Mathematics in Science and Engineering

Modeling and Analysis of Modern Fluid Problems

Mathematics in Science and Engineering

Modeling and Analysis of Modern Fluid Problems

Liancun Zheng
School of Mathematics and Physics,
University of Science and Technology Beijing,
Beijing, China

Xinxin Zhang
School of Energy and Environmental Engineering,
University of Science and Technology Beijing,
Beijing, China

Series Editor

Goong Chen

ELSEVIER

ACADEMIC PRESS
An imprint of Elsevier
elsevier.com

Academic Press is an imprint of Elsevier
125 London Wall, London EC2Y 5AS, United Kingdom
525 B Street, Suite 1800, San Diego, CA 92101-4495, United States
50 Hampshire Street, 5th Floor, Cambridge, MA 02139, United States
The Boulevard, Langford Lane, Kidlington, Oxford OX5 1GB, United Kingdom

ISBN: 978-0-12-811753-8

ISSN: 0076-5392

Publisher: Candice Janco
Senior Acquisition Editor: Graham Nisbet
Editorial Project Manager: Susan Ikeda
Production Project Manager: Vijayaraj Purushothaman
Designer: Alan Studholme

Typeset by TNQ Books and Journals

Contents

Preface

A significant feature of modern science and engineering technology is that we are witnessing revolutionary changes regarding problems involved in containing enhanced degrees of nonlinearity and the complexity of system dimensions. Such problems are becoming increasingly prominent.

With the rapid development of computational science and technology, numerical methods of partial differential equations have become a branch of numerical analysis that studies numerical solutions to partial differential equations. However, numerical methods have limitations. For example, when a nonlinear problem is singular or has multiple solutions, a numerical method may not converge. Sometimes it may provide only limited or even misleading information; i.e., numerical methods cannot completely replace analytical methods. There has been good progress in the development of analytical methods for solving nonlinear ordinary/partial/fractional differential equations. Many empirical studies have shown that the analytical method can be used not only for local analysis but also to help us understand global behavior, offer insights into the structure of problems being studied, and solve "real-world" problems.

The purpose of this book is to provide a useful approach to modeling and analyzing methods for modern fluid problems arising in scientific research and engineering technology. This book can be used as a reference for researchers working in applied mathematics, fluid mechanics, energy, the environment, power, metallurgy, the chemical industry, water resources and hydropower, and engineering technology. It can also be used as a textbook for a graduate-level course in applied mathematics, mechanics, and engineering.

Most of the contents of this book are taken from the authors' and their coauthors' publications. The authors thank Xuehui Chen, Yan Zhang, Chunying Ming, Xinhui Si, Jing Zhu, Botong Li, Yaqing Liu, Chunrui Li, Chaoli Zhang, Jize Sui, Jinhu Zhao, Lin Liu, Fangfang Zhao, Lijun Wang, Zhenlin Guo, Chunxia Chen, and Chengru Jiao for their contributions to this book. During the writing of this book, we obtained enthusiastic help from Professor Buxuan Wang (Tsinghua University, Beijing, China, Members of the Chinese Academy of Sciences) and Professor Goong Chen (Texas A&M University, College Station, TX). They offered many valuable suggestions. We express our sincerest thanks to Professors Buxuan Wang and Goong Chen for their assistance and encouragement.

The authors wish to thank the National Natural Science Foundation of China (Grant Nos. 51,076,012, 51,276,014, and 51,476,191) for supporting our investigations.

Finally, we are very grateful for the excellent cooperation's of Editorial Project Manager Susan Ikeda, Project Manager Vijayaraj Purushothaman, and Senior Acquisitions Editor Graham Nisbet, during the preparation of the manuscript at Elsevier.

Owing to the limitations of our knowledge, this book will inevitably have shortcomings and deficiencies. We sincerely hope our readers will offer timely remarks and comments to improve the quality of the book.

Liancun Zheng
School of Mathematics and Physics,
University of Science and Technology Beijing,
Beijing, China

Xinxin Zhang
School of Energy and Environmental Engineering,
University of Science and Technology Beijing,
Beijing, China

Chapter 1

Introduction

ABSTRACT
Nature is full of many nonlinear and random phenomena. The study of nonlinear phenomena is a matter of natural sciences, engineering, and even social economic problems. Many scientists and engineers spend a lot of time and effort to study mathematical modeling and solving methods for these problems. This chapter presents an introduction to modeling, the development of analytical methods for modern fluid problems, and the outline of this book.

1.1 BASIC IDEALS OF ANALYTICAL METHODS

1.1.1 Analytical Methods

Most problems in science and engineering are in essence nonlinear and can be modeled by nonlinear differential equations. However, because of the complexity of nonlinear differential equations, so far we have not found a suitable analytical method to arrive at general solutions for various nonlinear equations. For these nonlinear equations, most cannot be solved analytically and can be solved only by the approximate method. Many basic properties commonly used in linear equations, such as the superposition principle of solutions, are no longer held (Li, 1989; Wang, 1993; Zheng et al., 2003; Zheng and Zhang, 2013).

There is good progress in the development of approximate analytical methods to solve nonlinear partial/ordinary differential equations. Many approximate analytical methods have been proposed to solve nonlinear ordinary equations, partial equations, fractional differential equations (FDEs), and integral equations. The commonly used methods are perturbation, Adomian decomposition, homotopy analysis, variational iteration, and differential transformation. Based on those methods, this book presents some useful approaches to modeling and analytical methods for modern fluid problems.

To understand the essence of this book better, a short review of the concepts of Taylor series and Fourier series are presented here, which are the bases for exploring approximate analytical methods for solving nonlinear partial differential equations.

Mathematics in Science and Engineering. http://dx.doi.org/10.1016/B978-0-12-811753-8.00001-3

1.1.1.1 Taylor Series

Suppose $f(x) \in C^n[a,b]$, that $f^{(n+1)}(x)$ exists on $[a,b]$, $x_0 \in [a,b]$. For every $x \in [a,b]$, there exists a number $\xi = \xi(x)$ between x and x_0, such that

$$f(x) = P_n(x) + R_n(x), \tag{1.1}$$

where

$$P_n(x) = f(x_0) + f'(x_0)(x - x_0) + \cdots + \frac{f^{(n)}(x_0)}{n!}(x - x_0)^n, \tag{1.2}$$

is called the nth Taylor polynomial of $f(x)$ about $x_0 \in [a,b]$, and

$$R_n(x) = \frac{f^{(n+1)}(\xi)}{(n+1)!}(x - x_0)^{n+1},$$

is called the nth Lagrange remainder term (or truncation error) associated with $P_n(x)$.

The approximate error absolute value is $|R_n(x)| = |f(x) - P_n(x)|$.

If $|R_n(x)|$ decreases monotonously with an increase in n, one can improve the degree of approximation by increasing the number of items in the polynomial. If $f(x)$ has derivatives of all orders in the vicinity of x_0 and $\lim\limits_{n \to \infty} R_n(x) = 0$ holds, in terms of polynomial (1.1), we can obtain the infinite series

$$f(x_0) + f'(x_0)(x - x_0) + \frac{f''(x_0)}{2!}(x - x_0)^2 + \cdots + \frac{f^{(n)}(x_0)}{n!}(x - x_0)^n + \cdots. \tag{1.3}$$

Series (1.3) is called the Taylor series of $f(x)$ at the point $x = x_0$.

In the case of $x_0 = 0$, the Taylor polynomial is often called a Maclaurin polynomial and the Taylor series is called a Maclaurin series.

The term truncation error refers to the error involved in using a truncated, or finite, summation to approximate the sum of an infinite series.

1.1.1.2 Fourier Series

Let $f(x)$ be a periodic function of $2l$. If $f(x)$ satisfies these conditions:

1. it is continuous or has only a finite number of the first kind of discontinuity points in a period
2. it has at most a finite number of extreme points in a period

then $f(x)$ is able to expand into Fourier series

$$f(x) = \frac{a_0}{2} + \sum_{k=1}^{\infty}\left(a_n \cos\frac{k\pi}{l}x + b_n \sin\frac{k\pi}{l}x\right). \tag{1.4}$$

when x is the continuous point, the series converges to $f(x)$; when x is the discontinuity point, the series converges to

$$\frac{1}{2}[f(x^-)+f(x^+)]. \tag{1.5}$$

There is a relationship between coefficients a_0, a_1, b_1, ..., and the function of $f(x)$

$$\begin{cases} a_n = \dfrac{1}{l}\displaystyle\int_{-l}^{l} f(x)\cos\frac{n\pi}{l}x\mathrm{d}x \quad (n = 0,1,2,\ldots), \\ b_n = \dfrac{1}{l}\displaystyle\int_{-l}^{l} f(x)\sin\frac{n\pi}{l}x\mathrm{d}x \quad (n = 1,2,3,\ldots). \end{cases} \tag{1.6}$$

The coefficients determined by Eq. (1.6) are known as the Fourier coefficient of the function $f(x)$. The triangle series defined by the Fourier coefficient is called the Fourier series.

It can be seen that the essence of the Taylor series and the Fourier series is to select a suitable set of base functions (BFs)

$$\varphi_0(x),\ \ \varphi_1(x),\ \ \varphi_2(x),\ldots,\varphi_n(x),\ldots, \tag{1.7}$$

to express the unknown function $f(x)$ as:

$$f(x) = a_0\varphi_0(x) + a_1\varphi_1(x) + a_2\varphi_2(x) + \cdots, \tag{1.8}$$

The coefficients are determined by the selected basis functions and the given function $f(x)$ itself.

When the BF is selected by 1, x, x^2, ..., x^n,..., we obtain the Taylor series. If the BF is selected by 1, $\cos\frac{x}{l}$, $\sin\frac{x}{l}$, $\cos\frac{2x}{l}$, $\sin\frac{2x}{l}$, ..., we obtain the Fourier series.

It is obvious that the expressions of the different approximate series will be obtained when we select different sets of BFs. Therefore, the essence of the analytical approach to deal with the nonlinear problem is based on the following ideas:

1. Select the appropriate set of linear independent basis functions $\{\varphi_i(x)\}_{i=0}^{\infty} \in C[a,b]$;
2. Express unknown functions $\varphi(x)$ as the combination of the BFs

$$\varphi(x) = a_0\varphi_0(x) + a_1\varphi_1(x) + \cdots + a_n\varphi_n(x) + \cdots,\ \text{ i.e.,}$$

$$\varphi(x) \in \mathbf{\Phi} = \mathrm{span}\{\varphi_0(x), \varphi_1(x), \cdots, \varphi_n(x)\} \subset C[a,b];$$

3. Determine the coefficients such that it makes it possible to approximate the unknown solution function to the problems.

1.1.2 Padé Approximation

The Padé approximation method was developed in the process of studying the Taylor series expansion. The coefficients of Taylor series expansion and the value of the function are not only a profound mathematical problem but also an important practical problem (Baker, 1975; Baker and Graves-Morris, 1996; Thukral, 1999; Xu, 1990). Their investigation is based on the basis of mathematical analysis and physical and biological sciences (Coope and Graves-Morris, 1993; Graves-Morris, 1990, 1994; Graves-Morris and Jenkins, 1989). The interpretation criterion is that if the Taylor series expansion converges absolutely, it uniquely establishes the value of the function, which is differentiable an arbitrary number of times. On the other hand, if a function is differentiable an arbitrary number of times, it uniquely defines the Taylor series expansion, as shown in the subsequent theorem.

1.1.2.1 Weierstrass Approximation Theorem

Suppose $f(x) \in C[a,b]$; then for each $\varepsilon > 0$, there exists a polynomial $P(x)$ with the property that

$$|f(x) - P(x)| < \varepsilon, \quad \text{for all } x \text{ in } [a, b]$$

The proof of this theorem can be found in most elementary texts on real analysis.

In terms of the Weierstrass approximation theorem, it is possible to approximate a function as a polynomial by combining more and more BFs. However, this approach has undesirable limitations in practical operations.

Consider the following example:

$$f(x) = \left(\frac{1+2x}{1+x}\right)^{1/2} = 1 + \frac{1}{2}x - \frac{5}{8}x^2 + \frac{13}{16}x^3 - \frac{141}{128}x^4 + \cdots. \tag{1.9}$$

It is easy to see that the Taylor series expression is not convergent for any value of $x > 0.5$, even though for any value of $0 \le x < +\infty$, $f(x)$ is a smooth function located in the range of $1-\sqrt{2}$.

The standard method is to use the original expression to develop a new Taylor series expression for $f(x)$ from the old one by computing $f(x)$ and its derivatives at a new point x_0 ($0 < x_0 < 0.5$, this new Taylor series expression, can be met in a larger range of x but it does not include $x = \infty$). In fact, $x = \infty$ can never be achieved, any progress in this direction is lengthy, and it will be tedious to attempt to meet the very lengthy range.

It is supposed that a variable substitution is introduced as $x = w/(1 - 2w)$ or $w = x/(1 + 2x)$; then

$$f(x(w)) = (1-w)^{-1/2} = 1 + \frac{1}{2}w + \frac{3}{8}w^2 + \frac{5}{16}w^3 + \frac{35}{128}w^4 + \cdots. \tag{1.10}$$

In this substitution, $x = \infty$ is transformed into $w = 0.5$. The Taylor series expression of $f(x(w))$ converges into $w = 0.5$, i.e., $x = \infty$. Therefore, the sequence initial estimate values for $f(\infty)$ are:

$$1, \ 1.25, \ 1.34375, \ 1.38281, \ 1.39990, \ldots,$$

which converges into the value of $\sqrt{2} = 1.414\cdots$. In view of the initial variables of x, the successive approximations are expressed as:

$$1, \ \frac{1+(5/2)x}{1+2x}, \ \frac{1+(9/2)x+(43/8)x^2}{(1+2x)^2}, \ldots,$$

They are all the rational fractional functions of variable x.

The Padé approximation is a method for approximating the value of a known function using special fractional functions. The main idea is to match a Taylor series expression as quickly as possible (Graves-Morris, 1994; Graves-Morris and Jenkins, 1989).

For example, for this problem, we introduce an approximate function of the form

$$\frac{a+bx}{c+dx}, \tag{1.11}$$

The formula is bounded when x approaches infinity. When we approximate $f(x)$ using the first three coefficients of terms of the Taylor series, we can obtain the following expression:

$$\frac{1+(7/4)x}{1+(5/4)x} = 1 + \frac{1}{2}x - \frac{5}{8}x^2 + \frac{25}{32}x^3 - \frac{125}{128}x^4 + \cdots.$$

In view of this formula, when $x = \infty$, it has a value of 1.4. This result is better than any approximation results obtained previously. Similarly, we can compute a new approximate result as:

$$\frac{1+(13/4)x+(41/16)x^2}{1+(11/4)x+(29/16)x^2} \rightarrow \frac{41}{29} = 1.413793103.$$

This new result is close to $\sqrt{2} \approx 1.414213562$. Furthermore, if we continue to use this method to calculate the approximation, we find that the more coefficients are used, the better convergence result are obtained, which are shown as:

$$1.414201183, \ 1.414213198, \ 1.414213552, \ \ldots,$$

where the final value is obtained by 11 coefficients of Taylor series, which has an error of 10^{-8} compared with the exact value.

In the same way, we can arrive at an approximation for function $f(x(w))$, i.e., by substituting $x = w/(1 - 2w)$ or $w = x/(1 + 2x)$ into this formula, and we obtain:

$$1, \quad \frac{1-\frac{1}{4}w}{1-\frac{3}{4}w}, \quad \frac{1-\frac{3}{4}w+\frac{1}{16}w^2}{1-\frac{5}{4}w+\frac{5}{16}w^2},$$

The three results are obtained using the coefficients of the first, third, and fifth terms of the function $f(x(w))$.

When using $w = 0.5$ (i.e., $x = \infty$) in this formula, we obtain the values of 1, 1.4, 41/29, … . Those values are identical with the results obtained using the Taylor series polynomial expansion of x. This invariance property is a general and important property of Padé approximation methods and is the basis of their ability to sum the x series in our example; it gives excellent results, such that we can obtain ideal results even up to the value of $x = \infty$.

1.1.2.2 Definition of Padé Approximants

Definition: The Padé approximant of a function $A(x)$ is denoted as:

$$[L/M] = P_L(x)/Q_M(x), \tag{1.12}$$

where $P_L(x)$ is a polynomial of the highest degree of L and $Q_M(x)$ is a polynomial of the highest degree of M.

In view of the formal power series

$$A(x) = \sum_{j=0}^{\infty} a_j x^j, \tag{1.13}$$

we can determine the coefficients of polynomials $P_L(x)$ and $Q_M(x)$ by the following equation

$$A(x) - P_L(x)/Q_M(x) = O(x^{L+M+1}). \tag{1.14}$$

When the fraction with the numerator and denominator is multiplied by a nonzero constant the fractional values remain unchanged, we can define the standardized normalization condition as

$$Q_M(0) = 1.0. \tag{1.15}$$

Note that there are no public factors of the polynomial for functions $P_L(x)$ and $Q_M(x)$.

If we express the coefficients of polynomial functions $P_L(x)$ and $Q_M(x)$ as

$$\begin{aligned} P_L(x) &= p_0 + p_1 x + \cdots + p_L x^L, \\ Q_M(x) &= 1 + q_1 x + \cdots + q_M x^M, \end{aligned} \tag{1.16}$$

In view of Eq. (1.16), multiplying Eq. (1.14) by $Q_M(x)$, we can obtain the following linear systems of coefficients:

$$\begin{array}{lllllllll} a_0 & & & & & & & & = p_0, \\ a_1 & + & a_0 q_1 & & & & & & = p_1, \\ a_2 & + & a_1 q_1 & + & a_0 q_2 & & & & = p_2, \\ \vdots & & & & & & & & \vdots \\ a_L & + & a_{L-1} q_1 & + & \cdots & + & a_0 q_L & & = p_L, \\ a_{L+1} & + & a_L q_1 & + & \cdots & + & a_{L-M+1} q_M & & = 0, \\ \vdots & & & & & & & & \vdots \\ a_{L+M} & + & a_{L+M-1} q_1 & + & \cdots & + & a_L q_M & & = 0, \end{array} \tag{1.17}$$

and we may define

$$a_n \equiv 0 \text{ if } n < 0 \quad \text{and} \quad q_j \equiv 0 \text{ if } j > M, q_j \equiv 0. \tag{1.18}$$

Definition Padé approximants given here are different from the definition of classical approximants in several ways. First, the expression of symbols is the different in the classical definition:

$$[M,L] = [L/M]. \tag{1.19}$$

However, some authors use the notation:

$$[L,M] = [L/M]. \tag{1.20}$$

The formula employed here is used (Eq. 1.12) to avoid confusion. By convention, L denotes the degree of numerator and M denotes the degree of denominator and the following formula:

$$L + M = N, \ \ L - M = J, \tag{1.21}$$

to express the sum and difference of those degrees of the numerator and the denominator.

A important point in mathematics is that in the standardized condition, the new definition is clearly different from the classical definition in the normalization condition (Eq. 1.15). Frobenius (1881) and Padé (1892) required only that $Q_M(x) \neq 0$, as illustrated in the following example:

$$A(x) = 1 + x^2 + \cdots, \tag{1.22}$$

For $L = M = 1$, it is easy to see that

$$P_1(x) = Q_1(x) = x, \ \ P_1(x)/Q_1(x) = 1, \tag{1.23}$$

It satisfies

$$Q_M(x)A(x) - P_L(x) = O(x^{N+1}), \tag{1.24}$$

However, it does not satisfy Eq. (1.14). In fact, for this series, the expression [1/1] does not exist in the new definition.

1.1.2.3 Uniqueness Theorem of Padé Approximants

According to the definition of Frobenius (1881) and Padé (1892), we have the following theorem.

Theorem (uniqueness): If the power series of Padé approximants exists, $[L/M]$ of Padé is unique.

Proof. If the theorem is not true, without a loss of generality we may assume that two Padé approximants exist, i.e., $X(x)/Y(x)$ and $U(x)/V(x)$, in which the degree of X and U is less than or equal to L, and the degree of Y and V is less than or equal to M, respectively. Therefore, the following formula holds:

$$X(x)/Y(x) - U(x)/V(x) = O\left(x^{L+M+1}\right). \tag{1.25}$$

Multiplying both sides of Eq. (1.25) by $Y(x)V(x)$ yields

$$X(x)\cdot V(x) - U(x)\cdot Y(x) = O\left(x^{L+M+1}\right). \tag{1.26}$$

It is seen that the highest degree of polynomial on the left-hand of Eq. (1.26) is at most $L+M$, so both sides are identically zero. Because neither Y nor V is identically zero, we have

$$X/Y = U/V. \tag{1.27}$$

Therefore, by definition, X and Y, and U and V are relatively prime and $Y(0) = V(0) = 1.0$. We can assert that the assumption of two Padé approximants is actually the same. The proof of theorem is completed.

This theorem is held whether or not the defining equations are nonsingular. If the equations are nonsingular, we can solve them directly and obtain:

$$[L/M] = \frac{\begin{vmatrix} a_{L-M+1} & a_{L-M+2} & \cdots & a_{L+1} \\ \vdots & \vdots & \ddots & \vdots \\ a_L & a_{L+1} & \cdots & a_{L+M} \\ \sum_{j=M}^{L} a_{j-M}x^j & \sum_{j=M-1}^{L} a_{j-M+1}x^j & \cdots & \sum_{j=0}^{L} a_j x^j \end{vmatrix}}{\begin{vmatrix} a_{L-M+1} & a_{L-M+2} & \cdots & a_{L+1} \\ \vdots & \vdots & \ddots & \vdots \\ a_L & a_{L+1} & \cdots & a_{L+M} \\ x^M & x^{M-1} & \cdots & 1 \end{vmatrix}}. \tag{1.28}$$

1.1.2.4 Table of Padé Approximants

Frobenius (1881) initially expressed Padé approximants in an indexed array. Padé (1892) was the first to emphasize their importance and arrange the values of Padé approximants into a table

$$\begin{array}{cccccc} [0/0] & [0/1] & [0/2] & [0/3] & [0/4] & \cdots \\ [1/0] & [1/1] & [1/2] & [1/3] & [1/4] & \cdots \\ [2/0] & [2/1] & [2/2] & [2/3] & [2/4] & \cdots \\ [3/0] & [3/1] & [3/2] & [3/3] & [3/4] & \cdots \\ [4/0] & [4/1] & [4/2] & [4/3] & [4/4] & \cdots \\ \vdots & \vdots & \vdots & \vdots & \vdots & \cdots \end{array} \tag{1.29}$$

We can see from the table that the partial sums of the Taylor series occupy the first column of the table. This table transposes the original expressions of Padé (1892) and many subsequent workers.

Using this method to make Padé approximants to the function of e^x, we can obtain:

$$[1/1] = \frac{2+x}{2-x}, \tag{1.30}$$

$$[2/2] = \frac{12+6x+x^2}{12-6x+x^2}, \tag{1.31}$$

$$[3/3] = \frac{120+60x+12x^2+x^3}{120-60x+12x^2-x^3}, \tag{1.32}$$

$$[4/4] = \frac{1680+840x+180x^2+20x^3+x^4}{1680-840x+180x^2-20x^3+x^4}. \tag{1.33}$$

For $x = 1$, the values of approximants satisfy:

$$e \approx 3,\ \ 19/7,\ \ 193/71,\ \ 2721/1001. \tag{1.34}$$

The last value has an error to the exact value of 10^{-8}.

1.2 REVIEW OF ANALYTICAL METHODS

1.2.1 Perturbation Method

Perturbation theory is closely related to the methods used in numerical analysis. The earliest use of what astrophysical scientists now call perturbation theory was to deal with otherwise unsolvable mathematical problems of celestial mechanics.

Perturbation theory was first devised to solve otherwise intractable problems in calculating the motions of planets in the solar system. The gradually increasing accuracy of astronomical observations led to incremental demands in the accuracy of solutions to Newton's gravitational equations, which led several notable 18th- and 19th-century mathematicians to extend and generalize the methods of perturbation theory. The main idea of the perturbation method is to express the problem by analyzing the small or large parameter. The first several terms (often one or two) can reveal the important features of the solution to the problem; subsequent steps give only a little correction. Thus, the theory of perturbation theory has been widely applied (Li, 1999; Wasow, 1994).

Lindstedt et al. first established the perturbation when they studied the problem of planetary orbit (Sheng and Gui, 1996). The perturbation method is a kind of asymptotic analysis method that can analyze the global behavior of the solution of a differential equation. Its advantage is that it is not only able to give the correct approximate solution; it can also give the analytical structure for an qualitative and quantitative analysis of physical problems. The advantage is that the numerical solution cannot be reached.

After World War II, the mechanics analysis method obtained an important development in the generalized variational principle and singular perturbation theory. These well-developed perturbation methods were adopted and adapted to solve new problems arising in 20th-century atomic and subatomic physics. Combined with computers, the generalized variational principle provided a broad working field, and perturbation theory provided an effective means for mechanics to enter the nonlinear fields (Grasman and Matkowsky, 1977; Grasman et al., 1978; Kreiss and Parter, 1974; Lin and Segel, 1974; Matkowsky, 1975; Pearson, 1968; Stuff, 1972).

1.2.1.1 Regular Perturbation and Singular Perturbation

Perturbation theory is a theory for solving approximate solutions with small parameters. A differential equation with small parameters is formulated as:

$$P_\varepsilon : \begin{cases} L_\varepsilon[u(x,\varepsilon)] = f(x,\varepsilon), & x = (x_1, x_2, \cdots, x_m) \in \Omega, \\ B_{\varepsilon,j}[u(x,\varepsilon)] = \psi_j(x,\varepsilon), & j = 1, 2, \cdots, k, x \in \partial\Omega, \end{cases} \tag{1.35}$$

where L_ε is a differential operator containing small parameters ε and $B_{\varepsilon,j}$ is the differential operators defined on the boundary.

The general procedure for solving the problem of P_ε is:

1. Select an asymptotic series sequence $\{\delta_n(\varepsilon)\}$ in which $\{\varepsilon^n\}$ is the most commonly used asymptotic sequence, satisfying

$$\text{(a)} \lim_{\varepsilon \to \infty} \delta_n(\varepsilon) = 0, \quad \text{(b)} \lim_{\varepsilon \to \infty} \frac{\delta_{n+1}(\varepsilon)}{\delta_n(\varepsilon)} = 0 \ (n = 1, 2, \cdots). \tag{1.36}$$

2. Expand the unknown solution according to asymptotic series sequence

$$u(x,\varepsilon) = u_0(x) + \sum_{n=1}^{N} \delta_n(\varepsilon)u_n(x) + Z_N(x,\varepsilon), \tag{1.37}$$

3. Substitute the expanded formula into the original problem and compare the coefficients of $\delta_n(\varepsilon)$; one can obtain the recursive equation and boundary conditions as

$$L_0(u_i) = H_i[u_0, u_1, \cdots, u_{i+1}; x], \quad B_{0,j}(u_i) = \Phi_j[u_0, u_1, \cdots, u_{i-1}; x], \tag{1.38}$$

4. The solutions can be obtained by solving these equations, in which u_0 is the solution of degeneration problem ($\varepsilon = 0$).

In most practical physical problems, the solutions obtained are often compared with the experimental and numerical results, so as to judge the validity of the asymptotic expansion. If these steps can be used to obtain the uniformly valid solution in the region, the perturbation problem is called the regular perturbation problem in the region; otherwise it is known as the singular perturbation problem.

In a certain accuracy range, we can use the first few terms of the asymptotic expansion to approximate the original solution of $u(x; \varepsilon)$. Therefore, the perturbation method is widely used in practical problems. It makes up for the shortcomings of the pure numerical method and has the characteristics of the analytical solution, so it is called the semianalytical method.

1.2.1.2 Asymptotic Matching Method

The matching asymptotic expansion method is the main method used in dealing with the singular perturbation boundary layer problem. The boundary layer was first proposed by L. Prandtl in 1904 (presented at the third International Congress of Mathematicians in Heidelberg, Germany). It simplifies the Navier–Stokes equations of fluid flow by dividing the flow field into two parts: one is the part near the surface in the immediate vicinity of a bounding surface, called the boundary layer, where the effects of viscosity are significant. It is dominated by viscosity and creates most of the drag experienced by the boundary body. Another part is outside the boundary layer, where viscosity can be neglected without significant effects on the solution. Therefore, Prandtl derived the boundary layer equation and the solutions for two parts. Then the two solutions link up, leading to the boundary layer technique, or the method of matching, i.e., the Prandtl matching principle.

1.2.1.3 Poincare–Lighthill–Kuo Method

In general in nonlinear problems, the period or frequency is no longer a constant but is commonly associated with amplitude. When one makes asymptotic expansion a function, one should simultaneously make asymptotic expansion for the frequency, wave velocity, characteristic value, or coordinates, such as the independent variable of some parameters. The secular term elimination is used to determine whether the independent variable is perturbation expansion.

In this way, the uniformly valid asymptotic solution of the problem is obtained, which is the basic ideas of the Poincare–Lighthill–Kuo (PLK) method. Because the variables should make scale transform in this method, the PLK method is also called the coordinate deformation method.

Some Chinese scientists made important contributions to the development of perturbation theory. In 1948, Professor Qian Wei-Chang created a synthetic expansion method to solve the large deflection problem of a circular plate; good results were obtained. Guo Yong-Huai generalized the Poincaré–Lighthill method to the boundary layer problem of viscous flow in 1953. In 1956, Qian Xue-Sen (LLS. Tsien) pointed out the importance of this method and called it the PLK method. Lin Jia-Qiao in 1954 proposed the characteristics theory, which is usually referred to as an analytical method for the problem of hyperbolic differential equations, and which provides an effective way to study nonlinear wave problems. Many scholars have made contributions in this field (Kang and Gui, 1996; Li, 1999; Stuff, 1972).

1.2.1.4 Average Method

The average method originated from Vanderpol in 1926, in the study of the self-excited oscillation of the circuit. The main approach is to use the amplitude and phase characteristics of nonlinear vibration to derive the approximate equation, which is satisfied by the method of the average value in one period. The method is also called the KBM method; it is an important tool to study the periodic solution of a differential equation.

1.2.1.5 Multiple Scales Method

The multiple scales method has been the fastest development and most widely used singular perturbation method. For singular perturbation problems in solving nonlinear vibration or boundary layer theory, the solution of the independent variable is not necessarily the same order of magnitude everywhere, such as in nonlinear vibration problems, amplitude is often "slow" change the amount of, and as the fluid in the vicinity of the wall bounded the flow speed is fast changing the amount of. Therefore, it is necessary to deal with the independent variables in multiple scales to describe these kinds of mathematical and physical problems.

Professor Lin Jia-Qiao demonstrated that if the multiple scales are chosen properly, the zeroth-order asymptotic approximation of the solution will give the true solution to the problem. The multiple scales method contains problems that can be solved by means of the asymptotic matching method, PLK method, average method, and so on, but it is difficult to choose the form for multiple scales variables. It is necessary to have some practical knowledge and experience when using the method of multiple scales.

While pursuing a doctorate at the University of Science and Technology Beijing (USTB), Professor Zheng's doctoral student, Yan Zhang, proposed a novel technique, i.e., the embedding-parameters perturbation method. The basic idea of this method is by introducing special parameters transformation for both independent and dependent variables, we can embed special small parameters into nonlinear differential equations, which then can be solved by a standard perturbation method. Thereafter, the solutions to the original differential equations can be obtained by substituting the asymptotic series solutions into the reversed transformation. Moreover, this method can be applied to solve other nonlinear problems.

1.2.2 Adomian Decomposition Method

The Adomian decomposition method, also known as the inverse operator method, is a mathematical method for solving linear and nonlinear mathematical physics equations; it was proposed by George Adomian (1986, 1988, 1994).

The basic idea of the Adomian decomposition method is to decompose the unknown equation $Fu = g(t)$ (including linear and nonlinear equations, ordinary differential equations, partial differential equations, and integral equations) into the sum of the components of different degree solutions, and then to try to find the solutions of each order. The sum of these solutions is used to approximate the true solution to any desired accuracy. The basic steps are presented subsequently.

First, the whole equation is decomposed into several parts. The main ones are linear and nonlinear; the linear part is divided into the reversible and linear residual parts. Second, for the nonlinear part, the principle of splitting is used to produce an equivalent polynomial (called the Adomian polynomial). Third, the solutions are obtained by the recursive rule.

We show the steps as follows. Operator F is decomposed into the operator:

$$F = L + R + N, \tag{1.39}$$

Therefore

$$Fu = Lu + Ru + Nu = g(t), \tag{1.40}$$

i.e.,

$$Lu = g(t) - Ru - Nu. \tag{1.41}$$

Choose suitable operator L such that L^{-1} exists; then operator L^{-1} is used to act on both sides of this equation, obtaining

$$u = L^{-1}g - L^{-1}Ru - L^{-1}Nu + \Phi, \tag{1.42}$$

where the term of Φ satisfies $L\Phi = 0$.

If we decompose the true solution u of the equation into the sum of infinity components of different degree solutions, i.e., $u = \sum_{n=0}^{\infty} u_n$, the nonlinear term Nu is decomposed into the sum of infinite number Adomian polynomials $A_n(u_0, u_1, \ldots, u_n)$, and then Eq. (1.42) can be written as

$$\sum_{n=0}^{\infty} u_n = L^{-1}g - L^{-1}R\left(\sum_{n=0}^{\infty} u_n\right) - L^{-1}A_n(u_0, u_1, \cdots, u_n) + \Phi. \tag{1.43}$$

Adomian decomposition satisfies the following recurrence relation:

$$\begin{aligned} u_0 &= L^{-1}g + \Phi \\ u_1 &= -L^{-1}Ru_0 - L^{-1}A_0 \\ u_2 &= -L^{-1}Ru_1 - L^{-1}A_1 \\ u_{m+1} &= -L^{-1}Ru_m - L^{-1}A_m \end{aligned} \quad . \tag{1.44}$$

Let $Nu = F(u)$; the few terms of Adomian polynomials $A_0, A_1, \ldots, A_n$ can be written as:

$$\begin{aligned} A_0 &= F(u_0) \\ A_1 &= u_1 F'(u_0) \\ A_2 &= u_2 F'(u_0) + \frac{1}{2!} u_1^2 F''(u_0) \\ A_3 &= u_3 F'(u_0) + u_1 u_2 F''(u_0) + \frac{1}{3!} u_1^3 F'''(u_0) \end{aligned} \quad . \tag{1.45}$$

Adomian decomposition method is characterized by wide application, a simple calculation process, and a fast convergence rate with no approximate conditions. The high precision approximation of the equation can be obtained even with exact solutions.

In the process of calculation, the most important problem is to select the appropriate initial value to get the convergence of the solution. Adomian and Rach (1992) proposed a method of "noise terms in a decomposition series solution." In applications solved by the decomposition method, the appearance of noise terms sometimes makes it necessary to compute more terms to observe self-cancellations and separate solution terms from terms whose sum vanishes in the limit.

Based on the works of Adomian and Rach, Wazwaz (1995, 1998) studied the necessary conditions to ensure an appearance of the disturbance term in the nonhomogeneous equation. Wazwaz proposed improving the Adomian-decomposed method (Wazwaz, 1999a,b,c). Compared with the standard Adomian method, the improved method has a faster convergence rate than the standard method.

The Adomian decomposition method is an effective method for solving the strong nonlinear problem. It can be applied to both ordinary differential equations and partial differential equations. However, there are some limitations to the Adomian decomposition method, i.e., the asymptotic series given by the method usually contain polynomials that are a combination of the basis functions of power series for the approximation, because the convergence region of the power series is small. Thus, in some case the uniform effective solution of the whole region usually cannot be obtained.

In addition, the Adomian decomposition method does not allow the selection of different basis functions, and the method cannot adapt the convergence region and the convergence rate of asymptotic solutions.

1.2.3 Homotopy Analysis Method

The homotopy analysis method (HAM) was developed by Liao in 1992. It is different from the perturbation method and does not depend on small parameters. HAM has been applied successfully to many nonlinear problems. Liao (1992, 1995) proved that the series solution obtained by HAM convergence strongly depends on auxiliary parameter h. Some modified optimal HAMs were proposed. The way to construct the initial guess solutions for natural boundary conditions can be found in Liao (2003, 2004, 2012) and Yang and Liao (2006).

The basic idea of the homotopy method is illustrated as follows. Consider a differential equation:

$$A[u(t)] = 0, \tag{1.46}$$

where A is a nonlinear operator, t is time, and $u(t)$ is an unknown function.

Let $u_0(t)$ be an initial approximation of $u(t)$. L is an auxiliary linear operator, satisfying

$$Lf = 0, \ f = 0. \tag{1.47}$$

We construct the homotopy relation as

$$H[\phi(t;q);q] = (1-q)L[\phi(t;q) - u_0(t)] + qA[\phi(t;q)], \tag{1.48}$$

where $q \in [0,1]$ is the embed variable, and $\phi(t;\ q)$ is a function of t and q. When $q = 0$ and $q = 1$, we obtain

$$H[\phi(t;q);q]|_{q=0} = L[\phi(t;q) - u_0(t)], \tag{1.49}$$

and

$$H[\phi(t;q);q]|_{q=1} = A[\phi(t;1)]. \tag{1.50}$$

respectively.

It is clearly seen that, in view of Eq. (1.48), $\phi(t;\ 0) = u_0(t)$ is the solution of the equation $H[\phi(t;\ q);\ q]|_{q=0} = 0$ and $\phi(t;\ 1) = u(t)$ is the solution of equation $H[\phi(t;\ q);\ q]|_{q=1} = 0$. When embedded variable q changes from 0 to 1, the solution $\phi(t;\ q)$ of equation $H[\phi(t;\ q);\ q] = 0$, depending on q continuously changing from initial solution $u_0(t)$ to the solution $u(t)$ of Eq. (1.46). In topology, this continuous variation is called homotopy.

In brief, HAM is based on the topology homotopy concept, but it is different from the traditional one. By introducing a nonzero auxiliary parameter h and a nonzero auxiliary function H, one constructs a new homotopy:

$$\widetilde{H}(\phi; q, \hbar, H) = (1-q)L[\phi(t; q, \hbar, H) - u_0(t)] - q\hbar H(t)A[\phi(t; q, \hbar, H)], \tag{1.51}$$

This formula is more general than Eq. (1.48), because Eq. (1.48) is only a special case of Eq. (1.51) when $\hbar = -1$ and $H(t) = 1$, i.e.,

$$H(\phi; q) = \widetilde{H}(\phi; q, -1, 1) \tag{1.52}$$

Analogously, when q increases from 0 to 1, the function $\phi(t;\ q, \hbar, H)$ changes its value from the initial value $u_0(t)$ to the solution $u(t)$ of the original nonlinear problem $u(t)$. The solution to equation

$$\widetilde{H}[\phi(t; q, \hbar, H)] = 0. \tag{1.53}$$

depends not only on the embed parameter q but also on auxiliary parameter $\hbar$ and auxiliary function $H(t)$. When $q = 1$, the solution still depends on the auxiliary parameter $\hbar$ and auxiliary function $H(t)$.

The convergence radius of the obtained series approximate solution depends on the two auxiliary parameters; the convergence rate and the convergence region can be adjusted by the auxiliary parameter. HAM is completely free from the assumption of small parameters, which can overcome the limitation of the perturbation method:

1. The nonlinear problem is effective even if the problem is not contained in any small (or large) parameter.
2. The convergence region and the convergence rate of the asymptotic series solution can be conveniently adjusted.
3. It is free to choose different initial solutions to approximate the nonlinear problem.

HAM has limitations, including the introduction of auxiliary operators, auxiliary parameters, and auxiliary functions. So far, there is no rigorous mathematical theory to guide the selection of auxiliary operators, auxiliary parameters, or auxiliary functions.

1.2.4 Differential Transformation Method

In mathematics, a transformation is a variable or function (group) after a corresponding law become another variable or function (group). For example, the mathematical integral transformation is commonly used as Laplace or

Fourier integral transforms, etc. Integral transformations can change the differential equations into simple algebraic equations. The primitive function of a linear differential equation can be changed to corresponding algebraic operations. Therefore, integral transformations have been widely used.

The differential transformation method (DTM) was first proposed by Jakui Zhao (1988) to solve differential equations arising in circuits. With this method, Zhao successfully solved the linear ordinary differential equation of the circuit and some initial value problems of nonlinear ordinary differential equations.

DTM is based on the Taylor series expansion of an unknown function, but it is different from the traditional method of Taylor series. The Taylor series method needs to do differential operations for unknown functions. However, DTM does not need to do differential operations where the computing formula of differential transformation can be obtained directly according to the unknown function by an algebraic operation, which can be an iterative formula of the Taylor expansion coefficient. Compared with the traditional Taylor expansion method, DTM greatly reduces the amount of calculations and makes the Taylor expansion possible for some complex equations of unknown functions.

Assume a function $w(t)$ has kth-order derivatives for variable t; the differential transform of function $w(t)$ is defined as (Zhao, 1988)

$$W(k) = M(k)\left[\frac{\mathrm{d}^k q(t)w(t)}{\mathrm{d}t^k}\right]_{t=t_0}, \quad k = 0, 1, 2, 3, \ldots, \tag{1.54}$$

where $w(t)$ and $W(k)$ are called the original function and differential transform function, respectively. The inverse differential transform function of $W(k)$ is defined as

$$w(t) = \frac{1}{q(t)}\sum_{0}^{\infty}\frac{W(k)}{M(k)k!}(t - t_0)^k, \tag{1.55}$$

where $M(k) \neq 0$ denotes the transformation of the known function, and $q(t) \neq 0$ is the kernel of the transformation of a known function. If $q(t) = 1$, the proportional function $M(k) = H^k$ or $M(k) = \frac{H^k}{k!}$ (H is proportional constant). When $M(k) = \frac{H^k}{k!}$, the product operation of the transformation is relatively simple, so the scaling function is generally used in this form.

Furthermore, if the function $w(x, t)$ has continuous partial derivatives, the differential transformation of the function $w(x, t)$ is defined as

$$W(k,h) = M(k)N(h)\left[\frac{\partial^{k+h} q(t)p(x)w(x,t)}{\partial x^h \partial t^k}\right]_{x=x_0, t=t_0,}, \quad k = 0, 1, 2, 3, \ldots, \tag{1.56}$$

where $w(t,x)$ and $W(k,h)$ are the original function and differential transform functions, respectively. The inverse differential transform function of $W(k,h)$ is defined as

$$w(t,x) = \frac{1}{q(t)p(x)} \sum_{0}^{\infty} \frac{W(k,h)}{M(k)N(h)h!k!} (t - t_0)^k (x - x_0)^h. \tag{1.57}$$

It is analogous to the case of one variable function, $M(k) \neq 0$ and $N(h) \neq 0$, displayed for the transformation of the known function of the independent variable to be proportional to the integer; $q(t) \neq 0$ and $p(x) \neq 0$ are the kernels of the transformation of a known function, respectively. If $q(t) = 1$, $p(x) = 1$. Then the proportional functions $M(k) = H_1^k$ and $N(h) = H_2^h$ or $M(k) = \frac{H_1^k}{k!}$ and $N(h) = \frac{H_2^h}{h!}$(H_i ($i = 1,2$) are called proportional constants). When $M(k) = \frac{H_1^k}{k!}$ and $N(h) = \frac{H_2^h}{h!}$, the product operations of the transformation are relatively simple. If $q(t) = p(x) = 1$ is chosen, the differential transform of two variables function of $w(t,x)$ can be written as

$$W(k,h) = \frac{1}{k!h!} \left[\frac{\partial^{k+h} w(x,t)}{\partial x^h \partial t^k} \right]_{x=x_0,t=t_0}, \quad k = 0, 1, 2, 3, \cdots. \tag{1.58}$$

The inverse differential transform of function $W(k, h)$ is written as

$$w(t,x) = \sum_{0}^{\infty} W(k,h)(t - t_0)^k (x - x_0)^h. \tag{1.59}$$

In the same way, we can define the differential transform of three or more variables. The solutions obtained by DTM are in form of Taylor series, which has obvious advantages relative to other analytical methods. Since the DTM was put forward, scholars have given it much attention. Many types of commonly used functions of differential transformation formulas have been derived and applications have become more widely known (Chen and Ho, 1999). This method has gradually been applied to many linear and nonlinear ordinary differential equations, partial differential equations, FDEs, integral equations, etc. (Arikoglu, 2005; Arikoglu and Ozkol, 2009; Ayaz, 2004).

Although DTM has the advantages of high accuracy and a smaller amount of calculation, this method also has some limitations. DTM seems hard to solve for some differential equations on an unbounded domain because the results obtained by DTM are usually valid in a small region. This is because series solutions obtained by DTM are divergent when the variables of the problems go to infinity. Moreover, the power series in isolation are not useful for handling boundary value problems on an unbounded domain (Abazari and Borhanifar, 2010; Chen and Ho, 1996).

Boyd (1997) proved that under normal conditions, the power series solution of a single form cannot effectively express an unbounded region or even part of the bounded region in a boundary value problem. The reason is that the radius of a convergence of series solution cannot contain an infinite solving region of the boundary value problem. Thus, one needs to improve traditional DTM or combine it with other approximate methods of accelerating convergence.

For a boundary value problem of differential equations in a bounded area, Jang et al. (2010) proposed a method in which, the bounded area is divided into segments or fragments, performing the DTM in each subregion. However, this method cannot be performed for nonlinear boundary value problems on an unbounded region. Some scholars conducted many investigations and put forward some modified methods to improve the classical DTM so that the method can be used on an unbounded region. Ebaid (2011) combined the DTM and the posttreatment method of a sine or cosine function to solve the initial value problem in a vibration problem; the periodic analytical solution was obtained. Rashidi (2009) studied magnetohydrodynamic (MHD) boundary layer flow over a shrinking permeable sheet with the modified DTM. The basic thought mainly includes three steps: (1) introduce suitable similarity transformation to reduce boundary layer partial differential equations into a nonlinear boundary value problem of ordinary differential equations; (2) solve the corresponding initial value problems using DTM; and (3) determine the approximate solution of the algebraic equation under the corresponding parameter using the Padé approximants for infinity boundary conditions. The method is called differential transformation Padé approximants method (DTM-Padé).

DTM-Padé has obvious shortcomings. First, for some complex unbounded region, it is difficult to determine the introduced pending initial parameters to boundary value problems. Second, the solutions to the algebraic equation systems obtained by DTM-Padé may not be unique or complex roots may exist. To overcome these difficulties, while Professor Zheng's doctoral student, Xiaohong Su was at USTB, he proposed a novel analytical method called DTM-BF (2012) to solve differential equations on unbounded domains. The proposed method combines differential transformation and BFs, which is a valid and accurate method for solving nonlinear differential equations with infinity boundary conditions.

1.2.5 Variational Iteration Method and Homotopy Perturbation Method

The variational iteration method (VIM) was used to solve integer order differential equations (He, 1998, 1999a, 2006). Inokuti et al. (1978) proposed the Lagrange multiplier method to solve nonlinear problems, which was first applied to quantum mechanics. The characteristic of this method is that we can use the solution of the linearized equation as the approximate solution of the original equation. Lagrange multipliers are introduced to correct the value of some special points.

In He (1999a) improved Inokuti et al's method for solving the nonlinear FDE, and proposed the following iterative scheme:

$$y_{app}(x_0) = y_0(x_0) + \int_0^{x_0} \lambda(Ly_0 + Ny_0)\,dx, \tag{1.60}$$

where λ is the Lagrange multiplier. This formula is called the corrected functional; the second term on the right side is called the corrected term. The Lagrange multiplier should be understand such that for y_0 any small change in y_{app} should be taken as a stagnation point value; then the Lagrange multiplier can be optimally determined using the concept of the variable.

After many scholars' applications and developments (He and Wu, 2007; Molliq et al., 2009; Slota and Zielonka, 2009; Wazwaz, 2007), the variational iteration algorithm became a powerful mathematical tool for solving nonlinear differential equations. It has high accuracy and the approximate solution is valid in the entire domain.

The homotopy perturbation method (HPM) is another nonlinear analytical technique for solving nonlinear problems, which was proposed by He (1999b). In contrast to traditional perturbation methods, this technique does not require a small parameter in an equation. According to the homotopy technique, a homotopy with an imbedding parameter $p \in [0, 1]$ is constructed, and the imbedding parameter is considered as a small parameter, so the method is called the HPM. It can take the full advantage of traditional perturbation methods and homotopy techniques.

Many scientists and engineers have devoted effort to the applications and developments of HPM to solve linear and nonlinear problems (Abbasbandy, 2007; Cuce and Cuce, 2015; Ganji et al., 2009; Golbabai and Javidi, 2007; Grysa and Maciag, 2016; He, 2000, 2003, 2004, 2005; Rashidi and Ganji, 2009; Roy et al., 2015; Sushila et al., 2013; Yun and Temuer, 2015).

1.3 FRACTAL THEORY AND FRACTIONAL VISCOELASTIC FLUID

1.3.1 The Concept of Fractals

The noun "fractal" was first introduced into the field of natural science by Mandelbrot in the 1970s to characterize graphics and complex engineering problems. Its original intention was for irregular, fragmented objects (Bao, 2005; Chang et al., 2005a,b Havlin and Ben-Avraham, 2002; Li and Wang, 1997; Liu et al., 2003; Wang, 2000; Weeks et al., 1996).

Fractals can be divided into regular and irregular ones. Regular fractals such as the fractal Cantor set, Koch curve, Sierpinski gasket, carpet, and sponges have strict self-similarity. However, many phenomena in nature, science research, and engineering problems are complex and often random, such as winding coastline, transforming endless Brownian motion trajectory, and so on. This kind of curve of self-similarity is approximate or in a statistical sense. This type of self-similarity value exists in the scale-invariant area. Beyond the scale-invariant area, self-similarity ceased to exist and the curve is irregular.

The basic idea is of fractals is that objective things have a self-similar hierarchical structure. The parts, each of which is (at least approximately) a reduced-size copy of the whole in the form, function, information, time, space, etc., have the similarity in statistical significance, which is known as self-similarity. For this self-similar hierarchical structure, when it is enlarged or reduced to the appropriate geometrical size, the whole structure remains the same.

As an active and promising new field, the fractal theory has brought new perspectives, ideas, and tools for exploring nature and solving engineering problems. The universal nature of fractal theory provides a simple geometric language for scientists in various fields, which has contributed to enriching and deepening of scientific research in many areas.

The classical diffusion equation cannot be used to describe the diffusion phenomenon in media fractals, fractal crystals, and some impure media, known as anomalous diffusion. Anomalous diffusion must take the time dependence and spatial correlation of the motion process into account. In considering the time—space correlation, the diffusion motion of particles is no longer an integer order Brown motion, but fractional Brown motion. Instantaneous mean square displacement is a nonlinear function of time, $\langle r^2(t)\rangle \sim t^\gamma$, where $\gamma = 1$ is normal diffusion, and where $\gamma < 1$ or $\gamma > 1$ are respectively called anomalous subdiffusion or abnormal super diffusion (Havlin and Ben-Avraham, 2002; Li and Wang, 1997; Liu et al., 2003; Molliq et al., 2009; Slota and Zielonka, 2009; Wazwaz, 2007). In view of mathematics, anomalous diffusions come from the nonlocality of time and space. The main tools to describe the movement of particles are the generalized Langevin equation containing the non-Ohmic spectrum, the continuous time random walk, score activities—Planck equation, Lévy flight, Tsallis statistics, etc. (Bao, 2005; Chang et al., 2005b).

1.3.2 Fractional Order Calculus

The history of the study of fractional calculus is almost as long as the development of the theory of integral calculus. Early in 1695, l'Hospital wrote to Leibniz to discuss fractional derivative about a function, but it was not until 1819, that Lacroix first presented the results of a simple function of fractional derivative: $\mathrm{d}^{1/2}y/\mathrm{d}x^{1/2} = 2\sqrt{x}/\sqrt{y}$. Then, after hundreds years of development, Euler, Laplace, Fourier, Abel, Liouville, Riemann, Grunwald, Letnikov, Weyl, Lévy, Riesz, and other mathematicians carried out in-depth research and promoted the development of this discipline (Chang et al., 2005a; Podlubny, 1990; Povstenko, 2008; Wang, 2000; Weeks et al., 1996).

Owing to the lack of backgrounds in practical application, the development of fractional calculus theory was very slow at the beginning of the theory. In

1970s, Professor Mandelbrot first proposed "fractal" theory, and pointed out that nature and many science and technology problems have a lot of fractal dimensions, that there is a self-similar phenomenon, and that fractional Brownian motion and fractional calculus exist between the whole and the part. As the dynamic basis of fractal geometry and fractal dimension, fractional calculus theory and FDE theory have developed rapidly and have become a hot research topic in the world.

Fractional calculus provides a good tool to describe physical memory and heredity. Fractional order calculus has been applied to many fields such as flabby, oscillation, stochastic diffusion theory and wave propagation, biological materials, control and robotics, viscoelastic dynamics, and quantum mechanics. Those applications have also accelerated the development of the theory of fractional calculus. Fractional order calculus theory and the theory of chaos and dissipative structure is considered as the current theory of nonlinear science (Li et al., 2008; Zhu, 2006).

1.3.2.1 Definition of Fractional Order Derivatives

Fractional order calculus theory is used for dealing with any order of derivatives or integrals. It is the promotion of integer derivatives and integrals. There are many kinds of definitions for fractional order derivatives. The three most commonly used fractional derivatives are Riemann–Liouville, Caputo, and Grünwald–Letnikov.

Definition 1.3.1 (Riemann–Liouville fractional order derivative)

$$ {}_a^R\mathrm{D}_t^\alpha f(t) = \begin{cases} \dfrac{1}{\Gamma(m-\alpha)} \dfrac{\mathrm{d}^m}{\mathrm{d}t^m} \displaystyle\int_a^t (t-\tau)^{m-\alpha-1} f(\tau)\mathrm{d}\tau, & m-1<\alpha<m \\ \dfrac{\mathrm{d}^m}{\mathrm{d}t^m} f(t), & \alpha = m \in N \end{cases} . \tag{1.61} $$

Definition 1.3.2 (Caputo fractional order derivative)

$$ {}_a^C\mathrm{D}_t^\alpha f(t) = \begin{cases} \dfrac{1}{\Gamma(m-\alpha)} \displaystyle\int_a^t (t-\tau)^{m-\alpha-1} f^m(\tau)\mathrm{d}\tau, & m-1<\alpha<m \\ \dfrac{\mathrm{d}^m}{\mathrm{d}t^m} f(t), & \alpha = m \in N \end{cases} . \tag{1.62} $$

Definition 1.3.3 (Grünwald–Letnikov fractional order derivative)

$$ {}_a^G\mathrm{D}_t^\alpha f(t) = \lim_{\substack{h\to 0 \\ mh=t-a}} h^{-\alpha} \sum_{r=0}^{m} (-1)^r \binom{\alpha}{r} f(t-rh), \quad a \in R, \tag{1.63} $$

when $\alpha = n$ (positive integer), ${}_a^G\mathrm{D}_t^\alpha f(t)$ is an nth-degree derivative of $f(t)$; when $\alpha = -n$ (negative integer), ${}_a^G\mathrm{D}_t^\alpha f(t)$ is an nth-degree integration of $f(t)$; when $\alpha < 0$,

$${}_a^G\mathrm{D}_t^\alpha f(t) = \frac{1}{\Gamma(-\alpha)} \int_a^t f(t-\tau)^{-\alpha-1} f(\tau)\mathrm{d}\tau. \tag{1.64}$$

when $\alpha > 0$, if $f(t)$ has $n+1$ continuous derivative on $[a, t]$ and $n > \alpha - 1$, then

$${}_a^R\mathrm{D}_t^\alpha f(t) = \sum_{k=0}^{n} \frac{f^k(a)(t-a)^{-\alpha+k}}{\Gamma(-\alpha+k+1)} + \frac{1}{\Gamma(-\alpha+k+1)} \int_a^t (t-\tau)^{n-\alpha} f^{(n+1)}(\tau)\mathrm{d}\tau \tag{1.65}$$

where $\Gamma(\cdot)$ is Gamma function $\Gamma(z) = \int_0^\infty e^{-t} t^{z-1} \mathrm{d}t$, $Re(z) > 0$, and in the definition, integral lower limit $a = 0$.

Some main properties of fractional derivatives are:

Property 1.3.1 ${}_0\mathrm{D}_t^\alpha (Cf(t)) = C \cdot {}_0\mathrm{D}_t^\alpha f(t)$, $\alpha \in R$, C are constants.

Property 1.3.2 ${}_0\mathrm{D}_t^\varepsilon (f_1(t) + f_2(t)) = {}_0\mathrm{D}_t^\alpha f_1(t) + {}_0\mathrm{D}_t^\alpha f_2(t)$, $\alpha \in R$.

Property 1.3.3 ${}_0\mathrm{D}_t^{-\alpha} {}_0\mathrm{D}_t^{-\beta} (f(t)) = {}_0\mathrm{D}_t^{-\alpha-\beta} f(t)$, $\alpha > 0$, $\beta > 0$.

Particular, ${}_0\mathrm{D}_t^\alpha {}_0\mathrm{D}_t^{-\alpha} f(t) = f(t)$, $\alpha > 0$,

$${}_0\mathrm{D}_t^{-\beta} t^u = \frac{\Gamma(1+u)}{\Gamma(1+u+\beta)} t^{u+\beta}, \quad \beta > 0, \quad u > -1, \tag{1.66}$$

$${}_0\mathrm{D}_t^\alpha t^u = \frac{\Gamma(1+u)}{\Gamma(1+u-\alpha)} t^{u-\alpha}, \alpha > 0, \quad u > -1. \tag{1.67}$$

1.3.3 Fractional Integral Transformations and Their Properties

Integral transformations are commonly methods used to searching for the closed exact analytical solutions of fractional order partial differential equations. They mainly include Fourier transform, Hankel transform, Laplace transform, Weber transformation, and Mellin transform. Here, we introduce only some integral transforms relating to the contents of this book.

1.3.3.1 Fourier Transformation

Fourier transformation is established by integrating the specific form of the function. The corresponding relationship between Fourier transform has a clear physical meaning; it is also a useful mathematical tool.

From the point of view of the spectrum, Fourier transformation can be used to display the characteristic of the describing function (or signal), but it can also be used to simplify calculations to facilitate problem solving, especially in the field of signal processing. Fourier transformation is based on the Fourier integral operators, for classical sense of Fourier transformation, the function of doing Fourier transformation in addition to satisfying the Dirichlet conditions, but also needs absolute integral on the interval of $(-\infty, +\infty)$.

$$F(k) = F\{f(x);k\} = \int_{-\infty}^{+\infty} e^{ikx} f(x)\mathrm{d}x, k \in R, \tag{1.68}$$

$$f(x) = F^{-1}\{F(k);x\} = \frac{1}{2\pi}\int_{-\infty}^{+\infty} e^{-ikx} F(k)\mathrm{d}k, x \in R, \tag{1.69}$$

where $F(k)$ is called the Fourier transformation of function $f(x)$, and $f(x)$ is called the inverse Fourier transformation of $F(k)$.

The Fourier transform formula of the αth-order differential operator is

$$F\{{}_0\mathrm{D}_t^\alpha f(x);k\} = (-\mathrm{i}w)^\alpha F(k). \tag{1.70}$$

The convolution of the Fourier transformation is defined as

$$f(x) * g(x) = \int_{-\infty}^{+\infty} f(\varepsilon)g(x-\varepsilon)\mathrm{d}\varepsilon = \int_{-\infty}^{+\infty} f(x-\varepsilon)g(\varepsilon)\mathrm{d}\varepsilon, \tag{1.71}$$

The convolution of the Fourier transformation is important in solving FDEs

$$F\{f(x)\cdot g(x);k\} = F(k)\cdot G(k). \tag{1.72}$$

1.3.3.2 Laplace Transformation

Laplace transformation is based on Fourier transformation. It introduces the attenuation index function $e^{-\beta t}$ and the unit order jump function to ease the restrictions on function and make it more suitable for engineering

$$F(p) = L\{f(t);p\} = \int_0^{+\infty} e^{-pt} f(t)\mathrm{d}t, \quad R(p) > C_0, \tag{1.73}$$

$$f(t) = L^{-1}\{F(p);t\} = \frac{1}{2\pi i}\int_{c-i\infty}^{c+i\infty} e^{pt} F(p)\mathrm{d}t, \quad C = R(p) > C_0, \tag{1.74}$$

where $F(p)$ is called the Laplace transformation of function, $f(t)$, and $f(t)$ is called the inverse Laplace transformation of $F(p)$.

The convergence condition is that the integral function $f(t)$ growth cannot exceed the exponential term function e^{-pt} decrease rate with time t. When the integral converges, this implies the existence of Laplace transformation.

The Laplace transformation formula of an αth-degree R-L differential operator is

$$L\{{}_0\mathrm{D}_t^\alpha f(t);p\} = p^\alpha F(p) - \sum_{k=0}^{n-1} p^k \left[{}_0\mathrm{D}_t^{\alpha-k-1} f(t)\right]_{t=0}, \quad n-1 \le \alpha < n. \tag{1.75}$$

The Laplace transformation formula of an αth-degree Caputo differential operator is

$$L\{{}_0^c\mathrm{D}_t^\alpha f(t);p\} = p^\alpha F(p) - \sum_{k=0}^{n-1} p^{\alpha-k-1} f^{(k)}(0), \quad n-1<\alpha\le n. \tag{1.76}$$

The convolution of Laplace transformation is defined as

$$f(t)*g(t) = \int_0^t f(\tau)g(t-\tau)\mathrm{d}\tau = \int_0^t f(t-\tau)g(\tau)\mathrm{d}\tau. \tag{1.77}$$

It is similar to Fourier transformation; the convolution of Laplace transformation is important in solving FDEs

$$L\{f(t)\cdot g(t);p\} = F(p)\cdot G(p). \tag{1.78}$$

1.3.3.3 Mellin Transformation

The Mellin transformation is defined as

$$F(s) = M\{f(r);s\} = \int_0^{+\infty} f(r)r^{s-1}\mathrm{d}r, \quad \gamma_1 < R(s) < \gamma_2, \tag{1.79}$$

$$f(r) = M^{-1}\{F(s);r\} = \frac{1}{2\pi i}\int_{r-i\infty}^{r+i\infty} F(s)r^{-s}\mathrm{d}s, \quad \gamma_1<\gamma<\gamma_2, \tag{1.80}$$

where $F(s)$ is called the Mellin transformation of function, $f(r)$, and $f(r)$ is called the inverse Mellin transformation of $F(s)$.

The convolution of Mellin transformation is defined as

$$f(t)\cdot g(t) = \int_0^\infty f(t\tau)g(\tau)\mathrm{d}\tau. \tag{1.81}$$

Analogously, the convolution of Mellin transformation is important in solving FDEs

$$M\{f(t)\cdot g(t);s\} = F(s)\cdot G(1-s) \tag{1.82}$$

1.3.4 Fractional Viscoelastic Fluid

Interest in the flow and heat transfer of non-Newtonian fluids has grown considerably because of advances in technological applications. However, it is difficult to suggest a single parameter model that exhibits all of the properties of non-Newtonian fluids as is done for Newtonian fluids. For this reason, a number of constitutive equations have been proposed. Among them, the models of the differential type and the rate type have received much attention.

The fractional calculus theory has achieved many important achievements in the study of viscoelastic mechanics. Fractional derivatives have been found to be flexible in describing viscoelastic behavior. In general, the constitutive equations for generalized viscoelastic non-Newtonian fluids can be modified from well-known fluid models by replacing the time derivative of an integer order with so-called Riemann–Liouville fractional calculus operators. Some good experimental data are achieved when introducing the constitutive equation with fractional derivatives.

Many scholars have carried out investigations on various kinds of viscoelastic fluid. It is shown that the fractional derivatives are good tools for studying viscoelastic fluid (Li et al., 2012b, 2016a; Liu et al., 2011a,b; Pan et al., 2016a,b; Tan and Xian, 2002; Tan et al., 2003; Tong and Liu, 2005; Tong and Wang, 2004; Xu and Tan, 2001, 2006; Zhao et al., 2016; Zheng et al., 2010, 2011a,b,d,f, 2012b).

1.4 NUMERICAL METHODS

Numerical methods for partial differential equations have become a branch of numerical analysis. Many sophisticated numerical discretization schemes and methods have been developed, such as the finite difference method, finite element method, finite volume method, finite analysis method, and boundary element method (Ames, 1977; Chen et al., 2010; Ciarlet, 1978; Fletcher, 1988; Liu et al., 2009, 2013; Samarskii, 2001; Smith, 1985).

The finite difference method is the earliest and most widely used numerical simulation method. It is based on Taylor series expansion, to replace derivatives with the function value difference on the grid nodes and solve algebraic equations of unknown functions for grid nodes.

The finite element method is based on the variational principle and the weighted residual method. The basic solution procedure is to divide the computational domain into a finite number of nonoverlapping units by selecting some appropriate nodes in each cell. The differential equation of variables is rewritten by the variable or its derivative node values with the selected interpolation function of linear expression. By means of the variational principle or weighted residual method, one can obtain the discrete solution to differential equations.

The finite volume method is also known as the control volume method. The basic idea is to divide the calculation area into a series of repetitive control volumes and make each grid point in a control volume. Solve the differential equation for each control volume integral and a set of discrete equations is obtained.

The finite analysis method is a numerical method based on the finite element method. The basic idea of finite analysis method is to divide the calculation area into rectangular divisions of a finite number of rules in terms of unit subdomain boundary conditions, using the interpolation function approximation to obtain the local analytical solution in the unit.

The boundary element method is based on applying Green's function formula by selecting the appropriate weighting function to solve the space domain. The partial differential equation is converted into the boundary integral equation. The solution variables are associated with boundary conditions. By discretizing the continuous variables, the integral equation is derived boundary nodes unknown values of algebra equation, and then we can obtain the function values in interior by using the boundary integral equation.

1.5 MODELING AND ANALYSIS FOR MODERN FLUID PROBLEMS

The authors of this book and their coauthors have been working in the fields of "modeling and analysis for modern fluid problems" for more than 20 years. Based on the mathematical methodology and multidisciplinary approaches, they made a comprehensive study of complex flow, heat, and mass transfer arising in modern fluid problems. The problems that were involved include power law non-Newtonian (Ostwald–de Waele model) fluid, Bingham fluid, Maxwell fluid, Oldroyd-B fluid, Burgers' fluid, MHD fluid, fractional viscoelastic fluids, micropolar fluid, and nanofluids subject to various initial or boundary conditions: over an accelerating plate or stretching sheet; in a pipe; over an oscillating wall; on a rotating disk; on a vertical permeable plate; with velocity slip or a temperature jump; with natural convection, mixed convection, or Marangoni convection, etc. The fluid's physical prosperities are involved: variable viscosity or variable heat conductivity, with heat generation or chemical reaction, nonuniform heat source/sink, Soret and Dufour effects, Ohmic heating, etc. The analytical theory and methods include the Padé approximation theory, embedded-parameters perturbation method, Adomian decomposition method, HAM, modified DTM, fractal theory, fractional calculus, FDEs, and numerical techniques for solving nonlinear partial differential equations (Dong et al., 2014a,b; Jiao et al., 2016; Han et al., 2014; Li et al., 2011, 2012a, 2014, 2016b; Lin et al., 2014; 2015a,b,c,d; Liu et al., 2011a,b; Ming et al., 2011; Niu et al., 2014, 2015; Si et al., 2011a,b, 2012, 2013, 2014a,b,c, 2016a,b; Su et al., 2012a,b; Sui et al., 2015a,b, 2016a,b; Yin et al., 2015; Zhang et al., 2015; Zhang and Zheng, 2012; Zheng et al., 2010, 2011a,b,c,d,e,f, 2012a,c,d,e, 2013a,b,c,d,e,f).

There are some useful reference books associated with these fields (Bellout and Bloom, 2014; Deville and Gatski, 2012; Ruban, 2015; Shang, 2012; Weigand, 2015).

1.6 OUTLINE

Chapter 1 presents a review of relevant background and the development of analytical methods for solving nonlinear differential equations.

In Chapter 2, a novel method is proposed for solving nonlinear differential equations, i.e., the embedding-parameters perturbation method. By introducing parameter transformations, we can convert nonlinear differential equations without parameters to new ones with parameters. Solutions to the new equations can be obtained and expressed in the asymptotic series of the parameters. Then the original problem can be solved using the inverse transformation. The method is used to study two types of Marangoni convection problem. One is for a power law fluid caused by a temperature gradient and the other is for finite-thickness Marangoni convection problems caused by a quadratic relationship of the temperature gradient. The approximate analytical solutions are obtained successfully for both problems and the associated transport characteristics for velocity and temperature fields are analyzed in detail.

In Chapter 3, the Adomian decomposition method is introduced to study two boundary layer problems of a power law non-Newtonian fluid over a continuously moving surface with a constant speed and a free stream speed subjected to injection or suction velocity of an MHD power law fluid flow over a power law velocity wall. Similarity transformations and the generalized Crocco variable transformations are introduced to reduced problems to singular nonlinear boundary value problems. Approximate analytical solutions are obtained and the effects of involved parameters on the boundary layer behavior are discussed. Moreover, ADM is used to investigate Marangoni convection problem over a fluid—gas free surface caused by an imposed temperature gradient. The solutions are obtained and the effects of parameters on velocity and temperature fields are analyzed and discussed.

In Chapter 4, HAM is applied to study the unsteady boundary layer flow and radiation heat transfer of an MHD fluid over a horizontal continuously stretching surface in an otherwise quiescent fluid medium in the presence of a nonuniform magnetic field, the flow and heat transfer of nanofluids over a porous rotating disk with velocity slip and temperature jump with three different types of nanoparticles (Cu, Al_2O_3, and CuO) with water as the base fluid, and the mixed convection boundary layer heat transfer of a power law fluid over a moving conveyor along an inclined plate and MHD thermosolutal Marangoni convection heat and mass transfer in power law non-Newtonian fluids driven by a power law temperature and power law concentration. The governing equations are reduced to nonlinear ordinary differential equations using suitable local similarity transformations. Approximate analytical solutions are obtained using HAM. The influence of the parameters on the velocity, temperature, and concentration fields are analyzed in detail.

In Chapter 5, we introduce the applications of the DTM-Padé and DTM-BF methods. The two methods are used to solve to study Falkner—Skan boundary layer flow and its magnetic effects over a permeable wall, the mixed convective and radiation heat transfer of MHD fluid over a stretching wedge surface, and MHD boundary layer stagnation flow and radiation heat transfer of a nanofluid with uniform suction or blowing on a flat plate embedded in a porous medium. Approximate analytical solutions are obtained and the effects of pertinent parameters on the velocity and temperature fields are analyzed.

Numerical results are also presented to verify the reliability and validity of the proposed methods.

Chapter 6 introduces the applications of VIM and HPM to solving nonlinear partial differential equations. The method is applied to solve the fractional diffusion equation with variable coefficients and the fractional diffusion equation with time and space derivatives, and the unsteady flow and heat transfer of a power law fluid over an unsteadily stretching surface. Approximate analytical solutions are obtained and the effects of pertinent parameters on velocity and temperature fields are analyzed.

In Chapter 7, fractional calculus approaches are used to establish the constitutive relationship of viscoelastic non-Newtonian fluids. The fractional governing partial differential equations according to various initial or boundary conditions are formulated for fractional Maxwell fluid flow, the helical flows of fractional Oldroyd-B fluid in a porous medium, Oldroyd-B fluid between two infinite circular cylinders, etc. Closed-form analytical solutions are obtained using fractional integral transformation techniques, including Laplace transformation, Fourier transformation, Hankel transformation, and Merlin transformation. The solutions are presented in series form and as generalized functions. The effects of pertinent parameters on velocity and temperature and shear stress field distribution are analyzed graphically and discussed.

In Chapter 8, we present some numerical investigations for convection heat transfer of power law in a duct or over a rotating disk with modified Fourier's power law thermal conductivity models. The flow and heat transfer of fractional viscoelastic fluid with modified Fourier's thermal conductivity law and Darcy's diffusion law in porous medium, the fractional viscoelastic fluid unsteady natural convection, the fractional convection diffusion in a comb-like frame with Cattaneo—Christov flux etc., are studied, where the fractional nonlinear coupled governing equations are formulated. The solutions are obtained numerically and the influences of involved parameters on the velocity and temperature fields are analyzed.

REFERENCES

Abazari, R., Borhanifar, A., 2010. Numerical study of the solution of the Burgers and coupled Burgers equations by a differential transformation method. Computers and Mathematics with Applications 59 (8), 2711—2722.

Abbasbandy, S., 2007. Application of He's homotopy perturbation method to functional integral equations. Chaos, Solitons and Fractals 31, 1243—1247.

Adomian, G., Rach, R., 1992. Noise terms in decomposition series solution. Computers and Mathematics with Applications 24 (11), 61—64.

Adomian, G., 1986. Nonlinear Stochastic Operator Equations. Academic Press.

Adomian, G., 1988. Applications of Nonlinear Stochastic Systems Theory to Physics. Kulwer.

Adomian, G., 1994. Solving Frontier Problems of Physics, The Decomposition Method. Kluwer Academic Publisher, Boston, MA.

Ames, W.F., 1977. Numerical Methods for Partial Differential Equations, second ed. Academic, New York.

Arikoglu, A.I., Özkol, I., 2009. Solution of fractional integro-differential equations by using fractional differential transform method. Chaos, Solitons and Fractals 40, 521–529.

Arikoglu, A.I., 2005. Solution of boundary value problems for integro-differential equations by using differential transform method. Applied Mathematics and Computation 168 (2), 1145–1158.

Ayaz, F., 2004. Solutions of the systems of differential equations by differential transform method. Applied Mathematics and Computation 147 (2), 547–567.

Baker, G.A., Graves-Morris, P.R., 1996. Padé Aproximants. Cambridge University Press, Cambridge.

Baker, G.A., 1975. Essentials of Padé Approximants. Academic Press, New York.

Bao, J.D., 2005. Fractional Brown motion and anomalous diffusion. Progress in Physics 25 (4), 359–367.

Bellout, H., Bloom, F., 2014. Incompressible Bipolar and Non-Newtonian Viscous Fluid Flow, Advances in Mathematical Fluid Mechanics. Birkhauser/Springer.

Boyd, J., 1997. Padé approximant algorithm for solving nonlinear ordinary differential equation boundary value problems on an unbounded domain. Computers in Physics 11 (3), 299–303.

Chang, F.X., Chen, J., Huang, W., 2005a. The anomalous diffusion and fractional convection diffusion equation. Acta Physica Sinica 54 (3), 1113–1117.

Chang, F.X., Wu, J.C., Xue, Y.Q., et al., 2005b. The fractional dispersion solute transport in porous media problems. Research and Development of Water Dynamics 20 (1), 50–55.

Chen, C.K., Ho, S.H., 1996. Application of differential transformation to eigenvalue problems. Applied Mathematics and Computation 79 (2–3), 173–188.

Chen, C.K., Ho, S.H., 1999. Solving partial differential equations by two dimensional differential transform method. Applied Mathematics and Computation 106 (2), 171–179.

Chen, C., Liu, F.W., Anh, V., Turner, I., 2010. Numerical schemes with high spatial accuracy for a variable-order anomalous subdiffusion equation. SIAM Journal on Scientific Computing 32 (4), 1740–1760.

Ciarlet, P.G., 1978. The Finite Element Method for Elliptic Problems. North-Holland Publishing Company, Amsterdam.

Coope, I.D., Graves-Morris, P.R., 1993. The rise and fall of the vector epsilon algorithm. Numerical Algorithm 5, 275–286.

Cuce, E., Cuce, P.M., 2015. A successful application of homotopy perturbation method for efficiency and effectiveness assessment of longitudinal porous fins. Energy Conversion and Management 93, 92–99.

Deville, M., Gatski, T., 2012. Mathematical Modelling for Complex Fluids and Flows. Springer, Heidelberg.

Dong, S.L., Zheng, L.C., Zhang, X.X., Lin, P., 2014a. Improved drag force model and its application in simulating flow utilizing nanofluids. Microfluidics and Nanofluidics 17 (2), 253–261.

Dong, S.L., Zheng, L.C., Zhang, X.X., Wu, S.P., Shen, B.Y., 2014b. A New model for Brownian force and the application tosimulating nanofluid flow. Microfluidics and Nanofluidics 16, 131–139.

Ebaid, A.E., 2011. A reliable aftertreatment for improving the differential transformation method and its application to nonlinear oscillators with fractional nonlinearities. Communications in Nonlinear Science and Numerical Simulation 16, 528–536.

Fletcher, C.A.J., 1988. Computationl Technfor Fluid Dynamics. Spring-Verlag, New York.

Frobenius, G. (1881) Ueber Relationen zwischen den Näherungbrüchen von Poteneihen. J. für Math. (Crelle) 90, 1–17.

Ganji, D.D., Sahouli, A.R., Famouri, M., 2009. A new modification of He's homotopy perturbation method for rapid convergence of nonlinear undamped oscillators. Journal of Applied Mathematics and Computing 30, 181–192.

Golbabai, A., Javidi, M., 2007. New iterative methods for nonlinear equations by modified HPM. Applied Mathematics and Computation 191, 122–127.

Grasman, J., Matkowsky, B.J., 1977. A variational approach to singularly perturbed boundary value problems for ordinary and partial differential equations with turning points. SIAM Journal on Applied Mathematics 32, 588–597.

Grasman, J., Jansen, M.J.W., Veling, E.J.M., 1978. Asymptotic methods for relaxation oscillations. North-Holland Mathematics Studies 31, 225–238.

Graves-Morris, P.R., Jenkins, C.D., 1989. Degeneracies of generalised inverse, vector-valued Padé approximants. Constructive Approximation 5, 463–485.

Graves-Morris, P.R., 1990. Solution of integral equations using generalised inverse, function-valued Padé approximants. Journal of Computational and Applied Mathematics 32, 117–124.

Graves-Morris, P.R., 1994. A review of Padé methods for the acceleration of convergence of a sequence of vectors. Applied Numerical Mathematics 15, 153–174.

Grysa, K., Maciag, A., 2016. Temperature dependent thermal conductivity determination and source identification for nonlinear heat conduction by means of the Trefftz and homotopy perturbation methods. International Journal of Heat and Mass Transfer 100, 627–633.

Han, S.H., Zheng, L.C., Li, C.R., Zhang, X.X., 2014. Coupled flow and heat transfer in viscoelastic fluid with Cattaneo-Christov heat flux model. Applied Mathematics Letters 38, 87–93.

Havlin, S., Ben-Avraham, D., 2002. Diffusion in disordered media. Advances in Physics 51 (1), 187–292.

He, J.H., Wu, X.H., 2007. Variational iteration method: new development and applications. Computers and Mathematics with Applications 54, 881–894.

He, J.H., 1998. Approximate analytical solution for seepage flow with fractional derivatives in porous media. Computer Methods in Applied Mechanics and Engineering 167, 57–68.

He, J.H., 1999a. Variational iteration method-a kind of non-linear analytical technique: some examples. International Journal of Non-Linear Mechanics 34, 699–708.

He, J.H., 1999b. Homotopy perturbation technique. Computer Methods in Applied Mechanics and Engineering 178, 257–262.

He, J.H., 2000. A coupling method of Homotopy technique and perturbation technique for nonlinear problems. International Journal of Non-Linear Mechanics 35, 37–43.

He, J.H., 2003. Homotopy perturbation method: a new nonlinear analytical technique. Applied Mathematics and Computation 135, 73–79.

He, J.H., 2004. Comparison of Homotopy perturbation method and Homotopy analysis method. Applied Mathematics and Computation 156, 527–539.

He, J.H., 2005. Homotopy perturbation method for bifurcation of nonlinear problems. International Journal of Non-Linear Science 6, 207–208.

He, J.H., 2006. Non-perturbative Methods for Strongly Nonlinear Problems. De-Verlag im Internet GmbH, Berlin (Dissertation).

Inokuti, M., Sekine, H., Mura, T., 1978. General use of the Lagrange multiplier in nonlinear mathematical physics. In: Variational Methods in the Mechanics of Solids, pp. 156–162.

Jang, M.J., Yeh, Y.L., Chen, C.L., Yeh, W.C., 2010. Differential transformation approach to thermal conductive problems with discontinuous boundary condition. Applied Mathematics and Computation 216, 2339–2350.

Jiao, C.R., Zheng, L.C., Lin, Y.H., Ma, L.X., Chen, G., 2016. Marangoni abnormal convection heat transfer of power-law fluid driven by temperature gradient in porous medium with heat generation. International Journal of Heat and Mass Transfer 92, 700–707.

Kang, S., Gui, Z., 1996. Modern Methods in Physical and Mathematical Equations. Tongji University Press, Shanghai (in Chinese).

Kreiss, H.O., Parter, S.B., 1974. Remarks on singular perturbations with turning points. Society for Industrial and Applied Mathematics 5, 230–251.

Li, H.Q., Wang, F.Q., 1997. Fractal Theory and Its Application in Molecular Science. Science Press, Beijing.

Li, B., Chen, Z.J., Zhao, H.W., 2008. Application and development of rheology. Contemporary Chemical Industry 37 (2), 221–224.

Li, B.T., Zheng, L.C., Zhang, X.X., 2011. Heat transfer in pseudo-plastic non-Newtonian fluids with variable thermal conductivity. Energy Conversion and Management 52 (1), 355–358.

Li, B.T., Zheng, L.C., Zhang, X.X., 2012a. Comparison between thermal conductivity models on heat transfer in power-law non-Newtonian fluids. Journal of Heat Transfer 134, 041702.

Li, C.R., Zheng, L.C., Zhang, Y., Ma, L.X., Zhang, X.X., 2012b. Helical flows of a heated generalized Oldroyd-B fluid subject to a time-dependent shear stress in porous medium. Communications in Nonlinear Science and Numerical Simulation 17, 5026–5041.

Li, B.T., Chen, X., Zheng, L.C., Zhu, L.L., Zhou, J.L., Wang, T.T., 2014. Precipitation phenomenon of nanoparticles in power-lawfluids over a rotating disk. Microfluidics and Nanofluidics 17, 107–114.

Li, C.R., Zheng, L.C., Zhang, X.X., Chen, G., 2016a. Flow and heat transfer of a generalized Maxwell fluid modified fractional Fourier's law and Darcy's law. Computers and Fluids 125, 25–38.

Li, J., Zheng, L.C., Liu, L., 2016b. MHD viscoelastic flow and heat transfer over a vertical stretching sheet with Cattaneo-Christov heat flux effects. Journal of Molecular Liquids 221, 19–25.

Li, D.Q., 1989. Nonlinear Evolution Equations(in Chinese). Science Press, Beijing.

Li, J.C., 1999. Mathematical Methods in Mathematical Physics. Science Press, Beijing.

Liao, S.J., 1992. The Proposed Homotopy Analysis Technique for the Solution of Non-linear Problems. Shanghai Jiao Tong University (Ph.D. thesis).

Liao, S.J., 1995. An approximate solution technique not depending on small parameters: a special example. International Journal of Non-Linear Mechanics 303, 371–380.

Liao, S.J., 2003. Beyond Perturbation: Introduction to the Homotopy Analysis Method. Chapman & Hall/CRC Press, Boca Raton.

Liao, S.J., 2004. On the homotopy analysis method for non-linear problems. Applied Mathematics and Computation 47 (2), 499–513.

Liao, S.J., 2012. Homotopy Analysis Method in Nonlinear Differential Equations. Springer, Heidelberg.

Lin, C.C., Segel, L.A., 1974. Mathematics Applied to Deterministic Problems in the Nature Sciences. Macmillan, , New York.

Lin, Y.H., Zheng, L.C., Zhang, X.X., 2014. Radiation effects on Marangoni convection flow and heat transfer in pseudo-plastic non-Newtonian nanofluids with variable thermal conductivity. International Journal of Heat and Mass Transfer 77, 708–716.

Lin, Y.H., Zheng, L.C., Chen, G., 2015a. Unsteady flow and heat transfer of pseudo-plastic nanoliquid in a finite thin film on a stretching surface with variable thermal conductivity and viscous dissipation. Powder Technology 274, 324–332.

Lin, Y.H., Zheng, L.C., Li, B.T., Ma, L.X., 2015b. A new diffusion for laminar boundary layer flow of power law fluids past a flat surface with magnetic effect and suction or injection. International Journal of Heat and Mass Transfer 84, 1090–1097.

Lin, Y.H., Zheng, L.C., Zhang, X.X., 2015c. MHD Marangoni boundary layer flow and heat transfer of pseudo-plastic nanofluids over a porous medium with a modified model. Mechanics of Time-Dependent Materials 19 (4), 519–536.

Lin, Y.H., Zheng, L.C., Zhang, X.X., Ma, L.X., Chen, G., 2015d. MHD pseudo-plastic nanofluid unsteady flow and heat transfer in a finite thin film over stretching surface with internal heat generation. International Journal of Heat and Mass Transfer 84, 903–911.

Liu, S.D., Liang, F.M., Liu, S.S., et al., 2003. Chaos and Fractal in Natural Science. Peking University Press, Beijing.

Liu, F.W., Yang, C., Burrage, K., 2009. Numerical method and analytical technique of the modified anomalous subdiffusion equation with a nonlinear source term. Journal of Computational and Applied Mathematics 231, 160–176.

Liu, Y.Q., Zheng, L.C., Zhang, X.X., 2011a. Unsteady MHD Couette flow of a generalized Oldroyd-B fluid with fractional derivative. Computers and Mathematics with Applications 61 (2), 443–450.

Liu, Y.Q., Zheng, L.C., Zhang, X.X., 2011b. MHD flow and heat transfer of ageneralized Burgers' fluid due to an exponential accelerating plate with the effect of radiation. Computers and Mathematics with Applications 62, 3123–3131.

Liu, F.W., Meerschaert, M.M., McGough, R., Zhuang, P., Liu, Q., 2013. Numerical methods for solving the multi-term time fractional wave equations. Fractional Calculus and Applied Analysis 16 (1), 9–25.

Matkowsky, B.J., 1975. On boundary layer problems exhibiting resonance. Society for Industrial and Applied Mathematics 17, 82–100.

Ming, C.Y., Zheng, L.C., Zhang, X.X., 2011. Analysis of the power-law fluid over a rotating disk with heat transfer. International Journal of Heat and Mass Transfer 38, 280–284.

Molliq, R.Y., Noorani, M.S.M., Hashim, I., 2009. Variational iteration method for fractional heat- and wave-like equations. Nonlinear Analysis: Real World Application 10, 1854–1869.

Niu, J.J., Zheng, L.C., Yang, Y., Shu, C.W., 2014. Chebyshev spectral method for unsteady axisymmetric mixed convection heat transfer of power law fluid over a cylinder with variable transport properties. International Journal of Numerical Analysis and Modeling 1, 525–540.

Niu, J.J., Zheng, L.C., Zhang, X.X., 2015. Unsteady convective heat transfer of power-law fluid with variable fluid properties in a concentric annulus originated from a polymer flooding process. Numerical Heat Transfer, Part A: Applications 68, 1–16.

Padé, H., 1892. Sur la représentation approchée d'une fonction pour des fractions rationnelles (thesis). Annales Scientifiques de l'École Normale Supérieure 9, 1–93.

Pan, M.Y., Zheng, L.C., Liu, F.W., Zhang, X.X., 2016a. Lie group analysis and similarity solution for fractional Blasius flow. Communications in Nonlinear Science and Numerical Simulation 37, 90–101.

Pan, M.Y., Zheng, L.C., Liu, F.W., Zhang, X.X., 2016b. Modelling heat transport in nanouids with stagnation point ow using fractional calculus. Applied Mathematical Modelling 371, 1–11.

Pearson, C.E., 1968. On a different equation of boundary layer type. Journal of Mathematics and Physics 47, 134–154.

Podlubny, I., 1990. Fractional Differential Equations. Academic Press, San Diego-Boston-New York.

Povstenko, Y.Z., 2008. Fractional radial diffusion in a cylinder. Journal of Molecular Liquids 137, 46–50.

Rashidi, M.M., Ganji, D.D., 2009. Homotopy Perturbation Combined with Padé Approximation for Solving Two Dimensional Viscous Flow in the Extrusion Process. International Journal of Non-Linear Science 7, 387–394.

Rashidi, M.M., 2009. The modified differential transform method for solving MHD boundary layer equations. Computer Physics Communications 180 (11), 2210–2217.

Roy, P.K., Das, A., Mondal, H., Mallick, A., 2015. Application of homotopy perturbation method for a conductive–radiative fin with temperature dependent thermal conductivity and surface emissivity. Ain Shams Engineering Journal 6, 1001–1008.

Ruban, A.I., 2015. Fluid Dynamics, Asymptotic Problems of Fluid Dynamics. Oxford University, Press, Oxford.

Samarskii, A.A., 2001. The Theory of Difference Schemes. Marcel Dekker, New York.

Shang, D.Y., 2012. Free Convection Film Flows and Heat Transfer, second ed. Springer-Verlag Berlin Heidelberg.

Sheng, K.L., Gui, Z.P., 1996. Modern Methods of Mathematical Physics Equation. Tongji University Press, Shanghai.

Si, X.H., Zheng, L.C., Zhang, X.X., Si, X.Y., 2011a. Homotopy analysis method for the asymmetric laminar flow and heat transfer of viscous fluid between contracting rotating disks. Applied Mathematical Modelling 35, 4321–4329.

Si, X.H., Zheng, L.C., Zhang, X.X., Li, M., Yang, J.H., Chao, Y., 2011b. Multiple solutions for the laminar flow in a porous pipe with suction at slowly expanding or contracting wall. Applied Mathematics and Computation 218, 3515–3521.

Si, X.H., Zheng, L.C., Zhang, X.X., Si, X.Y., 2012. Homotopy analysis method for the asymmetric laminar flow and heat transfer of viscous fluid between contracting rotating disks. Applied Mathematical Modelling 36, 1806–1820.

Si, X.H., Zheng, L.C., Lin, P., Zhang, X.X., Zhang, Y., 2013. Flow and heat transfer of a micropolar fluid in a porous channel with expanding or contracting walls. International Journal of Heat and Mass Transfer 67, 885–895.

Si, X.H., Li, L., Zheng, L.C., 2014a. The exterior unsteady viscous flow and heat transfer due to a porous expanding stretching cylinder. Computers and Fluids 105, 280–284.

Si, X.H., Zheng, L.C., Chen, X.H., Zhang, X.X., Cao, L.M., Li, M., 2014b. The effects of slip velocity on a micropolar fluid through a porous channel with expanding or contacting walls. Computer Methods in Biomechanics and Biomedical Engineering 17 (4), 423–432.

Si, X.H., Zheng, L.C., Zhang, X.X., Li, M., 2014c. Asymmetric viscoelastic flow through a porous channel with expanding or contracting walls: a model for transport of biological fluids through vessels. Computer Methods in Biomechanics and Biomedical Engineering 17 (6), 623–631.

Si, X.H., Wang, C., Shen, Y.N.; Zheng, L.C., 2016a. Numerical method to initial-boundary value problems for fractional partial differential equations with time-space variable coefficients. Applied Mathematical Modelling 40 (7–8), 4397–4411.

Si, X.H., Yuan, L.L., Zheng, L.C., Shen, Y.N., Cao, L.M., 2016b. Lie group method for the modified model of MHD flow and heat transfer of a non-Newtonian fluid with prescribed heat flux over a moving porous plate. Journal of Molecular Liquids 220, 768–777.

Slota, D., Zielonka, A., 2009. A new application of He's variational iteration method for the solution of the one-phase Stefan problem. Computers and Mathematics with Applications 58, 2489–2494.

Smith, G.D., 1985. Numerical Solution of Partial Differential Equations: Finite Difference Methods. Clarendon Press, Oxford.

Stuff, R., 1972. Closed from solution for the sonic boom in a polytropic atmosphere. Journal of Aircraft 9, 556–562.

Su, X.H., Zheng, L.C., Zhang, X.X., 2012a. On DTM-BF method and dual solutions for an unsteady MHD flow over a permeable shrinking sheet with velocity slip. Applied Mathematics and Mechanics 33 (12), 1555–1568.

Su, X.H., Zheng, L.C., Zhang, X.X., Zhang, J.H., 2012b. MHD mixed convective heat transfer over a permeable stretching wedgewith thermal radiation and ohmic heating. Chemical Engineering Science 78, 1–8.

Sui, J.Z., Zheng, L.C., Zhang, X.X., 2015a. Convection heat transfer of power law fluids along the inclined non-uniformly heated plate with suction or injection. Journal of Heat Transfer-T AMSE 138, 021701.

Sui, J.Z., Zheng, L.C., Zhang, X.X., Chen, G., 2015b. Mixed convection heat transfer in power law fluids over a moving conveyor along an inclined plate. International Journal of Heat and Mass Transfer 85, 1023–1033.

Sui, J.Z., Zheng, L.C., Zhang, X.X., 2016a. Boundary layer Cattaneo-Christov double-diffusion model of heat and mass transfer in upper-convected Maxwell nanofluid past a stretching sheet with slip velocity. International of Thermal Science 104, 461–468.

Sui, J.Z., Zheng, L.C., Zhang, X.X., Chen, Y., Cheng, Z.D., 2016b. A novel equivalent agglomeration model for heat conduction enhancement in nanofluids. Scientific Reports 6, 19560.

Sushila, Singh, J., Shishodia, Y.S., 2013. An efficient analytical approach for MHD viscous flow over a stretching sheet via homotopy perturbation sumudu transform method. Ain Shams Engineering Journal 4, 549–555.

Tan, W.C., Xian, F., 2002. Whelan. Generalized second order fluid non steady Couette flow of exact solutions. Science Bulletin 47 (16), 1226–1228.

Tan, W.C., Pan, W.X., Xu, M.Y., 2003. A note on unsteady flows of a viscoelastic fluid with fractional Maxwell model between two parallel plates. International Journal of Non-Linear Mechanics 38 (5), 645–650.

Thukral, R., 1999. A family of Padé-type approximants for accelerating the convergence of sequences. Journal of Computational and Applied Mathematics 102, 287–302.

Tong, D.K., Liu, Y.S., 2005. Exact solutions for the unsteady rotational flow of non-Newtonian fluid in an annular pipe. International Journal of Engineering Science 43 (3), 281–289.

Tong, D.K., Wang, R.H., 2004. Fractal reservoir with non Newtonian viscoelastic liquid fractional flow analysis. Science in China (Series G) 34 (1), 87–101.

Wang, Y.M., 1993. Nonlinear Parabolic Equation (In Chinese). Science Press, Beijing.

Wang, Z.T., 2000. Singular diffusion in fractal porous media. Applied Mathematics and Mechanics 21 (10), 1033–1038.

Wasow, W., 1994. On the asymptotic solution of boundary value problems for ordinary differential equations containing aparameter. Journal of Mathematics and Physics 23, 173–183.

Wazwaz, A.M., 1995. A new approach to the nonlinear advection problem: an application of the decomposition method. Applied Mathematics and Computation 72, 175–181.

Wazwaz, A.M., 1998. A comparison between Adomian decomposition method and Taylor series method in the series solutions. Applied Mathematics and Computation 79, 37–44.

Wazwaz, A.M., 1999a. Analytic approximations and Padé approximants for Volterra's population model. Applied Mathematics and Computation 100, 13–25.

Wazwaz, A.M., 1999b. A reliable modification of Adomian's decomposition method. Applied Mathematics and Computation 102, 77–86.

Wazwaz, A.M., 1999c. The modified decomposition method and Padé approximants for solving Thomas-Fermi eqution. Applied Mathematics and Computation 105, 11–19.

Wazwaz, A.M., 2007. A comparison between the variational iteration method and Adomian decomposition method. Journal of Computational and Applied Mathematics 207, 129–136.

Weeks, E.R., Urbach, J.S., Harry, L.S., 1996. Anomalous diffusion in asymmetric random walks with a quasi-geostrophic flow example. Physica D: Nonlinear Phenomena 97, 291–310.

Weigand, B., 2015. Analytical Methods for Heat Transfer and Fluid Flow Problems. Springer, Heidelberg.

Xu, M.Y., Tan, W.C., 2001. Theoretical analysis of the two order fluid fractional anomalous diffusion velocity field. Stress Field and Vortex Layer 31 (7), 626–638.

Xu, M.Y., Tan, W.C., 2006. Intermediate processes, critical phenomena of fractional order operator theory, method and development and its application in modern mechanics. Science in China (Series G) 36 (3), 225–238.

Xu, X.Y., 1990. Padé Approximation Theory (In Chinese). Shanghai Science and Technology Press, Shanghai.

Yang, C., Liao, S.J., 2006. On the explicit, purely analytic solution of Von Karman swirling viscous flow. Communications in Nonlinear Science and Numerical Simulation 11, 83–93.

Yin, C.G., Zheng, L.C., Zhang, C.L., Zhang, X.X., 2015. Flow and heat transfer of nanofluids over a rotating porous disk with velocity slip and temperature jump. Zeitschrift für Naturforschung 70 (5), 351–358.

Yun, Y.S., Temuer, C., 2015. Application of the homotopy perturbation method for the large deflection problem of a circular plate. Applied Mathematical Modelling 39, 1308–1316.

Zhang, Y., Zheng, L.C., 2012. Analysis of MHD thermosolutal Marangoni convection with the heat generation and a first-order chemical reaction. Chemical Engineering Science 69, 449–455.

Zhang, C.L., Zheng, L.C., Zhang, X.X., Chen, G., 2015. MHD flow and radiation heat transfer of nanofluids in porous media with variable surface heat flux and chemical reaction. Applied Mathematical Modelling 39, 165–181.

Zhao, J.H., Zheng, L.C., Zhang, X.X., Liu, F.W., 2016. Unsteady boundary layer natural convection heat transfer of fractional Maxwell viscoelastic fluid over a vertical plate. International Journal of Heat and Mass Transfer 97, 760–766.

Zhao, J.K., 1988. Differential Transformation and its Application in Circuits. China University of science and Technology Press, Wuhan.

Zheng, L.C., Zhang, X.X., 2013. Modern Analysis Methods in Nonlinear Partial Differential Equations (In Chinese). Science Press, Beijing.

Zheng, L.C., Zhang, X.X., He, J.C., 2003. Singular Nonlinear Boundary Value Problems in Transfer Process (In Chinese). Science Press, Beijing.

Zheng, L.C., Zhao, F.F., Zhang, X.X., 2010. Exact solutions for generalized Maxwell fluid flow due to oscillatory and constantly accelerating plate. Nonlinear Analysis: Real World Applications 11, 3744–3751.

Zheng, L.C., Guo, Z.L., Zhang, X.X., 2011a. 3D Flow of a generalized Oldroyd-B fluid induced by a constantly pressure gradient between two side walls perpendicular to the plate. Nonlinear Analysis: Real World Applications 12, 3499–3508.

Zheng, L.C., Li, C.R., Zhang, X.X., Gao, Y.T., 2011b. Exact solutions for the unsteady rotating flows of a generalized Maxwell fluid with oscillating pressure gradient between coaxial cylinders. Computers and Mathematics with Applications 62, 1105–1115.

Zheng, L.C., Liu, Y.Q., Zhang, X.X., 2011c. Exact solutions for MHD flow of generalized Oldroyd-B fluid due to an infinite accelerating plate. Mathematical and Computer Modelling 54, 780–788.

Zheng, L.C., Zhang, X.X., Karem, B., UğurYücel, E.G., Yıldırım, A., 2011d. Similarity and BPES comparative solutions to the heat transfer equation for incompressible non-Newtonian fluids:Case of laminar boundary energy equation. European Physical Journal—Applied Physics 55, 21102.

Zheng, L.C., Wang, L.J., Zhang, X.X., 2011e. Analysis solutions of unsteady flow and heat transfer on a permeable stretching sheet with non-uniform heat source/sink. Communications in Nonlinear Science and Numerical Simulation 16, 731—740.

Zheng, L.C., Wang, K.N., Gao, Y.T., 2011f. Unsteady flow and heat transfer of a generalized Maxwell fluid due to a hyperbolic sine accelerating plate. Computers and Mathematics with Applications 61 (8), 2209—2212.

Zheng, L.C., Lin, Y.H., Zhang, X.X., 2012a. Marangoni convection of power law fluids driven by power-law temperature gradient. Journal of the Franklin Institute 349 (8), 2585—2597.

Zheng, L.C., Liu, Y.Q., Zhang, X.X., 2012b. Slip effects on MHD flow of a generalized Oldroyd-B fluid with fractional derivative. Nonlinear Analysis: Real World Applications 13, 513—523.

Zheng, L.C., Niu, J.J., Ma, L.X., Zhang, X.X., 2012c. Dual solutions forflow and radiative heat transfer of a micropolar fluid over stretching/shrinking sheet. International Journal of Heat and Mass Transfer 55, 7577—7586.

Zheng, L.C., Niu, J.J., Zhang, X.X., Gao, Y.T., 2012d. MHD flow and heat transfer over a porous shrinking surface with velocity slip and temperaturejump. Mathematical and Computer Modelling 56, 133—144.

Zheng, L.C., Wang, L.J., Zhang, X.X., Ma, L.X., 2012e. Unsteady flow and heat transfer over an stretching sheet with radiation in the presence of a magnetic field. Chemical Engineering Communications 99 (1), 1—16.

Zheng, L.C., Jin, X., Zhang, X.X., Zhang, J.H., 2013a. Unsteady heat and mass transfer in MHD flow over an oscillatory stretching surface with soret and dufour effects. Acta Mechanica Sinica 29 (5), 667—675.

Zheng, L.C., Li, B.T., Lin, P., Zhang, X.X., Zhang, C.L., Zhao, B., Wang, T.T., 2013b. Sedimentation and precipitation of nanoparticles in power-law fluids. Microfluidics and Nanofluidics 15, 11—18.

Zheng, L.C., Lin, Y.H., Zhang, X.X., 2013c. MHD thermosolutal Marangoni convection of non-Newtonian power law fluids driven by temperature gradient. Journal of Heat Transfer 135, 051702.

Zheng, L.C., Liu, N., Zhang, X.X., 2013d. Maxwell fluids unsteady mixed flow and radiation heat transfer over a stretching permeable plate with boundary slip and non-uniform heat source/sink. Journal of Heat Transfer 135, 031705.

Zheng, L.C., Liu, N., Niu, J.J., Zhang, X.X., 2013e. Slip and buoyancy lift effects on the mixed flow and radiation heat transfer of a micropolar fluid towards vertical permeable plate. Journal of Porous Media 16, 031705.

Zheng, L.C., Zhang, C.L., Zhang, X.X., Zhang, J.H., 2013f. Flow and radiation heat transfer of a Nanofluid over a stretching sheet with velocity slip and temperature jump in porous medium. Journal of the Franklin Institute 350, 990—1007.

Zhu, K.Q., 2006. Progress in the study of non Newtonian fluid mechanics. Mechanics and Practice 28 (4), 1—8.

Chapter 2

Embedding-Parameters Perturbation Method

ABSTRACT

In essence, the classical perturbation method depends on the availability of some small (or large) parameters, i.e., the so-called "perturbation variable." However, in reality not all nonlinear problems have such perturbation variables, so the classical perturbation method has serious limitations. In this chapter, we introduce a novel method for solving nonlinear differential equations, i.e., the embedding-parameters perturbation method. By introducing special embedding-parameter transformations for both the independent and dependent variables, we can embed the specified small parameters into nonlinear differential equations, which then can be solved by a standard perturbation method. Solutions to the original differential equations can be obtained by substituting the asymptotic series solutions into the reversed transformation.

The proposed methods are used to solving Marangoni convection boundary layer problems. Approximate analytical solutions are obtained and the effects of involved parameters on the velocity and temperature fields are graphically presented and analyzed.

2.1 BASICS OF PERTURBATION THEORY

In the real world, differential equations that model problems are too complicated to be solved exactly with usual mathematical methods. Therefore, various forms of approximate methods or numerical methods are sought instead. The perturbation method is a powerful mathematical method in science and engineering that has been applied successfully to solve many important problems and is known as the small parameter expansion method (Cole, 1968; Kevorkian and Cole, 1981; Li, 1999). When the perturbation method is used to solve differential equations, physical equations and initial or boundary conditions are usually required to be put into a dimensionless form. In the dimensionless equation, a dimensionless parameter that can reflect physical characteristics is often chosen as the perturbation parameter.

Mathematics in Science and Engineering. http://dx.doi.org/10.1016/B978-0-12-811753-8.00002-5

Assuming that the solution can be expanded in terms of a small parameter into a formal power series, it can be substituted in the form of a series into the dimensionless equation. On the basis of these equations, we can determine the coefficients of the power series. By truncating the series, we obtain an approximate solution to the original equations. Perturbation problems consist of two common types: regular and singular.

2.1.1 Perturbation Theory

Consider solving some certain classes of differential equations with small parameters (or large parameters):

$$P_\varepsilon : \begin{cases} L_\varepsilon[u_\varepsilon] = f(x;\varepsilon), & x = (x_1, x_2, \ldots, x_n) \in \Omega \\ B_{\varepsilon,j}[u_\varepsilon] = \varphi_j(x;\varepsilon), & j = 1, 2, 3, \ldots, k \end{cases} \tag{2.1}$$

where $0 < \varepsilon << 1$, L_ε is a differential operator with a small parameter ε ($L_\varepsilon \equiv L_0 + \varepsilon L_{1,\varepsilon}$) and $B_{\varepsilon,j}$ is a differential operator defined on the boundary (or part of the boundary) of Ω, respectively.

Perturbation theory studies how to find the solution u of the problem P_ε by using the solution u_0 of the degenerate problem P_0, and also the relationship between u and u_0. To begin with, we assume that P_ε has an asymptotic expansion u_ε in the following form

$$u_\varepsilon = u_0(x) + \sum_{i=1}^{N} \delta_i(\varepsilon) u_i(x) + Z_N(x;\varepsilon), \tag{2.2}$$

where $\{\delta_i(\varepsilon)\}_{i=1}^{+\infty}$ is an asymptotic sequence such that as $\varepsilon \to 0$

$$\lim_{\varepsilon\to 0} \delta_i(\varepsilon) = 0, \quad \lim_{\varepsilon\to 0} \frac{\delta_{i+1}(\varepsilon)}{\delta_i(\varepsilon)} = 0, \quad i = 1, 2, \cdots. \tag{2.3}$$

For example, $\{\delta_i(\varepsilon)\}_{i=1}^{+\infty}$ may be chosen in the form $\{\varepsilon^i\}_i^\infty$, $\{\varepsilon^{i/3}\}_i^\infty$, $\{(\ln \varepsilon)^{-i}\}_i^\infty$, etc. The work also involves estimating the truncation term $Z_N(x;\varepsilon)$.

2.1.2 Asymptotic Expansion of Solutions

If solution $u(x;\varepsilon)$ can be expanded as Eq. (2.2), and

$$|Z_N(x;\varepsilon)| \le M_N(x)\delta_{N+1}(\varepsilon), \tag{2.4}$$

where $M_N(x)$ is a bounded function of variable x and N is any positive integer, then Eq. (2.2) is called an asymptotic approximate expansion of $u(x;\varepsilon)$ with $\{\delta_n(\varepsilon)\}$. Its first $N+1$ term $u_0(x) + \sum_{i=1}^{N} \delta_i(\varepsilon) u_i(x)$ is the Nth-order asymptotic expansion.

If in expansion Eq. (2.2) the upper bound M of the remainder is uniformly bounded on $[a, b]$, we call Eq. (2.2) a uniformly effective asymptotic expansion of $u(x;\varepsilon)$ on $[a, b]$.

For example, an asymptotic expansion of *sin*$(x + \varepsilon)$ is

$$\sin(x+\varepsilon) = \begin{cases} \sin x + \varepsilon \cos x + \cdots + \varepsilon^N \dfrac{(-1)^{N/2}}{N!} \sin x + O\left(\varepsilon^{N+1}\right), & N = 2k, \ k \in Z \\ \sin x + \varepsilon \cos x + \cdots + \varepsilon^N \dfrac{(-1)^{(n-1)/2}}{N!} \cos x + O\left(\varepsilon^{N+1}\right), & N = 2k+1, \ k \in Z \end{cases},$$

which is uniformly effective on $(-\infty, \ \infty)$. Its first-order asymptotic approximation is

$$\sin(x+\varepsilon) \approx \sin x + \varepsilon \cos x. \tag{2.5}$$

2.1.3 Regular Perturbation and Singular Perturbation

If the solution of $u(x; \varepsilon)$ can be expressed in an asymptotic power series of ε

$$u(x,\varepsilon) = u_0(x) + \sum_{n=1}^{\infty} \varepsilon^n u_n(x), \tag{2.6}$$

which is uniformly effective on $[a, b]$, Eq. (2.6) corresponds to a regular perturbation problem; otherwise expansion Eq. (2.6) corresponds to a singular perturbation problem. In the regular case, the first few terms of the asymptotic expansions can approximate the original solution within a certain precision range. Practical problems are generally approximate in nature and require only some accuracy. The perturbation method can provide an approximate solution for such practical problems.

The perturbation method makes up for the shortcomings of solutions from a purely numerical method. It has the characteristics of an analytical method, so it is also called a semianalytical method. In its application to solve the asymptotic expansion problem, the coefficients in the expansion formula are often determined by some simple differential equations. Thus the method is an important tool for studying differential equations with nonlinear or higher-order or variable coefficients. Numerous applications can be found in optics, acoustics, chemistry, biology, mathematics, celestial mechanics, fluid mechanics, solid mechanics, and quantum mechanics (Chang and Cheng, 1983; Grasman and Matkowsky, 1977; Polyanin, 1984; Stein et al., 2002).

2.2 EMBEDDING-PARAMETER PERTURBATION

The classical perturbation theory works by finding an approximate solution to the problem with small parameters. However, in reality many nonlinear problems do not have such small parameter perturbation variables to overcome the shortcomings of the classical perturbation method. In this chapter, we introduce a novel method for solving nonlinear differential equations, i.e., the embedding-parameters perturbation method (EPPM), proposed by

Professor Zheng's doctoral student, Yan Zhang, during her doctoral work at the University of Science and Technology Beijing. The method can be used to solve many differential equations without small parameters.

Consider solving the following problem in which there are no parameters:

$$P_0:\begin{cases} L[u(x)]=f(x), & x\in\Omega \\ u^{(j)}(a_j)=b_j, & j=0,1,\ldots,n-1, a_j\in(\partial\Omega\cap\Omega) \end{cases}, \tag{2.7}$$

where L is an nth-order differential operator, a_j and b_j are both constants, and Ω is an interval including the origin.

First, a special transformation is introduced as

$$u(x)=\varepsilon^a g(\xi)+\sum_{k=0}^{n-1}\frac{u^{(k)}(0)}{k!}x^k, \quad x=\varepsilon^b\xi, \tag{2.8}$$

where ε is an artificial small parameter, ξ is an argument parameter, g is a dependent variable parameter, and a, b are both constants to be determined. Substituting Eq. (2.8) into Eq. (2.7), we obtain

$$P_1:\begin{cases} L_\varepsilon[g(\varepsilon,\xi)]=h(\xi), & \xi\in\Omega_1 \\ g^{(j)}(0)=0, & j=0,1,2,\ldots,n-1 \end{cases}, \tag{2.9}$$

where L_ε is also an nth-order differential operator, $h(\xi)$ is a function of ξ, and Ω_1 is another interval including the origin.

With the artificial small parameters properly determined and introduced, the double-parameters transformation includes the independent variable and the unknown function. Then, the problems for which without parameters are converted into an initial value problem involving an artificial small parameter.

Using a standard perturbation expanding method, we obtain the solution of P_1 as

$$g(\xi,\varepsilon)=\sum_{k=0}^{\infty}\varepsilon^k g_k\xi^k. \tag{2.10}$$

Once we substitute Eq. (2.10) into Eq. (2.8), the artificial small parameter is automatically eliminated, so we obtain the solution to the original differential Eq. (2.7). This method is called the EPPM.

In the following discussion, we verify the reliability and efficiency of the proposed method by applying it to studying the Blasius and Sakiadis flow problems.

2.2.1 Approximate Solution to Blasius Flow

Consider the solution to the classical Blasius equation (Howarth, 1938)

$$2f'''(\eta)+f(\eta)f''(\eta)=0, \quad \eta\in[0,+\infty), \tag{2.11}$$

subject to the boundary conditions

$$f(0)=0,\quad f'(0)=0,\quad f'(+\infty)=1 \tag{2.12}$$

Define a Crocco transformation as

$$y=f''(\eta),\quad x=f'(\eta), \tag{2.13}$$

where x is the dimensionless tangential velocity and $y(x)$ is the dimensionless shear force. Substituting Eq. (2.13) into Eqs. (2.11) and (2.12), we obtain the following singular nonlinear two-point boundary value problem

$$2yy''+x=0, \tag{2.14}$$

$$y'(0)=0,\quad y(1)=0. \tag{2.15}$$

An initial parameter value is assumed as:

$$f''(0)=y(0)=\alpha. \tag{2.16}$$

The embedding-parameters transformation is introduced as

$$x=\varepsilon^b\xi,\quad y=\varepsilon^a g+\alpha, \tag{2.17}$$

where ε is an artificial small parameter. Substituting Eq. (2.17) into Eq. (2.14) results in the following equation

$$\varepsilon^{2a-2b}2gg''+\varepsilon^{a-2b}2\alpha g''+\varepsilon^b\xi=0. \tag{2.18}$$

The initial conditions in Eqs. (2.15) and (2.16) become

$$g(0)=0,\quad g'(0)=0. \tag{2.19}$$

For ease of solving Eq. (2.18), we choose

$$a=3,\quad b=1. \tag{2.20}$$

Then Eq. (2.18) can be rewritten as

$$2\alpha g''+\xi+\varepsilon^3 2gg''=0. \tag{2.21}$$

The solution to Eq. (2.21) can be obtained by expanding g into a power series near $\varepsilon=0$ as follows:

$$g(\xi)=g_0+g_1\varepsilon+g_2\varepsilon^2+g_3\varepsilon^3+g_4\varepsilon^4+\cdots. \tag{2.22}$$

Substituting Eq. (2.22) into Eq. (2.21) and equating coefficients of the same power of ε, we obtain

$$\varepsilon^0: 2\alpha g_0''+\xi=0, \tag{2.23}$$

$$\varepsilon^1: g_1''=0, \tag{2.24}$$

$$\varepsilon^2: g_2''=0, \tag{2.25}$$

$$\varepsilon^3 : \alpha g_3'' + g_0 g_0'' = 0, \tag{2.26}$$

$\cdots$.

Solving the foregoing equations, we obtain

$$g(\xi) = -\frac{\xi^3}{12\alpha} - \varepsilon^3 \frac{\xi^6}{720\alpha^3} - \varepsilon^6 \frac{\xi^9}{6912\alpha^5} - \varepsilon^9 \frac{\xi^{12}}{138,240\alpha^7} + \cdots. \tag{2.27}$$

Therefore, by using the reversed transformation of Eq. (2.17), we obtain

$$y(x,\alpha) = \alpha - \frac{x^3}{12\alpha} - \frac{x^6}{720\alpha^3} - \frac{x^9}{6912\alpha^5} - \frac{x^{12}}{138,240\alpha^7} + \cdots. \tag{2.28}$$

Employing the right boundary condition $y(1) = 0$ in Eq. (2.15), we obtain the skin friction coefficient of $\alpha = 0.3351$. The reliability and efficiency of the proposed approximate value are verified by a numerical result of Howarth (1938) with $\alpha = 0.33206$; the absolute error is no more than 0.00304.

2.2.2 Approximate Solutions to Sakidias Flow

Now we consider the root-finding problem for the Sakiadis equation

$$f'''(\eta) + f(\eta)f''(\eta) = 0, \quad \eta \in [0, +\infty), \tag{2.29}$$

Subject to the boundary conditions

$$f(0) = 0, \ f'(0) = 1, \ f'(+\infty) = 0. \tag{2.30}$$

Introducing the Crocco variables of $y = f''(\eta), \ x = f'(\eta)$, Eqs (2.29) and (2.30) are converted into

$$yy'' + x = 0, \tag{2.31}$$

$$y(0) = 0, \ y'(1) = 0. \tag{2.32}$$

An initial parameter value is assumed as

$$f''(0) = y(1) = \alpha. \tag{2.33}$$

Let

$$t = x - 1, \ h(t) = y(x). \tag{2.34}$$

Substituting Eq. (2.34) into Eqs. (2.31) and (2.32), we obtain the following singular two-point boundary value problem:

$$hh'' + t + 1 = 0, \tag{2.35}$$

$$h(0) = \alpha, \ h'(0) = 0, \ h(-1) = 0. \tag{2.36}$$

The embedding-parameters transformation is introduced as

$$h = \varepsilon^a g + \alpha, \ t = \varepsilon^b \xi. \tag{2.37}$$

Substituting Eq. (2.37) into Eq. (2.35) results in the following equation

$$\varepsilon^{2a-2b} gg'' + \varepsilon^{a-2b} \alpha g'' + \varepsilon^b \xi + 1 = 0. \tag{2.38}$$

The initial conditions in Eq. (2.36) become

$$g(0) = 0, \;\; g'(0) = 0. \tag{2.39}$$

To use the proposed EPPM to solve Eq. (2.38), we choose

$$a = 2, \;\; b = 1. \tag{2.40}$$

Then, Eq. (2.38) is rewritten as

$$\varepsilon^2 gg'' + \varepsilon\xi + \alpha g'' + 1 = 0. \tag{2.41}$$

The solution to Eq. (2.41) can be obtained by expanding g into a power series near $\varepsilon = 0$ as:

$$g(\xi) = g_0 + g_1\varepsilon + g_2\varepsilon^2 + g_3\varepsilon^3 + g_4\varepsilon^4 + \cdots. \tag{2.42}$$

Substituting Eq. (2.42) into Eq. (2.41) and equating the coefficients of the same power of ε, yields

$$\varepsilon^0 : \alpha g_0'' + 1 = 0, \tag{2.43}$$

$$\varepsilon^1 : \alpha g_1'' + \xi = 0, \tag{2.44}$$

$$\varepsilon^2 : \alpha g_2'' + g_0 g_0'' = 0, \tag{2.45}$$

$$\varepsilon^3 : \alpha g_3'' + g_0 g_1'' + g_1 g_0'' = 0, \tag{2.46}$$

$\cdots$.

Solving the foregoing equations, we obtain

$$\begin{aligned} g(\xi) = &-\frac{1}{2\alpha}\xi^2 - \frac{1}{6\alpha}\xi^3 - \frac{1}{24\alpha^3}\xi^4 - \frac{1}{30\alpha^3}\xi^5 - \left(\frac{1}{180\alpha^3} + \frac{7}{720\alpha^5}\right)\xi^6 \\ &- \frac{59}{5040\alpha^5}\xi^7 - \left(\frac{1}{240\alpha^5} + \frac{127}{40,320\alpha^7}\right)\xi^8 - \left(\frac{1}{2160\alpha^5} + \frac{227}{45,360\alpha^7}\right)\xi^9 \\ &- \left(\frac{23}{8400\alpha^7} + \frac{4369}{3,628,800\alpha^9}\right)\xi^{10} + \cdots. \end{aligned} \tag{2.47}$$

Therefore,

$$\begin{aligned} h(t,\alpha) = \alpha &- \frac{1}{2\alpha}t^2 - \frac{1}{6\alpha}t^3 - \frac{1}{24\alpha^3}t^4 - \frac{1}{30\alpha^3}t^5 - \left(\frac{1}{180\alpha^3} + \frac{7}{720\alpha^5}\right)t^6 \\ &- \frac{59}{5040\alpha^5}t^7 - \left(\frac{1}{240\alpha^5} + \frac{127}{40,320\alpha^7}\right)t^8 - \left(\frac{1}{2160\alpha^5} + \frac{227}{45,360\alpha^7}\right)t^9 \\ &- \left(\frac{23}{8400\alpha^7} + \frac{4369}{3,628,800\alpha^9}\right)t^{10} + \cdots. \end{aligned} \tag{2.48}$$

Using the condition of $h(-1)=0$, we obtain the skin friction coefficient $\alpha=-0.6278$. The reliability and efficiency of the proposed approximate value are verified by a numerical result of Zheng et al. (2004) with $\alpha=-0.6254$, with an absolute error no more than 0.0024.

These examples show the effectiveness of the EPPM. In the following section, we shall apply the method to studying two types of Marangoni convection problems.

2.3 MARANGONI CONVECTION

Marangoni convection, which is induced by a surface tension gradient on the interface, is an important physical phenomenon under microgravity conditions. In the mid-1860s, Marangoni found the phenomenon that the natural convection dominated by liquid gravity gradually disappeared in a microgravity environment, whereas at the interface of liquid, the surface tension has a leading role and causes a surface tension gradient.

Marangoni convection appears in many applications. For example, it occurs around vapor bubbles during nucleation and the growth of vapor bubbles resulting from the variation in surface tension caused by temperature and/or concentration variations along the surface of the bubble. Experimental tests and numerical analysis of nucleate boiling have shown that heat transfer resulting from Marangoni flow can be significant under microgravity and may also be important in earth's gravity. Marangoni-induced flow is also important in crystal growth melts, in which the flow produces undesirable effects under microgravity in the same manner as buoyancy-induced natural convection.

The dissipative layers that occur along the liquid–gas or liquid–liquid interfaces are called Marangoni boundary layers. According to the different origins, the effects of Marangoni are divided into the thermal effect of Marangoni (EMT) and the solute Marangoni effect (EMS). EMT is caused by the thermal imbalance of the interfacial region; this imbalance is mainly caused by the heat source and temperature gradient. EMS is caused by the imbalance of the interfacial adsorption; this imbalance mainly results from the chemical reaction and the concentration gradient. The basic mechanism of EMT and EMS has been extensively investigated. For EMT, Pearson (1958) created the initial model and criterion. A thermal gradient along the surface results in a surface tension gradient along the direction of flow (Cazabat et al., 1990). Johnson and Narayanan (1999) provided a description of the various physics associated with convection in multiple fluid layers that are constrained by side walls. Experiments showed that the fluid surface tension is connected to the temperature; generally the surface tension decreases as the temperature increases (Arafune and Hirata, 1999). For EMS, Scriven and Sternling (1960) created the initial model and criterion. Bergman (1986) presented a numerical simulation of double-diffusive Marangoni convection. Chu and Velarde (1991) considered Korteweg–de Vries soliton excitation in Benard–Marangoni convection. Bergeon et al. (1998) considered a binary fluid contained in a rectangular cavity. Both the temperature and concentration gradient contribute to the initiation of convection. Boeck and Thess (1998) studied a lengthy direct numerical simulation of two-dimensional Benard–Marangoni convection at a low Prandtl number Pr driven exclusively by surface tension gradients. In general, the surface tension is assumed to vary linearly with temperature, or as a quadratic relationship:

$$\delta = \delta_0[1 - \gamma_T(T - T_\infty)] \tag{2.49}$$

The dependence of surface tension on both the temperature and concentration can be expressed as

$$\delta = \delta_0 - \gamma_T(T - T_\infty) - \gamma_C(C - C_\infty), \tag{2.50}$$

where

$$\gamma_T = \partial\delta/\partial T|_C, \ \ \gamma_C = \partial\delta/\partial C|_T. \tag{2.51}$$

In some special cases, it is also assumed that the surface tension is a quadratic function of the temperature:

$$\delta = \delta_0 + \frac{\gamma_T}{2}(T - T_\infty)^2, \quad \gamma_T = \frac{d^2\delta}{dT^2}\Big|_{T=T_\infty}, \tag{2.52}$$

where δ_0 is a positive constant and δ is the surface tension. A list of literature is given Boeck and Thess (2001), Chen (2003, 2007), Chistopher and Wang (2001), Hagstrom and Doering (2010), Hamid et al. (2011), Magyari and Chamkha (2007, 2008), Mikishev and Nepomnyashchy (2010), Mudhaf and Chamkha (2005), Naimi et al. (2000), Nepomnyashchy and Simanovskii (2009), Pereira and Kalliadasis (2008), Rongy and Wit (2006), Savino and Fico (2004), Svensson et al. (2006), Würger (2007), Zhang et al. (2007), and Zheng et al. (2006a, 2008a).

2.4 MARANGONI CONVECTION IN A POWER LAW NON-NEWTONIAN FLUID

In nature there is a large amount of non-Newtonian fluid, such as grease; paint; milk; toothpaste; mud; human and various types of animal blood; intraarticular synovial fluid; lymph, cell, and cerebrospinal fluid; bronchial secretion; electrorheological fluid, etc. Non-Newtonian fluid mechanics has been widely used in many fields such as the metallurgical, chemical, materials, petroleum, light, and food industries, national defense, and so on.

Non-Newtonian fluid does not obey the assumption of Newtonian fluid that the stress tensor is directly proportional to the deformation tensor which cannot be described by the conventional theory, various different constitutive equations were proposed to describe the flow and heat transfer mechanisms, among them the Ostwaald–de Waele model or the power law model is much attractive in which the relation between shear stress and velocity gradient can be described by a so-called power law. The flow of power law fluid is widely used in physics, chemistry and other industries, and the boundary layer assumption was adopted to simplify the Navier–Stokes equations.

Since the 1960s, considerable attention has been devoted to the problem of how to predict the behavior of non-Newtonian fluids. The main reason for this is probably that fluids (such as molten plastics, pulps, slurries, and emulsions), which do not obey the Newtonian law postulating that the stress tensor is directly proportional to the deformation tensor, are produced industrially in increasing quantities. Understanding the nature of this force by mathematical modeling with a view to predicting the associated behavior of fluid flow has been the focus of considerable research work. Pop et al. (1991) and Pop (1993) proposed a model in which the thermal conductivity of non-Newtonian fluids was power law–dependent on the velocity gradient. Based on this assumption, Gorla et al. (1992) and Ece and Buyuk (2002) performed a boundary-layer analysis for free convection flow over a vertical flat plate embedded in a porous medium saturated by a power law non-Newtonian fluid, and gave the similarity solution to the classical boundary layer equations of the power law wall plume problem.

Zheng and colleagues have performed research work on abnormal flow and the heat transfer of power law fluids, in which the effects of power law fluid viscosity on a temperature field were taken into account with a modified Fourier's law (Jiao et al., 2015, 2016; Li et al., 2011a,b; Lin et al., 2014; Zhang and Zheng, 2014; Zheng et al., 2006b, 2008b, 2011, 2013). The governing partial differential equations are reduced to ordinary differential equations by suitable similarity transformations. Similarity equations subject to variable initial boundary conditions are solved analytically or numerically and the transport characteristics for velocity and temperature fields were analyzed and discussed.

2.4.1 Marangoni Convection Caused by Temperature Gradient

We consider here the Marangoni convection heat transfer of a power law non-Newtonian fluid caused by a linear temperature gradient. It is assumed that the physical properties of the fluid are constants and the surface tension is a linear function of temperature:

$$\sigma = \sigma_0[1 - \gamma(T - T_0)], \tag{2.53}$$

where σ is the surface tension, σ_0 is positive constant, and $\gamma = -\frac{1}{\sigma_0}\frac{\partial \sigma}{\partial T}$ is the surface tension temperature coefficient. It is known that for most fluids, the surface tension is a decreasing function of temperature, so the value of γ is positive.

The shear force and velocity gradient satisfy the power law relation

$$\tau = K\left|\frac{\partial u}{\partial y}\right|^{n-1} \cdot \frac{\partial u}{\partial y}, \tag{2.54}$$

where K is the kinematic viscosity and n is the power law exponent. The case $n = 1$ corresponds to a Newtonian fluid, the case $0 < n < 1$ is a "power law" relation to model pseudoplastic non-Newtonian fluids, and $n > 1$ describes the dilatant fluid. It is assumed that the interfacial surface tension gradient is caused by the temperature gradient at the interface induced flow, satisfying

$$\frac{\partial \sigma}{\partial x} = \frac{\partial \sigma}{\partial T} \cdot \frac{\partial T}{\partial x}. \tag{2.55}$$

2.4.2 Mathematical Formulation

Consider a horizontal free surface; the Marangoni convection boundary layer flow of a power law fluid is induced by the surface temperature along the surface direction. Assuming that the temperature varies linearly along the surface, by neglecting the volume variation, the external pressure, and the energy dissipation, the dimensionless form of the mass conservation, momentum conservation, and energy conservation boundary layer equations can be expressed as, respectively,

$$\frac{\partial u}{\partial x} + \frac{\partial v}{\partial y} = 0, \tag{2.56}$$

$$u\frac{\partial u}{\partial x} + v\frac{\partial u}{\partial y} = \frac{K}{\rho}\frac{\partial}{\partial y}\left(\left|\frac{\partial u}{\partial y}\right|^{n-1}\frac{\partial u}{\partial y}\right), \tag{2.57}$$

$$u\frac{\partial T}{\partial x} + v\frac{\partial T}{\partial y} = \alpha\frac{\partial}{\partial y}\left(\left|\frac{\partial u}{\partial y}\right|^{n-1}\frac{\partial T}{\partial y}\right). \tag{2.58}$$

The corresponding boundary conditions are

$$y = 0: \quad K\left|\frac{\partial u}{\partial y}\right|^{n-1}\frac{\partial u}{\partial y} = \frac{\partial \sigma}{\partial T}\frac{\partial T}{\partial x}, \quad v = 0, \ \ T = T_0 + Ax, \tag{2.59a}$$

$$y = +\infty: \quad u = 0, \ \ T = T_0, \tag{2.59b}$$

where K and A are positive constants.

In view of the definition of the stream function, $u = \frac{\partial \psi}{\partial y}$ and $v = -\frac{\partial \psi}{\partial x}$, we introduce the similarity variables as

$$\eta = C_1 x^{-\frac{1}{3}} y, \ \ \psi(x, y) = C_2 x^{\frac{2}{3}} f(\eta), \ \ T = T_0 + Ax\theta(\eta), \tag{2.60}$$

where ψ is the dimensionless stream function and C_1 and C_2 are constants that can be established in terms of similarity solutions as

$$C_1 = \left(\frac{\rho^n}{K^2(A\sigma_0\gamma)^{n-2}}\right)^{\frac{1}{3n}}, \quad C_2 = \left(\frac{K(A\sigma_0\gamma)^{2n-1}}{\rho^{2n}}\right)^{\frac{1}{3n}}. \tag{2.61}$$

Substituting Eq. (2.60) into the boundary layer governing (Eqs. 2.56–2.58). Eq. (2.56) is satisfied automatically and Eqs. (2.57) and (2.58) are transformed as

$$[f'(\eta)]^2 - 2f(\eta)f''(\eta) = 3\left[|f''(\eta)|^{n-1}f''(\eta)\right]', \tag{2.62}$$

$$3Mf'(\eta)\theta(\eta) - 2Mf(\eta)\theta'(\eta) = 3\left[|f''(\eta)|^{n-1}\theta'(\eta)\right]', \tag{2.63}$$

where $M = \frac{AK}{\rho\alpha}$ is called the Marangoni number.

The boundary conditions are reduced to

$$f(0) = 0, \;\; f''(0) = -1, \;\; f'(\infty) = 0, \tag{2.64}$$

$$\theta(0) = 1, \;\; \theta(\infty) = 0. \tag{2.65}$$

2.4.3 Embedding-Parameters Perturbation Method Solutions

Boundary condition Eq. (2.64) implies that $f''(\eta) \le 0$; thus

$$|f''(\eta)|^{n-1}f''(\eta) = (-f''(\eta))^{n-1}f''(\eta) = -(-f''(\eta))^n. \tag{2.66}$$

Substituting Eq. (2.66) into Eqs. (2.62) and (2.63), we obtain the following nonlinear ordinary differential equations

$$f'^2 - 2ff'' = 3n(-f'')^{n-1}f''', \tag{2.67}$$

$$3Mf'\theta - 2Mf\theta' = -3(n-1)(-f'')^{n-2}f'''\theta' + 3(-f'')^{n-1}\theta''. \tag{2.68}$$

First, we solve the momentum Eq. (2.67) subject to boundary condition (Eq. 2.64). It is noted from Eq. (2.64) that two boundary values exist at $\eta = 0$. However, the value $f'(+\infty) = 0$ is presented at infinity $\eta = +\infty$, which cannot be used for computational purposes. To solve the problem, an initial parameter value is assumed to be $f'(0) = c$. Therefore, we can solve Eq. (2.62) subject to the initial conditions, and then determine the unknown initial parameter c using the boundary condition $f'(+\infty) = 0$.

Using the proposed EPPM in the form of Eq. (2.8), we obtain the embedding parameter transformation as

$$f(\eta) = \varepsilon g(\xi) + \varepsilon^{\frac{1}{3}}c\xi - \varepsilon^{\frac{2}{3}}\frac{1}{2}\xi^2, \quad \xi = \varepsilon^{-\frac{1}{3}}\eta, \tag{2.69}$$

where ε is an artificial small parameter. Substituting Eq. (2.69) into Eq. (2.67), in view of $f''(\eta) \le 0$, we arrive at

$$\left(\varepsilon^{\frac{2}{3}}g' + c - \varepsilon^{\frac{1}{3}}\xi\right)^2 - 2\left(\varepsilon g + \varepsilon^{\frac{1}{3}}c\xi - \varepsilon^{\frac{2}{3}}\frac{1}{2}\xi^2\right)\left(\varepsilon^{\frac{1}{3}}g'' - 1\right) = 3nc^{n-1}\left(1 - \varepsilon^{\frac{1}{3}}g''\right)^{n-1}g''', \tag{2.70}$$

with the boundary conditions

$$g(0) = 0, \quad g'(0) = 0, \quad g''(0) = 0. \tag{2.71}$$

Expand the nonlinear term $\left(1 - \varepsilon^{\frac{1}{3}}g''\right)^{n-1}$ into a power series such as $(1-x)^{n-1}$, as

$$\begin{aligned}&\left(\varepsilon^{\frac{2}{3}}g' + c - \varepsilon^{\frac{1}{3}}\xi\right)^2 - 2\left(\varepsilon g + \varepsilon^{\frac{1}{3}}c\xi - \varepsilon^{\frac{2}{3}}\frac{1}{2}\xi^2\right)\left(\varepsilon^{\frac{1}{3}}g'' - 1\right)\\ &= 3nc^{n-1}\left(1 - \varepsilon^{\frac{1}{3}}(n-1)g'' + \varepsilon^{\frac{2}{3}}\frac{(n-1)(n-2)}{2}g''^2\right.\\ &\qquad\left. - \varepsilon\frac{(n-1)(n-2)(n-3)}{6}g''^3 + \cdots\right)g'''.\end{aligned} \tag{2.72}$$

Furthermore, we expand the unknown function g into a Taylor series of ε, as

$$g(\xi) = g_0 + g_1\varepsilon^{\frac{1}{3}} + g_2\varepsilon^{\frac{2}{3}} + g_3\varepsilon + g_4\varepsilon^{\frac{4}{3}} + \cdots. \tag{2.73}$$

Substituting Eq. (2.73) into Eq. (2.72) and equating the coefficients of the same power of ε yields

$$\begin{aligned}f(\eta) &= c\eta - \frac{1}{2}\eta^2 + \frac{c^{3-n}}{18n}\eta^3 + \frac{1}{216n^2}c^{6-2n}(n-1)\eta^4\\ &+ \frac{1}{3240n^3}\left[2c^{9-3n}n^2 - 3c^{9-3n}n + c^{9-3n} - 6c^{5-2n}n\right]\eta^5\\ &+ \frac{1}{58,320n^4}\left[6c^{12-4n}n^3 - 11c^{12-4n}n^2 + 6c^{12-4n}n - 42c^{8-3n}n^2 - c^{12-4n}\right.\\ &\qquad\left. + 42c^{8-3n}n + 18c^{4-2n}n^2\right]\eta^6 + \cdots,\end{aligned} \tag{2.74}$$

$$\begin{aligned}f'(\eta) &= c - \eta + \frac{c^{3-n}}{6n}\eta^2 + \frac{1}{54n^2}c^{6-2n}(n-1)\eta^3\\ &+ \frac{1}{648n^3}\left[2c^{9-3n}n^2 - 3c^{9-3n}n + c^{9-3n} - 6c^{5-2n}n\right]\eta^4\\ &+ \frac{1}{9720n^4}\left[6c^{12-4n}n^3 - 11c^{12-4n}n^2 + 6c^{12-4n}n - 42c^{8-3n}n^2 - c^{12-4n}\right.\\ &\qquad\left. + 42c^{8-3n}n + 18c^{4-2n}n^2\right]\eta^5 + \cdots\end{aligned} \tag{2.75}$$

In the following discussion, we determine the unknown assumed initial parameter c using the boundary condition $f'(\infty) = 0$ and the Padé approximate technique.

The diagonal Padé approximation has been used by many researchers as the most accurate approximant; therefore, we will construct only the diagonal approximants in the following discussions. The diagonal approximants of [2/2] are computed for $f'(\eta)$ by choosing a_0, a_1, a_2, a_3, a_4, b_1, and b_2 as the coefficients of $\eta^i (i = 0, 1, 2, 3, 4)$ such that

$$f'(\eta)\left(1 + b_1\eta + b_2\eta^2\right) = a_0 + a_1\eta + a_2\eta^2 + \left(a_3\eta^3 + a_4\eta^4\right). \tag{2.76}$$

Substituting Eq. (2.75) into Eq. (2.76), using the condition of $f'(\infty) = 0$, and by equating the coefficients of each term η^n, we obtain

$$cb_2 - b_1 + \frac{c^{3-n}}{6n} = 0, \tag{2.77}$$

$$-b_2 + \frac{c^{3-n}}{6n} b_1 + \frac{1}{54n^2} c^{6-2n}(n-1) = 0, \tag{2.78}$$

$$\begin{aligned}&\frac{c^{3-n}}{6n} b_2 + \frac{1}{54n^2} c^{6-2n}(n-1)b_1 \\ &\quad + \frac{1}{648n^3}\left[2c^{9-3n}n^2 - 3c^{9-3n}n + c^{9-3n} - 6c^{5-2n}n\right] = 0.\end{aligned} \tag{2.79}$$

By solving Eqs. (2.77)–(2.79), we can determine the unknown initial parameter of c.

Analogously, we may solve the energy Eq. (2.68), subject to the conditions in Eq. (2.65). To solve an initial value problem first, we need to assume an initial value of $\theta'(0) = h$, using the EPPM. In the form of Eq. (2.8), we obtain the embedding parameter transformation as

$$\theta(\eta) = \varepsilon^{\frac{2}{3}}\varphi(\xi) + \varepsilon^{\frac{1}{3}}h\xi + 1, \quad \xi = \varepsilon^{-\frac{1}{3}}\eta. \tag{2.80}$$

Upon substituting Eq. (2.80) into Eq. (2.68), the governing equation is converted into

$$\begin{aligned}&3M\left(\varepsilon^{\frac{2}{3}}g' + c - \varepsilon^{\frac{1}{3}}\xi\right)\left(\varepsilon^{\frac{2}{3}}\varphi + \varepsilon^{\frac{1}{3}}h\xi + 1\right) \\ &\quad - 2M\left(\varepsilon g + \varepsilon^{\frac{1}{3}}c\xi - \varepsilon^{\frac{2}{3}}\frac{1}{2}\xi^2\right)\left(\varepsilon^{\frac{1}{3}}\varphi' + h\right) \\ &= -3(n-1)\left(1 - \varepsilon^{\frac{1}{3}}g''\right)^{n-2} + 3\varphi''\left(1 - \varepsilon^{\frac{1}{3}}g''\right)^{n-1}.\end{aligned} \tag{2.81}$$

Expanding $\left(1-\varepsilon^{\frac{1}{3}}g''\right)^{n-2}$ and $\left(1-\varepsilon^{\frac{1}{3}}g''\right)^{n-1}$ into power series, we obtain

$$
\begin{aligned}
&3M\left(\varepsilon^{\frac{2}{3}}g'+c-\varepsilon^{\frac{1}{3}}\xi\right)\left(\varepsilon^{\frac{2}{3}}\varphi+\varepsilon^{\frac{1}{3}}h\xi+1\right)-2M\left(\varepsilon g+\varepsilon^{\frac{1}{3}}c\xi-\varepsilon^{\frac{2}{3}}\frac{1}{2}\xi^{2}\right)\left(\varepsilon^{\frac{1}{3}}\varphi'+h\right)\\
&=-3(n-1)\left(\varepsilon^{\frac{1}{3}}\varphi'+h\right)\begin{bmatrix}1-\varepsilon^{\frac{1}{3}}(n-2)g''+\varepsilon^{\frac{2}{3}}\dfrac{(n-2)(n-3)}{2}g''^{2}\\ -\varepsilon\dfrac{(n-2)(n-3)(n-4)}{6}g''^{3}\\ +\varepsilon^{\frac{4}{3}}\dfrac{(n-2)(n-3)(n-4)(n-5)}{24}g''^{4}+\cdots\end{bmatrix}\\
&+3\varphi''\begin{bmatrix}1-\varepsilon^{\frac{1}{3}}(n-1)g''+\varepsilon^{\frac{2}{3}}\dfrac{(n-1)(n-2)}{2}g''^{2}\\ -\varepsilon\dfrac{(n-1)(n-2)(n-3)}{6}g''^{3}\\ +\varepsilon^{\frac{4}{3}}\dfrac{(n-1)(n-2)(n-3)(n-4)}{24}g''^{4}+\cdots\end{bmatrix}.
\end{aligned}
\tag{2.82}
$$

The initial conditions are

$$\varphi(0)=0,\quad \varphi'(0)=0. \tag{2.83}$$

To obtain the solution to Eq. (2.82), we expand the unknown function φ at $\varepsilon=0$ into a power series

$$\varphi(\xi)=\varphi_0+\varphi_1\varepsilon^{\frac{1}{3}}+\varphi_2\varepsilon^{\frac{2}{3}}+\varphi_3\varepsilon+\varphi_4\varepsilon^{\frac{4}{3}}+\cdots. \tag{2.84}$$

Upon substituting Eq. (2.84) into Eq. (2.82) and equating the coefficients of ε with the same exponent, we obtain:

$$
\begin{aligned}
\theta(\eta)&=1+h\eta+\frac{1}{2}(nh-h+Mc)\eta^{2}\\
&+\frac{1}{18n}\left(3n^{3}h-6n^{2}h+3n^{2}Mc+3nh-3nMc+nhc^{3-n}\right.\\
&\left.-hc^{3-n}+Mnc^{4-n}-Mc^{4-n}+Mcnh-3Mn\right)\eta^{3}+\cdots.
\end{aligned}
$$

Similarly, we can establish the initial parameter value of h by using the Padé approximation.

2.4.4 Results and Discussion

Some special parameters, approximate analytical solutions, which were obtained using the embedding-parameter method, are presented in Tables 2.1 and 2.2 and Figs. 2.1–2.5.

TABLE 2.1 Influence of Power Law Exponent n on Initial Surface Velocity $f'(0)$

$n = 0.6$	$n = 0.8$	$n = 1.0$	$n = 1.1$	$n = 1.2$
1.1211644	1.142355	1.144714	1.141509	1.13625

TABLE 2.2 Influence of Power Law Exponent n and Marangoni Number M on Initial Surface Temperature Gradient $-\theta'(0)$

$-\theta'(0)$	$M = 2$	$M = 3$	$M = 4$
$n = 0.8$	1.8100513	2.2817272	2.6821645
$n = 1$	1.6230846	2.1100871	2.5214073
$n = 1.1$	1.5408525	2.0262522	2.4377732
$n = 1.2$	1.4628330	1.9432602	2.3527847

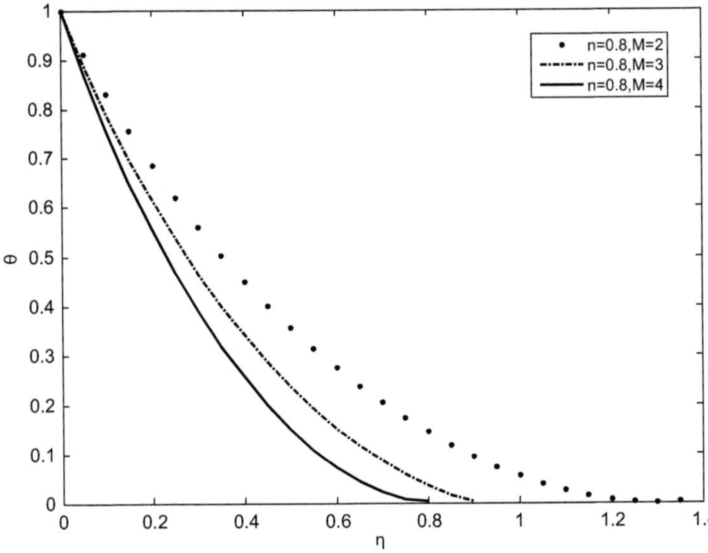

FIGURE 2.1 Effect of Marangoni number on temperature profile with $n = 0.8$.

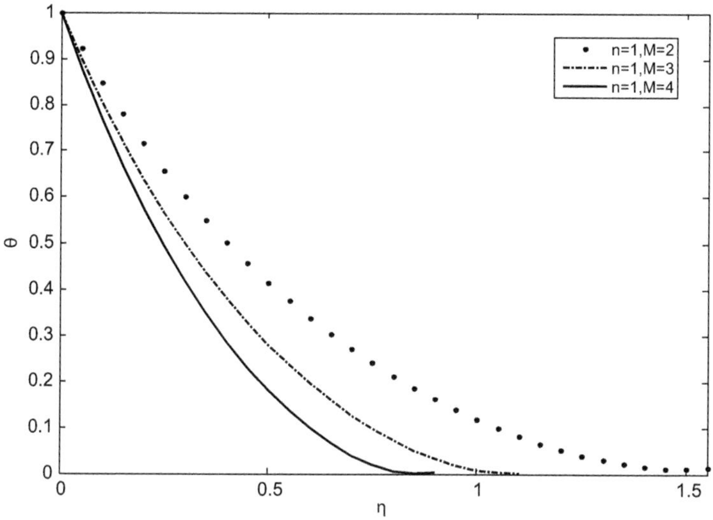

FIGURE 2.2 Effect of Marangoni number on temperature profile with $n = 1$.

Table 2.1 shows the influence of power law exponent n on initial surface velocity $f'(0)$. The results indicate that for $0 < n < 1$, the initial surface velocity $f'(0)$ increases with an increase in power law exponent n. For $n = 1$, the initial surface velocity achieves $f'(0)$ as its maximum value. However, for $n > 1$, the initial surface velocity $f'(0)$ decreases with an increase in power law

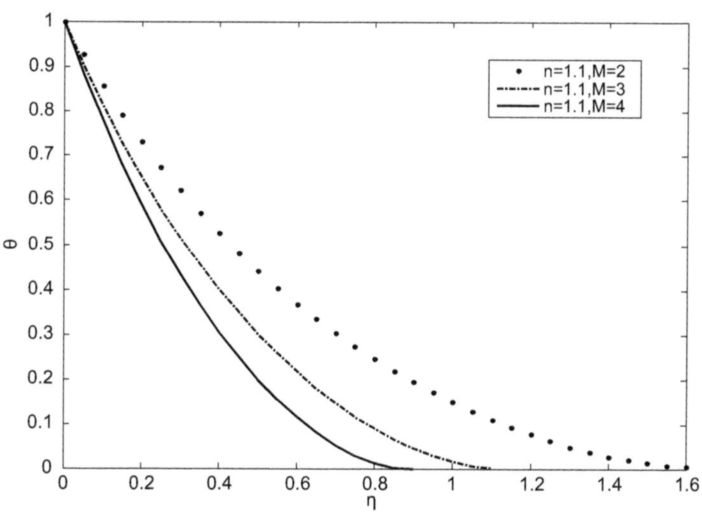

FIGURE 2.3 Effect of Marangoni number on temperature profile with $n = 1.1$.

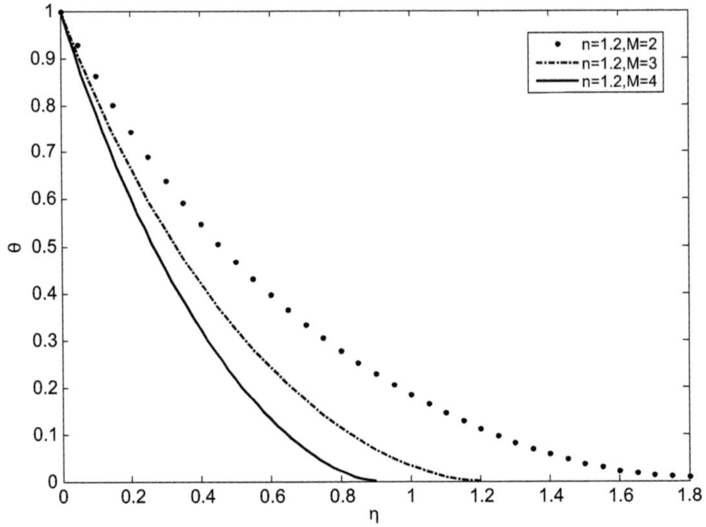

FIGURE 2.4 Effect of Marangoni number on temperature profile with $n = 1.2$.

exponent n. It implies that the Newtonian fluid ($n = 1$) has the largest surface velocity.

Table 2.2 shows the influence of power law exponent n and Marangoni number M on the initial surface temperature gradient (heat transfer coefficient) "$-\theta'(0)$".

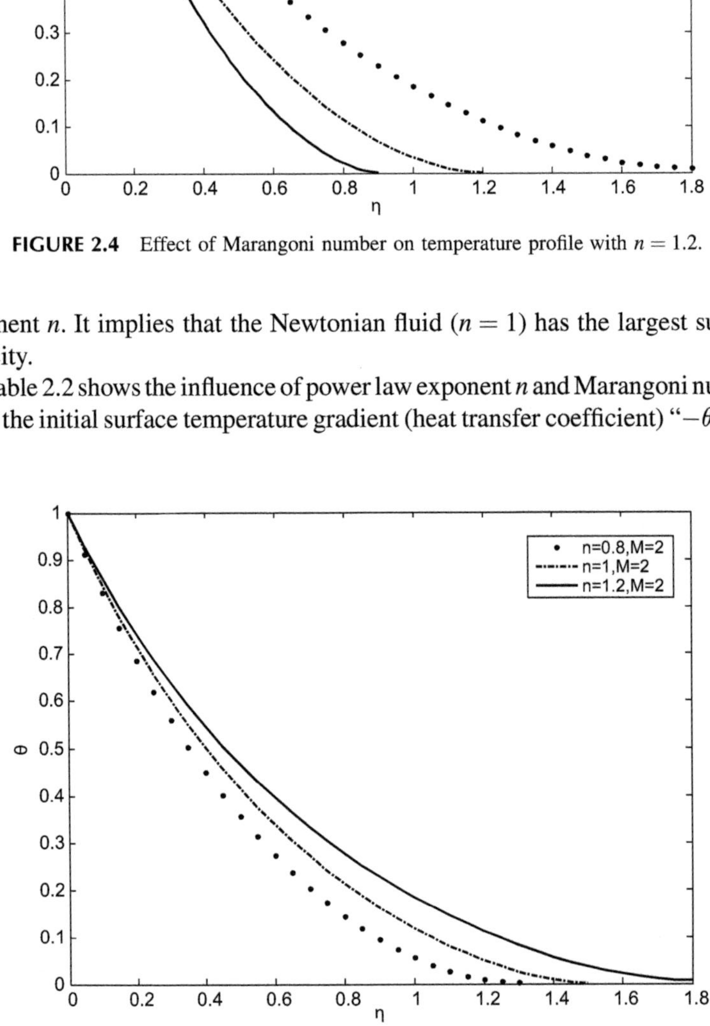

FIGURE 2.5 Effect of power law exponent n on temperature profile with Marangoni number for $M = 2.0$.

Obviously, the initial surface temperature gradient (heat transfer coefficient) is a decreasing function of power law exponent n, but it is an increasing function of Marangoni number.

Figs. 2.1–2.5 display the effects of the Marangoni number and power law exponent n on the dimensionless temperature distribution. Figs. 2.1–2.4 show the variation of the dimensionless temperature with the Marangoni number when the power law exponent is fixed. It can be seen that regardless of the values of power law exponent n ($0 < n < 1$, $n = 1$, or $n > 1$), the dimensionless temperature and the thickness of the boundary layer are a decreasing function of the Marangoni number. Fig. 2.5 shows how the dimensionless temperature varies with power exponent n when the Marangoni number is fixed. It can be observed that both the dimensionless temperature and the thickness of the boundary layer are increasing functions of the power law exponent of n.

2.5 MARANGONI CONVECTION IN FINITE THICKNESS

In the previous section, we studied the Marangoni convection of a power law fluid in a semiinfinite region caused by a temperature gradient in which the linear relationship between surface tension and temperature is assumed.

In practice, the thickness of the fluid is limited and the relationship between the surface tension and temperature are nonlinear. Because of the limited thickness of the liquid, the velocity boundary layer and the temperature boundary layer may not exist in the flow field. In this section, we study finite thickness Marangoni convection caused by a temperature gradient, and we assume that the surface tension has a quadratic relationship with the temperature gradient.

2.5.1 Background to the Problem

Until now, the Marangoni convection caused by heat transfer and mass transfer that we have considered was modeled so that the surface tension is a linear decreasing function of temperature. However, the surface tension of an aqueous solution such as fatty alcohol has a dependent relationship that first decreases and then increases as a function of temperature.

A boundary layer problem of two types of Marangoni convection caused by a temperature gradient was studied by Slavtchev and Dubovik (1992) and Slavtchev and Miladinova (1998). The surface tension of the fluid reaches a minimum value at a certain temperature, and the relationship between the fluid and the temperature can be expressed as a quadratic relationship

$$\sigma = \sigma_0 + \frac{\delta}{2}(T - T_0)^2, \tag{2.85}$$

where δ, σ_0, and T_0 are positive constants. For $T = T_0$, the surface tension takes a minimum value of σ_0, with δ as the second-order derivative of surface tension with respect to the temperature at $T = T_0$.

In this section, we consider a two-phase problem where a viscous liquid is located at the bottom whose width is not limited but whose thickness is limited, and where a static gas is located at the top. The physical parameters of the fluid are assumed to be constant. The properties and characteristics of Marangoni convection in a liquid are studied under several different temperatures. The emphasis in this study is on discussing the range of physical parameters such that the velocity and temperature boundary layers exist, and when there is a boundary layer, the thickness of the boundary layer changes with respect to the physical parameters.

2.5.2 Mathematical Model for Three Types of Conditions

Assume that the fluid is two-dimensional and incompressible. The lower boundary of the fluid is chosen as the x axis, with the y axis perpendicular to the lower boundary, as shown in Fig. 2.6.

The mass conservation, momentum conservation, and energy conservation equations of the boundary layer can be expressed as

$$\frac{\partial u}{\partial x} + \frac{\partial v}{\partial y} = 0, \tag{2.86}$$

$$q\left(u\frac{\partial}{\partial x} + v\frac{\partial}{\partial y}\right)(u, v) = -\left(\frac{\partial p}{\partial x}, \frac{\partial p}{\partial y}\right) + \mu\left(\frac{\partial^2 u}{\partial x^2}, \frac{\partial^2 u}{\partial y^2}\right) + q(0, -g), \tag{2.87}$$

$$u\frac{\partial T}{\partial x} + v\frac{\partial T}{\partial y} = \alpha\frac{\partial^2 T}{\partial y^2}. \tag{2.88}$$

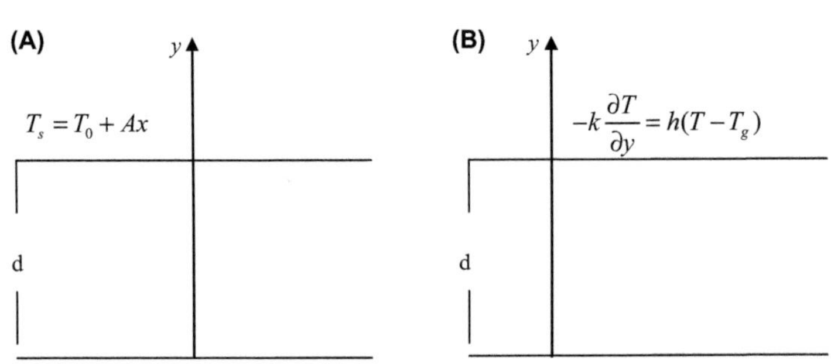

FIGURE 2.6 Boundary layer schematic with interface temperature: (A) at $y = 0$, $\frac{\partial T}{\partial y} = 0$ or $T = T_w$, at $y = d$, $T_s = T_0 + Ax$ and (B) at $y = 0$, $T = T_{w0} + Ax$, at $y = d$, $-k\frac{\partial T}{\partial y} = h\,(T - T_g)$.

The velocity field boundary conditions are

$$y = 0 : u(x, 0) = 0, \;\; v(x, 0) = 0, \tag{2.89}$$

$$y = d : v(x, d) = 0, \tag{2.90}$$

$$\mu\frac{\partial u}{\partial y} = \frac{\partial \sigma}{\partial x} = \delta(T - T_0)\frac{\partial T}{\partial x}. \tag{2.91}$$

Three types of temperature field boundary conditions will be considered:
Case I (Fig. 2.6A):

$$y = 0, \;\; \frac{\partial T}{\partial y} = 0; \quad y = d, \;\; T = T_0 + Ax, \tag{2.92}$$

Case II (Fig. 2.6A):

$$y = 0, \;\; T = T_w; \quad y = d, \;\; T = T_0 + Ax, \tag{2.93}$$

Case III (Fig. 2.6B):

$$y = 0, \;\; T = T_{w0} + Ax; \quad y = d, \;\; -k\frac{\partial T}{\partial y} = h(T - T_g). \tag{2.94}$$

Under Cases I and II, the lower surface of the liquid satisfies the condition that the heat exchange rate is zero or that the surface is of constant temperature, whereas on the upper surface, i.e., the free surface, there is a temperature gradient A ($A > 0$). Note that T_w is the temperature of the lower surface of the liquid, which is different from T_0.

The conditions in Case III are different from those in Cases I and II. A temperature gradient exists on the lower surface of the liquid, and heat exchange on the free surface satisfies Newton's law, in which T_{w0} is the initial temperature of the liquid lower surface, T_g is the temperature of the top gas, k is the thermal conductivity, and h is the heat transfer coefficient of the liquid surface.

We assume that the problem has a similarity solution. Thus, we introduce a similarity transformation as

$$\bar{x} = \frac{x}{d}, \;\; \eta = \frac{y}{d}, \;\; u = \frac{\delta A^2 d^2}{\mu}\bar{x}F'(\eta), \;\; v = -\frac{\delta A^2 d^2}{\mu}F(\eta),$$
$$p - p_g = \delta A^2 d\left[\frac{\lambda}{2}\bar{x}^2 + p(\eta)\right], \;\; T = T_1\theta_1(\eta) + Ad\,\bar{x}\;\theta(\eta). \tag{2.95}$$

where p_g is the pressure of gas and λ is an unknown constant. In Case I, $T_1 = T_0$; In Case II, $T_1 = T_w$; In Case III, $T_1 = T_{w0}$.

The momentum equation is transformed into a nonlinear ordinary differential equation

$$F''' + M\left(FF'' - F'^2\right) = \lambda, \tag{2.96}$$

subject to the boundary conditions

$$F(0) = F'(0) = 0,\ \ F(1) = 0,\ \ F''(1) = \theta^2(1), \tag{2.97}$$

where $M = q\delta A^2 d^3/\mu^2$ is the Marangoni number, which is related to the surface tension of the fluid. Parameter λ is the pressure coefficient of the fluid, which is a function of the Marangoni number.

The energy equation is transformed into two nonlinear ordinary differential equations

$$\theta_1'' + M\mathrm{Pr}F\theta_1' = 0, \tag{2.98}$$

$$\theta'' + M\mathrm{Pr}(F\theta' - F'\theta) = 0, \tag{2.99}$$

where θ_1 and θ are both functions of the variable y, and $\mathrm{Pr} = \frac{\mu}{qK}$ is the Prandtl number.

The boundary conditions for energy equations are:
Case I:

$$\theta_1'(0) = 0,\ \ \theta_1(1) = 1,\ \ \theta'(0) = 0,\ \ \theta(1) = 1, \tag{2.100}$$

Case II:

$$\theta_1(0) = 1,\ \ \theta_1(1) = \frac{T_0}{T_w},\ \ \theta(0) = 0,\ \ \theta(1) = 1, \tag{2.101}$$

Case III:

$$\theta_1(0) = 1,\ \ \theta_1'(1) + Bi\left[\theta_1(1) - \frac{T_g}{T_{w0}}\right] = 0,\ \ \theta(0) = 1,\ \ \theta'(1) + Bi\theta(1) = 0. \tag{2.102}$$

2.5.3 Embedding-Parameters Perturbation Method Solutions and Discussion

2.5.3.1 Solutions for Cases I and II

i) Solutions of the velocity field.

Eqs. (2.96) and (2.97) constitute a nonlinear boundary value problem. We treat it by solving an initial value problem. In doing so, we assume an initial value of $F''(0) = c$, in the form of Eq. (2.8), and we obtain an embedding-parameter transformation with an artificial small parameter ε as

$$F(\eta) = \varepsilon g(\xi) + \frac{c}{2}\eta^2 = \varepsilon g(\xi) + \varepsilon^{\frac{2}{3}}\frac{c}{2}\xi^2,\ \ \ \xi = \varepsilon^{-\frac{1}{3}}\eta. \tag{2.103}$$

Eq. (2.96) changes into the form

$$g''' + M\left(\left(\varepsilon g + \varepsilon^{\frac{2}{3}}\frac{c}{2}\xi^2\right)\left(\varepsilon^{\frac{1}{3}}g'' + c\right) - \left(\varepsilon^{\frac{2}{3}}g' + \varepsilon^{\frac{1}{3}}c\xi\right)^2\right) = \lambda, \tag{2.104}$$

with initial conditions

$$g(0) = 0, \;\; g'(0) = 0, \;\; g''(0) = 0. \tag{2.105}$$

Expanding the unknown function $g(\xi)$ at $\varepsilon = 0$ into a power series

$$g(\xi) = g_0 + g_1\varepsilon^{\frac{1}{3}} + g_2\varepsilon^{\frac{2}{3}} + g_3\varepsilon + g_4\varepsilon^{\frac{4}{3}} + \cdots. \tag{2.106}$$

Similar to the previous section, substituting Eq. (2.106) into Eq. (2.104) and comparing the coefficients of ε with the same power, and then by substituting back the obtained asymptotic series solution into the reversed transformation, we obtain

$$\begin{aligned} F(\eta) = {} & \frac{c}{2}\eta^2 + \frac{\lambda}{6}\eta^3 + \frac{Mc^2}{120}\eta^5 + \frac{M\lambda c}{360}\eta^6 + \frac{M\lambda^2}{2520}\eta^7 - \frac{M^2c^3}{40,320}\eta^8 - \frac{M^2\lambda c^2}{90,720}\eta^9 \\ & + \frac{M^3c^4}{1,478,400}\eta^{11} + \frac{181M^3c^3\lambda}{479,001,600}\eta^{12} - \frac{317M^4c^5}{29,059,430,400}\eta^{14}. \end{aligned} \tag{2.107}$$

The error of this solution is bounded by

$$\left|\frac{M^3\lambda^4}{2,476,656,000} - \frac{M^4c^4\lambda}{121,927,680}\right|. \tag{2.108}$$

Therefore

$$\begin{aligned} F' = {} & c\eta + \frac{\lambda}{2}\eta^2 + \frac{Mc^2}{24}\eta^4 + \frac{M\lambda c}{60}\eta^5 + \frac{M\lambda^2}{360}\eta^6 - \frac{M^2c^3}{5040}\eta^7 - \frac{M^2\lambda c^2}{10,080}\eta^8 \\ & - \frac{M^2c\lambda^2}{22,680}\eta^9 + \frac{M^3c^4}{134,400}\eta^{10} + \frac{181M^3c^3\lambda}{39,916,800}\eta^{11} - \frac{317M^4c^5}{2,075,673,600}\eta^{13}. \end{aligned} \tag{2.109}$$

Using the boundary conditions $F(1) = 0$ and $F''(1) = \theta^2(1) = 1$, we can obtain the assuming initial values of parameters c and λ.

ii) Solution of temperature fields.

Case I:

We first solve Eq. (2.98), which is similar to solving the momentum equation of the velocity field. Clearly, $\theta_1 = 1$. Then, we solve Eq. (2.99) in the form of Eq. (2.8) by assuming initial value $\theta(0) = \gamma$. We obtain the embedding-parameter transformation with artificial small parameter ε as

$$\theta(\eta) = \varepsilon^{\frac{2}{3}}\phi(\xi) + \gamma, \quad \xi = \varepsilon^{-\frac{1}{3}}\eta. \tag{2.110}$$

Analogously, we can obtain the asymptotic series solution, and then by substituting the obtained asymptotic series solution into the reversed transformation, we have

$$\begin{aligned}\theta(\eta) &= \gamma + \frac{M\Pr\gamma c}{6}\eta^3 + \frac{M\Pr\gamma\lambda}{24}\eta^4 + \left(-\frac{M^2\Pr^2\gamma c^2}{360} + \frac{M^2\Pr\gamma c^2}{720}\right)\eta^6 \\ &+ \left(-\frac{M^2\Pr^2\gamma\lambda c}{1008} + \frac{M^2\Pr\gamma\lambda c}{2520}\right)\eta^7 + \left(-\frac{M^2\Pr^2\gamma\lambda^2}{8064} + \frac{M^2\Pr\gamma\lambda^2}{20,160}\right)\eta^8 \\ &+ \left(\frac{M^3\Pr^3\gamma c^3}{12,960} - \frac{M^3\Pr\gamma c^3}{362,880}\right)\eta^9 + \cdots. \end{aligned} \tag{2.111}$$

Using the condition of $\theta(1) = 1$, we can obtain the parameter value γ.
Case II: Assuming two initial values

$$\theta_1'(0) = a, \;\; \theta'(0) = b. \tag{2.112}$$

By the embedding-parameter transformation similar that shown before, we obtain:

$$\begin{aligned}\theta_1(\eta) &= 1 + a\eta - \frac{M\Pr ac}{24}\eta^4 - \frac{M\Pr a\lambda}{120}\eta^5 + \left(\frac{M^2\Pr^2 ac^2}{504} - \frac{M^2\Pr ac^2}{5040}\right)\eta^7 \\ &+ \left(\frac{M^2\Pr^2 ac\lambda}{1152} - \frac{M^2\Pr ac\lambda}{20,160}\right)\eta^8 + \left(\frac{M^2\Pr^2 a\lambda^2}{10,368} - \frac{M^2\Pr a\lambda^2}{181,440}\right)\eta^9 \\ &+ \left(\frac{M^3\Pr^2 ac^3}{43,200} - \frac{M^3\Pr^3 ac^3}{12,960} + \frac{M^3\Pr ac^3}{3,628,800}\right)\eta^{10} + \cdots, \end{aligned} \tag{2.113}$$

and

$$\begin{aligned}\theta(\eta) &= b\eta + \frac{M\Pr bc}{24}\eta^4 + \frac{M\Pr b\lambda}{60}\eta^5 + \frac{M^2\Pr bc^2 - M^2\Pr^2 bc^2}{1008}\eta^7 \\ &+ \frac{1}{40,320}\left(12M^2\Pr bc\lambda - 23M^2\Pr^2 bc\lambda\right)\eta^8 \\ &+ \frac{1}{25,920}\left(M^2\Pr b\lambda - 2M^2\Pr^2 b\lambda^2\right)\eta^9 \\ &+ \left(\frac{1}{181,440}M^3\Pr^3 bc^3 - \frac{1}{60,4800}M^3\Pr^2 bc^3 + \frac{1}{453,600}M^3\Pr bc^3\right)\eta^{10} + \cdots. \end{aligned} \tag{2.114}$$

Using the boundary conditions $\theta_1(1) = T_0/T_w$ and $\theta(1) = 1$, we can obtain the assuming initial parameters c and b.

Fig. 2.7 shows the typical velocity profiles for various values of Marangoni number. As can be seen, a point exists on the interval that splits the whole

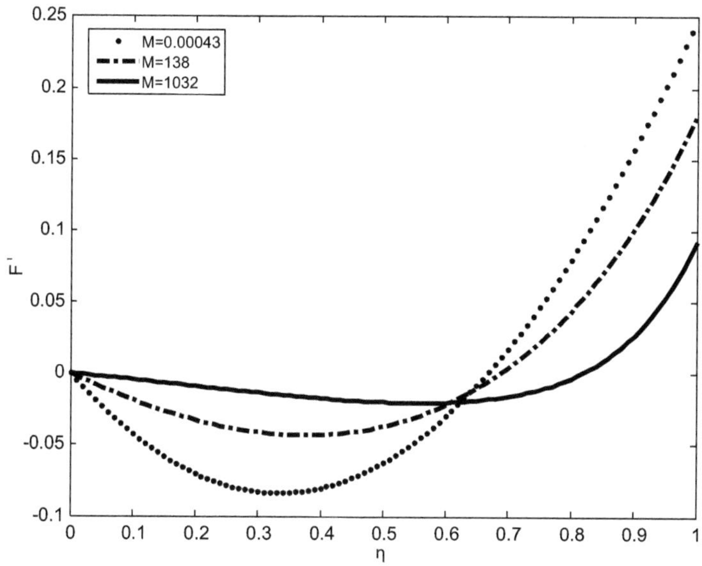

FIGURE 2.7 Effects of the Marangoni number on the velocity field.

interval into two subintervals: the fluid velocity profiles have negative and positive values in each subinterval. Physically, this implies that the whole boundary layer is split into two sublayers. The liquid on the free surface and in the sublayer below it moves in both directions, starting from point $x = 0$, to increase surface tension.

The results also indicate that, for relatively small Marangoni number, the velocity profiles changes very intensely on whole interval. When the Marangoni number is sufficiently large, the velocity profile varies slightly on the left subinterval (in the lower part of layer below the free surface), but it changes very intensely on the right subinterval (on the upper part of layer close to the free surface).

Fig. 2.8 shows the effects of the Marangoni number on a temperature field with $\mathrm{Pr} = 1$ under the temperature boundary condition of Case I. The temperature distribution behavior is different for small and large Marangoni numbers. For moderate or large Marangoni numbers, the temperature profile varies slightly in the left subinterval (in the lower part of layer below the free surface), but it changes intensely on right subintervals (on the upper part of layer close to the free surface), which is similar to the effects of the Marangoni number on velocity fields. However, when M is small enough, the temperature profile is a constant, which implies that the temperature boundary layer does not exist.

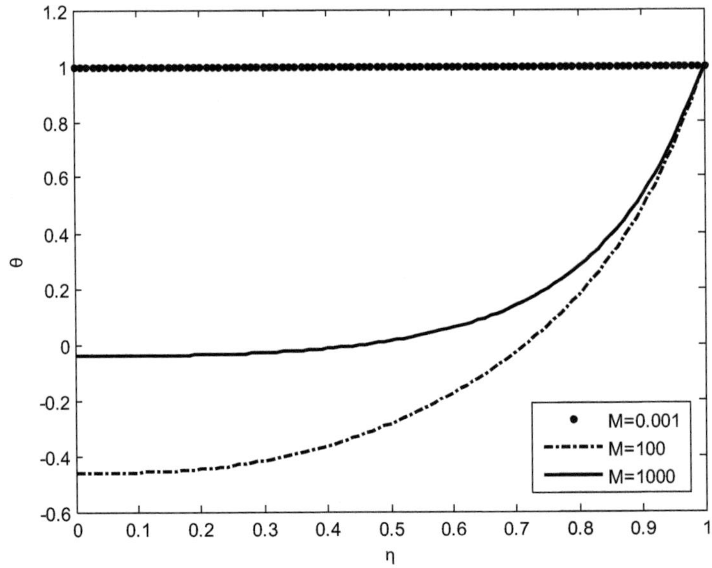

FIGURE 2.8 Effects of Marangoni number on temperature field in Case I with Pr = 1.

Fig. 2.9 shows the effects of the Prandtl number on temperature field distribution with a Marangoni number ($M = 100$) under the temperature boundary conditions of Case I. One can see that it is opposite the effects of the Marangoni number on the temperature field. For moderate or relatively large Prandtl numbers (Pr = 5, 10), the temperature profile varies slightly in the left subinterval (i.e., in the lower part of layer below the free surface) but it changes intensely on right subinterval (on the upper part of layer close to the free surface). However, when Pr is relatively small, the temperature profile changes very intensely on the whole interval, which implies that the temperature boundary layer exists on whole region.

Fig. 2.10 shows the effects of M on temperature distribution $\theta_1(\eta)$ under the temperature conditions in Case II with Pr = 1, whereas Fig. 2.11 shows the effects of Prandtl number on temperature distribution $\theta_1(\eta)$ under the temperature conditions in Case II with $M = 100$. We can find common features from the two graphics in that when M or Pr is small, the temperature change in the lower part of the liquid is almost linear, and when M or Pr is sufficiently large, temperature variation in the left subinterval (i.e., in the lower part of layer below the free surface) is not obvious. However, the temperature changes intensely on right subinterval (on the upper part of layer close to the free surface). Therefore, a temperature boundary layer is generated near the free surface. The comparison shows that the influence of Pr on the temperature boundary layer is more significant than that of Marangoni number.

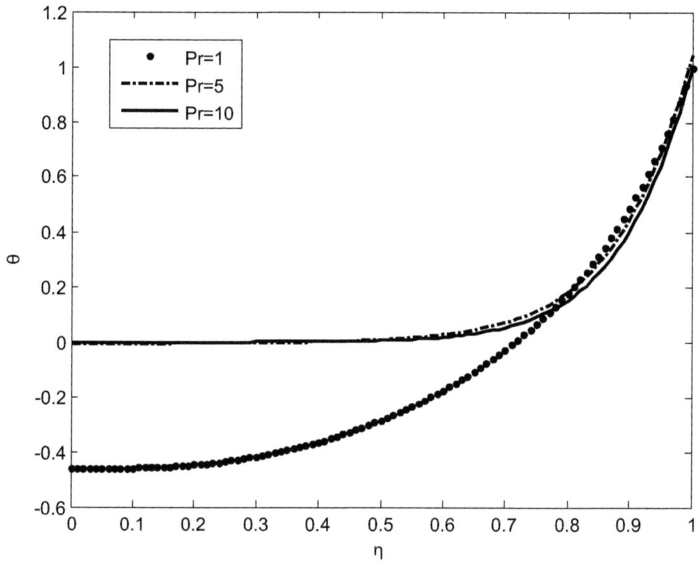

FIGURE 2.9 Effects of Prandtl number on temperature field with $M = 100$ in Case I.

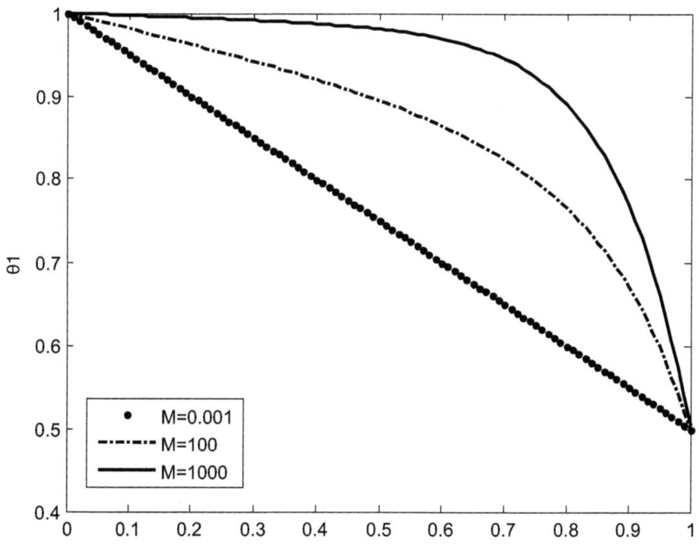

FIGURE 2.10 Effects of the Marangoni number on a temperature field with $\mathrm{Pr} = 1$ and $T_0/T_w = 0.5$ in Case II.

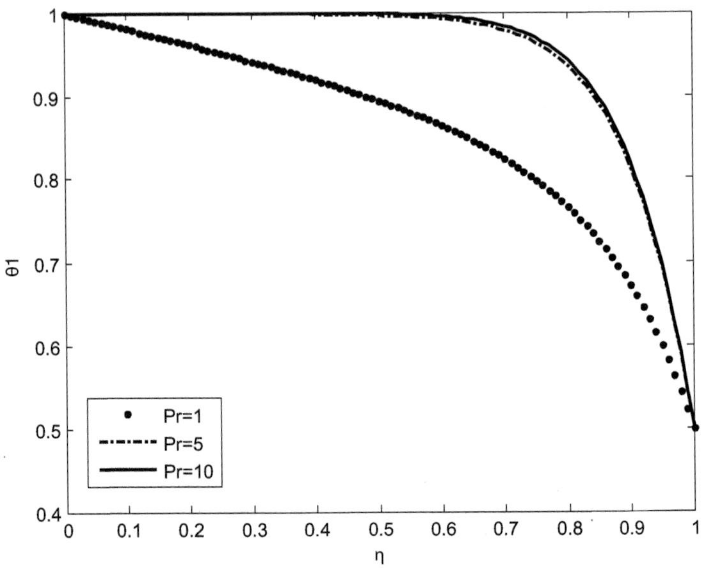

FIGURE 2.11 Effects of Prandtl number on temperature field with $M = 100$ and $T_0/T_w = 0.5$ in Case II.

Fig. 2.12 displays the distribution of temperature profiles of $\theta(\eta)$, subject to the temperature boundary conditions in Case II for $\mathrm{Pr} = 2$, for various values of Marangoni number; whereas Fig. 2.13 shows the distribution of temperature $\theta(\eta)$, subject to the boundary conditions in Case II for $M = 100$ with Pr taking on the different values. We can find some common features from the two graphics: When M or Pr is small, the temperature change in the liquid is almost linear, especially in the lower layer, whereas when M or Pr is sufficiently large, the temperature in the lower layer is not distinct but the temperature close the free surface becomes distinct, a temperature boundary layer is generated about the free surface.

Case III:

Now, we solve Eq. (2.99) in the form Eq. (2.8) by assuming initial value $\theta'(0) = s$ and introducing the embedding-parameter transformation with artificial small parameter ε as

$$\theta = \varepsilon\vartheta(\xi) + s\eta + 1, \quad \xi = \varepsilon^{-\frac{1}{2}}\eta. \tag{2.115}$$

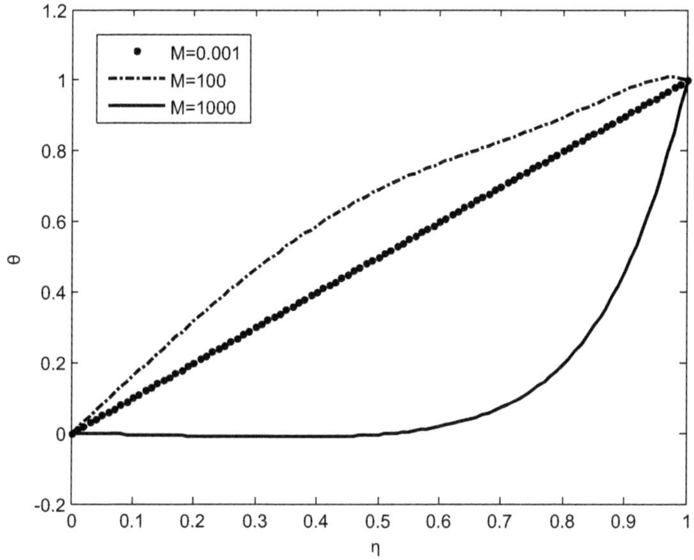

FIGURE 2.12 Effects of Marangoni number on temperature field with Pr = 2 in Case II.

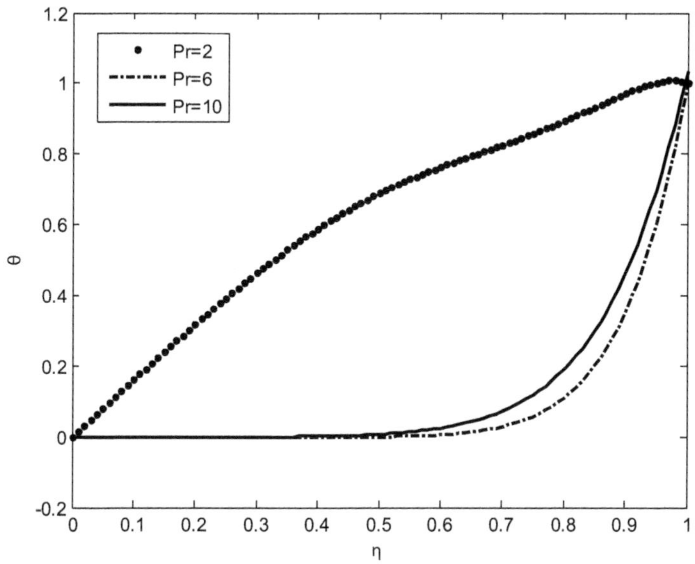

FIGURE 2.13 Effects of Prandtl number on temperature distribution behavior with $M = 100$ in Case II.

Substituting Eq. (2.115) into Eq. (2.98), similarly, we can obtain

$$\begin{aligned}\theta(\eta) &= 1 + s\eta + \frac{M\Pr c}{6}\eta^3 + \frac{1}{24}(M\Pr sc + M\Pr\lambda)\eta^4 + \frac{M\Pr s\lambda}{60}\eta^5 \\ &+ \frac{1}{720}\left(-2M^2\Pr^2 c^2 + M^2\Pr c^2\right)\eta^6 \\ &+ \frac{1}{5040}\left(4M^2\Pr sc^2 + 2M^2\Pr\lambda c - 5M^2\Pr^2 sc^2 - 5M^2\Pr^2 c\lambda\right)\eta^7 \\ &+ \frac{1}{40,320}\left(-23M^2\Pr^2 s\lambda c + 10M^2\Pr s\lambda c - 5M^2\Pr^2\lambda^2 + 2M^2\Pr\lambda^2\right)\eta^8 \\ &+ \frac{1}{3,62,880}\left(28M^3\Pr^3 c^3 - 28M^2\Pr^2 s\lambda^2 + 12M^2\Pr s\lambda^2 - M^3\Pr c^3\right)\eta^9 \\ &+ \frac{1}{3,628,800}\left(-66M^3\Pr^2 sc^3 + 2M^3\Pr^2\lambda c^2 + 100M^3\Pr^3 sc^3\right. \\ &\left.+ 156M^3\Pr^3 c^2\lambda - 7M^3\Pr sc^3 - 4M^3\Pr\lambda c^2\right)\eta^{10} + \cdots.\end{aligned} \tag{2.116}$$

For $\Pr = 1$, the error of this solution is bounded by

$$\left|\frac{2723}{279,417,600}M^3 c\lambda^2 + \frac{451}{39,916,800}M^3 s\lambda c^2\right|. \tag{2.117}$$

Using the boundary conditions of Case III, $\theta'(1) + Bi\theta(1) = 0$, $F(1) = 0$, and $F''(1) = \theta^2(1) = 1$, we can obtain the values of parameters S, c, and λ.

For example, for $\Pr = 1$, we solve Eq. (2.118)–(2.120):

$$\begin{aligned}&0.1 + 1.1s + \frac{31Mc}{60} + \frac{41}{240}(Msc + M\lambda) + \frac{51Ms\lambda}{600} - \frac{61}{7200}M^2c^2 \\ &- \frac{71}{50,400}\left(M^2sc^2 + 3M^2c\lambda\right) - \frac{81}{403,200}\left(13M^2s\lambda c + 3M^2\lambda^2\right) \\ &+ \frac{91}{3,628,800}\left(27M^3c^3 - 16M^2s\lambda^2\right) \\ &+ \frac{101}{36,288,000}\left(27M^3sc^3 + 154M^3c^2\lambda\right) = 0,\end{aligned} \tag{2.118}$$

$$\begin{aligned}&\frac{c}{2} + \frac{\lambda}{6} + \frac{Mc^2}{120} + \frac{M\lambda c}{360} + \frac{M\lambda^2}{2520} - \frac{M^2c^3}{40,320} - \frac{M^2\lambda c^2}{90,720} - \frac{M^2c\lambda^2}{226,800} \\ &+ \frac{M^3c^4}{1,478,400} + \frac{181M^3c^3\lambda}{479,001,600} - \frac{317M^4c^5}{29,059,430,400} = 0,\end{aligned} \tag{2.119}$$

$$c+\lambda+\frac{Mc^2}{6}+\frac{M\lambda c}{12}+\frac{M\lambda^2}{60}-\frac{M^2c^3}{720}-\frac{M^2\lambda c^2}{1260}-\frac{M^2c\lambda^2}{2520}+\frac{M^3c^4}{13,440}+\frac{181M^3c^3\lambda}{3,628,800}$$
$$-\frac{317M^4c^5}{159,667,200}=\left[1+s+\frac{Mc}{6}+\frac{1}{24}(Msc+M\lambda)+\frac{Ms\lambda}{60}-\frac{1}{720}M^2c^2\right.$$
$$-\frac{1}{5040}\left(M^2sc^2+3M^2c\lambda\right)-\frac{1}{40,320}\left(13M^2s\lambda c+3M^2\lambda^2\right)$$
$$+\frac{1}{3,62,880}\left(27M^3c^3-16M^2s\lambda^2\right)$$
$$\left.+\frac{1}{3,628,800}\left(27M^3sc^3+154M^3c^2\lambda\right)\right]^2. \tag{2.120}$$

For properly given value of Marangoni M, we can solve the values of s, c, and λ.

Now, we proceed to solve Eq. (2.98) in view of Eq. (2.8), by assuming an initial value $\theta_1'(0)=\alpha$ and introducing embedding-parameter transformation with artificial small parameter ε as

$$\theta_1=\varepsilon\psi(\xi)+\alpha\eta+1,\quad \xi=\varepsilon^{-\frac{1}{2}}\eta. \tag{2.121}$$

Analogously, we can obtain the solution

$$\theta_1(\eta)=1+\alpha\eta-\frac{M\mathrm{Pr}\alpha c}{24}\eta^4-\frac{M\mathrm{Pr}\alpha\lambda}{120}\eta^5$$
$$+\frac{1}{5040}\left(10M^2\mathrm{Pr}^2\alpha c^2-M^2\mathrm{Pr}\alpha c^2\right)\eta^7$$
$$+\frac{1}{40,320}\left(35M^2\mathrm{Pr}^2\alpha\lambda c-2M^2\mathrm{Pr}\alpha\lambda c\right)\eta^8$$
$$+\frac{1}{3,62,880}\left(35M^2\mathrm{Pr}^2\alpha\lambda^2-2M^2\mathrm{Pr}\alpha\lambda^2\right)\eta^9$$
$$+\frac{1}{3,628,800}\left(-280M^3\mathrm{Pr}^3\alpha c^3+84M^3\mathrm{Pr}^2\alpha c^3+M^3\mathrm{Pr}\alpha c^3\right)\eta^{10}$$
$$+\frac{1}{19,958,400}\left(-438M^3\mathrm{Pr}^3\alpha\lambda c^2+225M^3\mathrm{Pr}^2\alpha\lambda c^2+2M^3\mathrm{Pr}\alpha\lambda c^2\right)\eta^{11}+\cdots. \tag{2.122}$$

For $\mathrm{Pr}=1$, the error of the solution is bounded by $\left|\frac{4769}{470,001,600}M^3\alpha c\lambda^2\right|$. Using the boundary condition of $\theta(1)=1$, we can obtain the assuming initial value α.

Figs. 2.14 and 2.15 show the effects of the Marangoni number and the Prandtl number on the dimensionless velocity $F'(\eta)$. The results are similar to those in Fig. 2.7, in that a point exists on the interval that splits the whole interval into two subintervals. The fluid velocity profiles have negative and

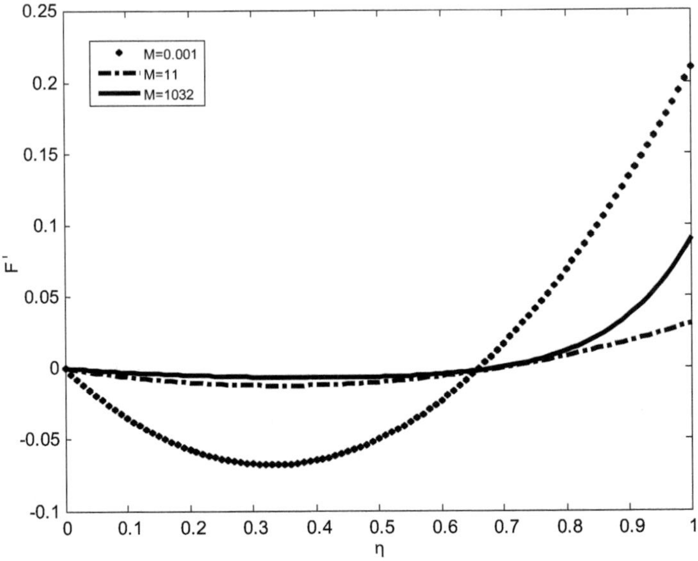

FIGURE 2.14 Effects of the Marangoni number on the velocity field with Pr = 1 in Case III.

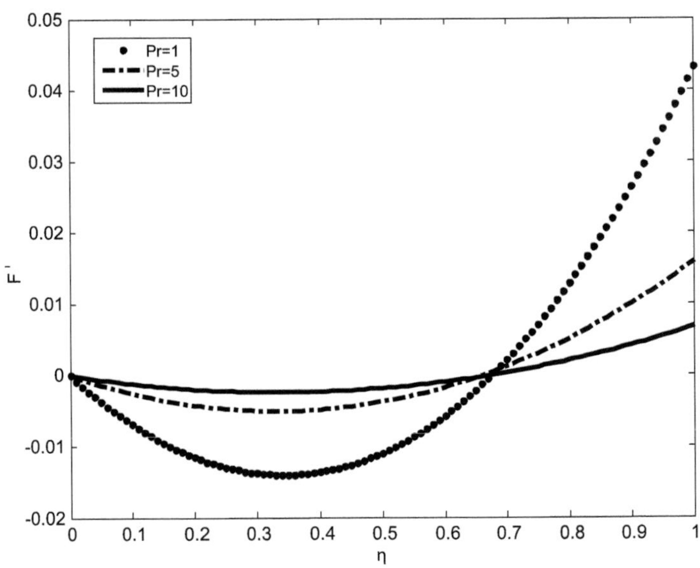

FIGURE 2.15 Effects of Prandtl number on velocity field with $M = 100$ in Case III.

positive values in left and right subinterval, respectively. The physical interpretation is that the whole boundary layer is split into two sublayers: the liquid is on the upper part close to the free surface, and on the sublayer below it moves in both directions, starting from point $x = 0$, to increase surface tension. The results also indicate that for relatively small Marangoni number or Prandtl number, the velocity profiles changes intensely throughout the whole interval.

When Marangoni number or the Prandtl number is sufficiently large, the velocity profile varies slightly on the left subinterval (on the lower part of layer below the free surface), but it changes intensely on the right subinterval (on the upper part of layer close to the free surface).

Fig. 2.16 shows the influence of the Marangoni number on temperature distribution $\theta(\eta)$ with $\Pr = 1$ and M taking on different values. The temperature distribution behavior is different for small and large values of the Marangoni number. For example, for $M = 0.000432$ or $M = 11$, temperature distribution function $\theta(\eta)$ is a constant on the whole region, so the temperature boundary layer does not exist. However, for suitable moderate or large values of M, such as $M = 138$ or $M = 1032$, the temperature distribution characteristics in the whole boundary layer is different and the temperature profile varies intensely on the left subinterval in the lower part of layer below the free surface.

Fig. 2.17 shows the temperature distribution of $\theta_1(\eta)$ when $\Pr = 1, T_g/T_{w0} = 0.5$, and $Bi = 0.1$, and M take on different values. For sufficiently small values of M, the

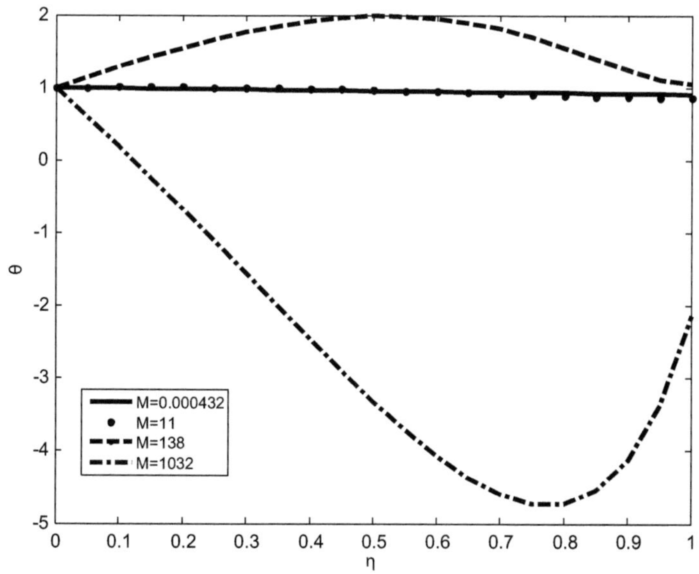

FIGURE 2.16 Effects of Marangoni number on temperature field with Pr = 1 in Case III.

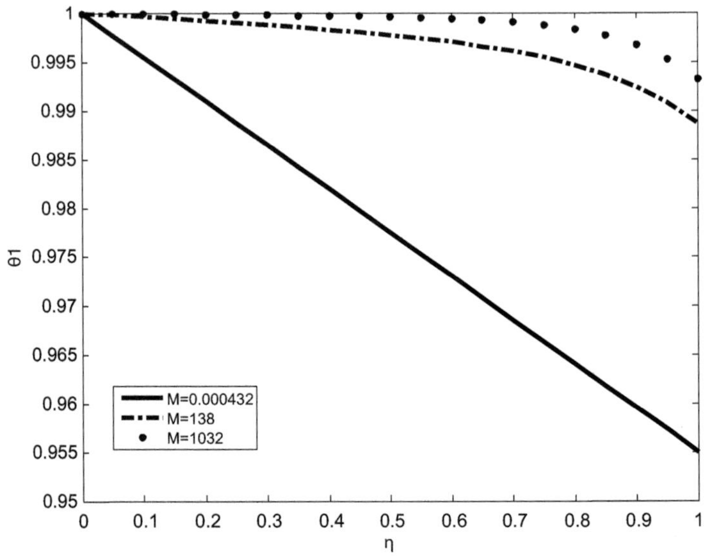

FIGURE 2.17 Effects of Marangoni number on temperature field with $\Pr = 1$, $T_g/T_{w0} = 0.5$ and $Bi = 0.1$ in Case III.

temperature distribution of $\theta_1(\eta)$ decreases almost linearly. However, for moderate or large values of $M(M = 138$ or $M = 1032)$, temperature distribution function $\theta_1(\eta)$ is about 1 in the lower part and decreases slightly in the upper one to reach a minimum at the free surface.

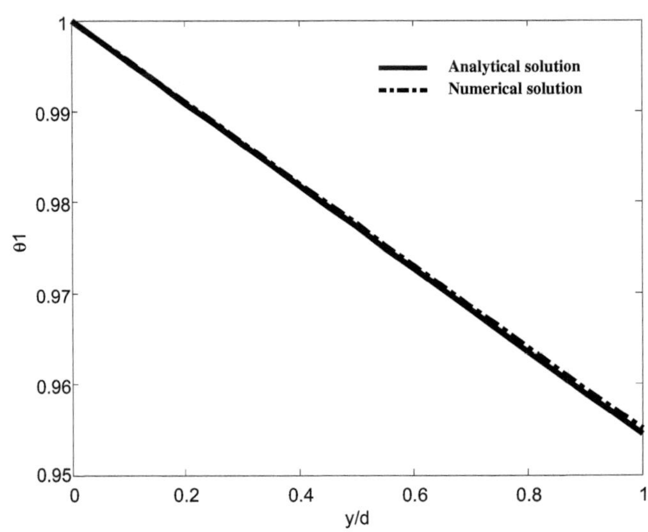

FIGURE 2.18 Comparison of analytical and numerical solutions with $M = 0.000432$ for temperature profiles θ_1.

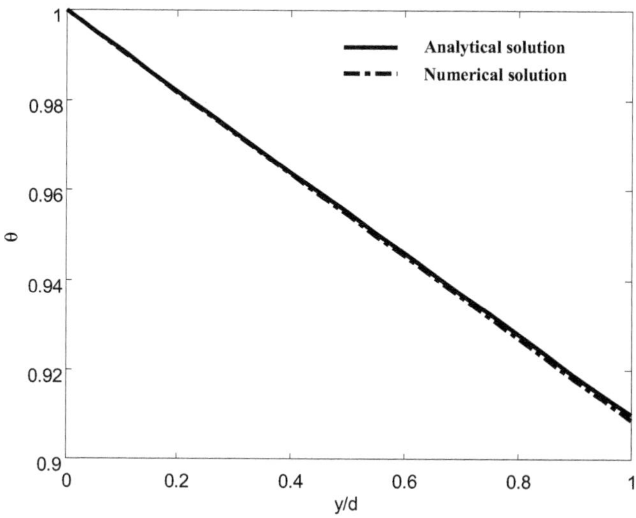

FIGURE 2.19 Comparison of analytical and numerical solutions with $M = 0.000432$ for temperature profiles θ.

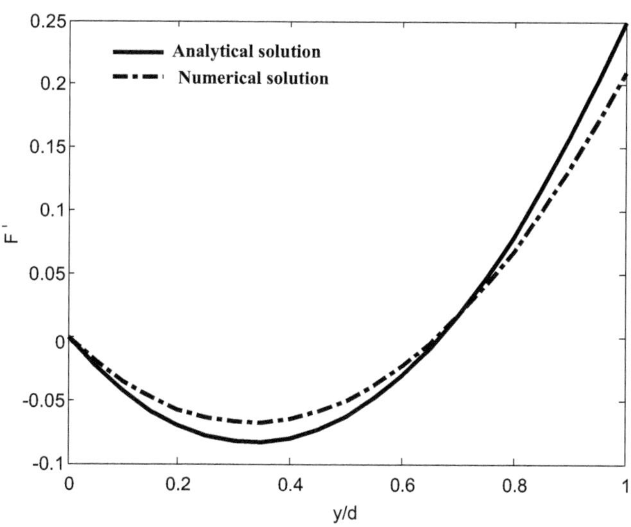

FIGURE 2.20 Comparison of analytical and numerical solutions with $M = 0.001$ for velocity profiles F'.

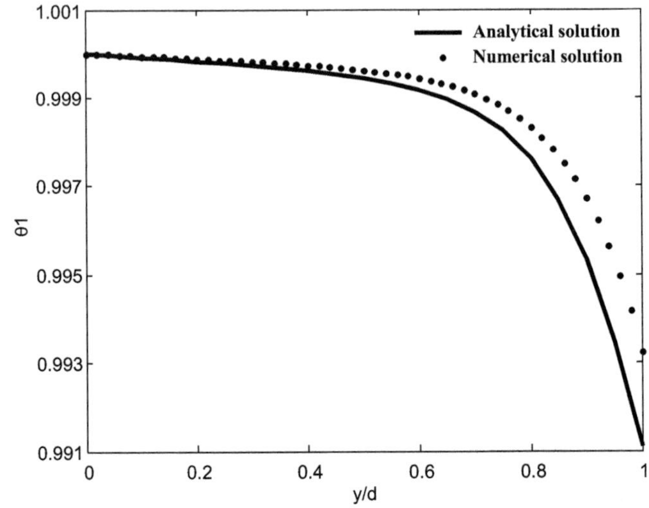

FIGURE 2.21 Comparison of analytical and numerical solutions with $M = 1032$ for temperature profiles θ_1.

Figs. 2.18–2.21 present some comparisons of approximate analytical results and numerical solutions for the dimensionless velocity and temperature profiles for some specific parameters. One can see that the results obtained by the two methods are basically consistent, which also proves the effectiveness of the method given in this book.

2.6 SUMMARY

Based on perturbation theory, this chapter introduces a method for solving nonlinear differential equations. By using the embedding-parameters transformation for both independent and dependent variables, we can convert nonlinear differential equations from ones without parameters to new ones with parameters. Solutions of new equations can be obtained and expressed in the asymptotic series of the parameters. Then the original problem can be solved using the inverse transformation.

Reliability and efficiency is verified by solving Blasius flow and Sakiadis flow. Then the method is used to study two types of Marangoni convection problems. One is for a power law fluid caused by temperature gradient; the other is for finite-thickness Marangoni convection problems caused by a quadratic relationship of the temperature gradient. The approximate analytical solutions are successfully obtained for both problems, and the associated transport characteristics for velocity and temperature fields are analyzed in detail.

REFERENCES

Arafune, K., Hirata, A., 1999. Thermal and solutal Marangoni convection in In-Ga-Sb system. Journal of Crystal Growth 197, 811–817.

Bergeon, A., Henry, D., Benhadid, H., Tuckerman, S., 1998. Marangoni convection in binary mixtures with Soret effect. Journal of Fluid Mechanics 375, 143–177.

Bergman, T.L., 1986. Numerical simulation of double-diffusive Marangoni convection. Physics of Fluids 29, 2103–2108.

Boeck, T., Thess, A., 1998. Turbulent Benard-Marangoni convection: results of two-dimensional simulations. Physical Review Letters 80, 1216–1219.

Boeck, T., Thess, A., 2001. Power-law scaling in Benard-Marangoni convection at large Prandtl numbers. Physical Review E 64, 027303.

Cazabat, A.M., Heslot, F., Troian, S.M., Carles, P., 1990. Fingering instability of thin spreading films driven by temperature gradients. Nature 346, 824–826.

Chang, I.D., Cheng, P., 1983. Matched asymptotic expansions for convection about an impermeable horizontal surface in a porous medium. International Journal of Heat and Mass Transfer 26, 163–174.

Chen, C.H., 2003. Heat transfer in a power-law fluid film over an unsteady stretching sheet. Heat Mass Transfer 39, 791–796.

Chen, C.H., 2007. Marangoni effects on forced convection of power-law liquids in a thin film over a stretching surface. Physics Letters A 370, 51–57.

Chistopher, D.M., Wang, B.X., 2001. Similarity simulation for Marangoni convection around a vapor bubble during nucleation and growth. International Journal of Heat and Mass Transfer 44, 799–810.

Chu, X.L., Velarde, M.G., 1991. Korteweg-de vries soliton excitation in Benard Marangoni convection. Physical Review A 43, 1094–1096.

Cole, J.D., 1968. Perturbation Methods in Applied Mathematics. Blaisdell Publishing Company, Waltham.

Ece, M.C., Buyuk, E., 2002. Similarity solutions for free convection to power-law fluids from a heated vertical plate. Applied Mathematics Letters 15, 1–5.

Gorla, R.S.R., Pop, I., Lee, J.K., 1992. Convective wall plume in power-law fluid: second-order correction for the adiabatic wall. Warme-und Stoffubertragung 27, 473–479.

Grasman, J., Matkowsky, B.J., 1977. A variational approach to singularly perturbed boundary value problems for ordinary and partial differential equations with turning points. SIAM Journal on Applied Mathematics 32, 588–597.

Hagstrom, G., Doering, C.R., 2010. Bounds on heat transport in Benard-Marangoni convection. Physical Review E 81, 047301.

Hamid, R.A., Arifin, N.M., Nazar, R., Ali, F.M., 2011. Radiation effects on Marangoni convection over a flat surface with suction and injection. Malaysian Journal of Mathematical Sciences 5, 13–25.

Howarth, L., 1938. On the solution of the laminar boundary layer equations. Series A Mathematical and Physical Sciences 164, 547–579.

Jiao, C.R., Zheng, L.C., Ma, L.X., 2015. MHD thermosolutal Marangoni convection heat and mass transport of power law fluid driven by temperature and concentration Gradient. AIP Advances 5, 087160.

Jiao, C.R., Zheng, L.C., Lin, Y.H., Ma, L.X., Chen, G., 2016. Marangoni abnormal convection heat transfer of power-law fluid driven by temperature gradient in porous medium with heat generation. International Journal of Heat and Mass Transfer 92, 700–707.

Johnson, D., Narayanan, R., 1999. A tutorial on the Rayleigh-Marangoni-Bebard problem with multiple layers and side wall effects. Chaos: An Interdisciplinary Journal of Nonlinear Science 9 (1), 124–140.

Kevorkian, J., Cole, J.D., 1981. Perturbation Methods in Applied Mathematics. Springer-Verlag Inc, New York.

Li, J.C., 1999. Mathematical Methods in Mathematical Physics. Science Press, Beijing.

Li, B.T., Zheng, L.C., Zhang, X.X., 2011a. A new model for flow and heat of a power law fluid in a pipe. Thermal Science 15, 127–130.

Li, B.T., Zheng, L.C., Zhang, X.X., 2011b. Heat transfer in pseudo-plastic non-Newtonian fluids with variable thermal conductivity. Energy Conversion and Management 52, 355–358.

Lin, Y.H., Zheng, L.C., Zhang, X.X., 2014. Radiation effects on Marangoni convection flow and heat transfer in pseudo-plastic non-Newtonian nanofluids with variable thermal conductivity. International Journal of Heat and Mass Transfer 77, 708–716.

Magyari, E., Chamkha, A.J., 2007. Exact analytical solutions for thermosolutal Marangoni convection in the presence of heat and mass generation or consumption. Heat Mass Transfer 43, 965–974.

Magyari, E., Chamkha, A.J., 2008. Exact analytical results for the thermosolutal MHD Marangoni boundary layers. International Journal of Thermal Sciences 47, 848–857.

Mikishev, A., Nepomnyashchy, A., 2010. Nonlinear large-scale Marangoni convection in a heated liquid layer with insoluble surfactant. Physical Review E 82, 046306.

Mudhaf, A.A., Chamkha, A.J., 2005. Similarity solutions for MHD thermosolutal Marangoni convection over a flat surface in the presence of heat generation. Heat Mass Transfer 42, 112–121.

Naimi, M., Hasnaoui, M., Platten, J.K., 2000. Marangoni convection of non-Newtonian power law fluids in a shallow rectangular cavity. Engineering Computations 17, 638–668.

Nepomnyashchy, A., Simanovskii, I., 2009. Instabilities and ordered patterns in nonisothermal ultrathin bilayer fluid films. Physical Review Letters 102, 164501.

Pearson, J.R.A., 1958. On convection cells induced by surface tension. Journal of Fluid Mechanics 4 (5), 489–500.

Pereira, A., Kalliadasis, S., 2008. Dynamics of a falling film with solutal Marangoni effect. Physical Review E 78, 036312.

Polyanin, A.D., 1984. An asymptotic analysis of some nonlinear boundary-value problems of convection mass and heat transfer of reacting particles with the flow. International Journal of Heat and Mass Transfer 27, 163–189.

Pop, I., 1993. Boundary layer flow at a three-dimensional stagnation point in power-law non-Newtonian fluids. International Journal of Heat and Mass Transfer 14, 408–412.

Pop, I., Rashidi, M., Gorla, R.S.R., 1991. Mixed convection to power-law type non-Newtonian fluids from a vertical wall. Polymer-Plastics Technology and Engineering 30, 47–66.

Rongy, L., Wit, A.D., 2006. Steady Marangoni flow traveling with chemical fronts. The Journal of Chemical Physics 124, 164705.

Savino, R., Fico, S., 2004. Transient Marangoni convection in hanging evaporating drops. Physics of Fluids 16 (10), 3738–3753.

Scriven, L.E., Sternling, C.V., 1960. The Marangoni effects. Nature 187, 186–188.

Slavtchev, S.G., Dubovik, K.G., 1992. Thermocapillary convection in a rectangular cavity at minimum of surface tension. Theoretical and Applied Mechanics Letters 23, 85–90.

Slavtchev, S.G., Miladinova, S., 1998. Thermocapillary flow in a liquid layer at minimum in surface tension. Acta Mechanica 127, 209–224.

Stein, C.F., Johansson, P., Bergh, J., 2002. An analytical asymptotic solution to a conjugate heat transfer problem. International Journal of Heat and Mass Transfer 45, 2485–2500.

Svensson, J., Bulgakova, N.M., Nerushev, O.A., Campbell, E.E.B., 2006. Marangoni effect in SiO_2 during field-directed chemical vapor deposition growth of carbon nanotubes. Physical Review B 73, 205413.

Würger, A., 2007. Thermophoresis in colloidal suspensions driven by Marangoni forces. Physical Review Letters 98, 138301.

Zhang, Y., Zheng, L.C., 2014. Similarity solutions of Marangoni convection boundary layer flow with gravity and external pressure. Chinese Journal of Chemical Engineering 22, 365–369.

Zhang, J., Behringer, R.P., Oron, A., 2007. Marangoni convection in binary mixtures. Physical Review E 76, 016306.

Zheng, L.C., Chen, X.H., Zhang, X.X., 2004. Analytical approximants for a boundary layer flow on a stretching moving surface with a power velocity. International Journal of Applied Mechanics 9, 795–802.

Zheng, L.C., Sheng, X.Y., Zhang, X.X., 2006a. Approximate analytical solutions for Marangoni convection boundary layer equations. Acta Physica Sinica 55, 5298–5304.

Zheng, L.C., Zhang, X.X., Lu, C.Q., 2006b. Heat transfer for power law non-Newtonian fluids. Chinese Physics Letters 23, 3301–3304.

Zheng, L.C., Zhang, X.X., Gao, Y.T., 2008a. Analytical solution for Marangoni convection over a liquid-vapor surface due to an imposed temperature gradient. Mathematical and Computer Modelling 48, 1787–1795.

Zheng, L.C., Zhang, X.X., Ma, L.X., 2008b. Fully developed convective heat transfer for power law fluids in a circular tube. Chinese Physics Letters 25, 195–197.

Zheng, L.C., Zhang, X.X., Boubaker, K., Yücel, U., Gargouze, E., Yıldırım, A., 2011. Similarity and Boubaker polynomials expansion scheme BPES comparative solutions to the heat transfer equation for incompressible non-Newtonian fluids: case of laminar boundary energy equation. The European Physical Journal Applied Physics 55, 21102.

Zheng, L.C., Lin, Y.H., Zhang, X.X., 2013. MHD thermosolutal Marangoni convection of non-Newtonian power law fluids driven by temperature gradient. Journal of Heat Transfer-T AMSE 135, 051702.

Chapter 3

Adomian Decomposition Method

ABSTRACT

In this chapter, Adomian decomposition method is introduced to solve two singular nonlinear boundary values problems of ordinary differential equations arising in flow and heat transfer of power law non-Newtonian fluids over a continuous moving surface with injection or suction velocity at surface, power law magnetohydrodynamic fluid flow over a horizontal wall with power law velocity, and Marangoni convection problem over a fluid-gas free surface due to an imposed temperature gradient. Approximate analytical solutions are obtained and the effects of involved parameters on the boundary layer behavior are analyzed and discussed.

3.1 INTRODUCTION

Adomian decomposition method (ADM), due to its many advantages, has obtained many scholars' attention and has been successfully applied to many problems of differential equations, integral equations, and differential-integral equations (Adomian, 1988, 1994, 1998). Especially for those problems with strong nonlinear, strong randomness and far from equilibrium, ADM can overcome the difficulties that the current approximation method finds it is difficult to overcome; many modified or optimized ADMs were developed (Wazwaz, 1995, 1998, 1999a,b, 2001).

In this chapter, the ADM is applied to study two types of boundary layer problems of power law non-Newtonian fluids: one is over a continuous moving surface with a constant speed U_w and a free stream speed U_∞ subjected to injection or suction velocity of $V_w(X) = V_0 X^{-\frac{N}{N+1}}$ at the wall and another is magnetohydrodynamic (MHD) power law fluid flow over a plate with a power law velocity. The governing partial differential equations are reduced to singular nonlinear boundary value problems of ordinary differential equations by Lie group similarity transformations and the generalized Crocco variable transformations. Approximate analytical solutions are obtained and the effects of

Mathematics in Science and Engineering. http://dx.doi.org/10.1016/B978-0-12-811753-8.00003-7

involved parameters on the boundary layer behavior are discussed. Moreover, ADM is used to investigate Marangoni convection problem over a fluid-gas free surface due to an imposed temperature gradient. The solutions are obtained and the effects of parameters on velocity and temperature fields are analyzed in detail.

The main advantage of ADM is that it can be used directly without using assumptions or transformation formula and the approximate analytical solution obtained by ADM may be presented in terms of a rapidly convergent power series with elegantly computable terms.

3.2 NONLINEAR BOUNDARY LAYER OF POWER LAW FLUID

3.2.1 Physical Background

The theoretical analysis of an external boundary layer flow and heat transfer of power law non-Newtonian fluid was first done by Schowalter (1960) and Acrivos et al. (1960). Later, Fox and Fan (1969) considered the boundary layer flow of power law non-Newtonian fluid over moving horizontal wall with a power law velocity. Jones and Atkinson (1974) studied similarity solutions in some nonlinear diffusion problems and in boundary layer flow of a pseudoplastic fluid. Nachman and Taliaferro (1979) and Nachman and Callegari (1980) studied theoretically the boundary layer problem of non-Newtonian fluid flow around semi-infinite plate. By introducing the similarity transformation and Crocco variable transformation technique, the boundary layer equation was reduced into a class singular nonlinear two-point boundary value problem of ordinary differential equations. The existence and uniqueness of positive solutions for boundary value problems were established by using the function analysis technique. Na and Hansen (1967) investigated similarity solutions of a class of laminar three-dimensional boundary layer equations of power law fluids. Timol and Kalthia (1986) also studied theoretically the existence of three-dimensional boundary layer similarity solutions for power law non-Newtonian fluids under normal conditions. Howell et al. (1997) studied the approximate solution of the boundary layer problem over horizontal moving plate.

It should be mentioned that Professor Pop and his coauthors made many excellent studies on flow and heat transfer of power law non-Newtonian fluids subject to various boundaries, for example, the free convection or mixed convection of power law—type non-Newtonian fluids along a vertical wall or along a vertical wavy surface, three-dimensional stagnation point flow in power law—type non-Newtonian fluid, etc (Crosan and Pop, 2001; Gorla et al., 1992; Kumari et al., 1997; Pop, 1993; Pop et al., 1991).

Recently, Professor Zheng and his coauthors have made systematic investigations of various unconventional flow and heat transfer of power law non-Newtonian fluids. Their work includes the qualitative problem of the boundary layer equations, i.e., the existence, uniqueness, nonuniqueness and analyticity of solutions of power law fluid boundary layers, analytical solutions or approximate analytical solutions, numerical solutions, Marangoni convection in power law non-Newtonian fluids, power law non-Newtonian base nanofluids, etc. (Li et al., 2012; Lin et al., 2014, 2015a,b,c; Sui et al., 2015a,b; Zheng et al., 2000, 2003, 2004c, 2007, 2008, 2011, 2012, 2013; Zheng and He, 1999).

In this chapter, the ADM is introduced to study two kinds of boundary layer problems of power law non-Newtonian fluids and Marangoni convection boundary layer problems, the suitable similarity transformations are introduced to reduce the problems into ordinary differential equations, the approximate analytical solutions are obtained, and the effects of pertinent parameters on velocity and temperature fields are analyzed and discussed.

3.2.2 Mathematical Formulation

Consider steady incompressible two-dimensional laminar boundary layer flow and heat transfer of power law non-Newtonian fluid over a continuous moving surface with a constant speed U_w and special injection or suction velocity of $V_w(X) = V_0 X^{-\frac{N}{N+1}}$ with a free stream of constant speed U_∞ (Fig. 3.1).

The boundary layer governing equations describing the conservation of mass, momentum, and energy can be written as (Acrivos et al., 1960; Lin et al., 2014, 2015a,b,c; Li et al., 2012; Schowalter, 1960; Sui et al., 2015a,b; Zheng et al., 2011, 2012, 2013):

$$\frac{\partial U}{\partial X} + \frac{\partial V}{\partial Y} = 0, \tag{3.1}$$

$$U\frac{\partial U}{\partial X} + V\frac{\partial V}{\partial Y} = \frac{1}{\rho}\frac{\partial \tau_{XY}}{\partial Y}, \tag{3.2}$$

$$U\frac{\partial T}{\partial X} + V\frac{\partial T}{\partial Y} = \frac{\partial}{\partial Y}\left(a\frac{\partial T}{\partial Y}\right), \tag{3.3}$$

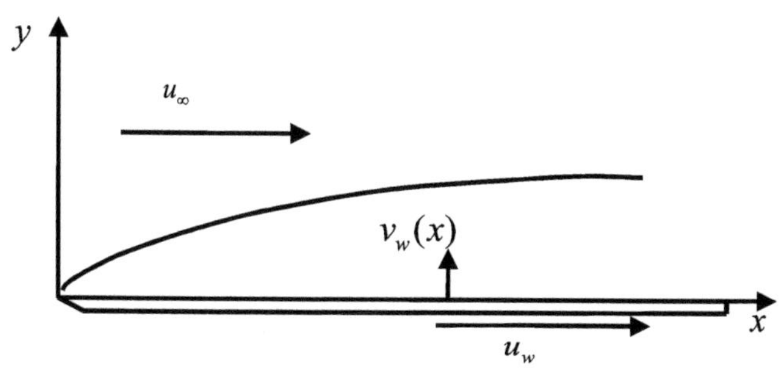

FIGURE 3.1 Boundary layer sketch over moving surface.

where the X- and Y-axes are taken along and perpendicular to the plate, U and V are the velocity components parallel and normal to the plate, and

$$\tau_{XY} = K\left|\frac{\partial U}{\partial Y}\right|^{N-1}\frac{\partial U}{\partial Y}, \tag{3.4}$$

is the shear stress, $\nu = \gamma\left|\frac{\partial U}{\partial Y}\right|^{N-1}$ $(\gamma = K/\rho)$ is the kinematic viscosity, and a is the thermal diffusivity. In terms of the relationship of Prandtl number N_{Pr} to the thermal diffusivity a and kinematic viscosity ν. The thermal diffusivity is proposed as $a = \omega\left|\frac{\partial U}{\partial Y}\right|^{N-1}$ for $U \neq 0$ such that $N_{\mathrm{Pr}} = \gamma/\omega$ with γ and ω as positive constant and $a = a_0$ for $U = 0$. The case $N = 1$ corresponds to a Newtonian fluid and the case $0 < N < 1$ is "power law" relation proposed as being descriptive of pseudoplastic non-Newtonian fluids and $N > 1$ describes the dilatant fluid. The appropriate boundary conditions are:

$$U|_{Y=0} = U_W, \quad V|_{Y=0} = V_w(X), \quad U|_{Y=+\infty} = U_\infty, \tag{3.5}$$

$$T|_{Y=0} = T_w, \quad T|_{Y=+\infty} = T_\infty. \tag{3.6}$$

3.2.3 Similarity Transformation

We introduce a dimensionless stream function $\psi(X,Y)$ and similarity variable η by the expressions

$$\psi = AX^{\sigma}f(\eta), \quad \eta = BX^{\beta}Y, \quad \varphi(\eta) = (T - T_w)/(T_\infty - T_w). \tag{3.7}$$

where A, B, α, and β are constants to be determined, and $f(\eta)$ denotes the dimensionless stream function. Thus the velocity components become:

$$U = \frac{\partial \psi}{\partial Y} = ABX^{\alpha+\beta}f'(\eta) = U_\infty f'(\eta), v = -\frac{\partial \psi}{\partial X} = -A\alpha\, X^{\alpha-1}[f(\eta) - \eta f'(\eta)],$$

Here we have chosen $\beta = -\alpha$ and $AB = U_\infty$. At this moment, Eq. (3.1) is satisfied automatically. Inserting u and v into (3.2), yields:

$$-U_\infty^2\ \alpha x^{-1}f(\eta)f''(\eta) = U_\infty^N B^{N+1}\gamma\, x^{-(N+1)\alpha}\left(|f''(\eta)|^{N-1}f''(\eta)\right)'.$$

After performing a series of similarity transformation and choosing suitable appropriate parameters,

$$\alpha := \frac{1}{N+1}, \quad B := \left(\frac{U_\infty^{2-N}}{(N+1)\gamma}\right)^{\frac{1}{N+1}},$$

we obtain:

$$\left(|f''(\eta)|^{N-1}f''(\eta)\right)' + f(\eta)f''(\eta) = 0, \tag{3.8}$$

$$f(0) = -C, \ \ f'(0) = \xi, \ \ f'(\eta)\Big|_{\eta=+\infty} = 1, \tag{3.9}$$

$$\left(|f''(\eta)|^{N-1}\varphi'(\eta)\right)' + N_{\mathrm{Pr}}f(\eta)\varphi'(\eta) = 0, \tag{3.10}$$

$$\varphi(0) = 0, \ \ \varphi(\eta)\Big|_{\eta=+\infty} = 1, \tag{3.11}$$

where $\xi = U_w/U_\infty$ is the velocity ratio parameter, $N_{\mathrm{Pr}} = \gamma/\omega$ is the Prandtl number, $C = (N+1)^{N/(N+1)}(u_\infty^{2-N}/\gamma)^{1/(N+1)}V_0/U_\infty$ is the injection or suction parameter, which can be positive, negative, or equals to zero. Physically $C > 0$ means that fluid is injected into the boundary layer and $C < 0$ implies that there is suction of fluid from the boundary layer. The special cases of $N = 1$, $C = 0$, $\xi = 0$ corresponds to the classical Blasius problems.

3.2.4 Crocco Variable Transformation

Eqs. (3.8) and (3.10) are strongly nonlinear and subject to the boundary conditions from 0 to ∞, and the problems are still very complex. In order to utilize the ADM to obtain approximate analytical solutions, we hope to transform the nonlinear boundary value problems of infinite range into a finite range. In doing so, a further transformation of independent variables and unknown functions, i.e., the generalized Crocco transformation, is introduced. We assume that the solutions of Eqs. (3.8) and (3.10) possess positive second derivative $f''(\eta)$ and $w''(\eta)$ in $(0,+\infty)$. Defining the general Crocco variable transformation as

$$g(z) = [f''(\eta)]^N, \ \ z = f'(\eta), \ \ w(z) = \phi(\eta), \ \ z \in [\xi, \ 1), \ \ 0 \le \xi < 1, \tag{3.12}$$

where the transformation has a very important physical meaning, $g(z)$ is the dimensionless shear stress, z is the dimensionless tangential velocity, and $w(z)$ is the dimensionless temperature.

Substituting (3.12) into (3.8)–(3.11) and applying the chain rule, we finally obtain the following singular nonlinear two-point boundary value problems of the form:

$$g''(z) = -zg^{-1/N}(z), \quad 0 \le \xi < z < 1, \tag{3.13}$$

$$g'(\xi) = C, \quad g(1) = 0, \tag{3.14}$$

$$w''(z)g(z) + (1 - N_{\mathrm{Pr}})w'(z)g'(z) = 0, \tag{3.15}$$

$$w(\xi) = 0, \quad w(1) = 1. \tag{3.16}$$

3.2.5 Adomian Decomposition Method Solutions

We now first solve the problem (3.13) and (3.14). In terms of the form (3.13) and (3.14) and the definition of ADM, we introduce an initial assumed parameter value for $g(\xi)$. Let $g(\xi) = \alpha$, the inverse operator ${L^{-1}}_{zz}$ is formally defined as

$$ {L^{-1}}_{zz} = \int_{\xi}^{z} dt \int_{\xi}^{t} ds, \tag{3.17}$$

By performing the operator ${L^{-1}}_{zz}$ on Eq. (3.13), we obtain:

$$ g(z) = \alpha + C(z - \xi) - \int_{\xi}^{z} \int_{\xi}^{t} s g^{-\frac{1}{N}}(s) ds dt \tag{3.18}$$

We assume that the solution of Eq. (3.18) can be expressed into the following form:

$$ g(z) = \sum_{m=1}^{\infty} g_m(z), \tag{3.19}$$

And the nonlinear term of Eq. (3.18) can be decomposed into the sum of Adomian polynomial,

$$ \sum_{m=0}^{\infty} A_m = z g^{-\frac{1}{N}}(z), \tag{3.20}$$

Let $p = -\frac{1}{N}$, ADM satisfies the following iteration recurrence relation:

$$ \begin{aligned} g_0(z) &= \alpha, \\ g_1(z) &= C(z - \xi) - L^{-1} A_0, \\ g_{k+1}(z) &= -L^{-1} A_k. \end{aligned} \tag{3.21}$$

A few first terms of Adomian polynomial are listed as following:

$$ \begin{aligned} A_0 &= z \cdot (g_0)^p \\ A_1 &= p \cdot z \cdot g_1 \cdot (g_0)^{p-1} \\ A_2 &= p \cdot z \cdot g_2 \cdot (g_0)^{p-1} + \frac{p \cdot (p-1)}{2} \cdot z \cdot (g_1)^2 \cdot (g_0)^{p-2} \\ A_3 &= p \cdot (p-1) \cdot z \cdot g_2 \cdot g_1 \cdot (g_0)^{p-2} \\ &\quad + \frac{p \cdot (p-1) \cdot (p-2)}{3} \cdot z \cdot (g_1)^3 (g_0)^{p-3} + p \cdot z \cdot g_3 \cdot (g_0)^{p-1}, \end{aligned} \tag{3.22}$$

Substituting (3.22) into (3.21), we obtain:

$$g_0 = \alpha,$$

$$g_1 = -\frac{\alpha^p}{6}z^3 + \frac{2C + \alpha^p \xi^2}{2}z - \frac{\alpha^p \xi^3 + 3C\xi}{3},$$

$$g_2 = \frac{p\alpha^{2p-1}}{180}z^6 - \frac{p(2C + \alpha^p \xi^2)\alpha^{p-1}}{24}z^4 - \frac{(-\alpha^p \xi^3 - 3C\xi)\alpha^{p-1}}{18}z^3,$$
$$-\frac{p\alpha^{p-1}\xi^3(\alpha^p \xi^2 + 5C)}{30}z + \frac{p\alpha^{p-1}\xi^4(\alpha^p \xi^2 + 6C)}{72}$$

$$\ldots, \ldots.$$

The approximate analytical solution of the problem (3.13) and (3.14) is to use the sum of a few previous terms of the series to approach the practical solution, i.e.,

$$\sigma_l = \sum_{i=0}^{l-1} g_i(z) = g_0(z) + g_1(z) + g_2(z) + \cdots + g_{l-1}(z), \quad l \geq 1, \tag{3.23}$$

Utilizing boundary conditions (3.14), for specified value of parameters C, p, and ξ, we can solve the assumed initial value of parameter α. By this way, the approximate analytical solution is obtained.

3.2.6 Results and Discussion

3.2.6.1 *Analysis of Velocity Field*

For some specified value of parameters C, p, and ξ, performing the computational work by ADM, the solutions are presented and analyzed. Figs. 3.2–3.4 present the approximate analytical solution of dimensionless shear stress distribution profiles with the dimensionless tangential velocity with $l = 5$.

Fig. 3.2 presents a comparison of shear stress distribution for parameters $C = 0.2$, $\xi = 0.0$, and $N = 0.5$, obtained by ADM and by numerical shooting technique. It is seen that the approximate analytical solution obtained by ADM by $l = 5$ is in good agreement with the solution obtained by numerical method, which demonstrated the validity of the analytical method.

Fig. 3.3 shows the effects of injection or suction parameter C on the shear stress distribution behavior for parameter $\xi = 0.0$, $N = 0.5$. It is seen that the skin friction coefficient profiles decrease with increasing in injection or suction parameter C. For $C \leq 0$ (which means that there exists a suction from wall to the fluid), the shear stress profile $g(z)$ is monotonous decreasing with the tangential velocity z in $[\xi,1]$, which implies that the largest shear stress is at the wall ($z = \xi$) and the smallest shear stress occurs at the free surface of the boundary layer ($z = 1$).

Fig. 3.4 displays the characteristics of shear stress distribution behavior for parameter $C = 0.5$, the suction or injection parameter changing from

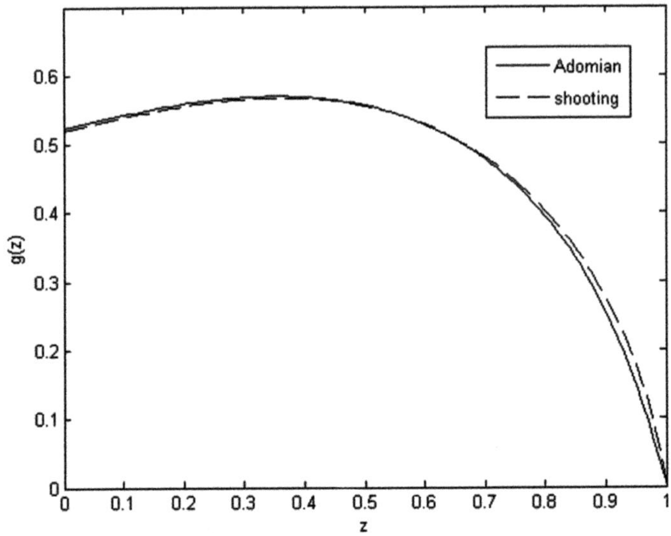

FIGURE 3.2 Shear stress distribution for $C = 0.2$, $\xi = 0.0$, $N = 0.5$.

$\xi = 0.3$ to 0.7, and the power law exponent changing from $N = 0.3$, 0.6, and 0.9. It is seen that for the case of $C > 0$, which means that there exists an injection of fluid from wall into the boundary layer. There is a point $z_0 \in (\xi, 1)$ such that the shear stress $g(z)$ increases with tangential velocity z in (ξ, z_0), and

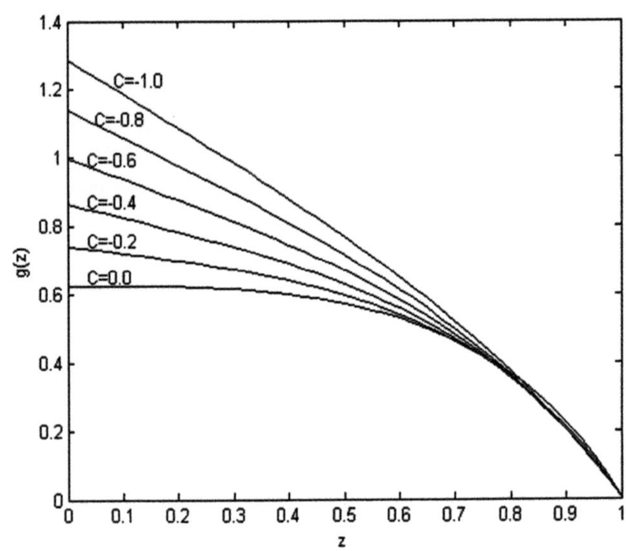

FIGURE 3.3 Effects of parameter C on the shear stress distribution for $\xi = 0.0$, $N = 0.5$.

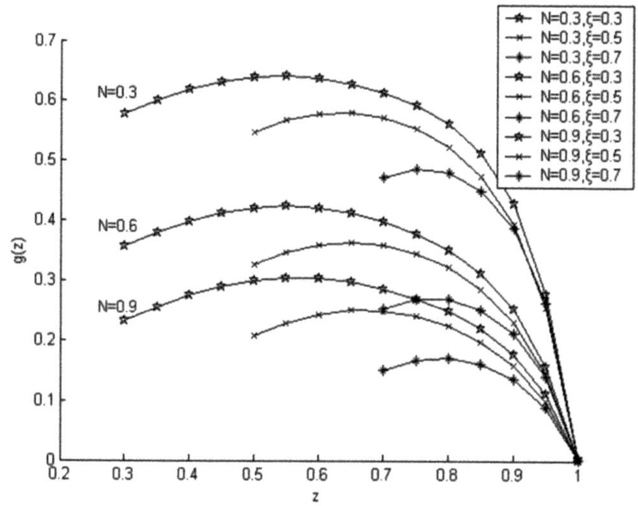

FIGURE 3.4 Shear stress distribution for $C=0.5$, $\xi=0.3$ to 0.7, $N=0.3$, 0.6, 0.9.

then $g(z)$ decreases with z in $(z_0,1)$ with $g'(z_0)=0$, where $z_0 \in (\xi,\ 1)$ is the largest value point of the shear stress in the boundary layer. It implies in physical meaning that for the case that there exists an injection of fluid from wall into the boundary layer, the largest shear stress occurs in the interior of the boundary layer.

Fig. 3.4 also indicates that, in the range of $\xi \in (0,1)$, the skin friction and shear stress $g(z)$ decrease with increasing in wall moving ratio parameter ξ. The skin friction and shear stress decrease rapidly with power law exponent, which physically means that a power law fluid with a smaller power exponent exerts a greater shear force on the wall.

3.2.6.2 Analysis of Temperature Field

Utilizing the solution of momentum equation, we rewrite Eq. (3.15) as

$$\frac{w''}{w'} = (N_{\mathrm{Pr}} - 1)\frac{g'}{g}. \tag{3.24}$$

Integrating both sides of (3.24) with respect to η over interval $[\xi,z]$, we obtain the approximate analytical solution of temperature field equation as following:

$$w(z) = \frac{\int_{\xi}^{z} g^{N_{\mathrm{Pr}}-1}(s)ds}{\int_{\xi}^{1} g^{N_{\mathrm{Pr}}-1}(s)ds}. \tag{3.25}$$

Substituting the approximate analytical solution of (3.13) and (3.14) into (3.25), yields:

$$w_l(z) = \frac{\int_{\xi}^{z}\left(\sum_{i=0}^{l-1} g_i\right)^{N_{\mathrm{Pr}}-1} ds}{\int_{\xi}^{1}\left(\sum_{i=0}^{l-1} g_i\right)^{N_{\mathrm{Pr}}-1} ds}, \tag{3.26}$$

where,

$$G_l(\xi) = \left(\sum_{i=0}^{l-1} g_i\right)^{N_{\mathrm{Pr}}-1}. \tag{3.27}$$

Expanding $G_l(\xi)$ into power law series as $G_l(\xi) = \sum_{n=0}^{\infty} c_n \xi^n$, for $N_{\mathrm{Pr}} > 2$, and approaching the approximate solution $G_l(z)$ by using the sum $G_{(l,q)}(z)$ of first $q+1$, obtains:

$$G_{(l,q)}(\xi) = \sum_{n=0}^{q} c_n \xi^n. \tag{3.28}$$

In view of (3.12), we have $w'(z) = \phi'(\eta)/f''(\eta)$, then

$$\phi'(0) = f''(0) w'(\xi). \tag{3.29}$$

We can determine the temperature gradient $\phi'(0)$ of the surface by using the value of $w'(\xi)$.

Fig. 3.5 shows the influence of power law exponent parameter N on distribution of the temperature field; it is seen that for pseudoplastic non-Newtonian fluid ($0 < N < 1$), the effects of power law exponent on

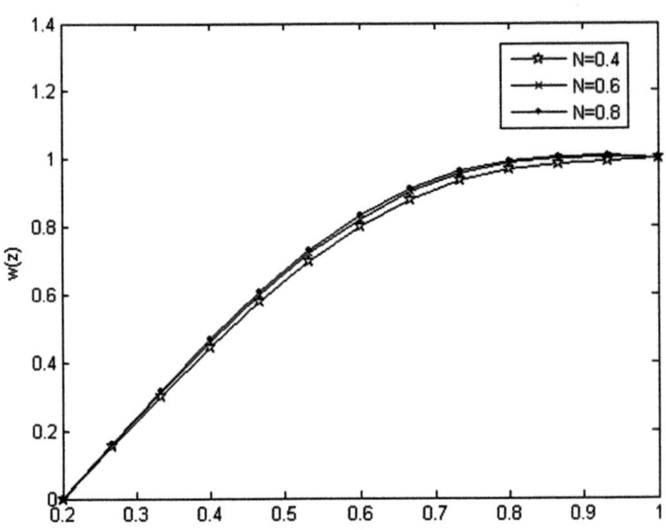

FIGURE 3.5 Temperature profiles for $C = 0$, $\xi = 0.2$, $N_{\mathrm{Pr}} = 5$.

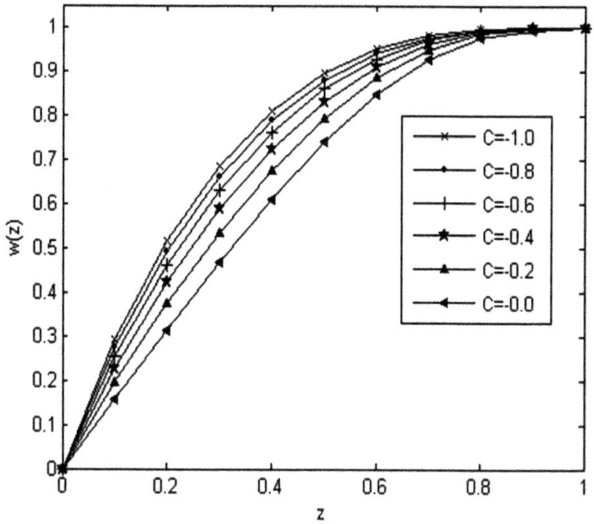

FIGURE 3.6 Temperature profiles for $\xi = 0$, $N = 0.5$, $N_{\rm Pr} = 4$.

temperature field are not severe. Fig. 3.6 displays the effects of suction or injection parameter on temperature field distributions, which indicates that when there exists suction or injection on the wall, the temperature gradient decreases with increasing in suction or injection parameter C.

Fig. 3.7 shows the effects of $N_{\rm Pr}$ on temperature field distributions; the results reveal that the temperature profile increases with increase in Prandtl

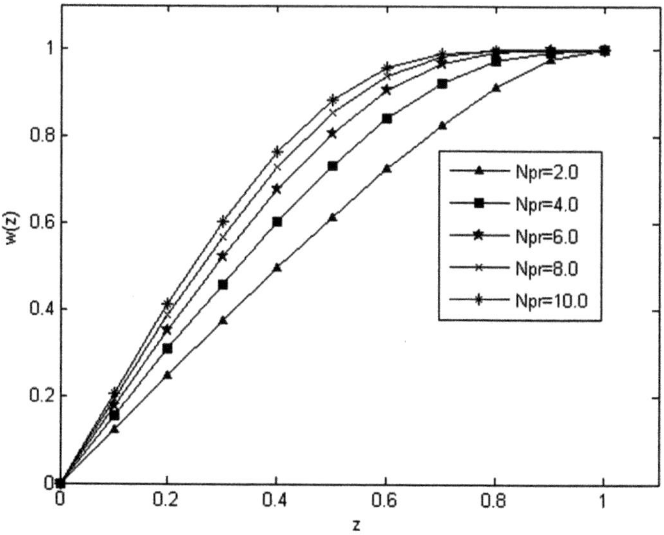

FIGURE 3.7 Temperature profiles for $C = 0$, $\xi = 0.0$, $N = 0.4$.

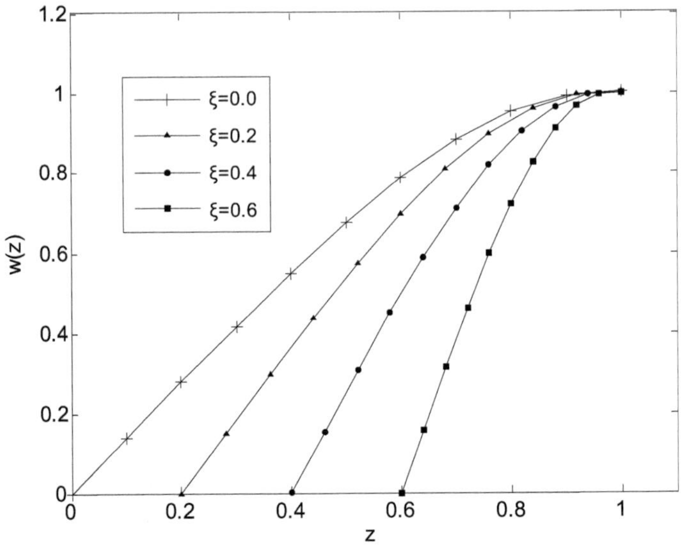

FIGURE 3.8 Temperature profiles for $C = 0$, $N_{Pr} = 3$, $N = 0.3$.

number N_{Pr}. Fig. 3.8 describes the effects of velocity ratio parameter $\xi (0 \leq \xi < 1)$ on temperature field distribution behavior, which indicates that the temperature gradient increases with increasing velocity ratio parameter.

3.3 POWER LAW MAGNETOHYDRODYNAMIC FLUID FLOW OVER A POWER LAW VELOCITY WALL

3.3.1 Physical Background

Magnetic fluid is a magnetic material in liquid form, which is a colloidal system formed by absorbing the molecules of surfactant and dispersing stably in an appropriate base medium. Magnetic fluid is widely used in the fields of mechanics, electronics, energy, chemicals, metallurgy, and so on.

The study of magnetic fluid originated in the 1950s in the United States. Papell obtained the first magnetic fluid patents, which had successfully been applied on NASA aerospace products seal in 1965 in United States. Since then, the research and development and application of this new type of material was initiated, and progress has been made. Pavlov (1974) studied the boundary layer flow of a Newtonian fluid in the magnetic field region, obtaining the exact solution of the boundary layer equation. On the basis of Pavlov, Chakrabarti and Gupta (1979) studied the temperature distribution in magnetic fluid boundary layer. Chiam (1995) studied the boundary layer flow and heat transfer of the magnetic fluid over a moving wall with a power law velocity. The partial differential equations were transformed to ordinary differential equations, and the asymptotic solutions of the magnetic field parameters were obtained. Mukhopadhyay et al. (2005) studied incompressible flow and heat transfer of magnetic fluid with variable properties; using the Lie group transformation, the partial differential equations were reduced into ordinary differential equations, and then a numerical solution was obtained by the Runge–Kutta method.

Most of the previous studies were focused on the problems of the boundary layer of Newton fluid. Sarpalaya (1961), Djukic (1973, 1974), Sapunkov (1967), and Vujanovic (1972) investigated flow and heat transfer of power law fluid in a magnetic field. In 1992, Andersson et al. (1992) studied boundary layer flow of non-Newtonian magnetic fluid on the moving surface, and a numerical solution was obtained by using the Runge–Kutta method and the Newton iteration method. In 2003, Liao (2003) studied MHD flows of non-Newtonian fluids over a stretching sheet, and the approximate analytical solutions were obtained for power law exponents n = 1, 2, 3 by using the homotopy analysis method. Chen et al. (2007) investigated MHD boundary layer flow of power law non-Newtonian fluid on a moving surface with a power law velocity, and the approximate analytical solutions were obtained by employing ADM.

In this section, we present a theoretical analysis for MHD boundary layer flow on a moving surface with the power law velocity. The boundary layer equation subject to corresponding boundary conditions are transformed to a singular, nonlinear two-point boundary value problem when suitable Lie group similarity transformations and generalized Crocco variables were introduced by Chen et al. (2007). An accurate expression of the skin friction coefficient is derived. The approximate analytical solution is obtained in terms of a rapid convergent power series by employing the ADMs, and the effects of pertinent physical parameters on the flow characteristics are analyzed in detail. Moreover, the reliability and efficiency of the analytical predictions are verified by the numerical results in the literature with good agreement.

For a sketch of magnetic fluid boundary layer, see Fig. 3.9.

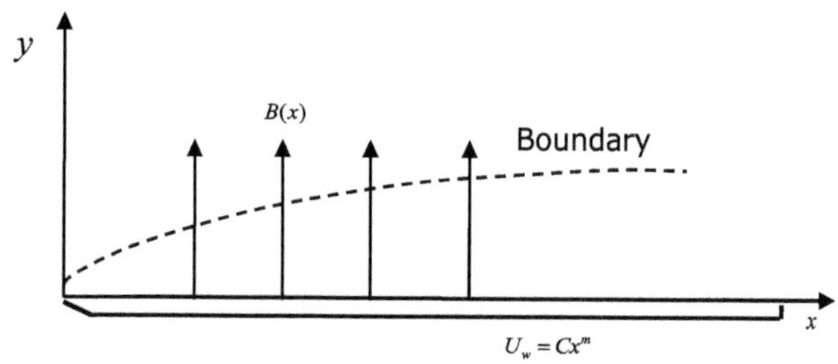

FIGURE 3.9 Sketch of magnetic fluid boundary layer.

3.3.2 Basic Governing Equations

The laminar boundary layer equations expressing the conservation of mass and momentum of MHD power law non-Newtonian fluid in dimensionless form can be written as follows:

$$\frac{\partial u}{\partial x}+\frac{\partial v}{\partial y}=0 \tag{3.30}$$

$$u\frac{\partial u}{\partial x}+v\frac{\partial u}{\partial y}=\frac{1}{\rho}\left(\frac{\partial \tau}{\partial y}\right)-\frac{\sigma B(x)^2}{\rho}u \tag{3.31}$$

where x-axis is taken along the surface and y-axis is perpendicular to it; u and v are the velocity components in the x and y directions, respectively; ρ is density of the fluid, σ is electrical conductivity, we consider here only the case of $\partial u/\partial y \leq 0$. The shear stress τ is defined as:

$$\tau=-K\left(-\frac{\partial u}{\partial y}\right)^N, \tag{3.32}$$

The corresponding boundary conditions are:

$$u|_{y=0}=Cx^m, \quad v|_{y=0}=0, \quad u|_{y=+\infty}=0, \tag{3.33}$$

where C is a known parameter.

3.3.3 Lie Group of Transformation

At present, most of the pertinent literature for moving velocity of the plate is considered to be the linear or constant velocity. In this section we assume the plate moving with a power law velocity. In order to solve the problem by employing the Adomian analytical decomposition method. The Li group transformation is introduced to reduce the boundary layer partial differential equations to ordinary differential equations.

In view of the stream function ψ, we have:

$$u=\frac{\partial \psi}{\partial y}, \quad v=-\frac{\partial \psi}{\partial x}. \tag{3.34}$$

Substituting (3.34) into boundary layer governing equations yields:

$$\frac{\partial \psi}{\partial y}\frac{\partial^2 \psi}{\partial x \partial y}-\frac{\partial \psi}{\partial x}\frac{\partial^2 \psi}{\partial y^2}=-\frac{K}{\rho}\frac{\partial}{\partial y}\left(-\frac{\partial^2 \psi}{\partial y^2}\right)^N-\frac{\sigma B(x)^2}{\rho}\frac{\partial \psi}{\partial y}. \tag{3.35}$$

Introducing Lie group of transformation as

$$\Gamma\ :\ x^*=xe^{\varepsilon\alpha_1}, y^*=ye^{\varepsilon\alpha_2}, \quad \psi^*=\psi e^{\varepsilon\alpha_3}, \quad u^*=ye^{\varepsilon\alpha_4}, \quad v^*=ve^{\varepsilon\alpha_5} \tag{3.36}$$

This transformation converts the coordinates of point (x,y,ψ,u,v) to the point of (x^*,y^*,ψ^*,u^*,v^*). Substituting (3.36) into Eq. (3.35), we obtain:

$$e^{\varepsilon(\alpha_1+2\alpha_2-2\alpha_3)}\left(\frac{\partial\psi^*}{\partial y^*}\frac{\partial^2\psi^*}{\partial x^*\partial y^*}-\frac{\partial\psi^*}{\partial x^*}\frac{\partial^2\psi^*}{\partial y^{*2}}\right)$$

$$=-\frac{K}{\rho}\frac{\partial}{\partial y^*}\left(-\frac{\partial^2\psi^*}{\partial y^{*2}}\right)^N e^{(2N+1)\varepsilon\alpha_2-N\varepsilon\alpha_3}-\frac{\sigma B(x^*)^2}{\rho}\frac{\partial\psi^*}{\partial y^*}e^{\varepsilon(\alpha_2-\alpha_3+2N\alpha_1)}, \tag{3.37}$$

The previous system has remained unchanged after the group transformation of Γ, and then we arrive at the following relations:

$$\alpha_1+2\alpha_2-2\alpha_3=(2N+1)\alpha_2-N\alpha_3=\alpha_2-\alpha_3+2N\alpha_1. \tag{3.38}$$

In view of boundary conditions:

$$\left.\frac{\partial\psi^*}{\partial y^*}\right|_{y=0}=Cx^{*m},\quad \left.\frac{\partial\psi^*}{\partial x^*}\right|_{y=0}=0,\quad \left.\frac{\partial\psi^*}{\partial y^*}\right|_{y=+\infty}=0. \tag{3.39}$$

We obtain the relation of $\alpha_4=\alpha_3-\alpha_2=\alpha_1 m$, and $\alpha_5=\alpha_3-\alpha_1$. Two variables are introduced in the form:

$$f(\eta)=A^{-1}x^r\psi,\quad \eta=Bx^s y, \tag{3.40}$$

Satisfying

$$A^{-1}x^r\psi=A^{-1}x^{*r}\psi^*\ ,\quad Bx^s y=Bx^{*s}y^*. \tag{3.41}$$

We may obtain $\alpha_2=-s\alpha_1$ and $\alpha_3=-r\alpha_1$.
By solving the linear systems of equations:

$$\begin{cases}\alpha_1+2\alpha_2-2\alpha_3=(2N+1)\alpha_2-N\alpha_3=\alpha_2-\alpha_3+2N\alpha_1 & \\ \alpha_4=\alpha_3-\alpha_2=\alpha_1 m & \alpha_5=\alpha_3-\alpha_1\ . \\ \alpha_2=-s\alpha_1 & \alpha_3=-r\alpha_1\end{cases}$$

The expressions for r and s are obtained as:

$$r=\frac{N+1}{2mN-m+1},\quad s=\frac{2m-mN-1}{N+1}.$$

Therefore, we establish the similarity transformation as following:

$$\psi(x,y)=\left[\frac{(N+1)C^{2N-1}K/\rho}{2mN-m+1}\right]^{1/(N+1)}x^{(2mN-m+1)/(N+1)}f(\eta),$$
$$\eta=y\left[\frac{(2mN-m+1)C^{2-N}}{(N+1)K/\rho}\right]^{1/(N+1)}x^{(2m-mN-1)/(N+1)}. \tag{3.42}$$

Taking (3.42) into (3.30) and (3.31), the boundary layer governing equations are reduced to an ordinary differential nonlinear boundary value problem:

$$N(-f'')^{N-1}f''' + ff'' - \beta(f')^2 - Mf' = 0. \tag{3.43}$$

the boundary conditions are:

$$f(0) = 0 \; f'(0) = 1 \; f'(+\infty) = 0, \tag{3.44}$$

where,

$$\beta = \frac{m(N+1)}{2mN - m + 1}, \quad M = \frac{\sigma B_0^2 (N+1)}{\rho C(2mN - m + 1)}. \tag{3.45}$$

The local skin friction coefficient C_f is defined as:

$$C_f = 2[-f''(0)]^N \left(\frac{2mN - m + 1}{N+1}\right)^{1/(N+1)} \left[\frac{(Cx^m)^{2-N} \cdot x^N}{K/\rho}\right]^{-1/(N+1)}, \tag{3.46}$$

where $(Cx^m)^{2-N}x^N/(K/\rho)$ describes the Reynolds number based on the velocity Cx^m of the wall. From the previous expression, we can see that the local skin friction coefficient C_f is closely related to the velocity gradient function $[-f''(0)]^N$ of the wall.

3.3.4 Generalized Crocco Variables Transformation

The following generalized Crocco variables are defined:

$$t = f'(\eta), \; w(t) = (-f''(\eta))^N, \quad 0 < t < 1. \tag{3.47}$$

Substituting (3.47) into Eqs. (3.43) and (3.44) and applying the chain rule, we obtain the following singular nonlinear two-point boundary value problems:

$$w''(t) = \left(\frac{\beta t^2 + Mt}{(w(t))^{1/N}}\right)' - \frac{t}{(w(t))^{1/N}}, \quad 0 < t < 1. \tag{3.48}$$

The boundary conditions are:

$$w(0) = 0, \quad w'(1) = (-1)^{1-N}\frac{\beta + M}{(w(1))^{1/N}}. \tag{3.49}$$

The local skin friction coefficient C_f is defined as:

$$C_f = 2w(1)\left(\frac{2mN - m + 1}{N+1}\right)^{1/(N+1)} \left[\frac{(Cx^m)^{2-N} \cdot x^N}{K/\rho}\right]^{-1/(N+1)}.$$

3.3.5 Adomian Decomposition Method Solutions

In order to apply ADM approximate method, in view of (3.48) and (3.49), we let $w(1) = \alpha$, the revised operator L_{tt}^{-1} is defined as:

$$L_{tt}^{-1} = \int_1^t dt' \int_1^{t'} dt'' \,, \tag{3.50}$$

Performing L_{tt}^{-1}on both sides of Eq. (3.48) yields:

$$w(t) = w(1) + \int_1^t \frac{\beta t'^2 + Mt'}{(w(t'))^{1/N}} dt' + \int_1^t \int_1^{t'} \frac{-t''}{(w(t''))^{1/N}} dt'' dt', \tag{3.51}$$

Decomposition method needs to write the solution of Eq. (3.51) into the sum of series:

$$w(t) = \sum_{n=0}^{\infty} w_n(t),$$

The nonlinear terms of Eq. (3.51) are decomposed to the sum of a series of Adomian polynomials as following:

$$\sum_{n=0}^{\infty} A_n = \frac{\beta t'^2 + Mt'}{(w(t'))^{1/N}}, \quad \sum_{n=0}^{\infty} B_n = \frac{-t''}{(w(t''))^{1/N}},$$

where A_n and B_n are, respectively, Adomian polynomials, which satisfy the following relationship:

$$\begin{aligned} w_0(t) &= \alpha \\ w_1(t) &= \int_1^t A_0 dt' + L_{tt}^{-1} B_0 \\ w_{k+1}(t) &= \int_1^t A_k dt' + L_{tt}^{-1} B_k, \quad k \geq 1 \end{aligned} ,$$

Let $p = -\frac{1}{N}$, we list several previous terms of Adomian polynomials as:

$$\begin{aligned} A_0 &= (\beta t'^2 + Mt') \cdot (w_0)^p, \\ A_1 &= (\beta t'^2 + Mt') p w_1(t')(w_0)^{p-1}, \\ A_2 &= (\beta t'^2 + Mt') \left(p w_2(t')(w_0)^{p-1} + \frac{p(p-1)}{2} w_1(t')^2 (w_0)^{p-2} \right), \\ &\ldots, \ldots \end{aligned} \tag{3.52}$$

and

$$
\begin{aligned}
B_0 &= -t''(w_0)^p, \\
B_1 &= -t'' p w_1(t'')(w_0)^{p-1}, \\
B_2 &= -t''\left(p w_2(t'')(w_0)^{p-1} + \frac{p(p-1)}{2} w_1(t'')^2 (w_0)^{p-2}\right), \\
B_3 &= -t''\left(p w_3(t'')(w_0)^{p-1} + \frac{p(p-1)(p-2)}{3} w_1(t'')^3 (w_0)^{p-3}\right. \\
&\quad \left. + p(p-1) w_1(t'') w_2(t'')(w_0)^{p-2}\right) \\
&\cdots, \cdots
\end{aligned}
\tag{3.53}
$$

The approximate analytical solution of the equation is to approximate the real solution $w(z)$ with the sum $\theta_l(z)$ of previous $l+1$ term of the solution, i.e.,

$$
\theta_l = \sum_{i=0}^{l} w_i(z) = w_0(z) + w_1(z) + w_2(z) + \cdots + w_l(z), \quad l \geq 1, \tag{3.54}
$$

Utilizing the values of boundary condition (3.49), for special values of parameters of N, β, and M, we can compute the value of parameter α. As soon as the value of α is established, we can obtain the approximation expression of the solution $w(z)$.

3.3.6 Results and Discussion

Table 3.1 presents the characteristics of skin friction coefficient $\alpha(\alpha = (-f''(0))^N)$ for specified fixed value of parameters $M = 0$ and $N = 1$ corresponding to parameter of β from $\beta = 0.5$ to $\beta = 5.0$. It is seen that the values of skin friction parameter increase with increasing of β. Table 3.2 shows the characteristics of skin friction coefficient α corresponding to the values of parameter M from $M = 0.0$ to $M = 5.0$ for fixed value of parameters $N = 1/3$ and $\beta = 1.0$. The results indicate that the skin friction coefficient increases with increasing in M.

TABLE 3.1 Values of α for $M = 0$ and $N = 1$, $\beta = 0.5$ to $\beta = 5.0$

l	α ($\beta = 0.5$)	α ($\beta = 1.0$)	α ($\beta = 1.5$)	α ($\beta = 2.0$)	α ($\beta = 5.0$)
6	0.81919	0.97899	1.11694	1.23991	1.81107
7	0.82152	0.98311	1.12271	1.24717	1.82492
8	0.82314	0.98607	1.12692	1.25250	1.83531
9	0.82432	0.98827	1.13010	1.25657	1.84339
10	0.82522	0.98997	1.13258	1.25977	1.84986

TABLE 3.2 Values of α for $N = 1/3$ and $\beta = 1.0$, $M = 0.0$ to $M = 5.0$

M	α ($l = 6$)	α ($l = 7$)	α ($l = 8$)	α ($l = 10$)
0.0	1.10811	1.11536	1.12083	1.12854
0.5	1.22611	1.23561	1.24284	1.25319
1.0	1.31717	1.32812	1.33648	1.34847
1.5	1.39243	1.40446	1.41367	1.42687
2.0	1.45713	1.47003	1.47991	1.49409
2.5	1.51418	1.52781	1.53827	1.55327
3.0	1.56542	1.57970	1.59064	1.60635
3.5	1.61206	1.62694	1.63829	1.65463
4.0	1.65498	1.67034	1.68211	1.69902
4.5	1.69479	1.71062	1.72275	1.74013
5.0	1.73197	1.74823	1.76070	1.77860

Table 3.3 presents a comparison of approximate analytical solutions for $f''(0)$ obtained by ADM with numerical results in literature, Chiam (1995); it is seen that the approximate analytical solutions are in close agreement with numerical results, which verify the reliable and effective solutions.

Figs. 3.10 and 3.11 present the distribution of shear stress profiles $w(z)$ with different power law exponents for $\beta = -1.0$ and $\beta = 1.0$, respectively. The figures show the distinct characteristics that the curve is convex for $\beta = -1.0$, but the curve is concave for $\beta = 1.0$. In both cases, the value of $w(z)$ decreases with increasing power law exponent parameter N.

TABLE 3.3 The Comparison of Analytical Solution $f''(0)$ With Numerical Results in Literature for $N = 1$ Chiam (1995)

β	M	$f''(0)_{app}$	$f''(0)_{com}$	Error
−1.0	0.5	−0.51787	−0.52395	6.08×10^{-3}
	1.0	−0.83424	−0.85111	1.69×10^{-2}
	5.0	−2.08665	−2.16287	7.62×10^{-2}
−1.5	1.0	−0.64101	−0.65298	1.20×10^{-2}
	5.0	−2.01214	−2.08524	7.31×10^{-2}
	10.0	−2.94042	−3.05623	1.15×10^{-1}

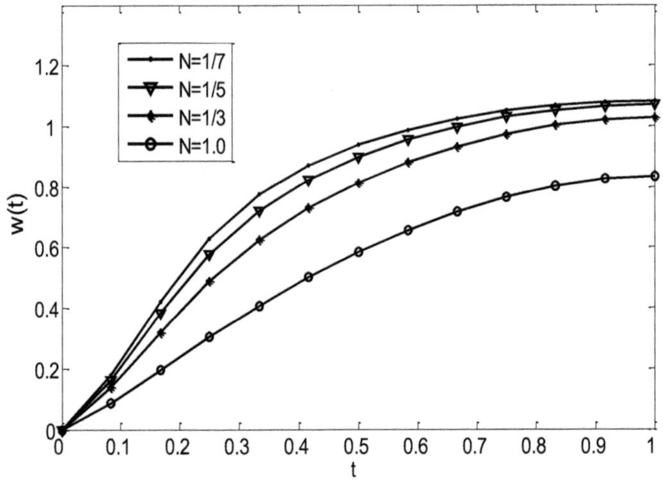

FIGURE 3.10 Shear stress profiles $w(z)$ for $\beta = -1.0$ and $M = 1$.

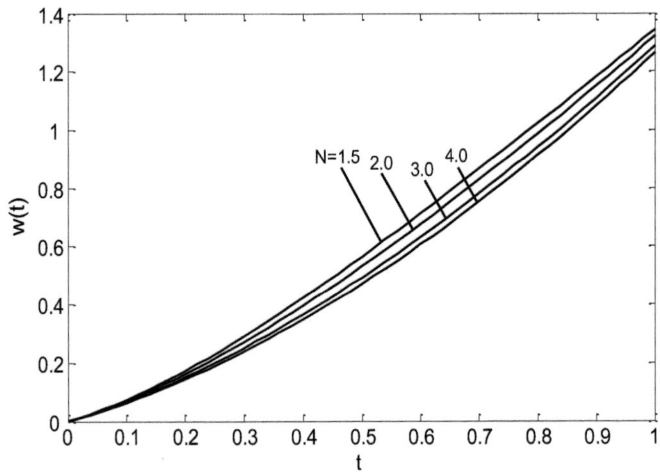

FIGURE 3.11 Shear stress profiles $w(z)$ for $\beta = 1.0$ and $M = 1$.

3.4 MARANGONI CONVECTION OVER A VAPOR–LIQUID SURFACE

We discussed in Chapter 2 that Marangoni convection may be induced by variations of surface tension along a liquid surface and appears in many natural and engineering problems. As buoyancy-induced natural convection, which causes undesirable effects in crystal growth of melts, the importance of Marangoni convection after carrying out single crystal growth under the microgravity environment has been recognized and the basic mechanism of Marangoni convection has been extensively investigated (Limbourg-Fontaine et al., 1985;

Villers and Platen, 1985; Legros et al., 1986; Arafune and Hirata, 1998). Many researchers have investigated Marangoni convection in a liquid bridge, which was modeled by the floating zone method, under the condition of microgravity utilizing a sounding rocket, an airplane, and drop shafts (Arafune and Hirata, 1999; Arafune et al., 1999; Croll et al., 1989; Christopher and Wang, 1998, 2001a,b; Okano et al., 1989; Straub, 1994; Slavtchev and Dubovik, 1992; Schwabe and Metzger, 1989; Slavtchev and Miladinova, 1998).

This section presents a similarity analysis for Marangoni convection–induced flow and heat transfer over a flat surface due to an imposed temperature gradient. The analysis assumes that the surface tension varies linearly with temperature but the temperature variation is a power law function of the location. In addition, the analysis assumes that a boundary layer develops along the surface due to the coupled Marangoni convection. The boundary layer equations for both the momentum equations and the energy equations are transformed to ordinary differential equations with a view to obtain the approximate analytical solutions (Adomian, 1994, 1988, 1998; Wazwaz, 1995, 2001; Zheng et al., 2004a,b).

3.4.1 Boundary Layer Governing Equations

Suppose that the flow is two dimensional and incompressible, then the laminar boundary layer equations expressing conservation of mass, momentum, and energy are written as follows:

$$\frac{\partial u}{\partial x}+\frac{\partial v}{\partial y}=0, \tag{3.55}$$

$$u\frac{\partial u}{\partial x}+v\frac{\partial u}{\partial y}=\nu\frac{\partial^2 u}{\partial y^2}, \tag{3.56}$$

$$u\frac{\partial T}{\partial x}+v\frac{\partial T}{\partial y}=\kappa\frac{\partial^2 T}{\partial y^2}, \tag{3.57}$$

where ν is the kinematic viscosity and κ thermal diffusivity which is the thermal conductivity divided by density and specific heat capacity at constant pressure, respectively. The boundary conditions at the liquid–vapor surface are:

$$\mu\left.\frac{\partial u}{\partial y}\right|_{y=0}=-\left.\frac{d\sigma}{dT}\frac{\partial T}{\partial x}\right|_{y=0}, \tag{3.58a}$$

$$v(x,0)=0, \tag{3.58b}$$

$$T(x,0)=T(0,0)+Ax^{k+1}, \tag{3.58c}$$

Far from the liquid–vapor surface, the boundary conditions are:

$$u(x,\infty)=0, \tag{3.59a}$$

$$\left.\frac{\partial T}{\partial y}\right|_{y=\infty}=0. \tag{3.59b}$$

By introducing similarity transformations (Lie group similarity transformations), the boundary layer governing Eqs. (3.55)–(3.59) can be transformed into the following nonlinear ordinary differential equations:

$$F''' = aF'^2 - bFF'', \tag{3.60a}$$

$$\theta'' = \Pr(-\,bF\theta' - tF'\theta), \tag{3.60b}$$

The boundary conditions are:

$$F(0)=0,\ F''(0)=-k-1,\ F'(\infty)=0, \tag{3.61}$$

$$\theta(0)=A, \quad \theta'(\infty)=0. \tag{3.62}$$

where we have chosen the stream function $\psi(x,y)$, temperature function $\theta(\eta)$, and the similarity variable η as:

$$\eta = C_1 x^{\alpha} y, \tag{3.63a}$$

$$F(\eta) = C_2 x^{\beta}\psi(x,y), \tag{3.63b}$$

$$\theta(\eta) = (T(x,y) - T(0,0))x^{t}, \tag{3.63c}$$

$F(\eta)$ is the dimensionless stream function and $\theta(\eta)$ denotes the dimensionless temperature function, and the parameters are chosen respectively:

$$C_1 = \sqrt[3]{\frac{(d\sigma/dT)A\rho}{\mu^2}}, \quad C_2 = \sqrt[3]{\frac{\rho^2}{(d\sigma/dT)A\mu}}, \tag{3.64}$$

$$a=\alpha-\beta=\frac{2k+1}{3}, \quad b=-\alpha=\frac{k+2}{3}, \quad t=-1-k. \tag{3.65}$$

3.4.2 Adomian Decomposition Method Solutions

The momentum equation and the energy equation are decoupled since the fluid is incompressible; the solution to the momentum equation may be discussed first. It is seen that the integration of Eq. (3.60a) cannot be started at $\eta=0$ because F' is not there, and the boundary conditions (3.61) provide only two of the three values, $F(0)$, $F''(0)$, that are required at $\eta=0$, but provide another value of F' at infinity. In order to solve the problem by employing the ADM, we need to introduce an auxiliary initial parameter, i.e., suppose $c=F'(0)$ and solve the equation with initial values firstly, then to determine the value c by using infinity condition $F'(\infty)=0$.

Integrate Eq. (3.60a) over $[0,\eta]$, we have:

$$F'' = t + \int \left(aF'^2 - bFF''\right). \tag{3.66}$$

Integrate (3.66) twice again, which yields:

$$F(\eta) = c\eta + \frac{1}{2}t\eta^2 + L^{-1}\left(\int \left(aF'^2 - bFF''\right)\right), \tag{3.67}$$

where L^{-1} is a twofold integral operator.

Define $F(\eta) = \sum_{n=0}^{\infty} F_n(\eta)$, then

$$\sum_{n=0}^{\infty} F_n(\eta) = c\eta + \frac{1}{2}t\eta^2 + L^{-1}\left(\int \left(aF'^2 - bFF''\right)\right). \tag{3.68}$$

Let $\sum_{n=0}^{\infty} A_n = aF'^2 - bFF''$, in view of Eq. (3.68), yields:

$$\sum_{n=0}^{\infty} F_n(\eta) = c\eta + \frac{1}{2}t\eta^2 + L^{-1}\left(\int \sum_{n=0}^{\infty} A_n d\eta\right), \tag{3.69}$$

where A_n are Adomian polynomials.

The modified decomposition technique suggests the recursive relation:

$$F_0 = c\eta, \tag{3.70}$$

$$F_1 = \frac{1}{2}t\eta^2 + L^{-1}\left(\int A_0 d\eta\right), \tag{3.71}$$

$$F_{k+1} = L^{-1}\left(\int A_k d\eta\right), \tag{3.72}$$

for the determination of the components $F_n(x)$. Here, we list the first few terms of Adomian polynomials that represent the nonlinear operator $aF'^2 - bFF''$:

$$A_0 = aF_0'^2 - bF_0F_0'', \tag{3.73}$$

$$A_1 = a\left(2F_0'F_1'\right) - b\left(F_1F_0'' + F_0F_1''\right), \tag{3.74}$$

$$A_2 = a\left(2F_0'F_2' + F_1'^2\right) - b\left(F_2F_0'' + F_1F_1'' + F_0F_2''\right), \tag{3.75}$$

$$A_3 = a\left(2F_0'F_3' + 2F_1'F_2'\right) - b\left(F_3F_0'' + F_2F_1'' + F_1F_2'' + F_0F_3''\right), \tag{3.76}$$

$$A_4 = a\left(2F_0'F_4' + 2F_1'F_3' + F_2'^2\right) - b\left(F_4F_0'' + F_3F_1'' + F_2F_2'' + F_1F_3'' + F_0F_4''\right), \tag{3.77}$$

$$A_5 = a\left(2F_0'F_5' + 2F_1'F_4' + 2F_2'F_3'\right) \\ - b\left(F_5F_0'' + F_4F_1'' + F_3F_2'' + F_2F_3'' + F_1F_4'' + F_0F_5''\right), \tag{3.78}$$

and so on for other polynomials. Substitute A_n into $F_n(x)$, we obtain:

$$F_0 = c\eta, \tag{3.79}$$

$$A_0 = ac^2, \tag{3.80}$$

$$F_1 = \frac{1}{2}t\eta^2 + \frac{1}{6}ac^2\eta^3, \tag{3.81}$$

$$A_1 = (2act - bct)\eta + \left(a^2c^3 - abc^3\right)\eta^2, \tag{3.82}$$

$$F_2 = \frac{2act - bct}{24}\eta^4 + \frac{a^2c^3 - abc^3}{60}\eta^5, \tag{3.83}$$

$$A_2 = \left(at^2 - \frac{1}{2}bt^2\right)\eta^2 + \left(\frac{5}{3}a^2c^2t - 2abc^2t + \frac{1}{2}b^2c^2t\right)\eta^3 \\ + \left(\frac{5}{12}a^3c^4 - \frac{2}{3}a^2bc^4 + \frac{1}{3}ab^2c^4\right)\eta^4, \tag{3.84}$$

$$F_3 = \frac{1}{60}\left(at^2 - \frac{1}{2}bt^2\right)\eta^5 + \frac{1}{120}\left(\frac{5}{3}a^2c^2t - 2abc^2t + \frac{1}{2}b^2c^2t\right)\eta^6 \\ + \frac{1}{210}\left(\frac{5}{12}a^3c^4 - \frac{2}{3}a^2bc^4 + \frac{1}{3}ab^2c^4\right)\eta^7, \tag{3.85}$$

$$A_3 = \left(\frac{5}{6}a^2ct^2 - \frac{4}{3}abct^2 + \frac{11}{24}b^2ct^2\right)\eta^4 \\ + \left(\frac{2}{3}a^3c^3t - \frac{83}{60}a^2bc^3t + \frac{103}{120}ab^2c^3t - \frac{1}{8}b^3c^3t\right)\eta^5 \\ + \left(\frac{1}{9}a^4c^5 - \frac{17}{60}a^3bc^5 - \frac{1}{15}ab^3c^5 + \frac{41}{180}a^2b^2c^5\right)\eta^6, \tag{3.86}$$

$$
\begin{aligned}
F_4 &= L^{-1}\left(\int A_3 d\eta\right) \\
&= \frac{1}{210}\left(\frac{5}{6}a^2ct^2 - \frac{4}{3}abct^2 + \frac{11}{24}b^2ct^2\right)\eta^7 \\
&+ \frac{1}{336}\left(\frac{2}{3}a^3c^3t - \frac{83}{60}a^2bc^3t + \frac{103}{120}ab^2c^3t - \frac{1}{8}b^3c^3t\right)\eta^8 \\
&+ \frac{1}{504}\left(\frac{1}{9}a^4c^5 - \frac{17}{60}a^3bc^5 - \frac{1}{15}ab^3c^5 + \frac{41}{180}a^2b^2c^5\right)\eta^9, \cdots, \cdots.
\end{aligned}
\tag{3.87}
$$

3.4.3 Results and Discussion

In terms of ADM, the series solution of the problem can be obtained in the form as:

$$
\begin{aligned}
F(\eta) &= c\eta + \frac{1}{2}t\eta^2 + \frac{1}{6}ac^2\eta^3 + \frac{2act - bct}{24}\eta^4 + \frac{a^2c^3 - abc^3 + at^2 - \frac{1}{2}bt^2}{60}\eta^5 \\
&+ \frac{1}{120}\left(\frac{5}{3}a^2c^2t - 2abc^2t + \frac{1}{2}b^2c^2t\right)\eta^6 \\
&+ \frac{1}{210}\left(\frac{5}{12}a^3c^4 - \frac{2}{3}a^2bc^4 + \frac{1}{3}ab^2c^4 + \frac{5}{6}a^2ct^2 - \frac{4}{3}abct^2 + \frac{11}{24}b^2ct^2\right)\eta^7, \\
&+ \frac{1}{336}\left(\frac{2}{3}a^3c^3t - \frac{83}{60}a^2bc^3t + \frac{103}{120}ab^2c^3t - \frac{1}{8}b^3c^3t\right)\eta^8 \\
&+ \frac{1}{504}\left(\frac{1}{9}a^4c^5 - \frac{17}{60}a^3bc^5 - \frac{1}{15}ab^3c^5 + \frac{41}{180}a^2b^2c^5\right)\eta^9 + \cdots
\end{aligned}
\tag{3.88}
$$

And then,

$$
\begin{aligned}
F'(\eta) &= c + t\eta + \frac{1}{2}ac^2\eta^2 + \frac{2act - bct}{6}\eta^3 \\
&+ \frac{a^2c^3 - abc^3 + at^2 - \frac{1}{2}bt^2}{12}\eta^4 + \frac{1}{20}\left(\frac{5}{3}a^2c^2t - 2abc^2t + \frac{1}{2}b^2c^2t\right)\eta^5 \\
&+ \frac{1}{30}\left(\frac{5}{12}a^3c^4 - \frac{2}{3}a^2bc^4 + \frac{1}{3}ab^2c^4 + \frac{5}{6}a^2ct^2 - \frac{4}{3}abct^2 + \frac{11}{24}b^2ct^2\right)\eta^6. \\
&+ \frac{1}{42}\left(\frac{2}{3}a^3c^3t - \frac{83}{60}a^2bc^3t + \frac{103}{120}ab^2c^3t - \frac{1}{8}b^3c^3t\right)\eta^7 \\
&+ \frac{1}{56}\left(\frac{1}{9}a^4c^5 - \frac{17}{60}a^3bc^5 - \frac{1}{15}ab^3c^5 + \frac{41}{180}a^2b^2c^5\right)\eta^8 + \cdots
\end{aligned}
\tag{3.89}
$$

Our concern now is to determine the constant c by using the condition $F'(\infty) = 0$ in Eq. (3.61). It may be seen that this condition cannot be applied directly to (3.89). This goal can be achieved by performing Padé approximant, which has the advantage of manipulating the polynomial approximation into a rational function. The Padé approximant of $F'(\eta)$ usually converges on the entire real axis algorithms if $F'(\eta)$ is free of singularities on the real axis.

The diagonal approximant has been used by many scholars as the most accurate approximant, therefore we will construct only the diagonal approximants in the following discussions. The diagonal approximants [2/2] and [3/3] are computed by using Maple built-in utilities. Substituting the boundary condition $F'(+\infty) = 0$ in each Padé approximant, the approximant will vanish if the coefficient of η with the highest power in the numerator vanishes. By solving the resulting polynomials of these coefficients, we obtain the values of c. Here, we construct the approximation for [2/2], [3/3], [4/4], etc., to obtain the value of parameter c. At this time, the error estimation of the approximation is $\sum_{i=0}^{\infty} c_i \eta^i - [L/M] = O(\eta^{L+M+1})$.

Firstly, we determine the [2/2] of Padé approximation. Using the same method we can also get [3/3], [4/4], ... For the approximation of [2/2], we choose a_0, a_1, a_2, b_1, and b_2 to be the coefficients of η^i ($i = 0, 1, 2$), let

$$F'(y)\left(1 + b_1 y + b_2 y^2\right) = a_0 + a_1 y + a_2 y^2 + \left(a_3 y^3 + a_4 y^4\right), \tag{3.90}$$

In view of definition of [2/2], we obtain $a_3 = a_4 = 0$,

$$F'(y) = \frac{a_0 + a_1 y + a_2 y^2}{1 + b_1 y + b_2 y^2} = \frac{\frac{a_0}{y^2} + \frac{a_1}{y} + a_2}{\frac{1}{y_2} + \frac{b_1}{y} + b_2}, \tag{3.91}$$

for $F'(\infty) = 0$, yields $a_2 = 0$.

Substituting (3.89) into (3.90) and equating the coefficients of y^i for $i = 0$, 1, 2,⋯, 4, yields:

$$\frac{1}{2}ac^2 + b_1 t + b_2 c = a_2 = 0. \tag{3.92}$$

In order to determine c, we have to determine the value of b_1, b_2.
From $a_3 = a_4 = 0$, we obtain:

$$\begin{cases} \dfrac{2act - bct}{6} + b_1 \cdot \dfrac{1}{2}ac^2 + b_2 \cdot t = 0 \\ \dfrac{a^2c^3 - abc^3 + at^2 - \frac{1}{2}bt^2}{12} + b_1 \cdot \dfrac{2act - bct}{6} + b_2 \cdot \dfrac{1}{2}ac^2 = 0 \end{cases}, \tag{3.93}$$

from (3.93), we can obtain the value of b_1, b_2. It is in the same way we can calculate [3/3], [4/4], ...

For specified values of parameters $k=-0.5$, $a=0$, $b=0.5$, $t=-0.5$, we obtain $c=0.98$, and

$$F=-\frac{1}{1199001}\eta^{10}-\frac{3}{281185}\eta^{9}+\frac{3}{224575}\eta^{8}+\frac{25}{187013}\eta^{7}-\frac{6}{11995}\eta^{6}-\frac{1}{960}\eta^{5}+\frac{49}{4800}\eta^{4}-\frac{1}{4}\eta^{2}+\frac{49}{50}\eta, \tag{3.94}$$

$$F'=-\frac{10}{1199001}\eta^{9}-\frac{15}{156214}\eta^{8}+\frac{7}{65501}\eta^{7}+\frac{48}{51295}\eta^{6}-\frac{36}{11995}\eta^{5}-\frac{1}{192}\eta^{4}+\frac{49}{1200}\eta^{3}-\frac{1}{2}\eta+\frac{49}{50}, \tag{3.95}$$

$$F''=-\frac{13}{173189}\eta^{8}-\frac{23}{29941}\eta^{7}+\frac{45}{60154}\eta^{6}+\frac{65}{11577}\eta^{5}-\frac{36}{2399}\eta^{4}-\frac{1}{48}\eta^{3}+\frac{49}{400}\eta^{2}-\frac{1}{2}, \tag{3.96}$$

For specified values of parameters $k=0$, $a=1/3$, $b=2/3$, $t=-1$, we obtain $c=1.25$, and

$$F=-\frac{1}{2716602}\eta^{11}+\frac{5}{572219}\eta^{10}-\frac{15}{273029}\eta^{9}+\frac{11}{55844}\eta^{8}-\frac{35}{49026}\eta^{7}+\frac{5}{2592}\eta^{6}-\frac{25}{6912}\eta^{5}+\frac{25}{288}\eta^{3}-\frac{1}{2}\eta^{2}+\frac{5}{4}\eta, \tag{3.97}$$

$$F'=-\frac{11}{2716602}\eta^{10}+\frac{50}{572219}\eta^{9}-\frac{135}{273029}\eta^{8}+\frac{22}{13961}\eta^{7}-\frac{245}{49026}\eta^{6}+\frac{5}{432}\eta^{5}-\frac{125}{6912}\eta^{4}+\frac{25}{96}\eta^{2}-\eta+\frac{5}{4}, \tag{3.98}$$

$$F''=-\frac{8}{197571}\eta^{9}+\frac{87}{110629}\eta^{8}-\frac{41}{10365}\eta^{7}+\frac{61}{5530}\eta^{6}-\frac{245}{8171}\eta^{5}+\frac{25}{432}\eta^{4}-\frac{125}{1728}\eta^{3}+\frac{25}{48}\eta-1. \tag{3.99}$$

For specified values of parameters $k=0.5$, $a=2/3$, $b=5/6$, $t=-1.5$, we obtain $c=1.55$, and

$$F=\frac{1}{546464}\eta^{11}+\frac{2}{72643}\eta^{10}+\frac{1}{75402}\eta^{9}-\frac{11}{168369}\eta^{8}-\frac{7}{420928}\eta^{7}+\frac{321}{461759}\eta^{6}+\frac{203}{81890}\eta^{5}-\frac{31}{640}\eta^{4}+\frac{961}{3600}\eta^{3}-\frac{3}{4}\eta^{2}+\frac{31}{20}\eta, \tag{3.100}$$

$$\begin{aligned} F' = & \frac{11}{546464}\eta^{10} + \frac{20}{72643}\eta^9 + \frac{1}{8378}\eta^8 - \frac{88}{168369}\eta^7 - \frac{49}{420928}\eta^6 \\ & + \frac{961}{230400}\eta^5 + \frac{203}{16378}\eta^4 - \frac{31}{160}\eta^3 + \frac{961}{1200}\eta^2 - \frac{3}{2}\eta + \frac{31}{20} \end{aligned}, \tag{3.101}$$

$$\begin{aligned} F'' = & \frac{7}{34775}\eta^9 + \frac{236}{95243}\eta^8 + \frac{4}{4189}\eta^7 - \frac{46}{12573}\eta^6 - \frac{11}{15749}\eta^5 \\ & + \frac{961}{46080}\eta^4 + \frac{153}{3086}\eta^3 - \frac{93}{160}\eta^2 + \frac{961}{600}\eta - \frac{3}{2} \end{aligned}. \tag{3.102}$$

In the following, we consider the solution of energy Eq. (3.60b). Define $\theta(\eta) = Ag(\eta)$, then Eq. (3.60b) becomes:

$$g = \Pr(-bFg' - tF'g), \tag{3.103}$$

The boundary condition (3.62) reduces:

$$g(0) = 1, \quad g'(\infty) = 0.$$

Let $g'(0) = \beta$, integrating Eq. (3.103) over $[0,\eta]$ yields:

$$g(\eta) = 1 + \beta\eta + \Pr L^{-1}(-bFg' - tF'g). \tag{3.104}$$

Noting that L^{-1} is a twofold integral operator, define

$$g(\eta) = \sum_{n=0}^{\infty} g_n(\eta), \quad \sum_{n=0}^{\infty} B_n(\eta) = -bFg' - tF'g, \tag{3.105}$$

where B_n are Adomian polynomials.

$$\begin{aligned} B_0(\eta) &= -bF_0g_0' - tF_0'g_0 \\ B_1(\eta) &= -b(F_0g_1' + F_1g_0') - t(F_0'g_1 + F_1'g_0) \\ B_2(\eta) &= -b(F_0g_2' + F_1g_1' + F_2g_0') - t(F_0'g_2 + F_1'g_1 + F_2'g_0) \\ & \ldots, \ \ldots, \end{aligned}. \tag{3.106}$$

Adomian decomposition satisfies the following relationship:

$$\begin{aligned} g_0(\eta) &= 1 \\ g_1(\eta) &= \beta\eta + \Pr L^{-1}B_0 \\ g_{k+1}(\eta) &= \Pr L^{-1}B_k \end{aligned}. \tag{3.107}$$

We may also determine the value of β by using Padé approximants technique. Choose $k = -0.5$, for example, we obtain the value of β for different Prandtl number. For values of parameters Pr, the polynomials are presented as follow:

$\Pr = 9$:

$$\begin{aligned} g(\eta) = & \frac{189}{3916}\eta^{10} + \frac{331}{5228}\eta^9 - \frac{269}{3044}\eta^8 - \frac{197}{1946}\eta^7 + \frac{345}{1379}\eta^6 \\ & + \frac{409}{2343}\eta^5 - \frac{301}{484}\eta^4 - \frac{3}{8}\eta^3 + \frac{441}{200}\eta^2 - \frac{1409}{701}\eta + 1 \end{aligned}, \tag{3.108}$$

Pr = 7:

$$g(\eta)=\frac{127}{9245}\eta^{10}+\frac{631}{27317}\eta^{9}-\frac{67}{2343}\eta^{8}-\frac{243}{5185}\eta^{7}+\frac{485}{4443}\eta^{6}+\frac{343}{3200}\eta^{5}-\frac{155}{427}\eta^{4}-\frac{7}{24}\eta^{3}+\frac{343}{200}\eta^{2}-\frac{560}{321}\eta+1, \quad (3.109)$$

$Pr = 5$:

$$g(\eta)=\frac{49}{19184}\eta^{10}+\frac{23}{3849}\eta^{9}-\frac{149}{25031}\eta^{8}-\frac{121}{7217}\eta^{7}+\frac{251}{7074}\eta^{6}+\frac{539}{9600}\eta^{5}-\frac{491}{2749}\eta^{4}-\frac{5}{24}\eta^{3}+\frac{49}{40}\eta^{2}-\frac{1035}{754}\eta+1, \quad (3.110)$$

The dimensionless velocity and shear stress are both the functions of the exponent parameter k. Figs. 3.12 and 3.13 show the approximate analytical solutions and numerical results of dimensionless velocity distributions for $k=-0.5$ and $k=0$. It may be seen that the dimensionless velocity F' decreases with the increasing η but increases with the increasing k. In physical meaning this indicates that both the velocity and the absolute value of shear stress are increasing functions of exponent parameter.

Figs. 3.14–3.16 show the comparisons of approximate analytical solutions obtained by ADM and numerical solutions of dimensionless temperature gradient distributions obtained by shooting technique coupled with the

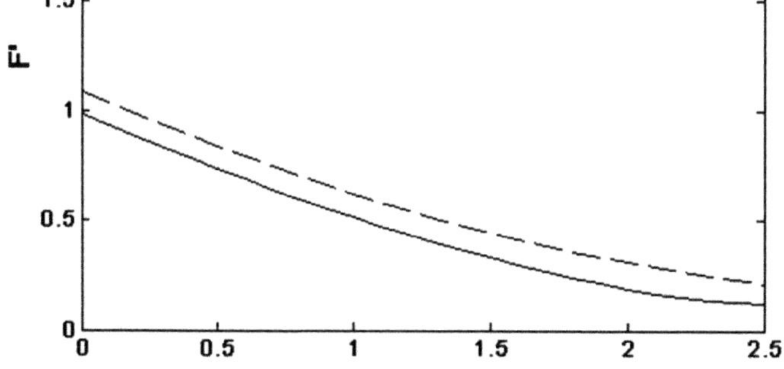

FIGURE 3.12 The dimensionless velocity profiles for $k=-0.5$.

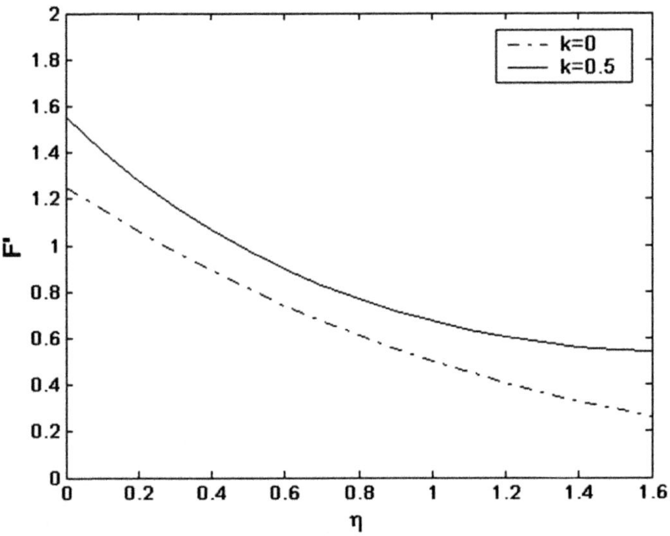

FIGURE 3.13 The dimensionless velocity profiles for $k = 0$ and $k = 0.5$.

Runge–Kutta method for $k = -0.5$, $\mathrm{Pr} = 5, 7, 9$, respectively. The approximate results are in close agreement for values of η near zero, and the errors increase with increasing η, which indicate the reliability and efficiency of the analytical decomposition and numerical technique near the surface due to the

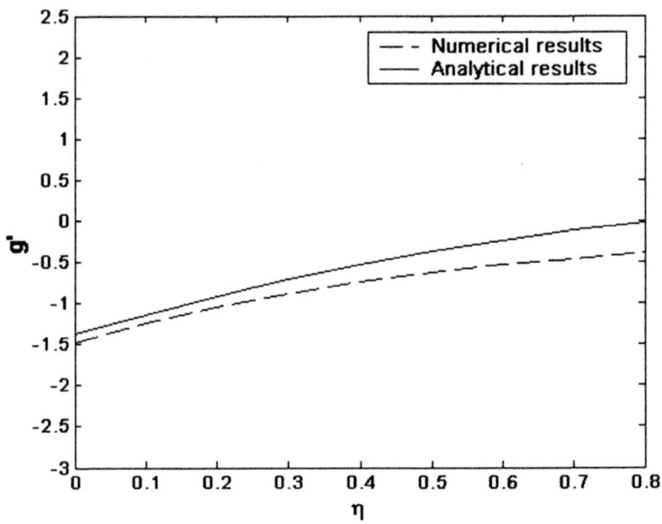

FIGURE 3.14 The dimensionless temperature gradient profiles for $\mathrm{Pr} = 5$.

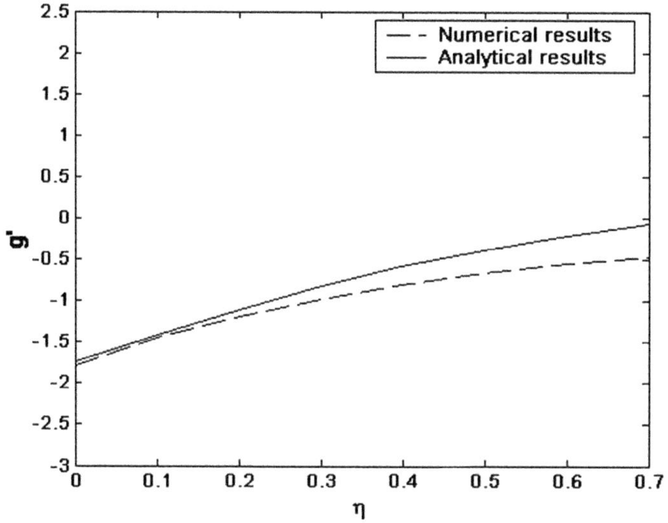

FIGURE 3.15 The dimensionless temperature gradient profiles for Pr = 7.

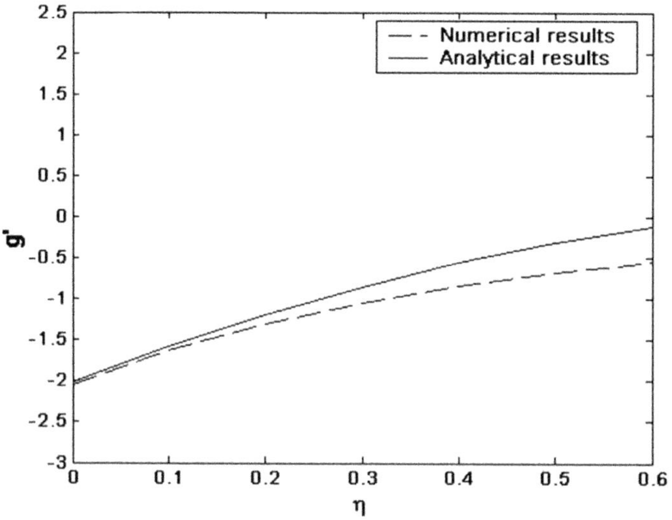

FIGURE 3.16 The dimensionless temperature gradient profiles for Pr = 9.

coupled Marangoni convection. The figures are also displaying that the heat transfer characteristics strongly depended on the wall temperature exponent and the Prandtl number. The magnitude of the slop of the temperature profiles at the surface sharply decreases with decreasing Prandtl number as the thermal

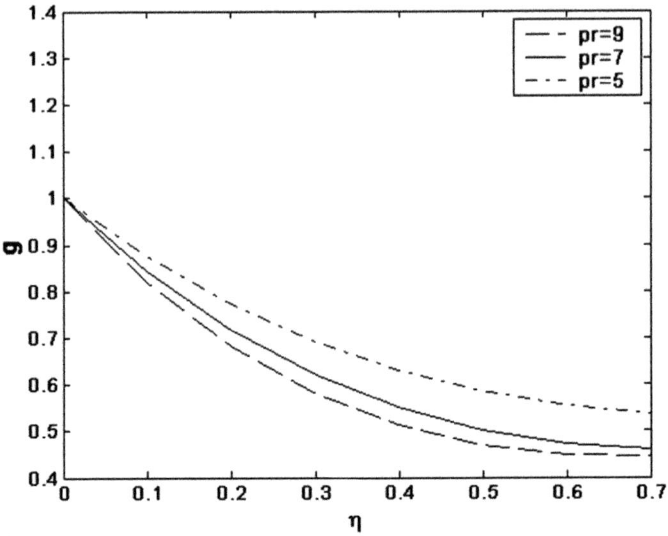

FIGURE 3.17 The dimensionless temperature profiles for Pr = 5, 7, 9.

boundary layer gets very thick for low Prandtl numbers and very thin for high Prandtl numbers.

Fig. 3.17 presents a comparison of the dimensionless temperature distributions for Pr = 5, 7, 9, the figure indicates that dimensionless temperature g decreases with increasing in Prandtl number, it implies that heat transfer rate increases and the thickness of thermal boundary is lost with increasing Prandtl number.

3.5 SUMMARY

In this chapter, ADM is introduced to study boundary layer problem flow and heat transfer of power law non-Newtonian fluids on a continuous moving surface with a constant speed and special injection or suction velocity, power law MHD boundary layer flow on the moving surface with the power law velocity and Marangoni convection–induced flow over a fluid-gas free surface due to an imposed temperature gradient is made. The analysis assumes that the surface tension varies linearly with temperature but the temperature variation is a power law function of the location. The partial differential equations are transformed into ordinary differential equations by suitable similarity transformation and generalized Crocco variables transformation. The approximate analytical solutions expressions are obtained by ADM in terms of a rapid convergent power series with elegantly computable terms, and the numerical results are also presented. Moreover, for specified pertinent parameters, the associated boundary layer transfer characteristics are presented graphically and analyzed.

REFERENCES

Acrivos, A., Shah, M.J., Petersen, E.E., 1960. Momentum and heat transfer in laminar boundary layer flows of non-newtonian fluids past external surface. AIChE Journal 6, 312–317.

Adomian, G., 1988. Applications of Nonlinear Stochastic Systems Theory to Physics. Kulwer.

Adomian, G., 1994. Solving Frontier Problems of Physics, the Decomposition Method. Kluwer Academic Publisher, Boston, MA.

Adomian, G., 1998. A review of the decomposition method in applied mathematics. Journal of Mathematical Analysis and Applications 135, 501–544.

Andersson, H.I., Bach, K.H., Dandapat, B.S., 1992. Magnetohydrodynamic flow of a power-law fluid over a stretching sheet. International Journal of Non-Linear Mechanics 27, 929–936.

Arafune, K., Hirata, 1998. A Interactive solutal and thermal Marangoni convection in a rectangular open boat. Numerical Heat Transfer Part A 34, 421–429.

Arafune, K., Sugiura, M., Hirata, A., 1999. AInvestigation of thermal Marangoni convection in low and high Prandtl number fluids. Journal of Chemical Engineering of Japan 32, 104–109.

Arafune, K., Hirata, A., 1999. Thermal and solutal Marangoni convection in In-Ga-Sb system. Journal of Crystal Growth 197, 811–817.

Chakrabarti, A., Gupta, A.S., 1979. Hydromagnetic flow and heat transfer over a stretching sheet. Quarterly of Applied Mathematics 37, 73–78.

Chiam, T.C., 1995. Hydromagnetic flow over a surface stretching with a power-law velocity. International Journal of Engineering Science 33, 429–435.

Chen, X.H., Zheng, L.C., Zhang, X.X., 2007. Mhd boundary layer flow of a non-newtonian fluid on a moving surface with a power-law velocity. Chinese Physics Letters 24 (7), 1989–1991.

Christopher, D.M., Wang, B.X., 1998. Marangoni convection around a bubble in microgravity, heat transfer. In: Proceedings of the 11th International Heat Transfer Conference, vol. 3. Taylor and Francis, Levittown, PA, pp. 489–494.

Christopher, D.M., Wang, B.X., 2001a. Similarity simulation for Marangoni convection around a vapor bubble during nucleation and growth. International Journal of Heat and Mass Transfer 44, 799–810.

Christopher, D.M., Wang, B.X., 2001b. Prandtl number effects for Marangoni convection over a flat surface. International journal of thermal sciences 40, 564–570.

Croll, A., Muller-Sebert, W., Nitsche, R., 1989. The critical Marangoni number for the onset of time-dependent convection in silicon. Materials Research Bulletin 24, 995–1004.

Crosan, T., Pop, I., 2001. Free Convection over a vertical flat plate with a variable wall temperature and internal heat generation in a porous medium saturated with a non-Newtonian Fluid. Technische Mechanik 4, 313–318.

Djukic, D.S., 1973. On the use of Crocco's equation for the flow of power-law fluids in a transverse magnetic field. AIChE Journal 19, 1159–1163.

Djukic, D.S., 1974. Hiemenz magnetic flow of power-law fluids. ASME Journal of Applied Mechanics 4, 822–823.

Fox, V.G., Fan, L.T., 1969. The laminar boundary layer on a moving continous flat sheet immersed in a non-Newtonian fluid. AIChE Journal 15 (3), 327–333.

Gorla, R.S.R., Pop, I., Lee, J.K., 1992. Convective wall plume in power-law fluid: second-order Correction for the Adiabatic Wall. Warme-und Stoffubertragung 27, 473–479.

Howell, T.G., Jeng, Dewitt, K.J., 1997. Momentum and heat transfer on a continuous moving surface in power law fluid. International Journal of Heat and Mass Transfer 40, 1853–1861.

Jones, C.W., Atkinson, C., 1974. Similarity solutions in some nonlinear diffusion problems and in boundary layer flow of a pseudo-plastic fluid. The Quarterly Journal of Mechanics and Applied Mathematics 27, 193–211.

Kumari, K., Pop, I., Takhar, H.S., 1997. Free convection boundary-layer flow of a non-Newtonian fluid along a vertical wavy surface. International Journal of Heat and Fluid Flow 18, 625–631.

Legros, J.C., Limbourg-Fontaine, M.C., Petre, G., 1986. Surface tension minimum and Marangoni convection, in: fluid dynamics and space proceedings of ESA Symp. Rhode-Saint-Genese 25–26, 137–143. June (ESA SP-265).

Li, B.T., Zheng, L.C., Zhang, X.X., 2012. Comparison between thermal conductivity models on heat transfer in power-law non-newtonian fluids. Journal of Heat Transfer-T ASME 134, 041702.

Liao, S.J., 2003. On the analytic solution of magnetodrodynamic flows of non-Newtionian fluids over a stretching sheet. Journal of Fluid Mechanics 488, 189–212.

Limbourg-Fontaine, M.C., Petre, G., Legros, J.C., 1985. Effects of a surface tension minimum on thermocapillary convection. PCH Physicochemical hydrodynamics 6, 301–310.

Lin, Y.H., Zheng, L.C., Zhang, X.X., Ma, L.X., Chen, G., 2015a. MHD pseudo-plastic nanofluid unsteady flow and heat transfer in a finite thin film over stretching surface with internal heat generation. International Journal of Heat and Mass Transfer 84, 903–911.

Lin, Y.H., Zheng, L.C., Zhang, X.X., 2015b. MHD Marangoni boundary layer flow and heat transfer of pseudo-plastic nanofluids over a porous medium with a modified model. Mechanics of Time-Dependent Materials 19 (4), 519–536.

Lin, Y.H., Zheng, L.C., Chen, G., 2015c. Unsteady flow and heat transfer of pseudo-plastic nanoliquid in a finite thin film on a stretching surface with variable thermal conductivity and viscous dissipation. Powder Technology 274, 324–332.

Lin, Y.H., Zheng, L.C., Zhang, X.X., 2014. Radiation effects on Marangoni convection flow and heat transfer in pseudo-plastic non-Newtonian nanofluids with variable thermal conductivity. International Journal of Heat and Mass Transfer 77, 708–716.

Mukhopadhyay, S., Layek, G.C., Samad, S.A., 2005. Study of MHD boundary layer flow over a heated stretching sheet with variable viscosity. International Journal of Heat and Mass Transfer 48, 4460–4466.

Na, T.Y., Hansen, A.G., 1967. Similarity solutions of a class of laminar three-dimensional. Boundary layer equations of power law fluids. International Journal of Non-Linear Mechanics 2, 373–385.

Nachman, A., Callegari, A., 1980. A nonlinear singular boundary value problem in the theory of pseudoplastic fluids. Society for Industrial and Applied Mathematics 38 (2), 275–281.

Nachman, A., Taliaferro, S., 1979. Mass transfer into boundary layers for power law fluids. Proceedings of the Royal Society A 365, 313–326.

Okano, Y., Itoh, M., Hirata, A., 1989. Natural and marangoni convections in a two-dimensional rectangular open boat. Journal of Chemical Engineering of Japan 22, 275–281.

Pavlov, K.B., 1974. Magnetohydrodynamic flow of an incompressible viscous fluid caused by deformation of a plane surface. Magnitnaya Gidrodinamika (U.S.S.R.) 4, 146–147.

Pop, I., Rashidi, M., Gorla, R.S.R., 1991. Mixed convection to power-law type non-Newtonian fluids from a vertical wall. Polymer-Plastics Technology and Engineering Journal 30, 47–66.

Pop, I., 1993. Boundary layer flow at a three-dimensional stagnation point in power-law non-Newtonian Fluids. International Journal of Heat and Fluid Flow 14, 408–412.

Sapunkov, Y.G., 1967. Self-similar solutions of non-Newtonian fluid boundary layer in MHD. Fluid Dynamics 2, 42–47.

Sarpkaya, T., 1961. Flow of non-Newtonian fluids in a magnetic field. AIChE Journal 7, 324–328.

Schowalter, W.R., 1960. The application of boundary-layer theory to power law pseudo plastic fluid: similar solutions. AIChE Journal 6, 24–28.

Schwabe, D., Metzger, J., 1989. Coupling and separation of buoyant and thermocapillary convection. Journal of Crystal Growth 97, 23–33.

Slavtchev, S.G., Miladinova, S.P., 1998. Thermocapillary flow in a liquid layer at minimum in surface tension. Acta Mechanica 127, 209–224.

Slavtchev, S.G., Dubovik, K.G., 1992. Thermocapillary convection in a rectangular cavity at minimum of surface tension. Theoretical & Applied Mechanics Letters 23, 85–90.

Straub, J., 1994. The role of surface tension for two-phase heat and mass transfer in the absence of gravity. Experimental Thermal and Fluid Science 9, 253–273.

Sui, J.Z., Zheng, L.C., Zhang, X.X., 2015a. Convection heat transfer of power law fluids along the inclined non-uniformly heated plate with suction or injection. Journal of heat transfer-T AMSE 138, 021701.

Sui, J.Z., Zheng, L.C., Zhang, X.X., Chen, G., 2015b. Mixed convection heat transfer in power law fluids over a moving conveyor along an inclined plate. International Journal of Heat and Mass Transfer 85, 1023–1033.

Timol, M.G., Kalthia, N.L., 1986. Similarity solutions of a class of laminar three-dimensional boundary layer equations of non-Newtonian fluids. International Journal of Non-Linear Mechanics 21 (6), 475–481.

Villers, D., Platen, J.K., 1985. Marangoni convection in systems presenting a minimum in surface tension. PCH Physicochemical hydrodynamics 6, 435–451.

Vujanovic, B., 1972. A variational solution of Rayleigh problem for power-law non-Newtonian conductive fluid. Ingenieur-Archiv 41, 381–386.

Wazwaz, A.M., 1995. A new approach to the nonlinear advection problem: an application of the decomposition method. Applied Mathematics and Computation 72, 175–181.

Wazwaz, A.M., 1998. A comparison between Adomian decomposition method and Taylor series method in the series solutions. Applied Mathematics and Computation 97, 37–44.

Wazwaz, A.M., 1999a. A reliable modification of Adomian's decomposition method. Applied Mathematics and Computation 102, 77–86.

Wazwaz, A.M., 1999b. The modified decomposition method and Padé approximants for solving Thomas-Fermi eqution. Applied Mathematics and Computation 105, 11–19.

Wazwaz, A.M., 2001. A reliable algorithm for solving boundary value problems for higher-order integro-differential equations. Applied Mathematics and Computation 118, 327–342.

Zheng, L.C., He, J.C., 1999. Existence and non-uniqueness of positive solutions to a non-linear boundary value problems in the theory of viscous fluids. Dynamic systems and applications 8, 133–145.

Zheng, L.C., Chen, X.H., Zhang, X.X., 2004a. Analytical approximants for a boundary layer flow on a stretching moving surface with a power velocity. International Journal of Applied Mechanics and Engineering 9, 795–802.

Zheng, L.C., Chen, X.H., Zhang, X.X., 2004b. An approximately analytical solution for the marangoni convection in an In-Ga-Sb system. Chinese Physics Letters 21, 1983–1985.

Zheng, L.C., Lin, Y.H., Zhang, X.X., 2012. Marangoni convection of power law fluids driven by power-law temperature gradient. Journal of the Franklin Institute 349 (8), 2585–2597.

Zheng, L.C., Lin, Y.H., Zhang, X.X., 2013. MHD thermosolutal Marangoni convection of non-Newtonian power law fluids driven by temperature gradient. Journal of heat transfer-T AMSE 135, 051702.

Zheng, L.C., Ma, L.X., He, J.C., 2000. Bifurcation solutions to a boundary layer problem arising in the theory of pseudo-plastic fluids. Acta Mathematica Scientica 20 (1), 19–26.

Zheng, L.C., Zhang, X.X., He, J.H., 2007. Existence and estimate of positive solutions to a nonlinear singular boundary value problem in the theory of dilatant non-Newtonian fluids. Mathematical Computers Modelling 45, 387–393.

Zheng, L.C., Zhang, X.X., Gao, Y.T., 2008. Analytical solution for marangoni convection over a vapor-liquid surface due to an imposed temperature gradient. Mathematical Computers Modelling 48, 1787–1795.

Zheng, L.C., Zhang, X.X., He, J.C., 2003. Bifurcation behavior of reverse flow boundary layer problem with special injection/suction. Chinese Physics Letters 20, 83–86.

Zheng, L.C., Zhang, X.X., He, J.C., 2004c. Suitable heat transfer model for Self-similar laminar boundary layer in power law fluids. Journal of Thermal Science 13, 150–154.

Zheng, L.C., Zhang, X.X., Karem, B., Yücel, U., 2011. Emna Gargouri-Ellouze, and Ahmet Yıldırım, Similarity and BPES comparative solutions to the heat transfer equation for incompressible non-Newtonian fluids: case of laminar boundary energy equation. European Physical Journal of Applied Physics 55, 21102.

Chapter 4

Homotopy Analytical Method

ABSTRACT

In this chapter, the homotopy analysis method (HAM) is introduced for solving nonlinear differential equations arising in modern fluid problems. Involved problems include unsteady flow and radiation heat transfer of magnetohydrodynamic fluid over a horizontal continuously stretching surface, flow and heat transfer of nanofluids over a porous rotating disk with velocity slip and temperature jump, mixed convection boundary layer heat transfer of power law fluid over a moving conveyor along an inclined plate, and the MHD thermosolutal Marangoni convection heat and mass transfer of power law fluids driven by a power law temperature and a power law concentration. The effects of power law fluid viscosity on temperature field are taken into account here by us by assuming that the temperature field is similar to a velocity field with a modified Fourier's law. Approximate analytical solutions are obtained by using HAM and the effects of involved parameters on the velocity, temperature, and concentration fields are analyzed in detail.

4.1 INTRODUCTION

The homotopy analysis method (HAM), developed by Professor Shijun Liao (1992, 2012), is a powerful mathematical tool for solving nonlinear problems. The method employs the concept of homotopy from topology to generate a convergent series solution for nonlinear systems. It is enabled by using a homotopy–Maclaurin series to deal with nonlinearities in the system. Liao made further modifications in 1997 by introducing a nonzero auxiliary parameter, referred to as the convergence–control parameter, to construct a homotopy on a differential system in general form. The convergence–control parameter is a nonphysical variable that provides a simple way to verify and enforce convergence of a solution series. The capability of HAM to show convergence of the series solution naturally is unusual in analytical and semianalytic approaches to nonlinear partial differential equations.

HAM has several advantages over many traditional analytical methods. It is a series expansion method that does not directly depend on small or large

Mathematics in Science and Engineering. http://dx.doi.org/10.1016/B978-0-12-811753-8.00004-9

physical parameters. Thus, it is applicable not only for weakly but also strongly nonlinear problems. HAM is a unified method for the Lyapunov artificial small parameter method, Delta expansion method, Adomian decomposition method, and homotopy perturbation method. It gives excellent flexibility to the expression of the solution and how the solution is explicitly obtained, and provides great freedom in choosing the base functions of the desired solution and the corresponding auxiliary linear operator of homotopy. HAM provides a simple way to ensure the convergence of the solution series and is able to combine with other techniques employed in nonlinear differential equations such as spectral methods, Padé approximants, and so on. In this chapter, HAM is employed for solving some nonlinear differential equations arising in modern fluid problems.

4.2 FLOW AND RADIATIVE HEAT TRANSFER OF MAGNETOHYDRODYNAMIC FLUID OVER A STRETCHING SURFACE

4.2.1 Description of the Problem

The study of flow and heat transfer of viscous fluid over a continuously stretching surface in an otherwise quiescent fluid medium has attracted considerable attention. This is owing to the applications of such a problem in many manufacturing processes such as hot rolling, wire drawing, metal extrusion, crystal growing, continuous casting, glass fiber production, and paper production. The resulting flow over a continuous stretching surface extruded from a slit may be modeled as a boundary layer developing away from a silt. Following the pioneering work of Sakiadis (1961), the great majority of theoretical investigations in this field described fluid flow and heat transfer in the vicinity of the continuous stretching surface with the aid of similarity solutions to the boundary layer equations. Later, an experimental and theoretical study was made by Tsou et al. (1967), who experimentally determined the heat transfer rates for certain Prandtl numbers and confirmed that the flow field obtained by Sakiadis (1961) was broadly realized experimentally. Crane (1970) investigated the problem for a plate moving with a linear surface velocity, and Chiam (1995) investigated the problem for a plate moving with a power law velocity in the presence of a magnetic field. Vajravelu and Hadjinicolaou (1997) studied the effect of free convection currents in the presence of temperature-dependent heat sources/sinks, with and without a magnetic field of an electrically conducting viscous fluid. Cortell (2008) analyzed boundary layer flow and heat transfer over a nonlinearly stretching sheet with the effects of viscous dissipation and radiation. Zheng et al. (2000, 2002, 2006) studied the flow and heat transfer of power law fluid over a moving plate when the plate moves with a linear or a power law velocity in the direction or opposite the direction of the flow; the existence, uniqueness, nonuniqueness, and analysis of the solutions were established and the associated transfer characteristics were analyzed. Moreover, Zheng et al. (2010, 2011) studied the flow and heat transfer of a power law fluid over an unsteadily stretching surface and unsteady flow and heat transfer on a permeable stretching sheet with a nonuniform heat source/sink. The approximate analytical solutions were obtained and the effects of pertinent parameters on the velocity and temperature fields were presented graphically and analyzed in detail.

All of these works were concerned with steady flows induced by a stretching surface. Little attention has been given to unsteady flows over a stretching surface. Wang (1990a) discussed the hydrodynamic viscous flow of a thin liquid film over an unsteady stretching sheet. Abbas et al. (2008) investigated the unsteady flow of a second-grade fluid film over an unsteady stretching sheet (Abel et al., 2009; Anderssona et al., 2000; El-Aziz, 2009; Takhar et al., 1999; Wang, 2006).

4.2.2 Mathematical Formulation

We consider here the flow and heat transfer of a magnetohydrodynamic (MHD) viscous fluid over an unsteady horizontal stretching surface, which issues from a slot at the origin and keeps the original fixed. The continuous sheet is aligned with the x axis at $y = 0$ moves in its own plane with a velocity $U_w(x,t)$ and with the temperature $T_w(x,t)$. The boundary layers governing equations are written as:

$$\frac{\partial u}{\partial x} + \frac{\partial v}{\partial y} = 0, \tag{4.1}$$

$$\frac{\partial u}{\partial t} + u\frac{\partial u}{\partial x} + v\frac{\partial u}{\partial y} = \nu\frac{\partial^2 u}{\partial y^2} - \frac{\sigma_0 B^2}{\rho}u, \tag{4.2}$$

$$\frac{\partial T}{\partial t} + u\frac{\partial T}{\partial x} + v\frac{\partial T}{\partial y} = \frac{k}{\rho c_p}\frac{\partial^2 T}{\partial y^2} - \frac{1}{\rho c_p}\frac{\partial q_r}{\partial y}, \tag{4.3}$$

subject to the boundary conditions

$$u = U_w(x,t), v = 0, T = T_w(x,t) \quad \text{at} \quad y = 0; \quad u \to 0, T \to T_\infty \quad \text{as} \quad y \to \infty. \tag{4.4}$$

where u and v are the velocity components along the x and y axes, respectively; T is the fluid temperature; $\nu(=\mu/\rho)$ is the kinematic coefficient of viscosity; $B(x,t)$ is the applied transverse magnetic field; c_p is the heat capacity at constant pressure; and q_r is the radiative heat flux at a constant pressure. The applied transverse magnetic field is chosen in its specified form as

$$B(x,t) = B_0(1 - \alpha t)^{\frac{1}{2}}. \tag{4.5}$$

The radiative heat flux q_r under Rossland approximation has the form

$$q_r = -\frac{4\sigma}{3k_1}\frac{\partial T^4}{\partial y}. \tag{4.6}$$

Assume that the temperature differences within the flow is small such that the term T^4 may be expressed as a linear function of temperature. Hence, expand T^4 in a Taylor series about T_∞, and by neglecting the higher-order terms, we obtain

$$T^4 \cong 4T_\infty^3 T - 3T_\infty^4. \tag{4.7}$$

The stretching velocity is assumed to be the form

$$u_w = bx/(1-\alpha t). \tag{4.8}$$

The adopted formulation of the sheet velocity in Eq. (4.8) is valid only for times $t < \alpha^{-1}$ unless $\alpha = 0$.

The temperature $T_w(x,t)$is in the form

$$T_w(x,t) = T_\infty + T_0\left[bx^2/2\nu\right](1-\alpha t)^{-3/2}, \tag{4.9}$$

where T_0 is a constant reference temperature number.

Introduce the similarity variable η and the dimensionless variables f and θ as (Abel et al., 2009; El-Aziz, 2009)

$$\eta = (b/\nu)^{1/2}(1-\alpha t)^{-1/2}y, \tag{4.10}$$

$$\psi = (\nu b)^{1/2}(1-\alpha t)^{-1/2}xf(\eta), \tag{4.11}$$

$$T = T_\infty + T_0\left[bx^2/2\nu\right](1-\alpha t)^{-3/2}\theta(\eta). \tag{4.12}$$

Then Eqs. (4.1)–(4.3) are transformed into a set of ordinary differential equations of the form

$$f''' + ff'' - (f')^2 - A\left(f' + \frac{\eta}{2}f''\right) - Mf' = 0, \tag{4.13}$$

$$\theta'' + k_0 P_r\left[f\theta' - 2f'\theta - \frac{A}{2}(3\theta + \eta\theta')\right] = 0, \tag{4.14}$$

The associated boundary conditions (4.4) are

$$f(0) = 0, \quad f'(0) = 1, \quad \theta(0) = 1, \tag{4.15}$$

$$f'(\infty) = 0, \quad \theta(\infty) = 0. \tag{4.16}$$

where $A = \alpha/b$ is the unsteady parameter, $M = \sigma B_0^2/\rho b$ is the magnetic parameter, $\theta = (T - T_\infty)/(T_w - T_\infty)$ is the nondimensional temperature, $k_0 = 3R/(3R+4)$ with $R = kk_1/4\sigma T_\infty^3$ is the radiation parameter, and $\Pr = \nu/\alpha$ is the Prandtl number.

Because R is physically more relevant than the similarity parameter k_0 introduced previously, we take R instead of k_0 as a governing parameter. If the thermal radiation's effect is not considered in the energy Eq. (4.14), we have $k_0 = 1$ in Eq. (4.14).

The local skin coefficient is given by

$$C_f = \frac{2\mu(\partial u/\partial y)_{y=0}}{\rho U_w^2} = 2\mathrm{Re}_x^{-1/2} f''(0), \tag{4.17}$$

and the local Nusselt number is given by

$$Nu_x = \frac{x}{T_0}(\partial T/\partial y)_{y=0} = -\frac{1}{2}(1-\alpha t)^{-1/2}\mathrm{Re}_x^{3/2}\theta'(0), \tag{4.18}$$

where $\mathrm{Re}_x = u_w x/\nu$ is the local Reynolds number.

4.2.3 Homotopy Analysis Method Solutions

4.2.3.1 Zero-Order Deformation Equations

To solve Eqs. (4.13)–(4.16) by using HAM, boundary conditions (4.15) and (4.16) show that the initial solutions may be chosen as

$$f_0(\eta) = 1 - \exp(-\eta), \tag{4.19}$$

$$\theta_0(\eta) = \exp(-\eta/3), \tag{4.20}$$

as the initial approximations of $f(\eta)$ and $\theta(\eta)$, respectively, and we may choose

$$L_f[f(\eta;q)] = \frac{\partial^3 F(\eta;q)}{\partial \eta^3} + \frac{\partial^2 F(\eta;q)}{\partial \eta^2}, \tag{4.21}$$

$$L_\theta[f(\eta;q)] = \frac{\partial^2 \Theta(\eta;q)}{\partial \eta^2} + \frac{1}{3}\frac{\partial \Theta(\eta;q)}{\partial \eta}, \tag{4.22}$$

as the auxiliary linear operators, which have the following properties

$$L_f[c_1 + c_2\eta + c_3\exp(-\eta)] = 0, \quad L_\theta[c_4 + c_5\exp(-\eta/3)] = 0, \tag{4.23}$$

where $c_i (i = 1-5)$ are arbitrary constants.

In terms of Eqs. (4.13) and (4.14), we define the nonlinear operators

$$\begin{aligned} N_f\left[F(\eta;q)\right] &= \frac{\partial^3 F(\eta;q)}{\partial \eta^3} + F(\eta;q)\frac{\partial^2 F(\eta;q)}{\partial \eta^2} - \left(\frac{\partial F(\eta;q)}{\partial \eta}\right)^2 \\ &\quad - A\left(\frac{\partial F(\eta;q)}{\partial \eta} + \frac{\eta}{2}\frac{\partial^2 F(\eta;q)}{\partial \eta^2}\right) - M\frac{\partial F(\eta;q)}{\partial \eta}. \end{aligned} \tag{4.24}$$

$$N_\theta[\Theta(\eta;q)] = \frac{\partial^2\Theta(\eta;q)}{\partial\eta^2} + k_0\mathrm{Pr}\left[F(\eta;q)\frac{\partial\Theta(\eta;q)}{\partial\eta} - 2\Theta(\eta;q)\frac{\partial F(\eta;q)}{\partial\eta}\right.$$
$$\left.-\frac{A}{2}\left(3\Theta(\eta;q) + \eta\frac{\partial\Theta(\eta;q)}{\partial\eta}\right)\right]. \tag{4.25}$$

Let h_f and h_θ denote the nonzero auxiliary parameters. Then we construct the zeroth-order deformation equations

$$(1-q)L_f[F(\eta;q) - f_0(\eta)] = qh_fH_f(\eta)N_f[F(\eta;q)], \tag{4.26}$$

$$(1-q)L_\theta[\Theta(\eta;q) - \theta_0(\eta)] = qh_\theta H_\theta(\eta)N_\theta[\Theta(\eta;q)], \tag{4.27}$$

subject to the boundary conditions

$$F(0;q) = 0, \quad \left.\frac{\partial F(\eta;q)}{\partial\eta}\right|_{\eta=0} = 1, \quad \left.\frac{\partial F(\eta;q)}{\partial\eta}\right|_{\eta=\infty} = 0, \tag{4.28}$$

$$\Theta(0;q) = 1, \quad \Theta(\infty;q) = 0, \tag{4.29}$$

where $q \in [0, 1]$ is an embedding parameter. When $q = 0$, it is straightforward that

$$F(\eta;0) = f_0(\eta), \quad \Theta(\eta;0) = \theta_0(\eta). \tag{4.30}$$

When $q = 1$, the zeroth-order deformation Eqs. (4.26)–(4.29) are equivalent to the original Eqs. (4.13)–(4.16), so that we have

$$F(\eta;q) = f(\eta), \quad \Theta(\eta;q) = \theta(\eta), \tag{4.31}$$

As q increases from 0 to 1, $F(\eta;q)$ and $\Theta(\eta;q)$ vary from the initial guess $f_0(\eta)$ and $\theta_0(\eta)$ to the solutions $f(\eta)$and $\theta(\eta)$ of the problem, respectively. Expanding $F(\eta;q)$ and $\Theta(\eta;q)$ in Taylor's series with respect to q yields

$$F(\eta;q) = F(\eta;0) + \sum_{m=1}^{\infty} f_m(\eta)q^m, \tag{4.32}$$

$$\Theta(\eta;q) = \Theta(\eta;0) + \sum_{m=1}^{\infty} \theta_m(\eta)q^m, \tag{4.33}$$

where

$$f_m(\eta) = \left.\frac{1}{m!}\frac{\partial^m F(\eta;q)}{\partial q^m}\right|_{q=0}, \tag{4.34}$$

$$\theta_m(\eta) = \left.\frac{1}{m!}\frac{\partial^m \Theta(\eta;q)}{\partial q^m}\right|_{q=0}. \tag{4.35}$$

Assuming that h_f and h_θ are properly chosen, so that the series (4.32) and (4.33) are convergent at $q = 1$, we have, in terms of (4.30) and (4.31), the solutions in series form

$$f(\eta) = f_0(\eta) + \sum_{m=1}^{\infty} f_m(\eta), \tag{4.36}$$

$$\theta(\eta) = \theta_0(\eta) + \sum_{m=1}^{\infty} \theta_m(\eta). \tag{4.37}$$

4.2.3.2 Higher-Order Deformation Equations

Differentiating the zero-order deformation Eqs. (4.26) and (4.27) m times with respect to q, then setting $q = 0$, and finally dividing them by $m!$, we obtain the mth-order deformation equations

$$L_f[f_m(\eta) - \chi_m f_{m-1}(\eta)] = h_f H_f(\eta) R_m(\eta), \tag{4.38}$$

$$L_\theta[\theta_m(\eta) - \chi_m \theta_{m-1}(\eta)] = h_\theta H_\theta(\eta) S_m(\eta), \tag{4.39}$$

with the boundary conditions

$$f_m(0) = f_m'(0) = f_m'(\infty) = 0, \tag{4.40}$$

$$\theta_m(0) = \theta_m(\infty) = 0, \tag{4.41}$$

where

$$\chi_m = \begin{cases} 0 & m \le 1 \\ 1 & m \ge 2 \end{cases}, \tag{4.42}$$

and

$$\begin{aligned} R_m(\eta) = {} & f_{m-1}'''(\eta) + \sum_{k=0}^{m-1} f_k(\eta) f_{m-1-k}''(\eta) - \sum_{k=0}^{m-1} f_k'(\eta) f_{m-1-k}'(\eta) \\ & - (A + M) f_{m-1}'(\eta) - \frac{\eta A}{2} f_{m-1}''(\eta), \end{aligned} \tag{4.43}$$

$$\begin{aligned} S_m(\eta) = \theta_{m-1}''(\eta) + k_0 \mathrm{Pr} \Bigg[& \sum_{k=0}^{m-1} f_k(\eta) \theta_{m-1-k}'(\eta) - 2 \sum_{k=0}^{m-1} f_k'(\eta) \theta_{m-1-k} \\ & - \frac{A}{2} (3\theta_{m-1}(\eta) + \eta \theta_{m-1}' \Bigg]. \end{aligned} \tag{4.44}$$

According to the solution expression and coefficient ergodic principle, we set

$$H_f(\eta) = \exp(-\eta), \ H_\theta(\eta) = \exp(-\eta/3). \tag{4.45}$$

The first-order deformation equations are

$$L_f[f_1(\eta)] = h_f H_f(\eta) R_1(\eta), \tag{4.46}$$

$$L_\theta[\theta_1(\eta)] = h_\theta H_\theta(\eta) S_1(\eta), \tag{4.47}$$

owing to the boundary conditions

$$f_1(0) = f_1'(0) = f_1'(\infty) = 0, \tag{4.48}$$

$$\theta_1(0) = \theta_1(\infty) = 0. \tag{4.49}$$

Then, we have

$$\begin{aligned} R_1(\eta) &= f_0''' + f_0 f_0'' - (f_0')^2 - (A+M)f_0' - \frac{\eta}{2} f_0'' \\ &= -(A+M)\exp(-\eta) + \frac{A\eta}{2}\exp(-\eta), \end{aligned} \tag{4.50}$$

$$\begin{aligned} S_1(\eta) = {} & \frac{1}{18}(3k_0 A\mathrm{Pr}\eta - 27k_0 A\mathrm{Pr} - 6k_0\mathrm{Pr} + 2)\exp\left(-\frac{1}{3}\eta\right) \\ & - \frac{5}{3}k_0\mathrm{Pr}\exp\left(-\frac{4}{3}\eta\right). \end{aligned} \tag{4.51}$$

From Eqs. (4.38) and (4.39) and boundary conditions (4.40) and (4.41), we have

$$\begin{aligned} f_1(\eta) = h_f \bigg[& -\left(\frac{1}{2}m + \frac{1}{8}A\right)\exp(-\eta) + \left(\frac{1}{4} - \frac{1}{8}A\eta\right)\exp(-2\eta) \\ & + \frac{1}{4}m + \frac{1}{8}A \bigg], \end{aligned} \tag{4.52}$$

$$\begin{aligned} \theta_1(\eta) = \frac{1}{8}h_\theta \exp\left(-\frac{1}{3}\eta\right) \bigg[& 18k_0\mathrm{Pr} + 27k_0 A\mathrm{Pr} - 4 + (6k_0\mathrm{Pr}A\eta - 27k_0 A\mathrm{Pr} \\ & - 12k_0\mathrm{Pr} + 4)\exp\left(-\frac{1}{3}\eta\right) - 6k_0\mathrm{Pr}\exp\left(-\frac{4}{3}\eta\right)\bigg], \end{aligned} \tag{4.53}$$

And second-order deformation equations

$$L_f[f_2(\eta) - f_1(\eta)] = h_f H_f(\eta) R_2(\eta), \tag{4.54}$$

$$L_\theta[\theta_2(\eta) - \theta_1(\eta)] = h_\theta H_\theta(\eta) S_2(\eta). \tag{4.55}$$

Similarly, we have

$$f_2(\eta) = -\frac{1}{1728} h_f\big[-216A - 432M - 348Mh_f - 214Ah_f - 120M^2h_f - 35A^2h_f - 138MAh_f$$
$$+ \big(288M^2h_f + 576Mh_f + 318Ah_f + 35A^2h_f + 216MAh_f + 864M + 216A\big)\exp(-\eta))$$
$$+ (216A\eta - 108Mh_f - 54Ah_f + 108MA\eta h_f - 432M - 54MAh_f + 27A^2\eta h_f$$
$$- 216M^2h_f)\exp(-2\eta) + (24A^2\eta^2h_f + 8A^2\eta h_f + 60A\eta h_f - 72MA\eta h_f$$
$$+ 48M^2h_f - 24MAh_f - 120Mh_f - 50Ah_f)\exp(-3\eta)\big], \tag{4.56}$$

$$\theta_2(\eta) = \frac{1}{1881600} h_\theta \exp\left(-\frac{1}{3}\eta\right)\Big[18963000k_0^2A^2\mathrm{Pr}^2h_\theta + 19072368k_0^2a\mathrm{Pr}^2h_\theta + 4809840k_0^2\mathrm{Pr}^2h_\theta$$
$$- 705600k_0A\mathrm{Pr}h_\theta + 1134000Mk_0\mathrm{Pr}h_f + 564975k_0A\mathrm{Pr}h_f + 6350400k_0A\mathrm{Pr}$$
$$+ 141120k_0\mathrm{Pr}h_\theta + 423360k_0\mathrm{Pr} - 156800h_\theta - 940800 + (4762800k_0^2A^2\mathrm{Pr}^2\eta h_\theta$$
$$- 21432600k_0^2A^2\mathrm{Pr}^2h_\theta + 3175200k_0^2A\mathrm{Pr}^2\eta h_\theta - 2381400k_0^2A\mathrm{Pr}^2h_\theta$$
$$- 6350400k_0^2\mathrm{Pr}^2h_\theta - 705600k_0A\mathrm{Pr}\eta h_\theta + 1411200k_0A\mathrm{Pr}\eta - 352800k_0A\mathrm{Pr}h_f$$
$$- 6350400k_0A\mathrm{Pr}h_\theta - 6350400k_0A\mathrm{Pr} - 705600Mk_0\mathrm{Pr}h_f + 352800k_0\mathrm{Pr}h_\theta$$
$$- 2822400k_0\mathrm{Pr} - 470400h_\theta + 940800)\exp\left(-\frac{1}{3}\eta\right) + (705600k_0^2A^2\mathrm{Pr}^2\eta^2h_\theta$$
$$- 3880800k_0^2A^2\mathrm{Pr}^2\eta h_\theta - 2822400k_0^2A\mathrm{Pr}^2\eta h_\theta + 1411200k_0A\mathrm{Pr}\eta h_\theta$$
$$+ 2469600k_0^2A^2\mathrm{Pr}^2h_\theta + 7761600k_0^2A\mathrm{Pr}^2h_\theta + 2822400k_0^2\mathrm{Pr}^2h_\theta$$
$$- 5644800k_0A\mathrm{Pr}h_\theta - 2822400k_0\mathrm{Pr}h_\theta + 627200h_\theta)\exp\left(-\frac{2}{3}\eta\right)$$
$$+ (-4762800k_0^2A\mathrm{Pr}^2h_\theta - 3175200k_0^2\mathrm{Pr}^2h_\theta - 176400k_0A\mathrm{Pr}h_f - 705600Mk\mathrm{Pr}h_f$$
$$+ 705600k_0\mathrm{Pr}h_\theta - 1411200k_0\mathrm{Pr})\exp\left(-\frac{4}{3}\eta\right) + (-917280k_0^2A\mathrm{Pr}^2\eta h_\theta$$
$$+ 1742832k_0^2A\mathrm{Pr}^2h_\theta + 1834560k_0^2\mathrm{Pr}^2h_\theta - 1552320k_0\mathrm{Pr}h_\theta)\exp\left(-\frac{5}{3}\eta\right)$$
$$+ (-138600k_0A\mathrm{Pr}\eta h_f - 35775k_0A\mathrm{Pr}h_f + 277200Mk_0\mathrm{Pr}h_f)\exp\left(-\frac{7}{3}\eta\right)$$
$$+ 58800k_0^2\mathrm{Pr}^2h_\theta \exp\left(-\frac{8}{3}\eta\right)\Big] \tag{4.57}$$

Solutions $f_3(\eta), f_4(\eta)$, and $\theta_3(\eta)$ are too long to be presented here; therefore, they are shown graphically.

4.2.4 Results and Discussion

The convergence and rate of approximation for the HAM solution strongly depend on the values of auxiliary parameters h_f and h_θ. Therefore, if we choose h_f and h_θ properly, the solutions we obtained will be convergent. This was verified by Liao (2012). In general, the so-called h curve provides the proper value of h to control the convergence of the approximation series. The range of admissible values of h_f and h_θ can be obtained from Figs. 4.1 and 4.2. Because the valid region of h_f and h_θ corresponds to the line segments nearly parallel to the horizontal axis, we choose the values of h_f and h_θ from the range $-1 \le h_f \le -0.1$ and $-1 \le h_\theta \le -0.3$ (Table 4.1).

For specified involved values of parameters, Table 4.1 presents a comparison of wall temperature gradient $-\theta'(0)$ obtained by HAM and in El-Aziz (2009). It is seen that the results are in close agreement.

The nondimensional velocity component in the x direction $f'(\eta)$ at $M = 3$ for the varying unsteadiness parameter is plotted in Fig. 4.3, from which we know that the velocity profiles in the boundary layer slightly increases with the increasing of A [boundary conditions (4.4) implies that the lower the velocity curve is, the bigger of the fluid flow velocity is, with a thinner thickness of momentum boundary layer]. Fig. 4.4 shows the velocity profiles at $A = 0.6$ for different values of M. It is clear that the nondimensional velocity remarkably increases with a thinner boundary layer with the increase in M for $0 \le M \le 4$; this may be why the increase in M causes the increase in the Lorentz force.

Fig. 4.5 shows the effects of magnetic parameters M on the temperature distribution for some specified parameters. It indicates that the temperature profiles increase as M increases from 0 to 10. In a physical meaning this can be

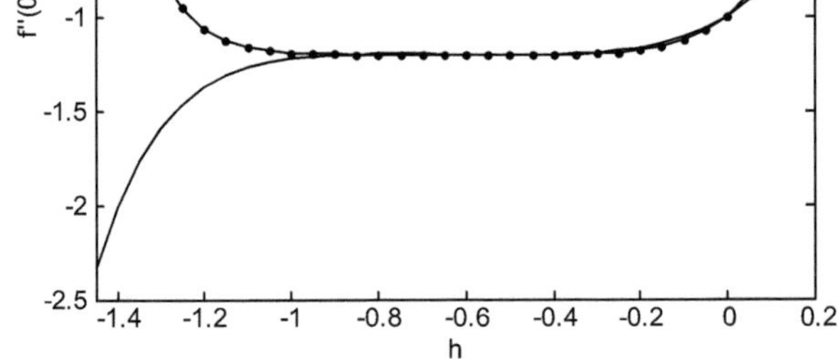

FIGURE 4.1 The h_f curves of $f''(0)$ obtained by the fourth-order approximation of HAM. When $A = 0.6$, $M = 2$.

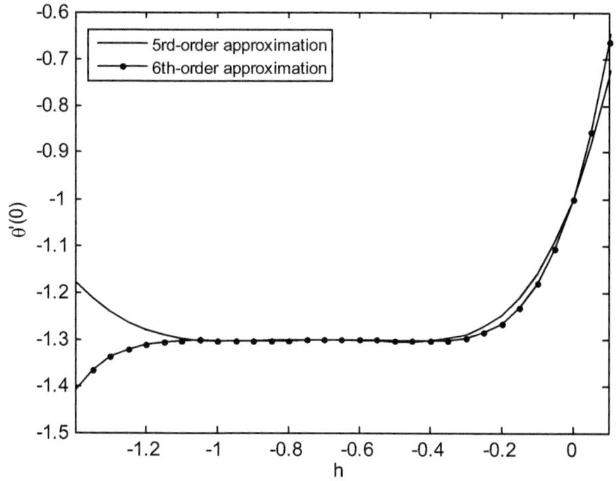

FIGURE 4.2 The h_θ curves of $f''(0)$ obtained by the fourth-order approximation of HAM when $A = 0.6$, $M = 2$, and $\Pr = 0.7$, $R = 1$.

explained as that the increase of magnetic parameter will induce the increasing of the thermal boundary layer thickness. The effects of radiation parameter R on the temperature field distributions are depicted in Fig. 4.6. The figure shows that the temperature profiles decrease with an increase in the radiation parameter, which implies that the increase in the radiation parameter will induce faster heat transfer efficiency and reduce the thermal boundary layer thickness. From Fig. 4.7, we know that the temperature profiles decrease as the Prandtl number increases. This signifies that the large Prandtl number will

TABLE 4.1 Comparison of Wall Temperature Gradient $-\theta'(0)$ With $k_0 = 1$, $M = 0$

Parameter	Parameter	El-Aziz (2009)	Present Results
A	Pr	$-\theta'(0)$	$-\theta'(0)$
0.8	0.1	0.4517	0.4367
0.8	1.0	1.6728	1.6755
1.2	0.1	0.5087	0.4765
1.2	1.0	1.818	1.8106
2.0	0.1	0.604013	0.533822
2.0	1.0	2.07841	2.07838

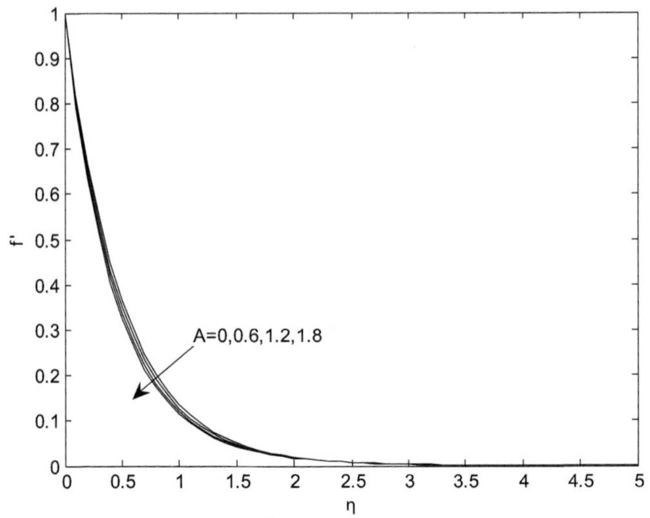

FIGURE 4.3 Velocity profiles for various values of A with $M=3$, $h_f=-0.4$.

induce faster heat transfer efficiency and reduce the thermal boundary layer thickness, which is similar to the effects of the radiation parameter.

Fig. 4.8 illustrates the temperature profiles for different unsteadiness parameter A. It is seen that the effects of unsteadiness parameter A on

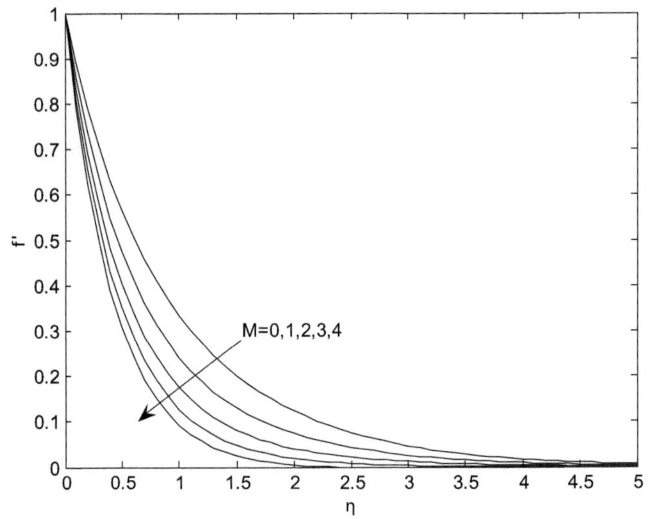

FIGURE 4.4 Velocity profiles for various values of M with $A=0.6$, $h_f=-0.4$.

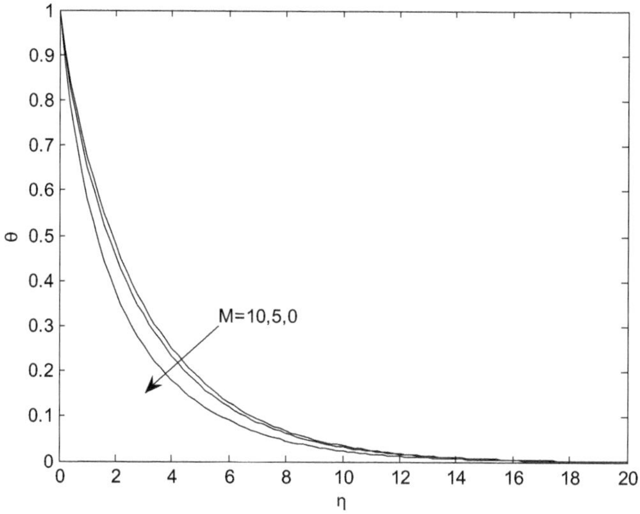

FIGURE 4.5 Temperature profiles for various values of M with $A = 0.2$, $\Pr = 0.72$, $R = 1$, $h_\theta = h_f = -0.4$.

temperature are similar to the Prandtl number and the radiation parameter; an increase in unsteadiness parameter A would induce faster heat transfer efficiency and a decrease in thermal boundary layer thickness. The wall temperature gradient profiles for small values of A are depicted in Fig. 4.9. It indicates that the increase in unsteadiness parameter can increase the local Nusselt number.

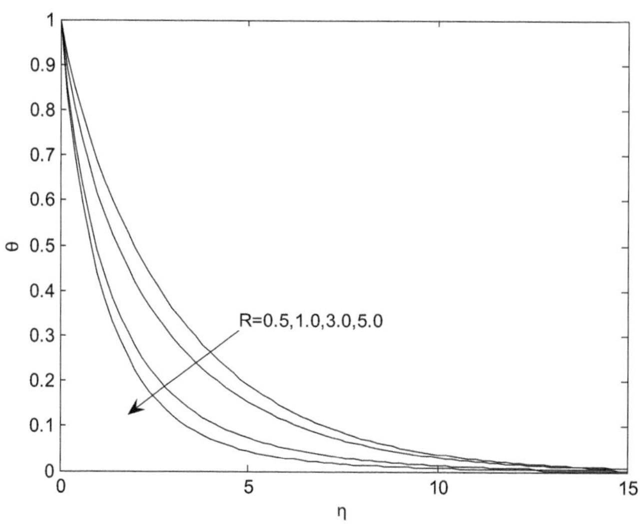

FIGURE 4.6 Temperature profiles for various values of R with $A = 0.2$, $M = 3$, $\Pr = 0.72$, $h_f = h_\theta = -0.4$.

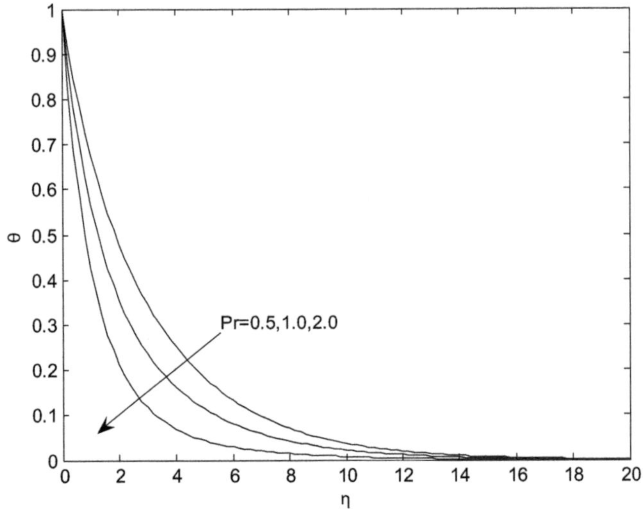

FIGURE 4.7 Temperature profiles for various values of Pr with $A = 0.6$, $M = 3$, $R = 0.5$, $h_f = h_\theta = -0.5$.

Fig. 4.10 depicts the wall temperature gradient distribution for different Prandtl numbers. The increase in Prandtl number will cause an increasing in the wall temperature gradient. The effect of *R* on the wall temperature gradient is shown in Fig. 4.11, which indicates that the wall temperature gradient increases with the increase in radiation parameter *R*.

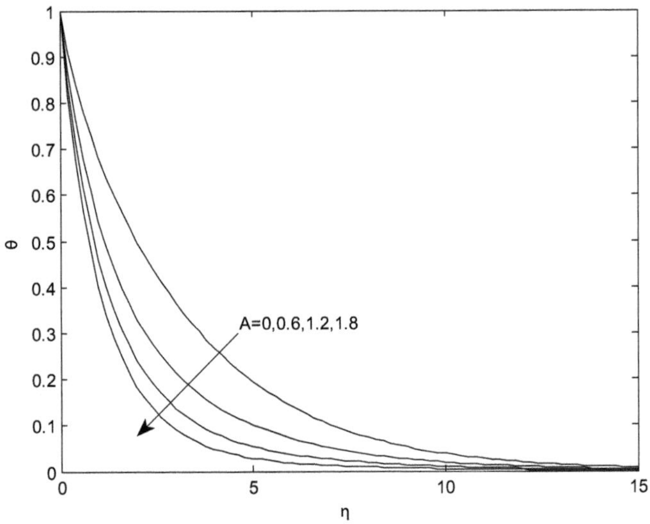

FIGURE 4.8 Temperature profiles for various values of *A* with $R = 1$, $M = 3$, $\mathrm{Pr} = 0.72$, $h_f = h_\theta = -0.5$.

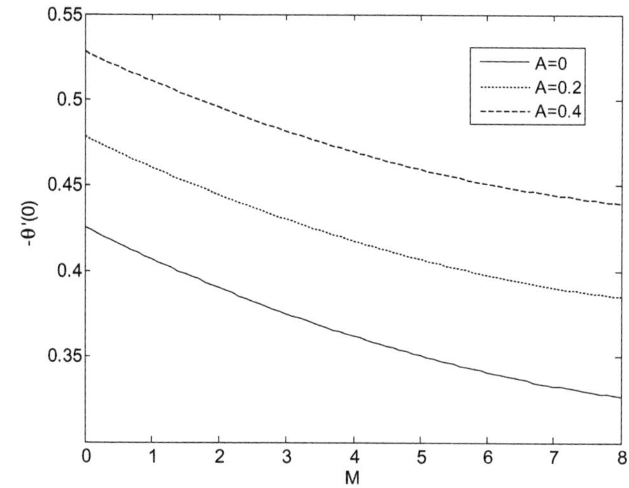

FIGURE 4.9 Wall temperature gradient $-\theta'(0)$ versus M for various values of A with $R = 0.5$, $\mathrm{Pr} = 0.7$, $h_\theta = h_f = -0.4$.

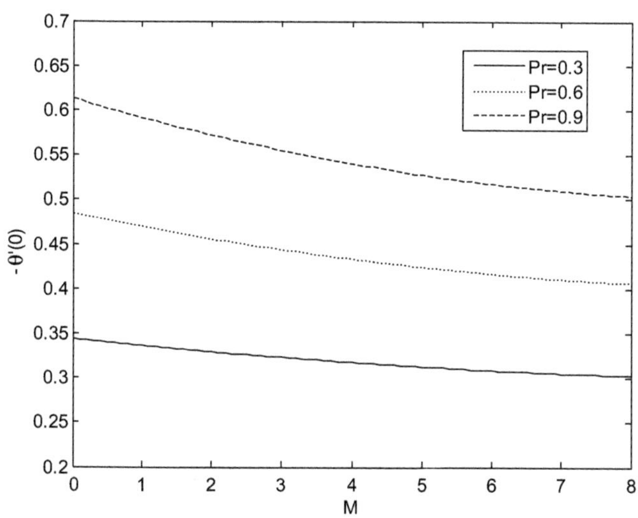

FIGURE 4.10 Wall temperature gradient $-\theta'(0)$ versus M for various values of Pr with $A = 0.4$, $R = 0.5$, $h_\theta = h_f = -0.4$.

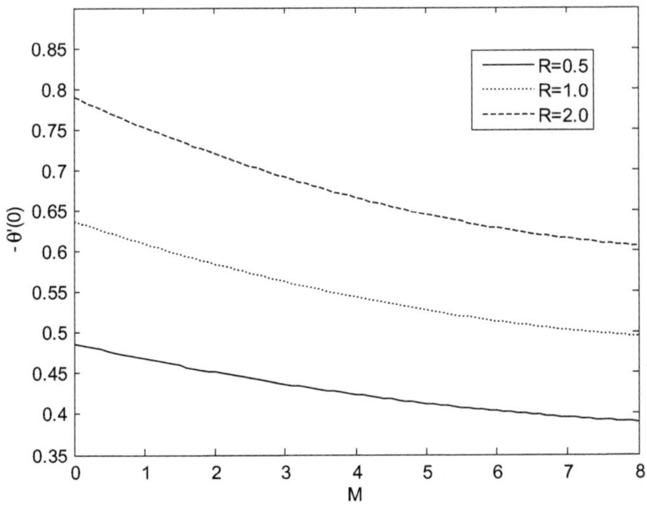

FIGURE 4.11 Wall temperature gradient $-\theta'(0)$ versus M for various values of R with $A = 0.2$, $\Pr = 0.72$, $h_\theta = h_f = -0.4$.

4.3 FLOW AND HEAT TRANSFER OF NANOFLUIDS OVER A ROTATING DISK

4.3.1 Background of the Problem

As one of the classical problems in fluid mechanics, fluid flow and heat transfer over a rotating disk have been studied by many researchers because of numerous practical applications in many important fields such as computer storage devices, electronic devices, and rotating machinery. Such flow is also significant in engineering processes. Von Kármán (1921) first investigated hydrodynamic flow over an infinite rotating disk. In his study, a famous similarity transformation was proposed to reduce the governing partial differential equations into ordinary differential equations. Cochran (1934) solved the steady hydrodynamic problem formulated by Von Karman and established the asymptotic solution. Benton (1966) considered the unsteady flow problem on the basis of Cochran's research. Various physical features were then explored (Attia & Aboul-Hassan, 2001; Kuiken, 1971; Lingwood, 1995). Shevchuk (2009) studied convective heat and mass transfer in rotating disk systems. Griffiths et al. (2014) investigated the neutral curve for stationary disturbances in rotating disk flow for power law fluids. Asghar et al. (2014) considered the Lie group analysis of flow and heat transfer over a stretching rotating disk. Turkyilmazoglu (2015) investigated Bödewadt flow and heat transfer over a stretching stationary disk.

Attia (2009) considered steady flow and heat transfer over a rotating disk in a porous medium; the effects of the porosity of the medium on the velocity and temperature fields were discussed. Rashidi et al. (2012) presented the approximate analytical solutions using the HAM.

In practical applications of science and engineering, partial slip may exist between the fluid and the moving surface, For example, in the situation where the fluid is particulate, such as emulsions, suspensions and rarefied gas, the proper boundary condition is the partial slip. At the same time, the presence of velocity slip on the wall may cause temperature jump, which must be taken into consideration in practical applications on a microscopic scale (Hong and Asako, 2010). The linear slip boundary condition was first proposed by Navier (1823). Rashidi et al. (2014) investigated slip flow resulting from a rotating infinite disk with variable properties of the fluid. Jiji and Ganatos (2010) considered steady laminar flow and heat transfer generated by two infinite parallel disks in the presence of velocity slip and temperature jump. Turkyilmazoglu and Senel (2013) studied the traditional Von Karman swirling flow problem in which the rotating disk surface admitted partial slip with a uniform suction or injection.

The term "nanofluids" was introduced by Choi (1995) at the American Society of Mechanical Engineers winter annual meeting. Nanofluid is a colloidal mixture made by adding nanoparticles ($<$100 nm) in a base fluid, which can considerably improve the heat transfer performance of the fluid (Bahiraei and Hangi, 2015; Kamatchi and Venkatachalapathy, 2015; Kandelousi, 2014a,b; Lomascolo et al., 2015; Pang et al., 2015; Sheikholeslami, 2014; Sheikholeslami & Ganji, 2014, 2015; Sheikholeslami et al., 2012; Sarkar et al., 2015).

Bachok et al. (2011) studied the steady flow of an incompressible viscous fluid over a rotating disk in a nanofluid. Rashidi et al. (2013) considered electrically conducting incompressible nanofluid flowing over a porous rotating disk with an externally applied uniform vertical magnetic field. Turkyilmazoglu (2014) investigated the flow and heat transfer characteristics resulting from a rotating disk immersed in different nanofluids. In this section we investigate the flow and heat transfer of nanofluid over a porous rotating disk with three types of nanoparticles: Cu, CuO, and Al_2O_3. The effects of porous parameter, velocity slip, temperature jump, and the types of nanofluid on velocity and temperature fields are also analyzed.

4.3.2 Formulation of the Problem

We consider an incompressible, steady, and axially symmetric nanofluid flow over a porous rotating disk. The disk is placed at $z=0$ and rotates with an angular velocity Ω through a porous medium, where the Darcy model is assumed. A physical model of a rotating disk is shown in Fig. 4.12. The governing equations in cylindrical coordinates are written as

$$\frac{\partial u}{\partial r}+\frac{u}{r}+\frac{\partial w}{\partial z}=0 \ , \tag{4.58}$$

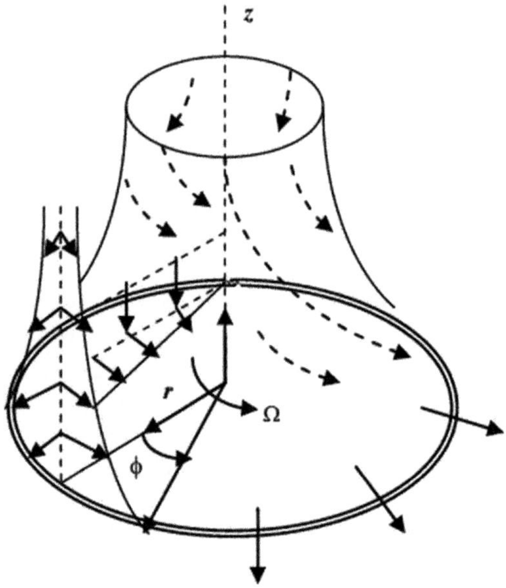

FIGURE 4.12 Physical model sketch of a rotating disk.

$$u\frac{\partial u}{\partial r}-\frac{v^2}{r}+w\frac{\partial u}{\partial z}+\frac{1}{\rho_{nf}}\frac{\partial p}{\partial r}=\frac{\mu_{nf}}{\rho_{nf}}\left(\frac{\partial^2 u}{\partial r^2}+\frac{1}{r}\frac{\partial u}{\partial r}-\frac{u}{r^2}+\frac{\partial^2 u}{\partial z^2}\right)-\frac{\mu_{nf}}{\rho_{nf}K}u, \quad (4.59)$$

$$u\frac{\partial v}{\partial r}+\frac{uv}{r}+w\frac{\partial v}{\partial z}=\frac{\mu_{nf}}{\rho_{nf}}\left(\frac{\partial^2 v}{\partial r^2}+\frac{\partial}{\partial r}\left(\frac{v}{r}\right)+\frac{\partial^2 v}{\partial z^2}\right)-\frac{\mu_{nf}}{\rho_{nf}K}v, \quad (4.60)$$

$$u\frac{\partial w}{\partial r}+w\frac{\partial w}{\partial z}+\frac{1}{\rho_{nf}}\frac{\partial p}{\partial z}=\frac{\mu_{nf}}{\rho_{nf}}\left(\frac{\partial^2 w}{\partial r^2}+\frac{1}{r}\frac{\partial w}{\partial r}+\frac{\partial^2 w}{\partial z^2}\right)-\frac{\mu_{nf}}{\rho_{nf}K}w, \quad (4.61)$$

$$u\frac{\partial T}{\partial r}+w\frac{\partial T}{\partial z}=\alpha_{nf}\left(\frac{\partial^2 T}{\partial r^2}+\frac{1}{r}\frac{\partial T}{\partial r}+\frac{\partial^2 T}{\partial z^2}\right). \quad (4.62)$$

The slip boundary conditions are assumed to be

$$z=0: u=\frac{2-\sigma_u}{\sigma_u}\lambda_0\frac{\partial u}{\partial z}, \quad v=\Omega r+\frac{2-\sigma_u}{\sigma_u}\lambda_0\frac{\partial v}{\partial z},$$
$$w=0, \quad T=T_w+\frac{2-\sigma_T}{\sigma_T}\frac{2\beta}{1+\beta}\frac{\lambda_0}{\Pr}\frac{\partial T}{\partial z}, \quad (4.63)$$

$$z\to\infty: u\to 0, \quad v\to 0, \quad T\to T_\infty, \quad P\to P_\infty, \quad (4.64)$$

where σ_u is the tangential momentum accommodation coefficient, σ_T is the thermal accommodation coefficient, λ_0 is the molecular mean free path, β is the specific heat ratio, T is the temperature of the nanofluid, T_∞ denotes the temperature of the ambient nanofluid, K is the Darcy permeability, the pressure is P, and the pressure of the ambient nanofluid is P_∞. Moreover, μ_{nf} and α_{nf} are the dynamic viscosity and thermal diffusivity of the nanofluid, respectively, and ρ_{nf} is the density of the nanofluid. These are defined as

$$\mu_{nf} = \frac{\mu_f}{(1-\varphi)^{2.5}}, \quad \alpha_{nf} = \frac{k_{nf}}{(\rho C_p)_{nf}}, \quad \rho_{nf} = (1-\varphi)\rho_f + \varphi\rho_s$$

$$(\rho C_p)_{nf} = (1-\varphi)(\rho C_p)_f + \varphi(\rho C_p)_s, \quad \frac{k_{nf}}{k_f} = \frac{(k_s+2k_f) - 2\varphi(k_f-k_s)}{(k_s+2k_f) + \varphi(k_f-k_s)}, \tag{4.65}$$

in which the nanoparticle volume fraction is denoted by φ, μ_f is the viscosity of the fluid fraction, and ρ_f and ρ_s are the densities of the fluid and of the solid fractions, respectively. The heat capacitance of the nanofluid is given by $(\rho C_p)_{nf}$ and k_{nf} stands for the effective thermal conductivity of the nanofluid approximated by the model given by Oztop and Abu-Nada (2008), which is restricted to spherical nanoparticles only. The thermophysical properties of water and different nanoparticles are given in Table 4.2.

4.3.3 Von Karman's Transformation

We introduce Von Karman's transformations as

$$\eta = (\Omega/v_f)^{1/2}z, \quad u = \Omega r F(\eta), \quad v = \Omega r G(\eta), \quad w = (\Omega v_f)^{1/2}H(\eta),$$
$$p - p_\infty = 2\mu_f\Omega p(\eta), \quad \theta(\eta) = (T-T_\infty)/(T_w-T_\infty). \tag{4.66}$$

Substituting Eqs. (4.65) and (4.66) into Eqs. (4.58)–(4.62), we can obtain the following ordinary differential equations:

$$H' + 2F = 0, \tag{4.67}$$

$$\left(\frac{1}{(1-\varphi)^{2.5}(1-\varphi+\varphi\rho_s/\rho_f)}\right)F'' - HF' - F^2 + G^2 - MF = 0, \tag{4.68}$$

TABLE 4.2 Thermophysical Properties of Water and Different Nanoparticles (Oztop and Abu-Nada, 2008)

Physical Properties	Pure Water	Cu	CuO	Al_2O_3
C_p (J/kg k)	4179	385	531.8	765
ρ (kg/m^3)	997.1	8933	6320	3970
k (W/mk)	0.613	400	76.5	40

$$\left(\frac{1}{(1-\varphi)^{2.5}(1-\varphi+\varphi\rho_s/\rho_f)}\right)G''-HG'-2FG-MG=0, \tag{4.69}$$

$$\frac{1}{\Pr}\frac{K_{nf}/K_f}{(\rho c_p)_{nf}/(\rho c_p)_f}\theta''-H\theta'=0, \tag{4.70}$$

The transformed boundary conditions become

$$\begin{aligned}&F(0)=\gamma F'(0), G(0)=1+\gamma G'(0), H(0)=0, \theta(0)=1+\delta\theta'(0)\\&F(\infty)=G(\infty)=\theta(\infty)=P(\infty)=0,\end{aligned} \tag{4.71}$$

where $M=\mu_{nf}/K\Omega\rho_{nf}$ is the porosity parameter, $\gamma=2-\sigma_u/\sigma_u\lambda_0\sqrt{\Omega/\nu_f}$ is the velocity slip parameter, $\delta=2-\sigma_T/\sigma_T(2\beta/\beta+1)\lambda_0/\Pr\sqrt{\Omega/\nu_f}$ is the temperature jump parameter, and Pr is the Prandtl number. The skin friction coefficient C_f and the Nusselt number N_u are physical quantities, which are introduced as

$$C_f=\frac{\sqrt{\tau_{wr}^2+\tau_{w\phi}^2}}{\rho_f(\Omega r)^2},\quad Nu=\frac{rq_w}{k_f(T_w-T_\infty)}, \tag{4.72}$$

where τ_{wr} and $\tau_{w\phi}$ are the radial and transversal shear stress at the surface of the disk, respectively, and q_w is the surface heat flux, defined as

$$\tau_{wr}=\left[\mu_{nf}(u_z+w_\phi)\right]_{z=0},\quad \tau_{w\phi}=\left[\mu_{nf}\left(v_z+\frac{1}{r}w_\phi\right)\right]_{z=0},\quad q_w=-k_{nf}(T_z)_{z=0}. \tag{4.73}$$

Substituting Eq. (4.65) in Eq. (4.73) and using Eq. (4.72), we obtain

$$\mathrm{Re}^{1/2}C_f=\frac{\sqrt{F'(0)^2+G'(0)^2}}{(1-\varphi)^{2.5}},\quad \mathrm{Re}^{1/2}Nu=-\frac{k_{nf}}{k_f}\theta'(0). \tag{4.74}$$

where $\mathrm{Re}=\Omega r^2/\nu_f$ is the local Reynolds number.

4.3.4 Homotopy Analysis Method Solutions

In this section, HAM is used to solve the nonlinear boundary value Eqs. (4.67)–(4.71). The initial approximations are chosen as

$$F_0(\eta)=0,\quad G_0(\eta)=e^{-\eta}/(1+\gamma), \tag{4.75}$$

$$H_0(\eta)=0,\quad \theta_0(\eta)=e^{-\eta}/(1+\delta). \tag{4.76}$$

The auxiliary linear operators are chosen as follows, respectively:

$$\ell_1[H]=H',\quad \ell_2[F]=F''+F',\quad \ell_3[G]=G''+G',\quad \ell_4[\theta]=\theta''+\theta', \tag{4.77}$$

satisfying the following properties

$$\ell_1[c_1] = 0,\ \ \ell_2[c_2e^{-\eta} + c_3] = 0,\ \ \ell_3[c_4e^{-\eta} + c_5] = 0,\ \ \ell_4[c_6e^{-\eta} + c_7] = 0, \tag{4.78}$$

where c_i, $i = 1–7$, are the arbitrary constants. The nonlinear operators are given by

$$N_1 = \frac{\partial H(\eta;q)}{\partial \eta} + 2F(\eta;q), \tag{4.79}$$

$$N_2 = C_f\frac{\partial^2 F(\eta;q)}{\partial \eta^2} - F(\eta;q)^2 + G(\eta;q)^2 - H(\eta;q)\frac{\partial F(\eta;q)}{\partial \eta} - MF(\eta;q), \tag{4.80}$$

$$N_3 = C_f\frac{\partial^2 G(\eta;q)}{\partial \eta^2} - H(\eta;q)\frac{\partial G(\eta;q)}{\partial \eta} - 2F(\eta;q)G(\eta;q) - MG(\eta;q), \tag{4.81}$$

$$N_4 = \frac{C_\theta}{\Pr}\frac{\partial^2 \theta(\eta;q)}{\partial \eta^2} - H(\eta;q)\frac{\partial \theta(\eta;q)}{\partial \eta}, \tag{4.82}$$

where $q \in [0,1]$ is the embedding parameter. The zero-order deformation equations are constructed as the following forms:

$$(1-q)\ell_1[H(\eta;q) - H_0(\eta)] = qh_H H_H(\eta)N_1[F(\eta;q), G(\eta;q), H(\eta;q)], \tag{4.83}$$

$$(1-q)\ell_2[F(\eta;q) - F_0(\eta)] = qh_F H_F(\eta)N_2[F(\eta;q), G(\eta;q), H(\eta;q)], \tag{4.84}$$

$$(1-q)\ell_3[G(\eta;q) - G_0(\eta)] = qh_G H_G(\eta)N_3[F(\eta;q), G(\eta;q), H(\eta;q)], \tag{4.85}$$

$$(1-q)\ell_4[\theta(\eta;q) - \theta_0(\eta)] = qh_\theta H_\theta(\eta)N_4[F(\eta;q), G(\eta;q), H(\eta;q), \theta(\eta;q)], \tag{4.86}$$

with the boundary conditions

$$F(\eta;q)|_{\eta=0} = \gamma\left.\frac{\partial F(\eta;q)}{\partial \eta}\right|_{\eta=0},\quad G(\eta;q)|_{\eta=0} = 1 + \gamma\left.\frac{\partial G(\eta;q)}{\partial \eta}\right|_{\eta=0},$$

$$H(\eta;q)|_{\eta=0} = 0,\ \ \theta(\eta;q)|_{\eta=0} = 1 + \gamma\left.\frac{\partial \theta(\eta;q)}{\partial \eta}\right|_{\eta=0}, \tag{4.87}$$

$$F(\eta;q)|_{\eta=\infty} = 0,\ \ G(\eta;q)|_{\eta=\infty} = 0,\ \ \theta(\eta;q)|_{\eta=\infty} = 0, \tag{4.88}$$

where h_H, h_F, h_G, and h_θ denote the auxiliary nonzero parameters and $H_H(\eta)$, $H_F(\eta)$, $H_G(\eta)$, and $H_\theta(\eta)$ are the auxiliary functions.

Expand $H(\eta;q)$, $F(\eta;q)$, $G(\eta;q)$, and $\theta(\eta;q)$ into Taylor series at $q=0$, as

$$H(\eta;q)=H_0(\eta)+\sum_{m=1}^{\infty}H_m(\eta)q^m,\quad F(\eta;q)=F_0(\eta)+\sum_{m=1}^{\infty}F_m(\eta)q^m, \tag{4.89}$$

$$G(\eta;q)=G_0(\eta)+\sum_{m=1}^{\infty}G_m(\eta)q^m,\quad \theta(\eta;q)=\theta_0(\eta)+\sum_{m=1}^{\infty}\theta_m(\eta)q^m, \tag{4.90}$$

where

$$H_m(\eta)=\frac{1}{m!}\frac{\partial^m H(\eta;q)}{\partial q^m}\bigg|_{q=0},\qquad F_m(\eta)=\frac{1}{m!}\frac{\partial^m F(\eta;q)}{\partial q^m}\bigg|_{q=0}, \tag{4.91}$$

$$G_m(\eta)=\frac{1}{m!}\frac{\partial^m G(\eta;q)}{\partial q^m}\bigg|_{q=0},\qquad \theta_m(\eta)=\frac{1}{m!}\frac{\partial^m \theta(\eta;q)}{\partial q^m}\bigg|_{q=0}, \tag{4.92}$$

Now we derive the high-order deformation equations as

$$\ell_1[H_m(\eta)-\chi_m H_{m-1}(\eta)]=h_H H_H(\eta)R_{1,m}(\eta), \tag{4.93}$$

$$\ell_2[F_m(\eta)-\chi_m F_{m-1}(\eta)]=h_F H_F(\eta)R_{2,m}(\eta), \tag{4.94}$$

$$\ell_3[G_m(\eta)-\chi_m G_{m-1}(\eta)]=h_G H_G(\eta)R_{3,m}(\eta), \tag{4.95}$$

$$\ell_4[\theta_m(\eta)-\chi_m \theta_{m-1}(\eta)]=h_\theta H_\theta(\eta)R_{4,m}(\eta). \tag{4.96}$$

subject to the following boundary conditions

$$F_m(0)=\gamma F'_m(0),\ \ G_m(0)=\gamma G'_m(0),\ \ \theta_m(0)=\delta\theta'_m(0),\ \ H_m(0)=0, \tag{4.97}$$

$$F_m(\infty)=0,\quad G_m(\infty)=0,\ \ \theta_m(\infty)=0, \tag{4.98}$$

where

$$R_{1,m}(\eta)=H'_{m-1}(\eta)+2F_{m-1}(\eta), \tag{4.99}$$

$$R_{2,m}(\eta)=C_f F''_{m-1}(\eta)-\sum_{k=0}^{m-1}F_k(\eta)F_{m-1-k}(\eta)$$
$$+\sum_{k=0}^{m-1}G_k(\eta)G_{m-1-k}(\eta)-\sum_{k=0}^{m-1}H_k(\eta)F'_{m-1-k}(\eta)-MF_{m-1}(\eta), \tag{4.100}$$

$$R_{3,m}(\eta) = C_f G''_{m-1}(\eta) - \left[\sum_{k=0}^{m-1} H_k(\eta) G'_{m-1-k}(\eta) + 2 \sum_{k=0}^{m-1} F_k(\eta) G_{m-1-k}(\eta) \right] - M G_{m-1}(\eta), \tag{4.101}$$

$$R_{4,m}(\eta) = \frac{C_\theta}{\Pr} \theta''_{m-1}(\eta) - \sum_{k=0}^{m-1} H_k(\eta) \theta'_{m-1-k}(\eta), \tag{4.102}$$

$$\chi_m = \begin{cases} 0 & m \leq 1 \\ 1 & m \geq 2 \end{cases}. \tag{4.103}$$

Finally, the auxiliary functions are chosen as

$$H_F(\eta) = H_G(\eta) = e^{-\eta}, \ \ H_H(\eta) = H_\theta(\eta) = 1. \tag{4.104}$$

4.3.5 Results and Discussion

The nonlinear ordinary differential Eqs. (4.67)–(4.70) subjected to boundary conditions (4.71) are solved analytically by HAM. The convergence of the HAM solutions strongly depends on the auxiliary parameter h. By means of the h curve, it is straightforward to choose a proper value of h to ensure the convergence of the solution series (Table 4.3).

The h curves of $H''(0)$, $F'(0)$, $G'(0)$, and $\theta'(0)$ obtained by 10th approximation are presented in Fig. 4.13. Moreover, the reliability of analytical results is verified with numerical ones obtained by the finite difference technique and the results published in the literature Rashidi et al. (2013) and Turkyilmazoglu (2014), which are also shown in Table 4.3.

4.3.5.1 Effects of Velocity Slip Parameter

Fig. 4.14 shows the effects of velocity slip parameter γ on the radial velocity profiles. It indicates that the slip parameter has a significant effect on radial

TABLE 4.3 Comparison of Results for $F'(0)$, $-G'(0)$, $-H(\infty)$, and $-\theta'(0)$ When $\varphi = 0$, $M = 0$, $\gamma = 0$, $\delta = 0$, and $\Pr = 6.2$

	Rashidi et al. (2013)	Turkyilmazoglu (2014)	Current Study
$F'(0)$	0.510186	0.51023262	0.51022941
$-G'(0)$	0.61589	0.61592201	0.61591990
$-H(\infty)$		0.88447411	0.88446912
$-\theta'(0)$		0.93387794	0.93387285

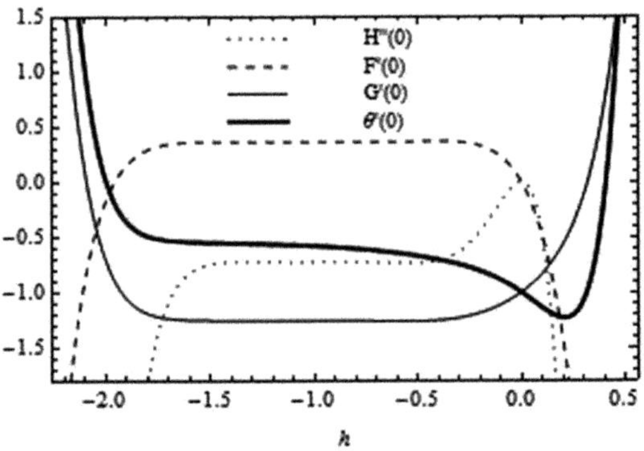

FIGURE 4.13 $\hbar$ curves of $H''(0)$, $F'(0)$, $G'(0)$, and $\theta'(0)$ obtained by the 10th-order approximation of the HAM solution for Cu-water nanofluid with $\varphi = 0.1$, $\gamma = 0$, $\delta = 0$, and $M = 1$.

velocity distributions; there is a peak for the radial velocity profiles (maximum) that decreases rapidly and moves to the disk as the slip parameter γ increases.

Figs. 4.15 and 4.16 present the profiles of the tangential and axial velocity, respectively. The results indicate that for different values of velocity slip parameter γ, the tangential velocity decreases but the axial velocity (negative) increases with the increase in γ.

In addition, Figs. 4.14–4.16 compare the analytical results obtained by HAM and the numerical solutions; the results are in a good agreement.

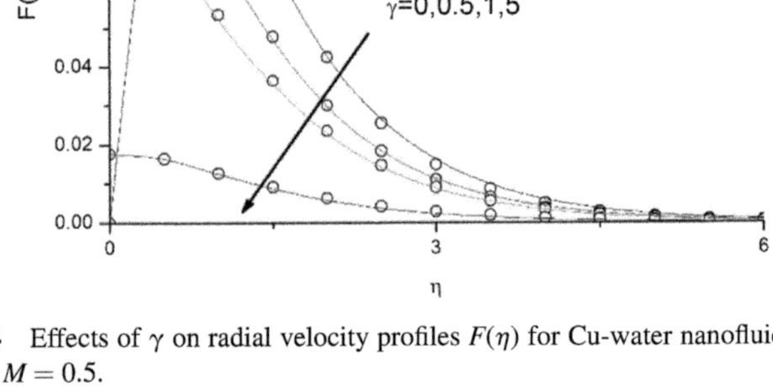

FIGURE 4.14 Effects of γ on radial velocity profiles $F(\eta)$ for Cu-water nanofluid with $\varphi = 0.1$, Pr = 6.2, and $M = 0.5$.

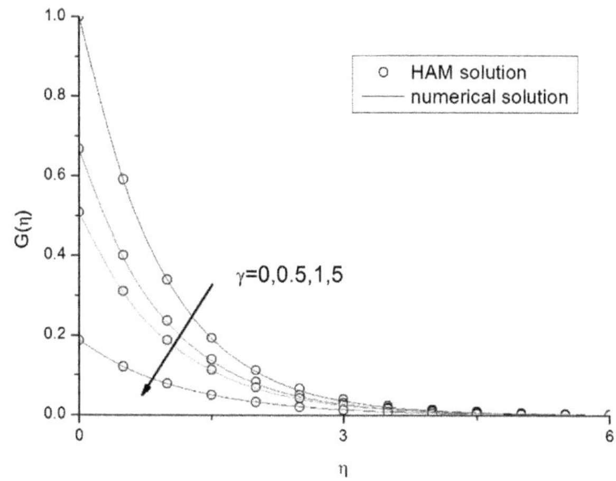

FIGURE 4.15 Effects of γ on tangential velocity profiles $G(\eta)$ for Cu-water nanofluid with $\varphi = 0.1$, $\Pr = 6.2$, and $M = 0.5$.

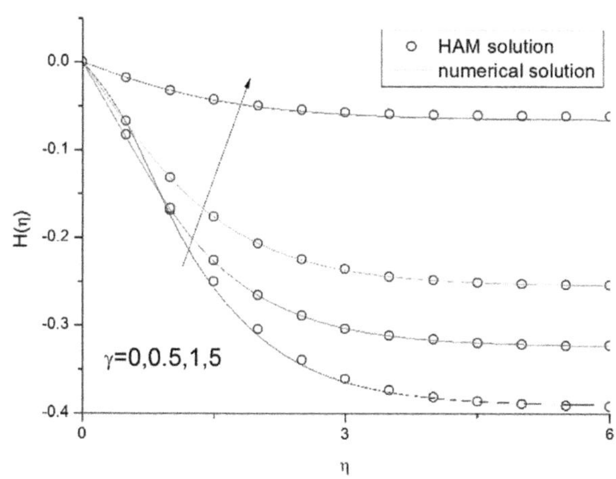

FIGURE 4.16 Effects of γ on axial velocity profiles $H(\eta)$ for Cu-water nanofluid with $\varphi = 0.1$, $\Pr = 6.2$, and $M = 0.5$.

4.3.5.2 Effects of Temperature Jump Parameter

The profiles of temperature distribution for various jump parameters δ are shown in Fig. 4.17; the results reveal that the surface temperature and thickness of the thermal boundary layer decrease with increasing values of δ.

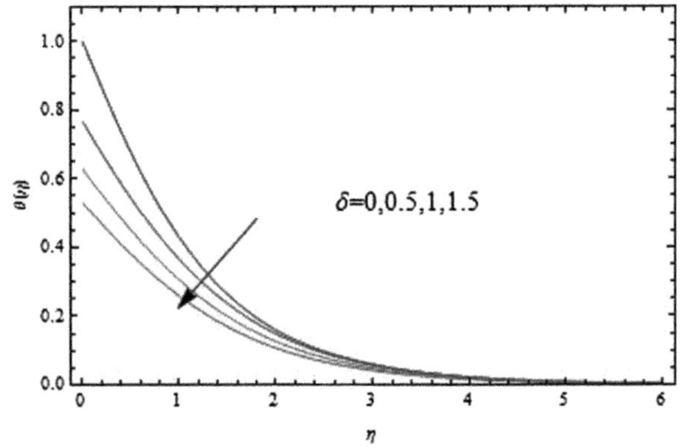

FIGURE 4.17 Effects of δ on temperature profiles $\theta(\eta)$ for Cu-water nanofluid with $\varphi = 0.1$, $Pr = 6.2$, and $M = 0.5$.

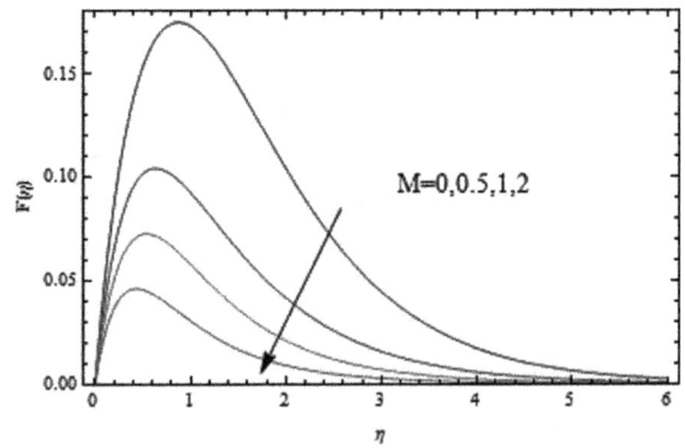

FIGURE 4.18 Effects of M on radial velocity profiles $F(\eta)$ for Cu-water nanofluid with $\varphi = 0.1$, $Pr = 6.2$, and $\gamma = 0$.

4.3.5.3 Effects of Porosity Parameter

Figs. 4.18–4.21 demonstrate the effects of porosity parameter M on the velocity components in radial, tangential, and axial directions and temperature distribution. With increasing M, the velocity profiles in the radial, tangential, and axial directions decrease, whereas the thermal boundary layer thickness increases.

4.3.5.4 Effects of Types of Nanoparticle

Analytical results for the skin friction coefficient $Re^{1/2}C_f$ and the local Nusselt number $Re^{-1/2}Nu$ for a wide range of nanoparticle volume fractions and three different types of nanoparticles in the presence of velocity slip and temperature jump are presented in Figs. 4.22 and 4.23. The values of skin

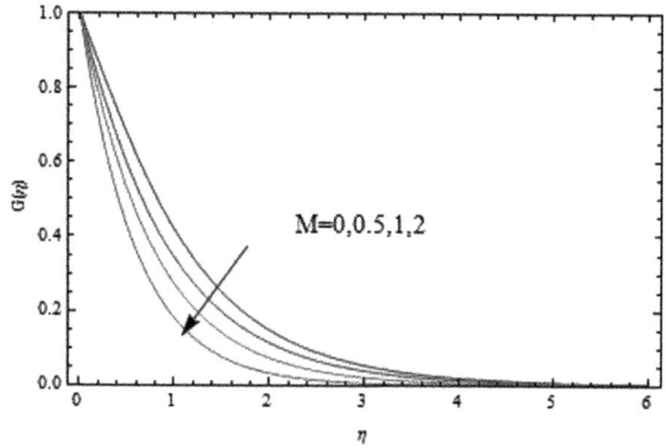

FIGURE 4.19 Effects of M on tangential velocity profiles $G(\eta)$ for Cu-water nanofluid with $\varphi = 0.1$, $\Pr = 6.2$, and $\gamma = 0$.

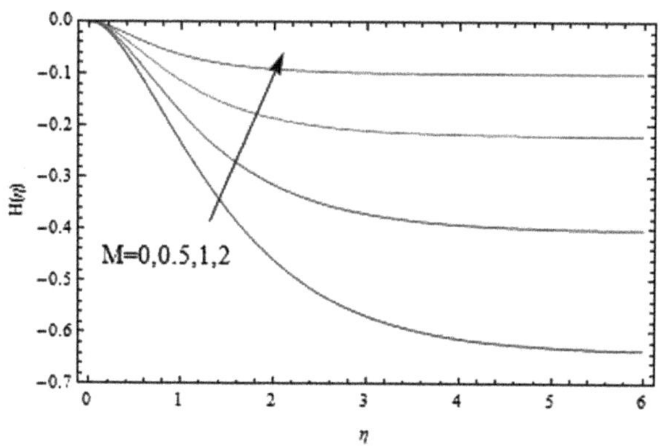

FIGURE 4.20 Effects of M on axial velocity profiles $H(\eta)$ for Cu-water nanofluid with $\varphi = 0.1$, $\Pr = 6.2$, and $\gamma = 0$.

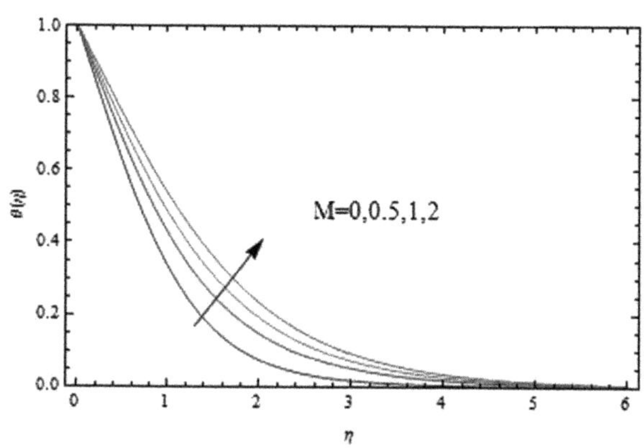

FIGURE 4.21 Effects of M on temperature profiles $\theta(\eta)$ for Cu-water nanofluid with $\varphi = 0.1$, $\Pr = 6.2$, and $\delta = 0$.

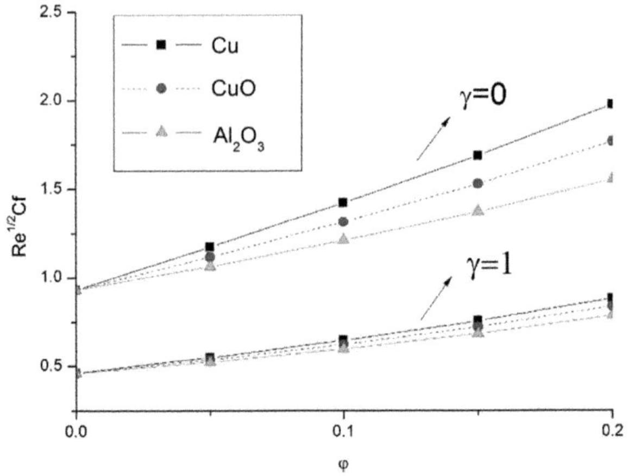

FIGURE 4.22 Variations in skin friction coefficient with φ for different nanoparticles and velocity slip with Pr = 6.2 and M = 0.5.

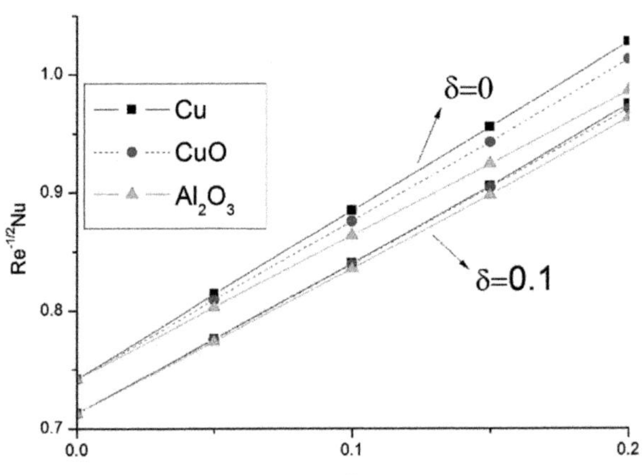

FIGURE 4.23 Variations in Nusselt number with φ for different nanoparticles and temperature jump with Pr = 6.2 and M = 0.5.

friction coefficient and local Nusselt number both increased nearly linearly with the nanoparticle volume fraction. The Cu-nanofluid had the largest skin friction coefficient and heat transfer rate owing to its largest thermal conductivity value. On the contrary, Al_2O_3-nanofluid had the lowest ones. Fig. 4.22 indicates that the increase in velocity slip parameter γ led to a reduction in the values of the skin friction coefficient.

The temperature jump parameter has also special effects on the local Nusselt number. Fig. 4.23 shows that the local Nusselt number decreases with an increase in the temperature jump parameter δ.

4.4 MIXED CONVECTION IN POWER LAW FLUIDS OVER MOVING CONVEYOR

4.4.1 Physical Background of the Problem

Mixed convection heat transfer has attracted considerable attention in many fields of science and technology because of its wide applications: for example, circulation caused by different densities along the vertical direction in a lake owing to seasonal variation, different temperature atmospheric flow, and heat exchanger in fuel in nuclear reactors. The most prominent feature of mixed convection is buoyancy force caused by varying density and temperature. Mathematically speaking, under Boussinesq's hypothesis, the momentum and energy equations describing mixed convection are highly coupled. According to Prandtl's boundary layer theory, a fluid with mixed convection will induce a boundary layer close to the vertical plate owing to the viscosity of the fluid. Through dimensional analysis, Navier—Stokes equations can be simplified to boundary layer equations. Air—heat convection around the vertical plate has been measured and the results of existence of momentum and thermal boundary layers have been proved (Brackenridge et al., 1965).

It is well recognized that non-Newtonian fluids are important in science research and engineering. Many models have been proposed to describe the behaviors of such fluids. Among those models, the power law model (Schlichting, 1979), in which shear stress varies according to a power law function of strain rate, has gained considerable acceptance. The boundary layer equations for a power law fluid were studied (Acrivos et al., 1960; Andersson and Irgens, 1988; Denier and Dabrowski, 2004; Hussaini et al., 1987; Schowalter, 1960). The flow and heat transfer in power law fluid over a stretching sheet were investigated (Donald Ariel, 2002; Hassanien et al., 1998). Zhang et al. (2007) and Bharti et al. (2007) numerically solved the thermal boundary layer on a continuously moving surface in a power law fluid. They asserted that boundary layer temperature distribution depends not only on plane velocity but also on the power law exponent n and generalized Prandtl number N_{pr}.

In classical works for a power law fluid, power law kinematic viscosity was introduced in the momentum equation, but the energy equation was treated the same as in Newtonian fluids. Obviously, this is inconsistent with the fact that changing viscosity should affect both momentum and heat transfer. Some researchers have paid special attention to this inconsistency because the rough assumption of thermal conductivity for non-Newtonian fluid does not meet sophisticated industrial requirements. Pop et al. (1991) and Pop (1993) proposed a model for heat transfer in which the thermal conductivity of non-Newtonian fluids has power law dependence on the velocity gradient. Zheng et al. (2006) and Li et al. (2011) took the effects of power law viscosity on temperature field into account by assuming that the temperature field is similar to the velocity field. The Navier–Stokes equation and energy equation are formulated with modified Fourier-law heat conduction law. They assumed that the thermal diffusivity varies as a function of velocity gradient or temperature gradient in the energy equation of a power law fluid. A comparison for both models was presented. Rosca and Pop (2013) investigated the steady mixed convection stagnation point flow over a vertical flat plate with a second-order slip and maintained heat flux. Two branches of solutions, upper and lower, were found in a certain range of mixed convection and velocity slip parameters. In addition, many other mixed convection heat transfer problems with different boundary conditions, such as convection in an inclined rectangle (Elsherbiny & Ragab, 2013) and mixed convection heat transfer with the effects of a magnetic field (Ashorynejad et al., 2013; Samanta and Guha, 2013), were also considered in porous media and nanofluids (Alves et al., 2014; Ece and Büyük, 2002; Kherbeet et al., 2014).

In this section, we study mixed convection heat transfer of a power law fluid over a moving conveyor along an inclined plate. Unlike most classical studies of Newtonian fluids, we take the effects of power law fluid viscosity on a temperature field into account here by assuming that the temperature field is similar to the velocity field with a modified Fourier's law. The effects of a moving conveyor and inclined plate on the overall convection system, i.e., the velocity ratio coefficient $\gamma_u = \frac{U_w}{U_m}$ and incline angle φ, the generalized Prandtl number N_{pr}, and power law exponent n on velocity and temperature field are graphically assessed, respectively. The generalized local Grashof number Gr_n and local Nusselt number Nu for power law fluids are also derived and discussed (Sui et al., 2015).

4.4.2 Mathematical Formulation

Consider a steady laminar mixed convection boundary layer flow and heat transfer in a power law fluid over a moving conveyor along an inclined plate, where the essential features of such a flow are illustrated in Fig. 4.24. It is assumed that the body force is $X = -\rho g$; the boundary layer governing equations describing the conservation of mass continuity, momentum, and energy can be written as:

$$\frac{\partial U}{\partial X} + \frac{\partial V}{\partial Y} = 0, \tag{4.105}$$

$$U\frac{\partial U}{\partial X} + V\frac{\partial U}{\partial Y} = \frac{\overline{\mu}}{\rho}\frac{\partial}{\partial Y}\left(\left|\frac{\partial U}{\partial Y}\right|^{n-1}\frac{\partial U}{\partial Y}\right) + g\beta(T - T_\infty)\sin\varphi \ , \tag{4.106}$$

$$U\frac{\partial T}{\partial X} + V\frac{\partial T}{\partial Y} = \frac{\overline{\lambda}}{\rho c_p}\frac{\partial}{\partial Y}\left(\left|\frac{\partial U}{\partial Y}\right|^{n-1}\frac{\partial T}{\partial Y}\right) \ , \tag{4.107}$$

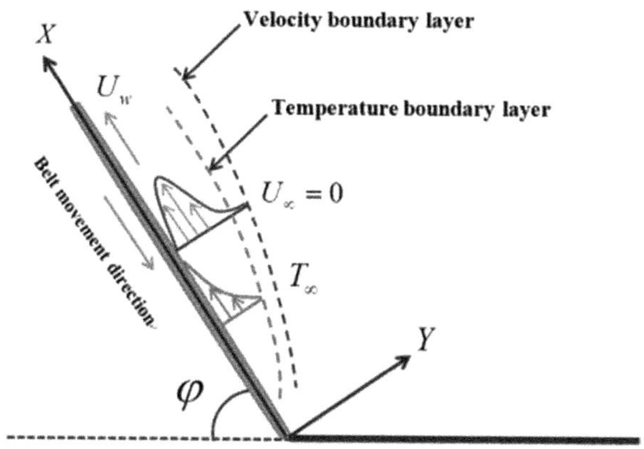

FIGURE 4.24 Schematic physical model and coordinate system of an inclined moving conveyor.

$$Y = 0 : U = U_w, V = 0, T = T_w, \tag{4.108}$$

$$Y \to \infty : U = 0, T = T_\infty, \tag{4.109}$$

where the shear stress is characterized as $\tau = \overline{\mu}\left(\left|\frac{\partial U}{\partial Y}\right|^{n-1}\frac{\partial U}{\partial Y}\right)$, the kinematic viscosity is $\overline{\nu} = \omega\left|\frac{\partial U}{\partial Y}\right|^{n-1}$ ($\overline{\mu}$ and $\omega = \overline{\mu}/\rho$ are positive constant), $\overline{\lambda}\left|\frac{\partial U}{\partial Y}\right|^{n-1}$ is the generalized thermal conductivity in terms of a modified Fourier's law, β is the expansion coefficient of power law fluid, U_w is the velocity of moving conveyor, T_w is the surface temperature of the inclined plate, and $0 < \varphi \leq \frac{\pi}{2}$ is the angle between the inclined plate with respect to the horizontal direction. The case $n = 1$ corresponds to a Newtonian fluid, $0 < n < 1$ is pseudoplastic non-Newtonian fluids, whereas $n > 1$ describes dilatant fluids, respectively.

The following dimensionless variables are introduced:

$$u = \frac{U}{U_m}, \quad v = \frac{V}{U_m}, \quad x = \frac{X}{L\sin\varphi}, \quad y = \frac{Y}{L\sin\varphi}, \quad \theta = \frac{T - T_\infty}{T_w - T_\infty}, \tag{4.110}$$

According to the Boussinesq's approximation, the boundary layer equations become:

$$\frac{\partial u}{\partial x} + \frac{\partial v}{\partial y} = 0, \tag{4.111}$$

$$u\frac{\partial u}{\partial x} + v\frac{\partial u}{\partial y} = \frac{1}{\sqrt{Gr_n}}\frac{\partial}{\partial y}\left(\left|\frac{\partial u}{\partial y}\right|^{n-1}\frac{\partial u}{\partial y}\right) + \theta\sin\varphi, \tag{4.112}$$

$$u\frac{\partial\theta}{\partial x}+v\frac{\partial\theta}{\partial y}=\frac{1}{N_{\mathrm{pr}}\sqrt{Gr_n}}\frac{\partial}{\partial y}\left(\left|\frac{\partial u}{\partial y}\right|^{n-1}\frac{\partial\theta}{\partial y}\right), \tag{4.113}$$

The boundary conditions are:

$$y=0: u=\frac{\gamma_u}{\sqrt{sin\varphi}},\quad v=0,\quad \theta=1, \tag{4.114}$$

$$y\to\infty: u=0,\quad \theta=0, \tag{4.115}$$

$$x=0:\quad u=0,\quad \theta=0, \tag{4.116}$$

where $N_{\mathrm{pr}}=\frac{\overline{\mu}c_p}{\overline{\lambda}}$ is the generalized Prandtl number. When $n=1$, N_{pr} reduces to the Pr for the Newtonian fluid, $Gr_n=\frac{(L\sin\varphi)^{n+2}}{\omega^2(g\beta(T_w-T_\infty))^{n-2}}$ with $U_m=\sqrt{g\,\beta\,L\sin\varphi\,(T_w-T_\infty)}$ are the generalized local Grashof number derived first for a power law fluid and the mean velocity of the whole boundary layer along the inclined plate, $\gamma_u=\frac{U_w}{U_m^{\perp}}$ is the velocity ratio coefficient where the conveyor velocity is $U_w=\gamma_u\sqrt{g\,\beta\,X\,(T_w-T_\infty)}$, and $U_m^{\perp}$ is the mean velocity of fluid along the vertical plate.

Eqs. (4.111)–(4.116) are strongly nonlinear with a moving boundary. The problem is complex to solve analytically and numerical simulation is just as challenging. Here, we approach the problem by similarity solutions.

4.4.3 Nonlinear Boundary Value Problems

Historically, the boundary layer flow past a flat plate was a first example posed by Blasius in illustrating the application of the Prandtl boundary layer theory. The similarity equations of expressing conservation of momentum and energy for a Newtonian fluid can be written as Falkner–Skan's equations (Schlichting, 1979):

$$f'''(\eta)+f(\eta)f''(\eta)=0, \tag{4.117}$$

$$T''(\eta)+\mathrm{Pr}f(\eta)T'(\eta)=0. \tag{4.118}$$

We introduce the stream function ψ (x,y), similarity variable η, and dimensionless temperature function $w(\eta)$ as

$$\eta=AX^aY,\quad \psi=BX^bf(\eta),\quad \theta(X,Y)=w(\eta),$$

where A, B, a, and b are constants to be determined, and $f(\eta)$ denotes the dimensionless stream function. Thus, the velocity components U and V are respectively,

$$U = \frac{\partial \psi}{\partial Y} = ABX^{a+b} f'(\eta), \tag{4.119}$$

$$V = -\frac{\partial \psi}{\partial X} = -BbX^{b-1} f - ABaX^{a+b-1} Yf'. \tag{4.120}$$

Substituting U and V defined by Eqs. (4.119) and (4.120) into Eqs. (4.105)–(4.109), by choosing

$$a = -\frac{n}{2(n+1)}, \quad b = \frac{2n+1}{2(n+1)}, \quad A = \left[\frac{[g\beta(T_w - T_\infty)]^{2-n}}{\omega^2}\right]^{\frac{1}{2+2n}},$$

$$\eta = \left[\frac{Gr_{nx}}{(\sin\varphi)^{n+2}}\right]^{\frac{1}{2(n+1)}} \frac{Y}{X}.$$

we obtain the similarity equations:

$$\left(|f''|^{n-1} f''\right)' + \left(\frac{2n+1}{2(n+1)} ff'' - \frac{1}{2}(f')^2\right) + w \ \sin\varphi = 0, \tag{4.121}$$

$$\left(|f''|^{n-1} w'\right)' + N_{\text{pr}} \frac{2n+1}{2(n+1)} fw' = 0. \tag{4.122}$$

The boundary conditions are:

$$f(0) = 0, \ f'(0) = \gamma_u, \ w(0) = 1, \tag{4.123}$$

$$f'(+\infty) = 0, \quad w(+\infty) = 0. \tag{4.124}$$

All derivatives are with respect to the similarity variable η for $0 < \varphi \leq \frac{\pi}{2}$. The similarity form of velocity is $U = \sqrt{g\beta(T_w - T_\infty)X} f'(\eta) = U_m^{\frac{1}{m}} f'(\eta)$. According to the similarity transformation, the heat flux density and shear stress on surface of inclined plate are obtained as

$$q_w = -\overline{\lambda} \left|\frac{\partial U}{\partial Y}\right|^{n-1} \frac{\partial T}{\partial Y}\bigg|_{Y=0} = -\overline{\lambda}(t_w - t_\infty) \sqrt{g\beta(T_w - T_\infty)}^{\,n-1} A^n X^{\frac{-1}{2(n+1)}} |f''(0)|^{n-1} w'(0), \tag{4.125}$$

$$\tau_w = \overline{\mu}\left(\left|\frac{\partial U}{\partial Y}\right|^{n-1} \frac{\partial U}{\partial Y}\right)\bigg|_{Y=0} = \overline{\mu}\left[\frac{(g\beta(T_w - T_\infty))^3 X}{\omega^2}\right]^{\frac{n}{2n+2}} |f''(0)|^{n-1} f''(0), \tag{4.126}$$

We derive the local Nusselt number as $Nu = \frac{q_w X}{(T_w - T_\infty)\overline{\lambda}\left|\frac{\partial U}{\partial Y}\right|^{n-1}} = -(Gr_{nx}^{\frac{1}{n}})^{\frac{1}{2n+2}} w'(0)$. The similarity expressions of the local Nusselt number and nondimensional skin friction at the surface of the plate are $Nu \sim -w'(0)$ and

$C_f \sim |f''(0)|^{n-1} f''(0)$, respectively. $Gr_{nx}^{\perp} = Gr_n|_{\varphi=\pi/2}$ is the generalized local Grashof number for the case of a vertical plate.

4.4.4 Homotopy Analysis Method Solutions

In this section, we show an idea of how to solve the strong nonlinear coupled Eqs. (4.121)–(4.124) by HAM (Liao, 1992, 2012). The main structure of the solution is presented as follows.

Assume that $f(\eta)$ and $w(\eta)$ can be expressed in the form

$$f(\eta) = f_0 + \sum_{i=1}^{\infty} a_i e^{-i\eta}, \quad w(\eta) = w_0 + \sum_{i=1}^{\infty} b_i e^{-i\eta}, \tag{4.127}$$

where a_i and b_i are coefficients. The initial guess solution satisfies the boundary conditions (4.123) and (4.124) and auxiliary linear operators is defined as follows:

$$f_0(\eta) = \gamma_u(1 - e^{-\eta}), \quad w_0(\eta) = e^{-\eta} \tag{4.128}$$

$$L_f(\eta) = \eta''' - \eta', \quad L_w(\eta) = \eta'' - \eta \tag{4.129}$$

and the auxiliary linear operators have the following properties

$$L_f[C_1 + C_2 e^{\eta} + C_3 e^{-\eta}] = 0, \quad L_w[C_4 e^{\eta} + C_5 e^{-\eta}] = 0, \tag{4.130}$$

where $C_{1,2,3,4,5}$ are constants.

We construct the zeroth-order deformation equations

$$(1-p)L_f[F(\eta,p) - f_0(\eta)] = p h_f H_f(\eta) N_f[F(\eta,p), W(\eta,p)], \tag{4.131}$$

$$(1-p)L_w[W(\eta,p) - w_0(\eta)] = p h_w H_w(\eta) N_w[F(\eta,p), W(\eta,p)], \tag{4.132}$$

subject to the boundary conditions

$$\begin{aligned} &F(0,p) = 0, \quad \left.\frac{\partial F(\eta,p)}{\partial \eta}\right|_{\eta=0} = \gamma_u, \quad \left.\frac{\partial F(\eta,p)}{\partial \eta}\right|_{\eta=+\infty} = 0, \\ &W(0,p) = 1, \quad W(\infty,p) = 0, \end{aligned} \tag{4.133}$$

where $N_f[F(\eta,p),W(\eta,p)]$ and $N_w[F(\eta,p),W(\eta,p)]$ are two nonlinear differential operators defined by

$$N_f = \frac{\partial}{\partial \eta}\left(\left|\frac{\partial^2 F}{\partial \eta^2}\right|^{n-1} \frac{\partial^2 F}{\partial \eta^2}\right) + \frac{2n+1}{2(n+1)} F \frac{\partial^2 F}{\partial \eta^2} - \frac{1}{2}\left(\frac{\partial F}{\partial \eta}\right)^2 + W \sin\varphi, \tag{4.134}$$

$$N_w = \frac{\partial}{\partial \eta}\left(\left|\frac{\partial^2 F}{\partial \eta^2}\right|^{n-1} \frac{\partial W}{\partial \eta}\right) + N_{\mathrm{pr}} \frac{2n+1}{2(n+1)} F \frac{\partial W}{\partial \eta}, \tag{4.135}$$

and $p \in [0,1]$ is an embedding parameter, h_f and h_w are auxiliary nonzero parameters, and $H_f(\eta)$ and $H_w(\eta)$ are auxiliary functions. We expand $F(\eta,p)$ and $W(\eta,p)$ into Taylor series at $p = 0$, as

$$F(\eta,p) = f_0(\eta) + \sum_{m=1}^{\infty} f_m(\eta)p^m, \quad W(\eta,p) = w_0(\eta) + \sum_{m=1}^{\infty} w_m(\eta)p^m, \quad (4.136)$$

where

$$f_m(\eta) = \frac{1}{m!}\frac{\partial^m F(\eta,p)}{\partial p^m}\bigg|_{p=0}, \quad w_m(\eta) = \frac{1}{m!}\frac{\partial^m W(\eta,p)}{\partial p^m}\bigg|_{p=0}. \quad (4.137)$$

Furthermore, the high-order deformation equations are written as:

$$L_f[f_m(\eta) - \chi_m f_{m-1}(\eta)] = h_f H_f(\eta) R_m^f(\eta), \quad (4.138)$$

$$L_w[w_m(\eta) - \chi_m w_{m-1}(\eta)] = h_w H_w(\eta) R_m^w(\eta), \quad (4.139)$$

subject to boundary conditions

$$f_m(0) = 0, \quad f_m'(0) = 0, \quad f_m'(\infty) = 0, \quad w_m(0) = 0 \quad w_m(\infty) = 0. \quad (4.140)$$

where

$$\begin{aligned} R_m^f(\eta) = & \frac{1}{(m-1)!}\frac{\partial^{m-1}}{\partial p^{m-1}}\left(\frac{\partial}{\partial \eta}\left(\left|\frac{\partial^2 F}{\partial \eta^2}\right|^{n-1}\frac{\partial^2 F}{\partial \eta^2}\right)\right)\bigg|_{p=0} \\ & + \frac{2n+1}{2(n+1)}\sum_{j=0}^{m-1} f_j(\eta) f''_{m-j-1}(\eta) \\ & - \frac{1}{2}\sum_{j=0}^{m-1} f'_j(\eta) f'_{m-1-j}(\eta) + w_{m-1}(\eta)\sin\varphi, \end{aligned} \quad (4.141)$$

$$\begin{aligned} R_m^w(\eta) = & \frac{1}{(m-1)!}\frac{\partial^{m-1}}{\partial p^{m-1}}\left(\frac{\partial}{\partial \eta}\left(\left|\frac{\partial^2 F}{\partial \eta^2}\right|^{n-1}\frac{\partial W}{\partial \eta}\right)\right)\bigg|_{p=0} \\ & + N_{pr}\frac{2n+1}{2(n+1)}\sum_{j=0}^{m-1} f_j(\eta) w'_{m-1-j}(\eta), \end{aligned} \quad (4.142)$$

$$\chi_m = \begin{cases} 0 \ldots\ldots m = 1 \\ 1 \ldots\ldots m > 1 \end{cases} \quad (4.143)$$

4.4.5 Results and Discussion

Analytical computation is carried out for various values of parameters of the power law exponent n, inclined angle φ, and velocity ratio parameter γ_u. Results are presented graphically. Certain important physical quantities such as the local Nusselt number Nu and the skin friction parameter C_f are analyzed. It is well known that the Prandtl number Pr has important effects on the coupled velocity and temperature fields for a Newtonian fluid (Acrivos et al., 1960; Schowalter, 1960). Our results indicate that the generalized Prandtl number N_{pr} has similar effects on the temperature and velocity fields. Here we omit the discussions for N_{pr} and focus on the study for parameters n, φ, and γ_u emphatically.

4.4.5.1 *Effects of Power Law Exponent* n

The power law exponent n characterizes a type of power law fluids. The case $n = 1$ corresponds to a Newtonian fluid, $0 < n < 1$, which is a pseudoplastic fluid (shear thinning), whereas $n > 1$ describes a dilatant fluid (shear thickening).

Fig. 4.25 shows the velocity profiles for different values of power law exponent n. It indicates that the velocity maximum and mixed convection region close to the inclined plate increase as the power law exponent n increases from 0.7 to 1.5, which leads to velocity profiles intersecting each other (especially with the curve of $n = 1$, Newtonian fluid) at some points and the intersection point moves from a pseudoplastic fluid to a Newtonian fluid and then to a dilatant fluid as n increases. Also, we can draw the conclusion that heat transfer efficiency will be enhanced owing to mixed convection flow with the increase in n, which can be seen in Fig. 4.26, i.e., the thickness of the

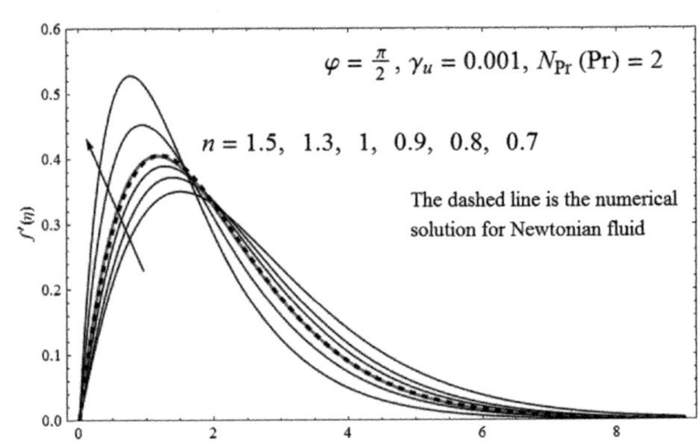

FIGURE 4.25 Velocity profiles for various values of power law index n. the 10th homotopy analysis method solution (*red line*) has good agreement with the numerical solution presented by shooting method (*dashed line*).

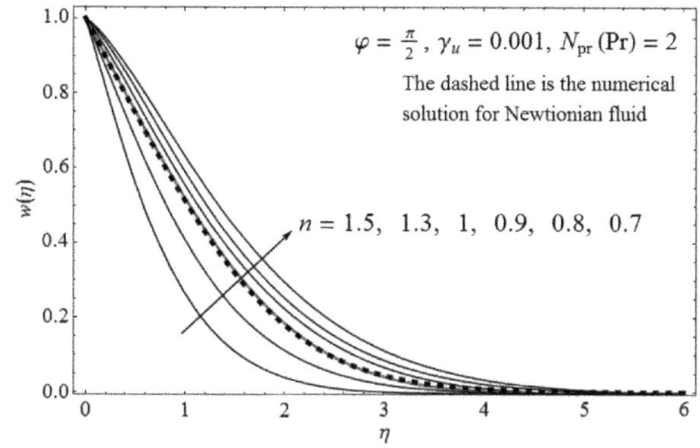

FIGURE 4.26 Temperature profiles for various values of power law index n. The 10th homotopy analysis method solution (*red line*) has good agreement with the numerical solution presented by shooting method (*dashed line*).

temperature boundary layer decreases, which agrees with the corresponding flow behavior in Fig. 4.25.

Figs. 4.25 and 4.26 demonstrate that a dilatant fluid (shear thickening) has more intense mixed convection (there is a larger peak of velocity and the convection area is closer to the plate), resulting in better heat transfer efficiency contrary to a pseudoplastic fluid (shear thinning).

Fig. 4.27 displays the skin fiction C_f and local Nu for various values of n (from 0.7 to 1.5). The heat transfer efficiency becomes gradually better from a pseudoplastic fluid to a dilatant fluid. Meanwhile, the skin friction also increases with the increase in n for a small velocity ratio $\gamma_u = 0.001$, which is significantly different from a classical Blasius flow in which the skin friction decreases with the increase in n. What leads to this difference is that the whole process of natural convection occurs without the continuous pressure from outside but depends on thermal buoyancy with the fixed moving conveyor. Therefore, a fluid with stronger convection flow results in larger skin friction. We also find that skin friction increases rapidly, especially after $n > 1.3$, whereas the local Nu increases quickly when $n < 1$ and slowly when $n > 1$ in Fig. 4.27, and the growth of C_f and Nu are nonlinear.

4.4.5.2 Effects of Incline Angle φ

The inclined angle φ has a significant impact on the mixed convection system. Here we analyze the cases of power law fluid for $n = 0.7$, $n = 1$, and $n = 1.3$, respectively.

Figs. 4.28 and 4.29 illustrate the velocity profiles of a power law fluid with $n = 0.7$ and $n = 1.3$ for values of $\varphi = \frac{\pi}{6}, \frac{\pi}{4}, \frac{\pi}{3}$, and $\frac{\pi}{2}$. As pointed out earlier, the results show that the decrease in φ causes a loss of velocity boundary layer, i.e., the velocity peak value decreases (maximum peak value with vertical plate

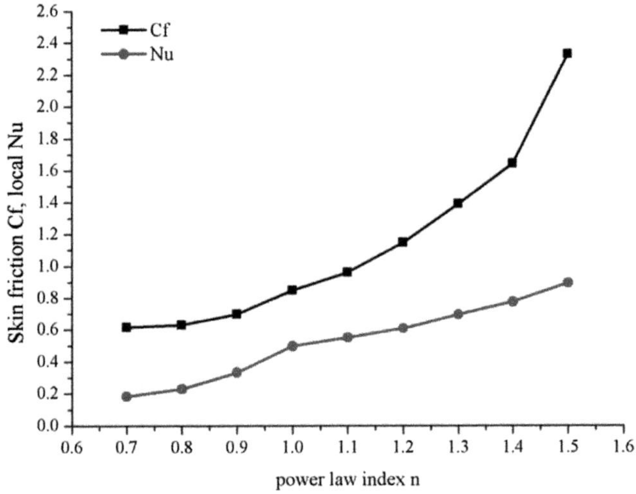

FIGURE 4.27 Skin friction parameter C_f and local Nusselt number Nu for various values of n with $\gamma_u = 0.001$ and $N_{pr}(Pr) = 2$, $\varphi = \frac{\pi}{2}$. Apparent nonlinear growth of C_f and Nu is presented, especially the large increasing rate of C_f when $n \geq 1$ and the large increasing rate of Nu when $n \leq 1$.

$\varphi = \frac{\pi}{2}$) and its horizontal position moves backward (far from the inclined plate). In addition, the velocity profiles intersect because velocity falls back slowly with the decrease in φ and the obvious intersection for $n = 0.7$ but the inconspicuous intersection for $n = 1.3$.

The slowing of velocity will lead to an increase in the thermal boundary layer, which is presented in Figs. 4.30 and 4.31. A similar characteristic for velocity and temperature can be obtained, showing that the velocity and

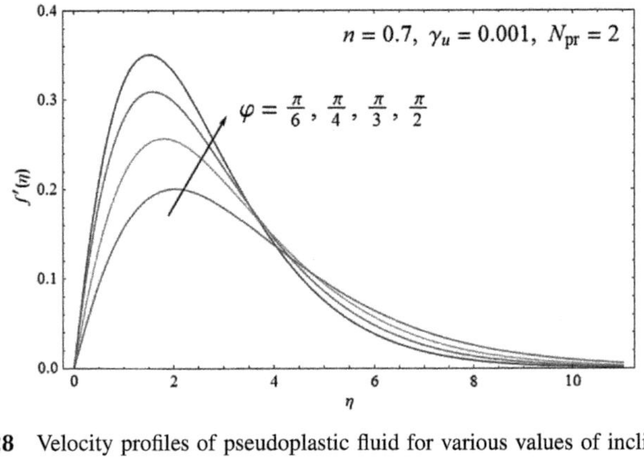

FIGURE 4.28 Velocity profiles of pseudoplastic fluid for various values of inclined angle φ.

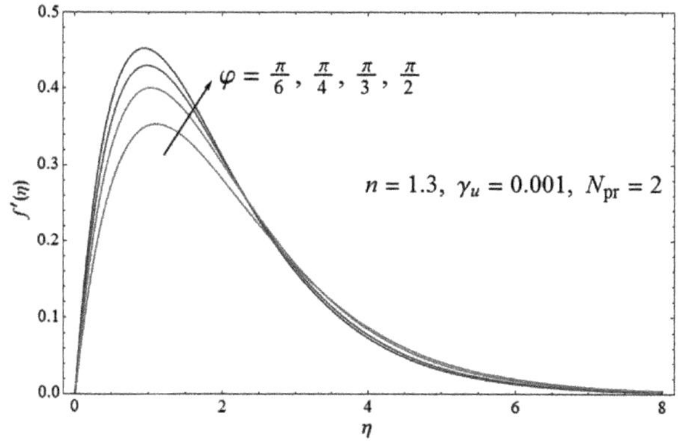

FIGURE 4.29 Velocity profiles of dilatant fluid for various values of inclined angle φ.

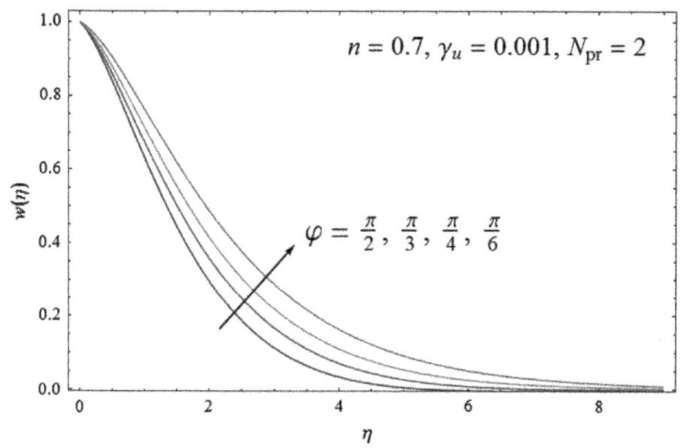

FIGURE 4.30 Temperature profiles of pseudoplastic fluid for various values of inclined angle φ.

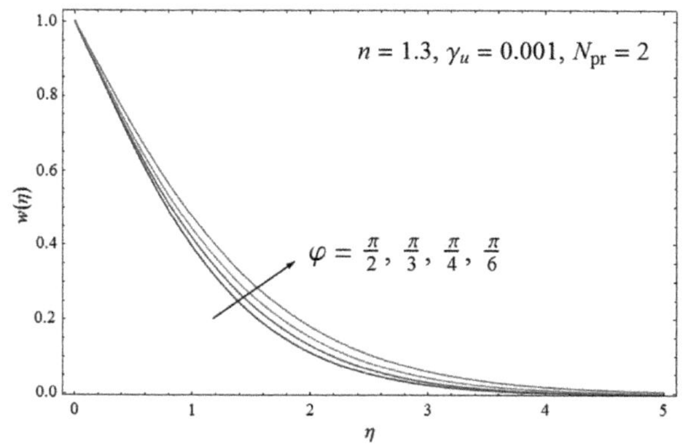

FIGURE 4.31 Temperature profiles of dilatant fluid for various values of inclined angle φ.

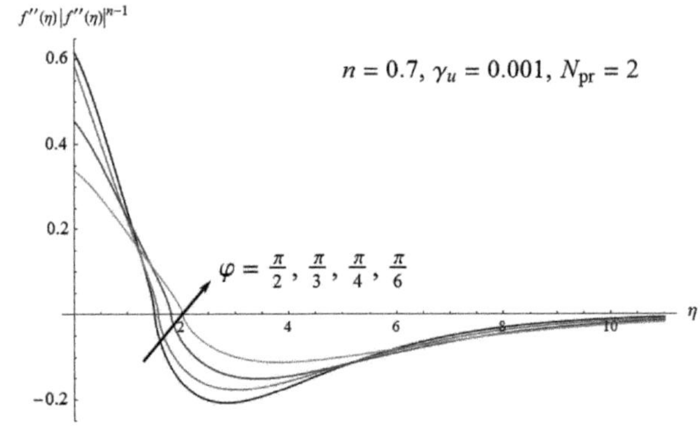

FIGURE 4.32 Shear stress profiles of pseudoplastic fluid for various values of inclined angle φ. Shear stress is reduced in positive orientation (above the η axis) and increases in negative orientation (below the η axis) within the whole boundary layer.

temperature boundary layers of dilatant fluid are thinner than those of pseudoplastic fluid for each fixed angle φ.

The shear stress profiles of power law fluid ($|f''|^{n-1}f''$) with $n = 0.7$ and $n = 1.3$ for various angles of φ are presented in Figs. 4.32 and 4.33. The shear stress changes its direction, i.e., the shear stress can be positive or negative (the plus or minus represents the direction). It implies that the velocity increases first to arrive at its maximum and then falls back. Meanwhile, the horizontal position where the shear stress equals zero moves forward (close to the plate) with the increase in φ, which is consistent with the moving direction of the horizontal position of velocity peak. As $\eta \to \infty$, both velocity and shear stress profiles tend toward zero, which indicates the behavior nearing the edge of the

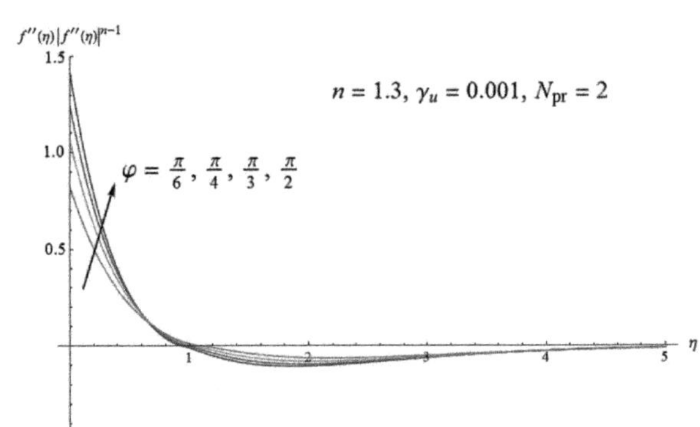

FIGURE 4.33 Shear stress profiles of dilatant fluid for various values of inclined angle φ. Obviously the reducing rate is larger than that of pseudoplastic fluid in positive orientation but the increasing rate is smaller than that of pseudoplastic fluid in negative orientation.

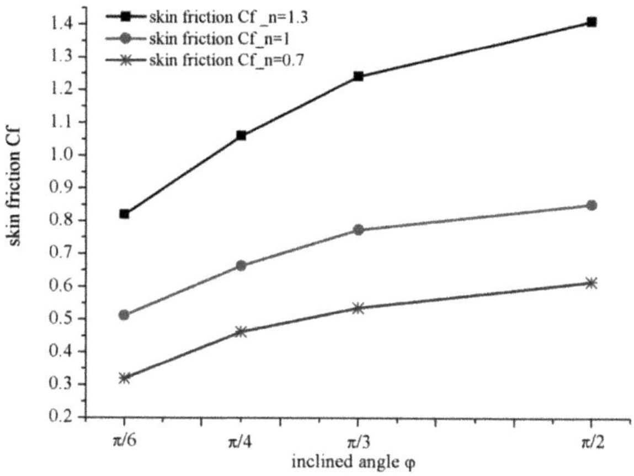

FIGURE 4.34 Skin friction profiles of different power law fluids for various values of φ with fixed parameters $\gamma_u = 0.001$ and $N_{pr} = 2$.

boundary layer. The skin friction C_f and local Nu of a power law fluid are presented in Figs. 4.34 and 4.35, respectively.

We see that both parameters increase as the value of φ increases. The increasing rate of C_f and Nu for fluid $n = 1.3$ is larger than the case of $n = 0.7$ for every fixed angle φ.

4.4.5.3 Effects of Velocity Ratio Coefficient γ_u

According to the assumptions for velocity of a moving conveyor $U_w = \gamma_u \sqrt{g\,\beta\,X\,(T_w - T_\infty)} = \gamma_u U_m^{\perp}$, the velocity ratio $\gamma_u = U_w / U_m^{\perp}$ is an

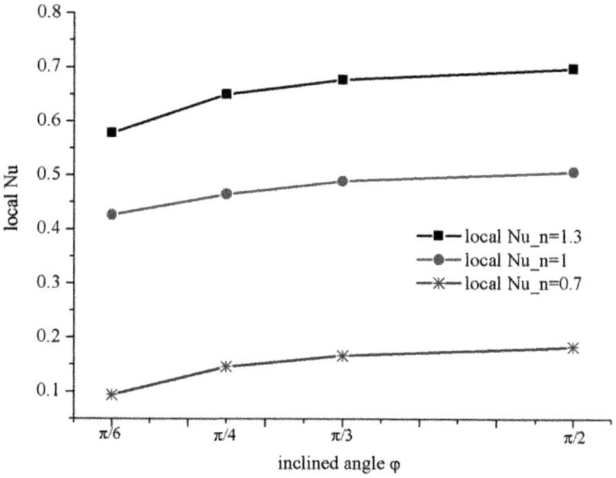

FIGURE 4.35 Local Nusselt number profiles of different power law fluids for various values of φ with fixed parameters $\gamma_u = 0.001$ and $N_{pr} = 2$.

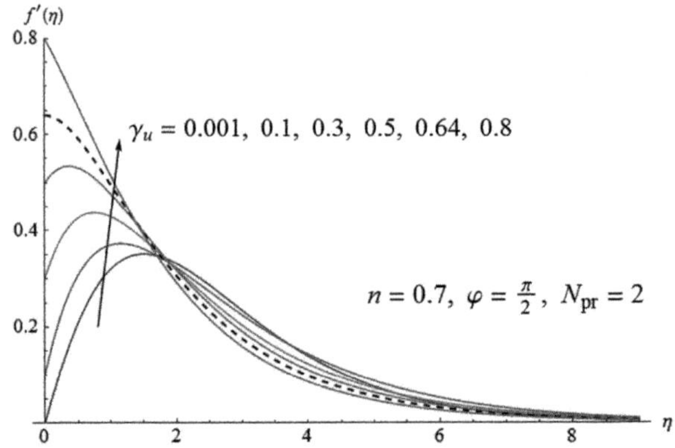

FIGURE 4.36 Velocity profiles of pseudoplastic fluid for various values of the velocity ratio. The critical velocity ratio is $\gamma_c = 0.64$ corresponding to dashed line, which means the skin friction is zero. If the velocity ratio continues to increase as $\gamma_u = 0.8$, the moving conveyor will govern the convection system instead of thermal buoyancy.

important parameter associating the control variable of the velocity of a conveyor with the mean velocity of fluid flow within the boundary layer. We discuss the effects of γ_u on different power law exponents for $n = 0.7$ and $n = 1.3$ along the vertical plate, such as the velocity profiles for various γ_u in Figs. 4.36 and 4.37, respectively.

It is noticed that the increase in ratio γ_u can gradually break up the convection boundary layer dominated by thermal buoyancy, making the effects of moving conveyor become a principal factor. Observing the behavior of power law fluid with $n = 0.7$ and $n = 1.3$ in Figs. 4.36 and 4.37, we find a different

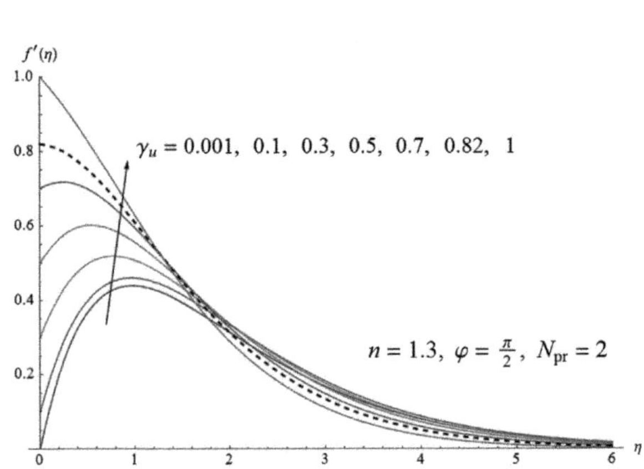

FIGURE 4.37 Velocity profiles of dilatant fluid for various values of the velocity ratio. The critical velocity ratio is $\gamma_c = 0.82$ corresponding to the dashed line, which means the skin friction is zero. If the velocity ratio continues to increase as $\gamma_u = 1$, the moving conveyor will govern the convection system instead of thermal buoyancy.

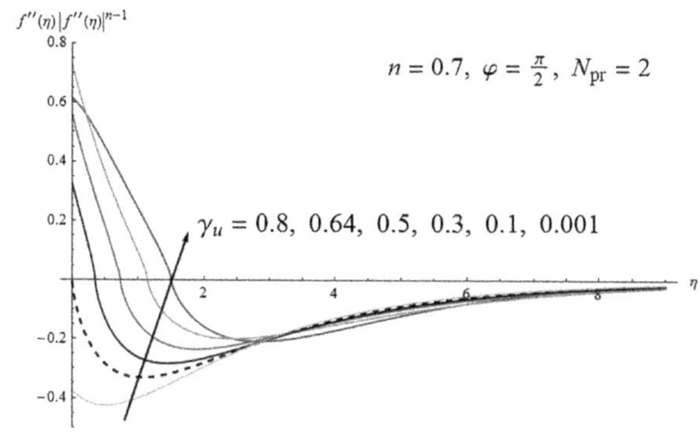

FIGURE 4.38 Shear stress profiles of pseudoplastic fluid for various values of the velocity ratio. The dashed line (skin friction is zero) corresponds to the critical velocity ratio, $\gamma_c = 0.64$.

transition ratio interval but the same small ratio, $\gamma_u = 0.001$, between these two kinds of power law fluid. For a pseudoplastic fluid ($n = 0.7$), the transition ratio is $0.1 \leq \gamma_u \leq 0.64$, and for a dilatant fluid ($n = 1.3$), the transition ratio is $0.1 \leq \gamma_u \leq 0.82$. It is obvious that transition ratio interval expands with an increase in n from 0.7 to 1.3. We are interested in investigating the critical values of velocity ratio so that thermal buoyancy is no longer dominant for a power law fluid.

Figs. 4.38 and 4.39 show the shear stress profiles for a power law fluid with $n = 0.7$ and $n = 1.3$, respectively. The figures indicate that shear stress (the positive value) decreases gradually until it becomes negative with an increase in γ_u from a transition ratio to a large ratio.

The critical ratio corresponds to the case in which shear stress at the surface (skin friction) equals zero, which indicates that the shear stress always keeps a negative orientation (the whole shear stress profile just falls below the

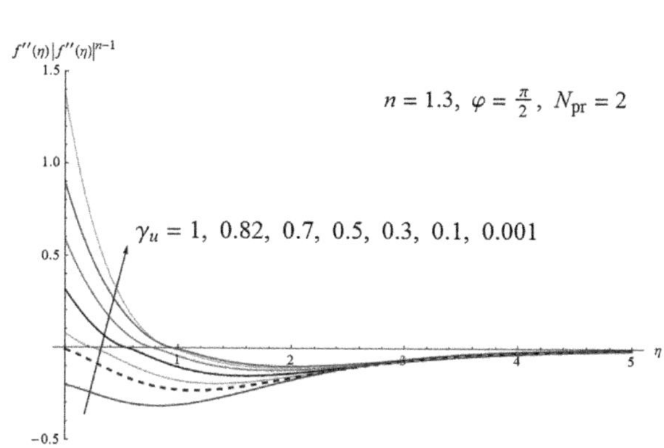

FIGURE 4.39 Shear stress profiles of a dilatant fluid for various values of the velocity ratio. The dashed line (skin friction is zero) corresponds to the critical velocity ratio $\gamma_c = 0.82$.

η axis) as shown in Figs. 4.38 and 4.39, i.e., the velocity falls back directly rather than increases first then falls back (the obvious peak). We obtain a critical ratio about $\gamma_c = 0.64$, 0.79, and 0.82 for power law fluid $n = 0.7$, $n = 1$, and $n = 1.3$, respectively, along the vertical plate, which shows that critical ratios increase with an increase in power law exponent n. It means that mixed convection of a pseudoplastic fluid ($n = 0.7$) tends to be affected more easily than a Newtonian fluid because of the relatively small critical ratio ($\gamma_c = 0.64$), which will be able to weaken the influences of thermal buoyancy completely; conversely, it is relatively difficult for a dilatant fluid ($n = 1.3$) during this process owing to its relatively large critical ratio ($\gamma_c = 0.82$).

Fig. 4.40A–C illustrates the effects of γ_u on skin friction with $n = 0.7$, $n = 1$, and $n = 1.3$, respectively. Skin friction first decreases to zero and then increases in a reverse manner (the minus means orientation) with an increase in γ_u from the transition ratio to a large one. We also find that the skin friction of fluid with $n = 0.7$ decreases faster than a Newtonian fluid and it is slowest

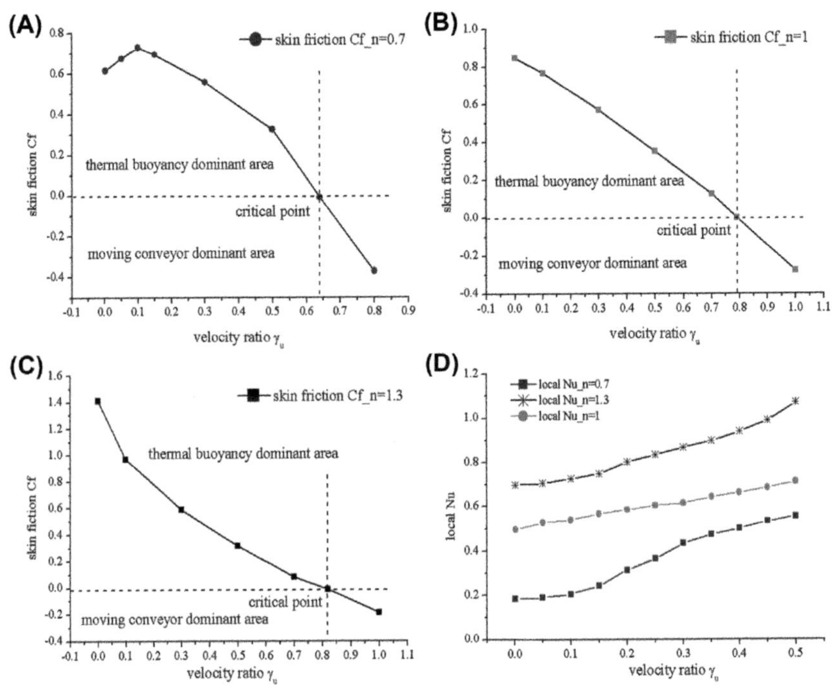

FIGURE 4.40 (A–C) represent skin friction C_f of different power law fluids for various values of γ_u, the critical point (the skin friction is zero corresponding to critical velocity ratio), which means that the thermal buoyancy has dominant effects above it but the moving conveyor has dominant effects below it within the whole boundary layer. (D) shows the local Nusselt number of different power law fluids for various values of γ_u, which demonstrates the evident nonlinear growth for a non-Newtonian fluid with $n = 0.7$ and $n = 1.3$, but linear growth for a Newtonian fluid.

for fluid with $n = 1.3$. Fig. 4.40D shows that the local Nu increases with an increasing in γ_u for a different power law fluid with power law exponent $n = 0.7$, $n = 1$, and $n = 1.3$, and has approximate linear growth for a Newtonian fluid but nonlinear growth for a power law fluid, $n = 0.7$ and $n = 1.3$.

4.5 MAGNETOHYDRODYNAMIC THERMOSOLUTAL MARANGONI CONVECTION IN POWER LAW FLUID

This section studies MHD thermosolutal Marangoni convection heat and mass transfer in power law non-Newtonian fluids driven by power law temperature and power law concentration in which we assume that the surface tension varies linearly with both temperature and concentration, and the heat diffusion depends nonlinearly on both the temperature and the temperature gradient and the mass diffusion depends nonlinearly on both the concentration and the concentration gradient with modified Fourier's heat conduction law and Fick's diffusion law for a power law fluid. The governing equations are reduced to nonlinear ordinary differential equations by using suitable local similarity transformations. Approximate analytical solution is obtained using HAM. The transport characteristics of velocity, temperature, and concentration fields are analyzed in detail.

4.5.1 Background of the Problem

It is known that magnetic fields are widely used in many engineering applications and have received much attention from scholars. Magyari and Chamkha (2007, 2008) analytically and numerically investigated MHD thermosolutal Marangoni convection along a flat surface in the presence of heat generation or absorption effects. Kandasamy et al. (2005) studied nonlinear MHD flow with the heat and mass transfer characteristics of an incompressible, viscous, electrically conducting fluid on a vertical stretching surface with chemical reaction and thermal stratification effects. Al-Mudhaf and Chamkha (2005) investigated the problem of steady, laminar, thermosolutal Marangoni convection flow of an electrically conducting fluid along a vertical permeable surface in the presence of a magnetic field, heat generation or absorption, and first-order chemical reaction effects. Lin et al. (2013) investigated MHD thermocapillary Marangoni convection heat transfer of an electrically conducting power law fluid driven by temperature gradient.

Philip (1961) proposed a model for some special diffusion process as $\partial C/\partial t = \nabla \cdot (AB)$, where C is the generalized concentration, t is time, A is a constant, and B is a function of the concentration gradient ∇C. When $B = \nabla C$, we get the classical diffusion equation (Fick's law). Let $B = |\nabla C|^{N-1}\nabla C (N > 0)$; we obtain the so-called N-diffusion equation. Wu (1985) and Wang (1990b) investigated a free boundary nonlinear problem for the N-diffusion equation; existence, uniqueness, and analyticity results were established. Later,

Pascal (1996) and Pascal and Pascal (1997a,b) presented a new convection-diffusion model by consider nonlinear molecular diffusion for mass transfer in a two-phase system. In the model, the molecular diffusion depends nonlinearly on both the concentration and the concentration gradient $B = C^m|\nabla C|^{N-1}\nabla C(N \geq 1)$.

This section focuses on Marangoni convection caused by surface tension that varies linearly with both temperature and concentration. We propose a new model for the constitutive relation of $B_1 = (T - T_\infty)^m|\nabla T|^{n-1}\nabla T$ and $B_2 = (C - C_\infty)^m|\nabla C|^{n-1}\nabla C$ $(0 < n \leq 1)$. The effects of power law fluid viscosity on temperature and concentration fields are taken into account by assuming that the temperature and concentration fields are similar to the velocity field (Zheng et al., 2000, 2002); the governing partial differential equations are transformed into ordinary differential equations using local similarity transformations. Approximate analytical solutions are obtained using HAM (Liao, 1992, 1995, 2012). The effects of power law index, temperature power law index, Marangoni number, Hartmann number, and thermosolutal surface tension ratio on the velocity and the temperature fields are graphically illustrated and analyzed.

4.5.2 Mathematical Formulation

Consider the two-dimensional, steady, laminar boundary layer flow of an incompressible power law fluid over a plate surface in the presence of surface tension resulting from temperature and concentration gradients. The Marangoni effect acts as a boundary condition on the governing equations for the flow. A magnetic field is applied in the Y direction. The laminar boundary layer equations of a viscous and incompressible fluid describing mass, momentum, energy, and concentration can be written as

$$\frac{\partial U}{\partial X} + \frac{\partial V}{\partial Y} = 0, \tag{4.144}$$

$$U\frac{\partial U}{\partial X} + V\frac{\partial U}{\partial Y} = \frac{\partial}{\partial Y}\left(\nu\left|\frac{\partial U}{\partial Y}\right|^{n-1}\frac{\partial U}{\partial Y}\right) - \frac{\Delta B^2}{\rho}U, \tag{4.145}$$

$$U\frac{\partial T}{\partial X} + V\frac{\partial T}{\partial Y} = \frac{\partial}{\partial Y}\left(\omega(T - T_\infty)^m\left|\frac{\partial T}{\partial Y}\right|^{n-1}\frac{\partial T}{\partial Y}\right), \tag{4.146}$$

$$U\frac{\partial C}{\partial X} + V\frac{\partial C}{\partial Y} = \frac{\partial}{\partial Y}\left(\lambda(C - C_\infty)^m\left|\frac{\partial C}{\partial Y}\right|^{n-1}\frac{\partial C}{\partial Y}\right), \tag{4.147}$$

where U and V are the velocity components in the X and Y directions, respectively. T is the temperature and C is the concentration. $\nu|\partial U/\partial Y|^{n-1}$ is the kinematic viscosity, $\alpha = \omega(T - T_\infty)^m|\partial T/\partial Y|^{n-1}$ is the thermal diffusivity, $D=\lambda(C-C_\infty)^m|\partial C/\partial Y|^{n-1}$ is the mass diffusivity, and $\Delta B=B_0X^{-b/2}(b=(3mn+n(2n-1))/(3mn+(n-1)(2n-1)))$ is the magnetic field. T_∞ and C_∞ are the constant, and denote the temperature and concentration of species far from the surface, respectively. The dependence of surface tension on temperature and concentration can be expressed as

$$\delta=\delta_0-\gamma_T(T-T_\infty)-\gamma_C(C-C_\infty), \tag{4.148}$$

where

$$\gamma_T=\partial\delta/\partial T|_C,\ \ \gamma_C=\partial\delta/\partial C|_T. \tag{4.149}$$

The boundary conditions for this problem can be written as

$$Y=0: \nu\left|\frac{\partial U}{\partial Y}\right|^{n-1}\frac{\partial U}{\partial Y}\bigg|_{Y=0}=-\frac{\partial\delta}{\partial T}\frac{\partial T}{\partial X}\bigg|_{Y=0}-\frac{\partial\delta}{\partial C}\frac{\partial C}{\partial X}\bigg|_{Y=0},\quad V(X,0)=0,$$
$$T(X,0)=T_\infty+AX^{1-n/(3mn+(n-1)(2n-1))},$$
$$C(X,0)=C_\infty+A^*X^{1-n/(3mn+(n-1)(2n-1))}, \tag{4.150}$$

$$Y\to\infty: U(X,\infty)=0,\quad T(X,\infty)=T_\infty,\quad C(X,\infty)=C_\infty. \tag{4.151}$$

Dimensionless variables are:

$$u=U/U_\Delta,\ \ v=V\nu^{-1/n+1}U_\Delta^{1-2n/n+1}L_\Delta^{n/n+1}, \tag{4.152}$$

$$x=X/L_\Delta,\ \ \ y=Y\nu^{-1/n+1}U_\Delta^{2-n/n+1}L_\Delta^{-1/n+1}, \tag{4.153}$$

$$t=(T-T_\infty)/T_\infty,\ \ \varphi=(C-C_\infty)/C_\infty. \tag{4.154}$$

Boundary layer Eqs. (4.144)–(4.147) then become:

$$\frac{\partial u}{\partial x}+\frac{\partial v}{\partial y}=0, \tag{4.155}$$

$$u\frac{\partial u}{\partial x}+v\frac{\partial u}{\partial y}=\frac{\partial}{\partial y}\left(\left|\frac{\partial u}{\partial y}\right|^{n-1}\frac{\partial u}{\partial y}\right)-\frac{B_0^2L_\Delta^{1-b}}{U_\Delta\rho}ux^{-b}, \tag{4.156}$$

$$u\frac{\partial t}{\partial x}+v\frac{\partial t}{\partial y}=\frac{1}{\Pr}\frac{\partial}{\partial y}\left(t^m\left|\frac{\partial t}{\partial y}\right|^{n-1}\frac{\partial t}{\partial y}\right), \tag{4.157}$$

$$u\frac{\partial\varphi}{\partial x}+v\frac{\partial\varphi}{\partial y}=\frac{1}{Sc}\frac{\partial}{\partial y}\left(\varphi^m\left|\frac{\partial\varphi}{\partial y}\right|^{n-1}\frac{\partial\varphi}{\partial y}\right), \tag{4.158}$$

$$\Pr=\frac{\nu}{\omega U_\Delta^{1-n}T_\infty^{m+n-1}},\ \ Sc=\frac{\nu}{\lambda U_\Delta^{1-n}C_\infty^{m+n-1}}. \tag{4.159}$$

Boundary conditions (4.150) and (4.151) are converted into:
At $y = 0$:

$$\left|\frac{\partial u}{\partial y}\right|^{n-1}\frac{\partial u}{\partial y}\bigg|_{y=0} = -q(\gamma_T A + \gamma_C A^*)\nu^{-1/n+1}U_\Delta^{-3n/n+1} \cdot L_\Delta^{-3n(m+n-1)/\left(3mn(n+1)+\left(n^2-1\right)(2n-1)\right)}x^{n(2-2n-3m)/(3mn+(n-1)(2n-1))}$$

$$v|_{y=0} = 0, \quad t|_{y=0} = AL_\Delta^{(1-n)/(3mn+(n-1)(2n-1))}\Big/T_\infty x^{(1-n)/(3mn+(n-1)(2n-1))},$$

$$\varphi|_{y=0} = A^* L_\Delta^{(1-n)/(3mn+(n-1)(2n-1))}\Big/C_\infty x^{(1-n)/(3mn+(n-1)(2n-1))}, \tag{4.160}$$

At y $= \infty$:

$$u|_{y=+\infty} = 0, \; t|_{y=+\infty} = 0, \; \varphi|_{y=+\infty} = 0. \tag{4.161}$$

Using the standard definition of the stream function $u = \partial\psi/\partial y$ and $v = -\partial\psi/\partial x$, the following similarity variables are introduced:

$$\psi(x,y) = C_1 x^{m/(3mn+(n-1)(2n-1))}f(\eta), \quad \eta = C_2 x^{1-n-2m/(3mn+(n-1)(2n-1))}y, \tag{4.162}$$

$$t(x,y) = AL_\Delta^{(1-n)/(3mn+(n-1)(2n-1))}\Big/T_\infty x^{(1-n)/(3mn+(n-1)(2n-1))}\theta(\eta), \tag{4.163}$$

$$\varphi(x,y) = A^* L_\Delta^{(1-n)/(3mn+(n-1)(2n-1))}\Big/C_\infty x^{(1-n)/(3mn+(n-1)(2n-1))}\phi(\eta), \tag{4.164}$$

where

$$C_1 = \left(\frac{1-n}{3mn+(n-1)(2n-1)}\gamma_T A\nu^{-1/(n+1)}U_\Delta^{-3n/(n+1)} \times L_\Delta^{-3n(m+n-1)/\left(3mn(n+1)+\left(n^2-1\right)(2n-1)\right)}\right)^{\frac{2n-1}{3n}}, \tag{4.165}$$

$$C_2 = \left(\frac{1-n}{3mn+(n-1)(2n-1)}\gamma_T A\nu^{-1/(n+1)}U_\Delta^{-3n/(n+1)} \times L_\Delta^{-3n(m+n-1)/\left(3mn(n+1)+\left(n^2-1\right)(2n-1)\right)}\right)^{\frac{2-n}{3n}}. \tag{4.166}$$

The boundary layer Eqs. (4.155)–(4.158) reduce to the following local similarity solution ordinary differential equations:

$$\left(|f''|^{n-1}f''\right)' + \frac{m+n-1}{3mn+(n-1)(2n-1)}f'^2 + \frac{m}{3mn+(n-1)(2n-1)}ff'' - B^2 f'(\eta) = 0, \tag{4.167}$$

$$\left(\theta^m|\theta'|^{n-1}\theta'\right)' + \frac{(n-1)M}{3mn+(n-1)(2n-1)}f'\theta + \frac{mM1}{3mn+(n-1)(2n-1)}f\theta' = 0, \tag{4.168}$$

$$\left(\phi^m|\phi'|^{n-1}\phi'\right)' + \frac{(n-1)M2}{3mn+(n-1)(2n-1)}f'\phi + \frac{mM2}{3mn+(n-1)(2n-1)}f\phi' = 0, \tag{4.169}$$

$$\begin{aligned} M_1 &= \nu T_\infty^{m+n-1}C_1 \Big/ \omega U^{1-n}T_\infty^{m+n-1}A^{m+n-1}L_\Delta^{(m+n-1)q}C_2^n \\ &= C_1C_2^{-n}A^{1-m-n}L_\Delta^{(n-1)(m+n-1)/(3mn+(n-1)(2n-1))}U^{n-1}\nu \Big/ \omega, \end{aligned} \tag{4.170}$$

$$\begin{aligned} M_2 &= \nu C_\infty^{m+n-1}C_1 \Big/ \lambda U^{1-n}C_\infty^{m+n-1}A^{*m+n-1}L_\Delta^{(m+n-1)q}C_2^n \\ &= C_1C_2^{-n}A^{*1-m-n}L_\Delta^{(n-1)(m+n-1)/(3mn+(n-1)(2n-1))}U^{n-1}\nu \Big/ \lambda, \end{aligned} \tag{4.171}$$

$$B^2 = B_0^2L_\Delta^{1-b} \Big/ C_1C_2U_\Delta\rho, \quad b = (3mn + n(2n-1))/3mn + (n-1)(2n-1). \tag{4.172}$$

The boundary conditions (4.160)–(4.161) are converted into the following:

$$\begin{aligned} &f''(0) = -(1+r)^{1/n},\ f(0) = 0,\ f'(\infty) = 0,\ \theta(0) = 1,\ \theta(\infty) = 0, \\ &\quad \phi(0) = 1,\ \phi(\infty) = 0, \end{aligned} \tag{4.173}$$

where $r = A^*\gamma_C/A\gamma_T$ is the thermosolutal surface tension ratio.

We now determine the surface velocity $u(x,0)$ as

$$u(x,0) = C_1C_2xf'(0), \tag{4.174}$$

The local Nusselt number Nu_x and Sherwood number Sh_x are defined as

$$Nu_x = xq_w(x)/\alpha[t(x,0) - t(x,+\infty)], \quad Sh_x = xc_w(x)/D[c(x,0) - c(x,+\infty)], \tag{4.175}$$

where α is the thermal conductivity of the fluid, D is mass diffusivity, and $q_w(x)$ and $c_w(x)$ are the heat flux and mass flux from the interface and are given by

$$q_w(x) = -\alpha(\partial t/\partial y)|_{y=0}, \quad c_w(x) = -D(\partial c/\partial y)|_{y=0}. \tag{4.176}$$

We obtain the generalized local Nusselt number and the generalized Sherwood number

$$Nu = Nu_x/C_2x = -\theta'(0), \quad Sh = Sh_x/C_2x = -\phi'(0). \tag{4.177}$$

4.5.3 Homotopy Analysis Method Solutions

The velocity, temperature, and concentration distribution, $f(\eta)$, $\theta(\eta)$, and $\phi(\eta)$, can be expressed by the set of base functions

$$\{\eta^k \exp(-n\eta) | k \geq 0, n \geq 0\}, \tag{4.178}$$

in the form

$$f(\eta) = \sum_{n=0}^{\infty}\sum_{k=0}^{\infty} a_{m,n}^k \eta^k \exp(-n\eta), \tag{4.179}$$

$$\theta(\eta) = \sum_{n=0}^{\infty}\sum_{k=0}^{\infty} b_{m,n}^k \eta^k \exp(-n\eta), \tag{4.180}$$

$$\phi(\eta) = \sum_{n=0}^{\infty}\sum_{k=0}^{\infty} c_{m,n}^k \eta^k \exp(-n\eta), \tag{4.181}$$

in which $a_{m,n}^k$, $b_{m,n}^k$, and $d_{m,n}^k$ are the coefficients. Based on the rule of solution expressions by Eqs. (4.36)–(4.38), it is straightforward to choose

$$f_0(\eta) = (1+r)^{1/n}(1-\exp(-\eta)), \quad \theta_0(\eta) = \exp(-\eta), \quad \phi_0(\eta) = \exp(-\eta), \tag{4.182}$$

as initial approximations of $f(\eta)$, $\theta(\eta)$, and $\phi(\eta)$. Besides that, we select

$$L_f(\eta) = f''' + f'', \quad L_\theta(\eta) = f'', \quad L_\phi(\eta) = f'', \tag{4.183}$$

as auxiliary linear operators satisfying the following properties:

$$L_f(C_1 + C_2\eta + C_3\exp(-\eta)) = 0, \quad L_\theta(C_4 + C_5\eta) = 0, \quad L_\phi(C_6 + C_7\eta) = 0. \tag{4.184}$$

where $C_i(i = 1, 2, \cdots, 7)$ are arbitrary constant. If $p(\in [0,1])$, $\hbar_i(i = f, \theta, \phi)$ and $H_i(i = f, \theta, \phi)$ are the embedding, nonzero auxiliary parameters and nonzero auxiliary functions, respectively, then the zeroth-order deformation problems are

$$(1-p)L_f(F(\eta;p) - f_0(\eta)) = p\hbar_1 N_f[F(\eta;p)], \tag{4.185}$$

$$(1-p)L_\theta(\Phi(\eta;p) - \theta_0(\eta)) = p\hbar_2 N_\theta[\Phi(\eta;p)], \tag{4.186}$$

$$(1-p)L_\phi(\Psi(\eta;p) - \phi_0(\eta)) = p\hbar_3 N_f[\Psi(\eta;p)], \tag{4.187}$$

$$F(0;p) = 0, \quad F''(0;p) = -(1+\mathrm{r})^{1/n}, \quad F'(\infty;p) = 0, \tag{4.188}$$

$$\Phi(0;p) = 1, \quad \Phi(\infty;p) = 0, \quad \Psi(0;p) = 1, \quad \Psi(\infty;p) = 0. \tag{4.189}$$

The nonlinear operators N_f, N_θ, and N_ϕ are

$$N_f[F(\eta;p)] = \frac{\partial}{\partial\eta}\left[\left|\frac{\partial^2 F(\eta;p)}{\partial^2\eta}\right|^{n-1}\frac{\partial^2 F(\eta;p)}{\partial^2\eta}\right] + \frac{m+n-1}{3mn+(n-1)(2n-1)}\left[\frac{\partial F(\eta;p)}{\partial\eta}\right]^2$$
$$+\frac{m}{3mn+(n-1)(2n-1)}F(\eta;p)\frac{\partial^2 F(\eta;p)}{\partial^2\eta} - B^2\frac{\partial F(\eta;p)}{\partial^2\eta}, \tag{4.190}$$

$$N_\theta[F(\eta;p),\Phi(\eta;p)] = \frac{\partial}{\partial\eta}\left[\Phi^m(\eta;p)\left|\frac{\partial\Phi(\eta;p)}{\partial\eta}\right|^{n-1}\frac{\partial\Phi(\eta;p)}{\partial\eta}\right]$$
$$+\frac{n-1}{3mn+(n-1)(2n-1)}M1\frac{\partial F(\eta;p)}{\partial\eta}\Phi(\eta;p)$$
$$+\frac{m}{3mn+(n-1)(2n-1)}M1F(\eta;p)\frac{\partial\Phi(\eta;p)}{\partial\eta}, \tag{4.191}$$

$$N_\phi[F(\eta;p),\Psi(\eta;p)] = \frac{\partial}{\partial\eta}\left[\Psi^m(\eta;p)\left|\frac{\partial\Psi(\eta;p)}{\partial\eta}\right|^{n-1}\frac{\partial\Psi(\eta;p)}{\partial\eta}\right]$$
$$+\frac{n-1}{3mn+(n-1)(2n-1)}M2\frac{\partial F(\eta;p)}{\partial\eta}\Psi(\eta;p)$$
$$+\frac{m}{3mn+(n-1)(2n-1)}M2F(\eta;p)\frac{\partial\Psi(\eta;p)}{\partial\eta}. \tag{4.192}$$

Obviously, for $p=0$ and $p=1$, the above zeroth-order deformation equations have the solutions

$$F(\eta;0) = f_0(\eta)\ \ F(\eta;1) = f(\eta), \tag{4.193}$$

$$\Phi(\eta;0) = \theta_0(\eta),\ \ \Phi(\eta;1) = \theta(\eta), \tag{4.194}$$

$$\Psi(\eta;0) = \phi_0(\eta),\ \ \Psi(\eta;1) = \phi(\eta). \tag{4.195}$$

Expanding $F(\eta;p)$, $\Phi(\eta;P)$, and $\Psi(\eta;P)$ in Tailor's series with respect to p, we can write

$$F(\eta;p) = f_0(\eta) + \sum_{m=1}^{\infty} f_m(\eta)\eta^m,\ \ f_m(\eta) = \frac{1}{m!}\frac{\partial^m F(\eta;p)}{\partial^m p}\bigg|_{p=0}, \tag{4.196}$$

$$\Phi(\eta;P) = \theta_0(\eta) + \sum_{m=1}^{\infty} \theta_m(\eta)\eta^m,\ \ \theta_m(\eta) = \frac{1}{m!}\frac{\partial^m \Phi(\eta;p)}{\partial^m p}\bigg|_{p=0}, \tag{4.197}$$

$$\Psi(\eta;P)=\phi_0(\eta)+\sum_{m=1}^{\infty}\phi_m(\eta)\eta^m,\ \ \phi_m(\eta)=\frac{1}{m!}\frac{\partial^m\Psi(\eta;p)}{\partial^m p}\bigg|_{p=0}. \tag{4.198}$$

The convergence of the series in Eqs. (4.185)–(4.187) depends on $\hbar_f$, $\hbar_\theta$, and $\hbar_\phi$. Assuming that $\hbar_f$, $\hbar_\theta$, and $\hbar_\phi$ are selected in such a way that the series in Eqs. (4.185)–(4.187) converge at $p=1$, then owing to Eqs. (4.193) and (4.194), we have

$$f(\eta)=F(\eta;1)=f_0(\eta)+\sum_{m=1}^{\infty}f_m(\eta), \tag{4.199}$$

$$\theta(\eta)=\Phi(\eta;1)=\theta_0(\eta)+\sum_{m=1}^{\infty}\theta_m(\eta), \tag{4.200}$$

$$\phi(\eta)=\Psi(\eta;1)=\phi_0(\eta)+\sum_{m=1}^{\infty}\phi_m(\eta). \tag{4.201}$$

Differentiating the zeroth-order deformation Eqs. (4.185)–(4.187) k times with respect to p, then setting $p=0$, and finally dividing by $k!$, the kth-order deformation equations can be expressed as

$$L_f(f_k(\eta)-\chi_k f_{k-1}(\eta))=\hbar_f H_f(\eta)R_{f,k}(\eta), \tag{4.202}$$

$$L_\theta(\theta_k(\eta)-\chi_k\theta_{k-1}(\eta))=\hbar_\theta H_\theta(\eta)R_{\theta,k}(\eta), \tag{4.203}$$

$$L_\phi(\phi_k(\eta)-\chi_k\phi_{k-1}(\eta))=\hbar_\phi H_\phi(\eta)R_{\phi,k}(\eta), \tag{4.204}$$

$$f_k(0)=f_k''(0)=f_k'(\infty)=0,\ \ \theta_k(0)=\theta_k(\infty)=0,\ \ \phi_k(0)=\phi_k(\infty)=0. \tag{4.205}$$

$$R_{f,k}(\eta)=\frac{1}{(k-1)!}\frac{\partial^{k-1}N_f[F(\eta;p)]}{\partial^{k-1}p}\bigg|_{p=0}, \tag{4.206}$$

$$R_{\theta,k}(\eta)=\frac{1}{(k-1)!}\frac{\partial^{k-1}N_\theta[F(\eta;p),\Phi(\eta;p)]}{\partial^{k-1}p}\bigg|_{p=0}, \tag{4.207}$$

$$R_{\phi,k}(\eta)=\frac{1}{(k-1)!}\frac{\partial^{k-1}N_\phi[F(\eta;p),\Phi(\eta;p)]}{\partial^{k-1}p}\bigg|_{p=0}, \tag{4.208}$$

$$\chi_k=\begin{cases}0, k\le 1\\ 1, k>1\end{cases}. \tag{4.209}$$

The general solutions of Eqs. (4.179) and (4.180) are

$$f_k(\eta) = f_k^*(\eta) + C_1 + C_2 \exp(\eta) + C_3 \exp(-\eta), \tag{4.210}$$

$$\theta_k(\eta) = \theta_k^*(\eta) + C_4 \exp(\eta) + C_5 \exp(-\eta), \tag{4.211}$$

$$\phi_k(\eta) = \phi_k^*(\eta) + C_6 \exp(\eta) + C_7 \exp(-\eta), \tag{4.212}$$

where $f_k^*(\eta)$, $\theta_k^*(\eta)$, and $\phi_k^*(\eta)$ denote the solutions of Eqs. (4.202)–(4.204), and the integral constants $C_i(i = 1, \ldots, 7)$ are determined by the boundary condition (4.205). Therefore it is easy to solve the linear nonhomogeneous Eqs. (4.202)–(4.204) by using Mathematica one after the other in the order $k = 1, 2, 3 \ldots$.

4.5.4 Results and Discussion

Figs. 4.41 and 4.42 show the effects of various parameters on the velocity, temperature, and concentration fields. Fig. 4.41 illustrates the effects of temperature and concentration power law index m on dimensionless stream function f, dimensionless velocity f', and dimensionless shear stress $|f''|^{n-1}f''$ when $n = 1/3$, $M_1 = M_2 = 1$, $r = 0$, and $B = 0$.

We can see that f increases and converges more slowly as m increases, while f' and the absolute value of $|f''|^{n-1}f''$ decrease as the location similarity variable η increases (the values of $|f''|^{n-1}f''$ are negative). As m increases, $f'(0)$ increases. The effects of temperature and concentration power law index m on dimensionless temperature θ for $n = 1/3$, $M_1 = M_2 = 1$, $r = 0$, and $B = 0$ are shown in Fig. 4.42. The figure shows that the surface temperature gradient induces surface flow and the temperature profiles decreases with the distance

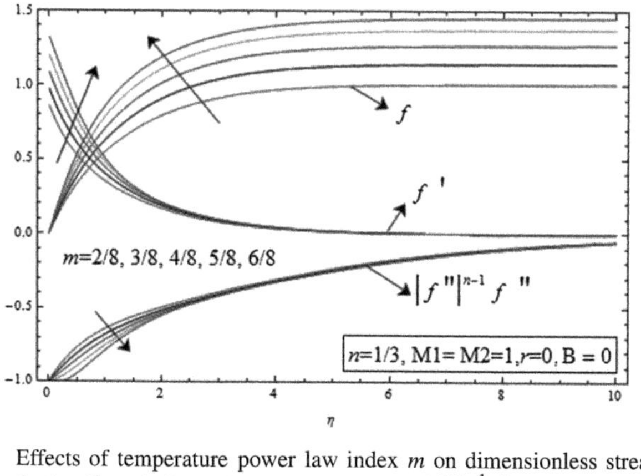

FIGURE 4.41 Effects of temperature power law index m on dimensionless stream function f, dimensionless velocity f', and dimensionless shear stress $|f''|^{n-1}f''$ for $n = 1/3$, $M_1 = M_2 = 1$, $r = 0$, and $B = 0$.

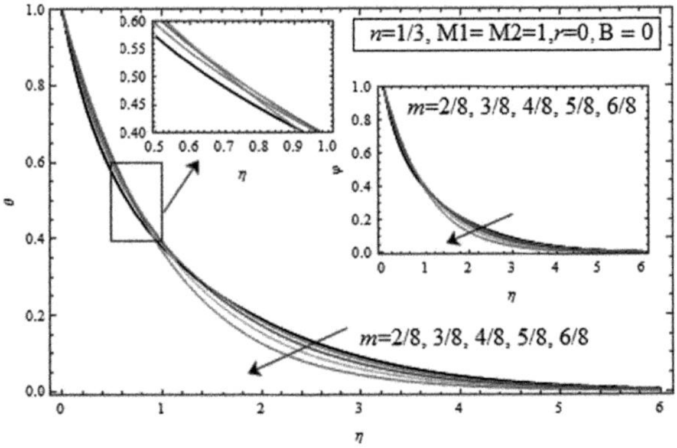

FIGURE 4.42 Effects of temperature power law index m on dimensionless temperature θ and dimensionless concentration ϕ for $n = 1/3$, $M_1 = M_2 = 1$, $r = 0$, and $B = 0$.

away from the contact surface. As m increases, the dimensionless temperature θ decreases. The effects of m on ϕ is similar to the effects of m on θ.

Fig. 4.43 exhibits the effects of power law index n on dimensionless stream function f, dimensionless velocity f', and dimensionless shear stress $|f''|^{n-1}f''$ when $m = 1/2$, $M_1 = M_2 = 1$, $r = 0$, and $B = 0.6$. It can been seen that f increases and converges more slowly as n increases, while f' and the absolute value of $|f''|^{n-1}f''$ (the values of $|f''|^{n-1}f''$ are negative) decrease as the location similarity variable η increases. $f'(0)$ increases as n increases. $|f''|^{n-1}f''$ decreases as n increases and the values of $|f''|^{n-1}f''$ are not positive. The effects of power law index n on dimensionless temperature θ when $m = 1/2$, $M_1 = M_2 = 1$, $r = 0$, and $B = 0.6$ are shown in Fig. 4.44. The figure represents

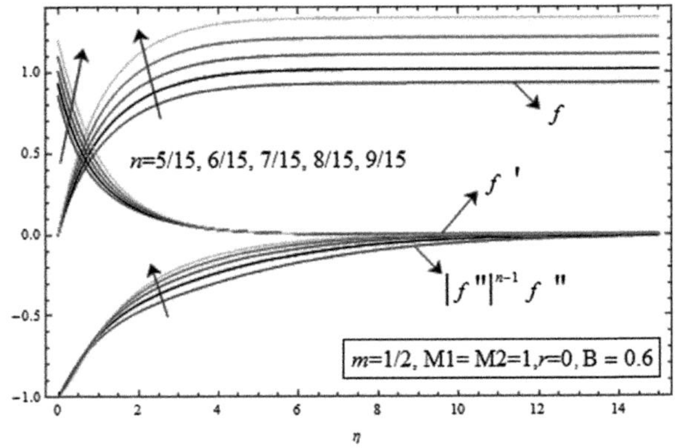

FIGURE 4.43 Effects of power law index n on dimensionless stream function f, dimensionless velocity f', and dimensionless shear stress $|f''|^{n-1}f''$ for $m = 1/2$, $M_1 = M_2 = 1$, $r = 0$, and $B = 0.6$.

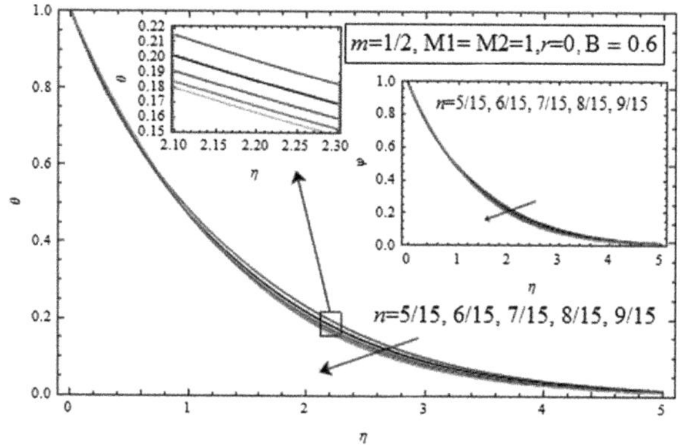

FIGURE 4.44 Effects of power law index n on dimensionless temperature θ and dimensionless concentration ϕ for $m = 1/2$, $M_1 = M_2 = 1$, $r = 0$, and $B = 0.6$.

that the surface temperature gradient induces surface flow; the temperature profiles decreases with the distance away from the contact surface. It can be observed that θ decreases as n increases. Obviously, the effects of n on ϕ are similar to the effects of n on θ.

Fig. 4.45 illustrates the effects of Hartmann number B on dimensionless stream function f, dimensionless velocity f', and dimensionless shear stress $|f''|^{n-1}f''$ when $m = 1/2$, $n = 1/3$, $M_1 = M_2 = 1$, and $r = 0$. From the figures we can see that f decreases and converges faster as B increases, and f' decreases as the location similarity variable η increases. It can also be observed that $|f''|^{n-1}f''$ decreases as η increases and the values of $|f''|^{n-1}f''$ are not positive. The effects of Hartmann number Bon dimensionless temperature θ when $m = 1/2$, $n = 1/3$, $M_1 = M_2 = 1$, and $r = 0$ and $B = 0.6$ are shown in Fig. 4.46. We

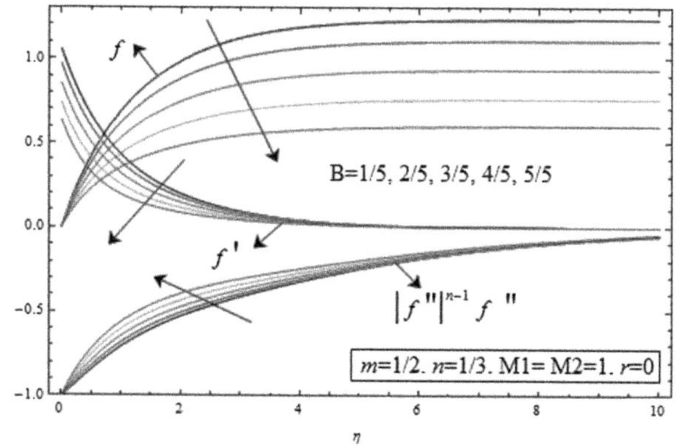

FIGURE 4.45 Effects of Hartmann number B on dimensionless stream function f, dimensionless velocity f', and dimensionless shear stress $|f''|^{n-1}f''$ for $m = 1/2$, $n = 1/3$, $M_1 = M_2 = 1$, and $r = 0$.

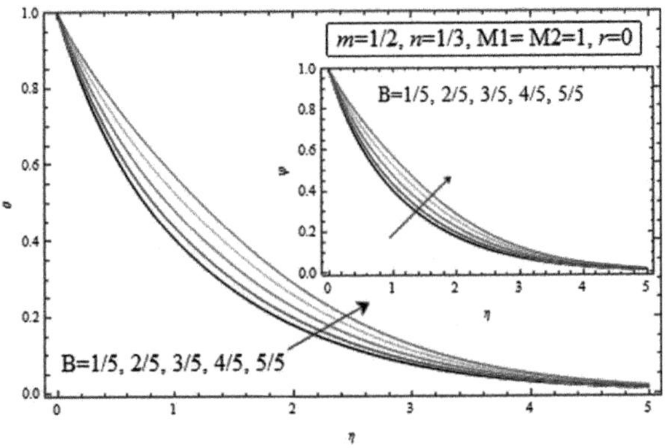

FIGURE 4.46 Effects of Hartmann number B on dimensionless temperature θ and dimensionless concentration ϕ for $m = 1/2$, $n = 1/3$, $M_1 = M_2 = 1$, and $r = 0$.

can see that θ decreases as η increases. We also observe that θ increases as B increases. Obviously, the effects of B on ϕ are similar to the effects of B on θ.

Fig. 4.47 illustrates the thermosolutal surface tension ratio r on dimensionless stream function f, dimensionless velocity f', and dimensionless shear stress $|f''|^{n-1}f''$ when $m = 1/2$, $n = 1/3$, $M_1 = M_2 = 1$, and $B = 0.6$. Fig. 4.48 shows that f and f' increase as r increases. It can be observed that f' and $|f''|^{n-1}f''$ decrease as η increases and the values of $|f''|^{n-1}f''$ are not positive. The effects of thermosolutal surface tension ratio$r$ on dimensionless temperature θ when $m = 1/2$, $n = 1/3$, $M_1 = M_2 = 1$, and $B = 0.6$ are shown in Fig. 4.48. It is seen that θ decreases as η increases. We also observe that θ decreases as r increases. Obviously, the effects of r on ϕ are similar to the

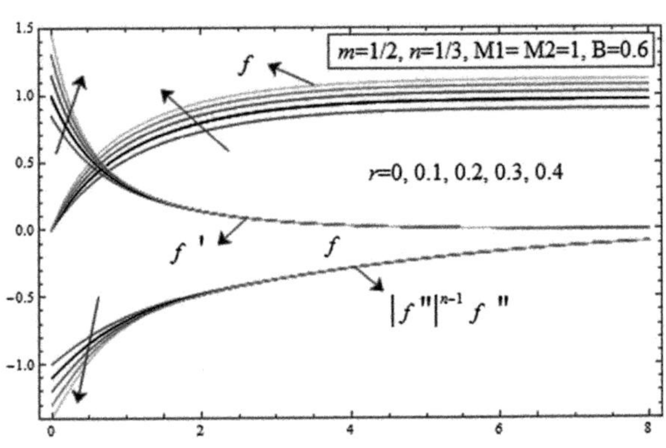

FIGURE 4.47 Effects of thermosolutal surface tension ratio r on dimensionless stream function f, dimensionless velocity f', and dimensionless shear stress $|f''|^{n-1}f''$ for $m = 1/2$, $n = 1/3$, $M_1 = M_2 = 1$, and $B = 0.6$.

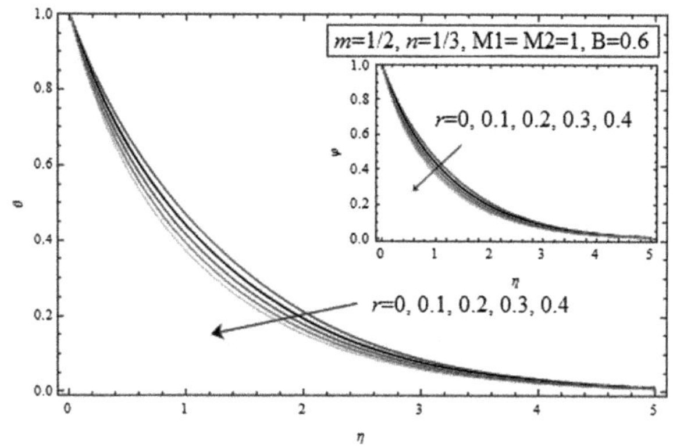

FIGURE 4.48 Effects of thermosolutal surface tension ratio r on dimensionless temperature θ and dimensionless concentration ϕ for $m = 1/2$, $n = 1/3$, $M_1 = M_2 = 1$, and $B = 0.6$.

effects of r on θ. Fig. 4.49 illustrates the effects of heat Marangoni number M_1 on temperature θ when $m = 1/2$, $n = 1/3$, $B = 0.6$, and $r = 0$. The temperature profiles strongly depend on M_1. We also observe that θ decreases as η increases, and θ decreases as r increases. The effects of mass Marangoni number M_2 on ϕ are similar to the effects of M_1 on θ.

Figs. 4.50–4.52 present the effects of various parameters on the generalized Nusselt number and generalized Sherwood number, respectively. Because the effects of various parameters on the generalized Sherwood number are similar to those on the generalized Nusselt number, here we will consider only the effects of various parameters on the generalized Nusselt number.

Fig. 4.50 illustrates the effects of Hartmann number B on generalized Nusselt number Nu with $n = 1/3$, $M_1 = 1$, and $r = 0$ and the effects of heat

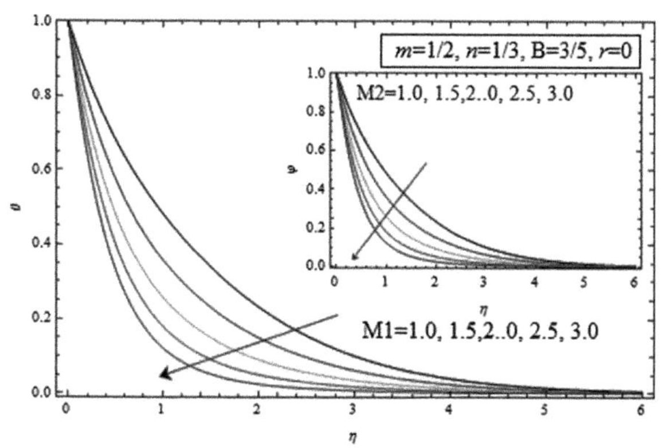

FIGURE 4.49 Effects of heat Marangoni number M_1 on dimensionless temperature θ and mass Marangoni number M_2 on dimensionless concentration ϕ for $m = 1/2$, $n = 1/3$, $r = 0$, and $B = 0.6$.

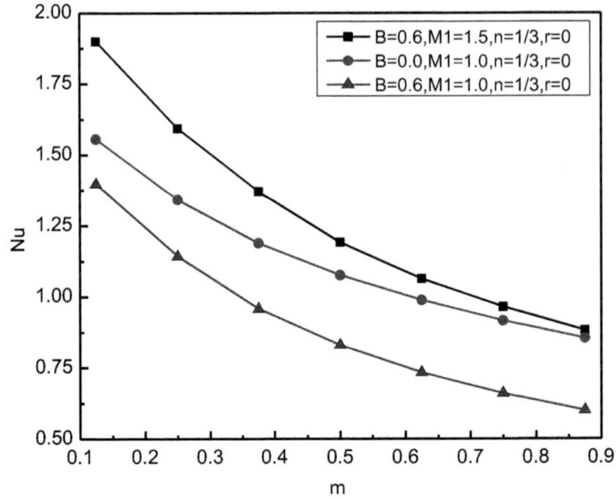

FIGURE 4.50 Generalized Nusselt number profiles of different Hartmann numbers and different Marangoni numbers for various values of temperature power law index m.

Marangoni number M_1 on generalized Nusselt number Nu with $n = 1/3$, $B = 0.6$, and $r = 0$ for various values of temperature power law index m. It is seen that the generalized Nusselt number of fluid with $B = 0.6$ decreases faster than the generalized Nusselt number of fluid with $B = 0.0$ as the temperature and concentration power law index m increases, which means that the

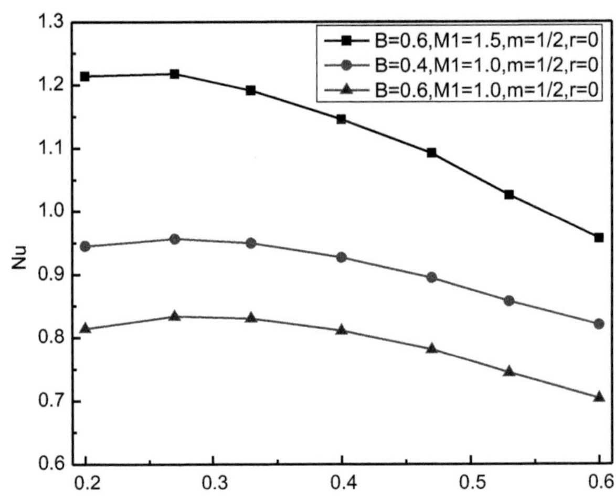

FIGURE 4.51 Generalized Nusselt number profiles of different Hartmann numbers and different Marangoni numbers for various values of power law index n.

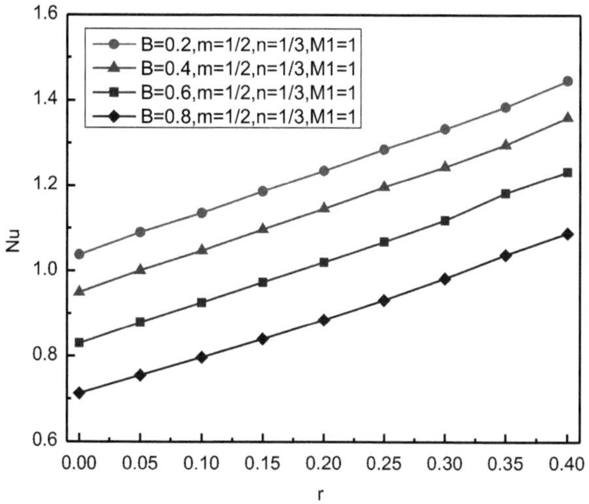

FIGURE 4.52 Generalized Nusselt number profiles of different Hartmann numbers for various values of thermosolutal surface tension ratio *r*.

Hartmann number will make the generalized Nusselt number fall faster. We also find the generalized Nusselt number of fluid with $M_1 = 1$ and $B = 0.6$ less than the generalized Nusselt number of fluid with $M_1 = 1.5$ for different temperature power law index m; their descent speed is similar.

Fig. 4.51 shows the effects of Hartmann number B on generalized Nusselt number Nu with $m = 1/2$, $M_1 = 1$, and $r = 0$ and the effects of heat Marangoni number M_1 on generalized Nusselt number Nu with $m = 1/2$, $B = 0.6$, and $r = 0$ for various values of power law index n. It is seen that the generalized Nusselt number of fluid first increases and then decreases with the increase in the power law index n. The generalized Nusselt number of fluid with $B = 0.4$ variation has a similar trend, with $B = 0.6$. We also find that the generalized Nusselt number of fluid with $M_1 = 1.5$ decreases faster than $M_1 = 1$ as the power law index n increases, which means that heat Marangoni number M_1 will make the generalized Nusselt number fall faster.

Fig. 4.52 presents the effects of Hartmann number B on generalized Nusselt number Nu with $m = 1/2$, $n = 1/3$, and $M_1 = 1$ for various values of thermosolutal surface tension ratio r. We can see that the generalized Nusselt number Nu increases with the increase in thermosolutal surface tension ratio r for different magnetofluids with $B = 0.2$, $B = 0.4$, $B = 0.6$, and $B = 0.8$. Their increment speeds are similar and have approximate linear growth. We also find that the generalized Nusselt number decreases as the Hartmann number increases for different temperature power law index m.

4.6 SUMMARY

In this chapter, using HAM, we present investigation for unsteady boundary layer flow and radiation heat transfer of an MHD fluid over a horizontal continuously stretching surface in an otherwise quiescent fluid medium in the presence of a nonuniform magnetic field, the flow and heat transfer of nanofluids over a porous rotating disk with velocity slip and temperature jump with three different types of nanoparticles (Cu, Al_2O_3, and CuO) with water as the base fluid, the mixed convection boundary layer heat transfer of a power law fluid over a moving conveyor along an inclined plate, and MHD thermosolutal Marangoni convection heat and mass transfer in power law non-Newtonian fluids driven by a power law temperature and power law concentration. The governing equations are reduced to nonlinear ordinary differential equations using suitable local similarity transformations. Approximate analytical solutions are obtained using HAM. The influence of the parameters on the velocity, temperature, and concentration fields are analyzed in detail.

REFERENCES

Abbas, Z., Hayat, T., Sajid, M., Asghar, S., 2008. Unsteady flow of a second grade fluid film over an unsteady streaching sheet. Mathematical and Computer Modelling 48, 518–526.

Abel, M.S., Mahesha, N., Tawade, J., 2009. Heat transfer in a liquid film over an unsteady stretching surface with viscous dissipation in presence of external magnetic field. Applied Mathematical Modelling 33, 3430–3441.

Acrivos, A., Shah, M.J., Petersen, E.E., 1960. Momentum and heat transfer in laminar boundary-layer flows of non-Newtonian fluids past external surfaces. AIChE Journal 6, 312–317.

Al-Mudhaf, A., Chamkha, A.J., 2005. Similarity solutions for MHD thermosolutal Marangoni convection over a flat surface in the presence of heat generation or absorption effects. Heat and Mass Transfer 42, 112–121.

Alves, L.S.B., Barletta, A., Hirata, S., Ouarzazi, M.N., 2014. Effects of viscous dissipation on the convective instability of viscoelastic mixed convection flows in porous media. International Journal of Heat and Mass Transfer 70, 586–598.

Anderssona, H.I., Aarseth, J.B., Dandapat, B.S., 2000. Heat transfer in a liquid film on an unsteady stretching surface. International Journal of Heat and Mass Transfer 43, 69–74.

Andersson, H.I., Irgens, F., 1988. Gravity-driven laminar film flow of power-law fluids along vertical walls. Journal of Non-Newtonian Fluid Mechanics 27, 153–172.

Asghar, S., Jalil, M., Hussan, M., Turkyilmazoglu, M., 2014. Lie group analysis of flow and heat transfer over a stretching rotating disk. International Journal of Heat and Mass Transfer 69, 140–146.

Ashorynejad, H.R., Mohamad, A.A., Sheikholeslami, M., 2013. Magnetic field effects on natural convection flow of a nanofluid in a horizontal cylindrical annulus using Lattice Boltzmann method. International Journal of Thermal Sciences 64, 240–250.

Attia, H.A., 2009. Steady flow over a rotating disk in porous medium with heat transfer. Nonlinear Analysis: Modelling and Control 14 (1), 21–26.

Attia, H.A., Aboul-Hassan, A.L., 2001. Effect of Hall current on the unsteady MHD flow due to a rotating disk with uniform suction or injection. Applied Mathematical Modelling 25 (12), 1089–1098.

Bachok, N., Ishak, A., Pop, I., 2011. Flow and heat transfer over a porous rotating disk in a nanofluid. Physica B: Condensed Matter 406 (9), 1767–1772.

Bahiraei, M., Hangi, M., 2015. Flow and heat transfer characteristics of magnetic nanofluids: a review. Journal of Magnetism and Magnetic Materials 374, 125–138.

Benton, E.R., 1966. On the flow due to a rotating disk. Journal of Fluid Mechanics 24 (4), 781–800.

Bharti, R.P., Chhabra, R.P., Eswaran, V., 2007. Steady forced convection heat transfer from a heated circular cylinder to power-law fluids. International Journal of Heat and Mass Transfer 50, 977–990.

Brackenridge, J.B., Gilbert, W.P., Interferometry, S., 1965. An optical method for determining temperature and velocity distributions in liquids. Journal of Applied Optics 4 (7), 819–821.

Chol, S.U.S., 1995. Enhancing thermal conductivity of fluids with nanoparticles. ASME-Publications-Fed 23, 199–206.

Chiam, T.C., 1995. Hydromagnetic flow over a surface stretching with a power-law velocity. International Journal of Engineering Science 33, 429–435.

Cochran, W.G., 1934. The flow due to a rotating disc. In: Mathematical Proceedings of the Cambridge Philosophical Society, vol. 30(3). Cambridge University Press, pp. 365–375.

Cortell, R., 2008. Effects of viscous dissipation and radiation on the thermal boundary layer over a nonlinearly stretching sheet. Physics Letters A 372, 631–636.

Crane, L.J., 1970. Flow past a stretching plate. Zeitschrift für Angewandte Mathematik und Physik 21, 645–647.

Denier, J.P., Dabrowski, P.P., 2004. On the boundary-layer equations for power-law fluids. Mathematical Physical and Engineering Sciences, the Royal Society 460 (2051), 3143–3158.

Donald Ariel, P., 2002. On the flow of power law fluid over a stretching sheet-techniques and solutions. Acta Mechanica 156, 13–27.

Ece, M.C., Büyük, E., 2002. Similarity solutions for free convection to power-law fluids from a heated vertical plate. Applied Mathematics Letters 15, 1–5.

El-Aziz, M.A., 2009. Radiation effect on the flow and heat transfer over an unsteady stretching sheet. International Communications in Heat and Mass Transfer 36, 521–524.

Elsherbiny, S.M., Ragab, E.H., 2013. Laminar natural convection in inclined rectangular cavities with a localized heat source. Alexandria Engineering Journal 52, 249–257.

Griffiths, P.T., Garrett, S.J., Stephen, S.O., 2014. The neutral curve for stationary disturbances in rotating disk flow for power-law fluids. Journal of Non-Newtonian Fluid Mechanics 213, 73–81.

Hassanien, I.A., Abdullah, A.A., Gorla, R.S.R., 1998. Flow and heat transfer in a power law fluid over a non-isothermal stretching sheet. Mathematical and Computer Modelling 28, 105–116.

Hong, C., Asako, Y., 2010. Some considerations on thermal boundary condition of slip flow. International Journal of Heat and Mass Transfer 53 (15), 3075–3079.

Hussaini, M.Y., Lakin, W.D., Nachman, A., 1987. On similarity solutions of a boundary layer problem with an upstream moving wall. SIAM Journal on Applied Mathematics 47, 699–709.

Jiji, L.M., Ganatos, P., 2010. Microscale flow and heat transfer between rotating disks. International Journal of Heat and Fluid Flow 31 (4), 702–710.

Kamatchi, R., Venkatachalapathy, S., 2015. Parametric study of pool boiling heat transfer with nanofluids for the enhancement of critical heat flux: a review. International Journal of Thermal Sciences 87, 228–240.

Kandasamy, R., Periasamy, K., Prabhu, K.K.S., 2005. Chemical reaction, heat and mass transfer on MHD flow over a vertical stretching surface with heat source and thermal stratification effects. International Journal of Heat and Mass Transfer 48, 4557–4561.

Kandelousi, M.S., 2014a. Effect of spatially variable magnetic field on ferrofluid flow and heat transfer considering constant heat flux boundary condition. The European Physical Journal Plus 129 (11), 1–12.

Kandelousi, M.S., 2014b. KKL correlation for simulation of nanofluid flow and heat transfer in a permeable channel. Physics Letters A 378 (45), 3331–3339.

Kármán, T., 1921. Über laminare und turbulente Reibung. Zeitschrift für Angewandte Mathematik und Mechanik 1 (4), 233–252.

Kherbeet, A.S., Mohammedb, H.A., Munisamya, K.M., Salman, B.H., 2014. The effect of step height of microscale backward-facing step on mixed convection nanofluid flow and heat transfer characteristics. International Journal of Heat and Mass Transfer 68, 554–566.

Kuiken, H.K., 1971. The effect of normal blowing on the flow near a rotating disk of infinite extent. Journal of Fluid Mechanics 47 (4), 789–798.

Li, B.T., Zheng, L.C., Zhang, X.X., 2011. Heat transfer in pseudo-plastic non-Newtonian fluids with variable thermal conductivity. Energy Conversion and Management 52, 355–358.

Liao, S.J., 1992. The Proposed Homotopy Analysis Technique for the Solution of Nonlinear Problems (Ph.D. thesis). Shanghai Jiao Tong University.

Liao, S.J., 1995. An approximate solution technique not depending on small parameters: a special example. International Journal of Non-Linear Mechanics 30 (3), 371–380.

Liao, S.J., 2012. Homotopy Analysis Method in Nonlinear Differential Equations. Higher Education Press, Beijing and Springer-Berlag (Berlin, Heidelberg).

Lin, Y.H., Zheng, L.C., Zhang, X.X., 2013. Magnetohydrodynamics thermocapillary Marangoni convection heat transfer of power-law fluids driven by temperature gradient. ASME Journal of Heat Transfer 135, 051702.

Lingwood, R.J., 1995. Absolute instability of the boundary layer on a rotating disk. Journal of Fluid Mechanics 299, 17–33.

Lomascolo, M., Colangelo, G., Milanese, M., Risi, A., 2015. Review of heat transfer in nanofluids: conductive, convective and radiative experimental results. Renewable and Sustainable Energy Reviews 43, 1182–1198.

Magyari, E., Chamkha, A.J., 2007. Exact analytical solutions for thermosolutal Marangoni convection in the presence of heat and mass generation or consumption. Heat and Mass Transfer 43, 965–974.

Magyari, E., Chamkha, A.J., 2008. Exact analytical results for the thermosolutal MHD Marangoni boundary layers. International Journal of Thermal Sciences 47, 848–857.

Navier, C., 1823. Mémoire sur les lois du mouvement des fluides. Mémoires de l'Académie Royale des Sciences de l'Institut de France 6, 389–440.

Oztop, H.F., Abu-Nada, E., 2008. Numerical study of natural convection in partially heated rectangular enclosures filled with nanofluids. International Journal of Heat and Fluid Flow 29 (5), 1326–1336.

Pascal, J.P., 1996. Effects of nonlinear diffusion in a two-phase system. Physica A 223, 99–112.

Pascal, J.P., Pascal, H., 1997a. Free convection in a non-Newtonian fluid saturated porous medium with lateral mass flux. International Journal of Non-Linear Mechanics 32 (3), 471–482.

Pascal, J.P., Pascal, H., 1997b. Non-linear effects on some unsteady non-Darcian flows through porous media. International Journal of Non-Linear Mechanics 32 (2), 361–376.

Pang, C., Lee, J.W., Kang, Y.T., 2015. Review on combined heat and mass transfer characteristics in nanofluids. International Journal of Thermal Sciences 87, 49–67.

Philip, J.R., 1961. n-diffusion. Australian Journal of Physics 14, 1–13.

Pop, I., 1993. Boundary layer flow at a three-dimensional stagnation point in power-law non-Newtonian fluids. International Journal of Heat and Fluid Flow 14, 408–412.

Pop, I., Rashidi, M., Gorla, R.S.R., 1991. Mixed convection to power-law type non-Newtonian fluids from a vertical wall. Polymer-Plastics Technology and Engineering Journal 30, 47–66.

Rashidi, M.M., Abelman, S., Freidooni Mehr, N., 2013. Entropy generation in steady MHD flow due to a porous rotating disk in a nanofluid. International Journal of Heat and Mass Transfer 62, 515–525.

Rashidi, M.M., Kavyani, N., Abelman, S., 2014. Investigation of entropy generation in MHD and slip flow over a porous rotating disk with variable properties. International Journal of Heat and Mass Transfer 70, 892–917.

Rashidi, M.M., Mohimanian Pour, S.A., Hayat, T., Obaidat, S., 2012. Analytic approximate solutions for steady flow over a rotating disk in porous medium with heat transfer by homotopy analysis method. Computers & Fluids 54, 1–9.

Rosca, N.C., Pop, I., 2013. Mixed convection stagnation point flow past a vertical flat plate with a second order slip: heat flux case. International Journal of Heat and Mass Transfer 65, 102–109.

Samanta, S., Guha, A., 2013. Magnetohydrodynamic free convection flow above an isothermal horizontal plate. Communications in Nonlinear Science and Numerical Simulation 18, 3407–3422.

Sakiadis, B.C., 1961. Boundary-layer behavior on continuous solid surfaces: I. Boundary-layer equations for two-dimensional and axisymmetric flow. AIChE Journal 7, 26–28.

Sarkar, J., Ghosh, P., Adil, A., 2015. A review on hybrid nanofluids: recent research, development and applications. Renewable and Sustainable Energy Reviews 43, 164–177.

Schlichting, H., 1979. Boundary Layer Theory. McGraw-Hill, New York, New York.

Schowalter, W.R., 1960. The application of boundary-layer theory to power-law pseudoplastic fluids: similar solutions. AIChE Journal 6, 24–28.

Sheikholeslami, M., 2014. Effect of uniform suction on nanofluid flow and heat transfer over a cylinder. Journal of the Brazilian Society of Mechanical Sciences and Engineering 1–11.

Sheikholeslami, M., Ganji, D.D., 2014. Ferrohydrodynamic and magnetohydrodynamic effects on ferrofluid flow and convective heat transfer. Energy 75, 400–410.

Sheikholeslami, M., Ganji, D.D., 2015. Nanofluid flow and heat transfer between parallel plates considering Brownian motion using DTM. Computer Methods in Applied Mechanics and Engineering 283, 651–663.

Sheikholeslami, M., Gorji-Bandpay, M., Ganji, D.D., 2012. Magnetic field effects on natural convection around a horizontal circular cylinder inside a square enclosure filled with nanofluid. International Communications in Heat and Mass Transfer 39 (7), 978–986.

Shevchuk, I.V., 2009. Convective Heat and Mass Transfer in Rotating Disk Systems. Springer Science & Business Media.

Sui, J.Z., Zheng, L.C., Zhang, X.X., Chen, G., 2015. Mixed convection heat transfer in power law fluids over a moving conveyor along an inclined plate. International Journal of Heat and Mass Transfer 85, 1023–1033.

Takhar, H.S., Chamkha, A.J., Nath, G., 1999. Unsteady flow and heat transfer on a semi-infinite flat plate with an aligned magnetic field. International Journal of Engineering Science 37, 1723–1736.

Tsou, F.K., Sparrow, E.M., Goldstein, R.J., 1967. Flow and heat transfer in the boundary-layer on a continuous moving surface. International Journal of Heat and Mass Transfer 10, 219–235.

Turkyilmazoglu, M., 2014. Nanofluid flow and heat transfer due to a rotating disk. Computers & Fluids 94, 139–146.

Turkyilmazoglu, M., 2015. Bödewadt flow and heat transfer over a stretching stationary disk. International Journal of Mechanical Sciences 90, 246–250.

Turkyilmazoglu, M., Senel, P., 2013. Heat and mass transfer of the flow due to a rotating rough and porous disk. International Journal of Thermal Sciences 63, 146–158.

Vajravelu, K.A., Hadjinicolaou, A., 1997. Convective heat transfer in an electrically conducting fluid at a strectching surface with a uniform free stream. International Journal of Engineering Science 35, 1237–1244.

Wang, C., 2006. Analytic solutions for a liquid film on an unsteady stretching surface. Heat and Mass Transfer 42, 759–766.

Wang, C.Y., 1990a. Liquid film on an unsteady stretching surface. Quarterly of Applied Mathematics 48, 601–610.

Wang, J.Y., 1990b. A free boundary problem for a generalized diffusions equation. Nonlinear Analysis 14 (8), 691–700.

Wu, Z.Q., 1985. A free boundary problem for degenerate quasilinear parabolic equations. Nonlinear Analysis: Theory, Methods & Applications 19 (9), 937–951.

Zhang, H., Zhang, X.X., Zheng, L.C., 2007. Numerical study of thermal boundary layer on a continuous moving surface in power law fluids. Journal of Thermal Science 16 (3), 243–247.

Zheng, L.C., Chen, C.X., Zhang, X.X., Gao, Y.T., 2010. Analyzing the flow and heat transfer of a power-law fluid over an unsteadily stretched surface using a modified homotopy perturbation method. International Journal of Nonlinear Science and Numerical Simulation 11 (10), 843–849.

Zheng, L.C., Ma, L.X., He, J.C., 2000. Bifurcation solutions to a boundary layer problem arising in the theory of power law fluids. Acta Mathematica Scientia 20, 19–26.

Zheng, L.C., Wang, L.J., Zhang, X.X., 2011. Analysis solutions of unsteady flow and heat transfer on a permeable stretching sheet with non-uniform heat source/sink. Communications in Nonlinear Science and Numerical Simulation 16, 731–740.

Zheng, L.C., Zhang, X.X., He, J.C., 2002. Momentum and heat transfer in laminar boundary layer behind shock wave. Journal of Thermal Science 11, 255–258.

Zheng, L.C., Zhang, X.X., Lu, C.Q., 2006. Heat transfer for power law non-Newtonian fluids. Chinese Physics Letters 23 (12), 3301–3304.

Chapter 5

Differential Transform Method

ABSTRACT

In this chapter, two approximate analytical methods, i.e., differential transform Padé and differential transform basis function, are introduced. Using the two methods, we study magnetohydrodynamics(MHD) Falkner–Skan boundary layer flow, MHD fluid mixed convective and radiation heat transfer over the stretching wedge surface, MHD nanofluid boundary layer stagnation flow, and radiation heat transfer in a porous medium. Approximate analytical solutions are obtained for the problems and all results are verified by numerical solutions in good agreement. The effects of pertinent parameters on velocity and temperature fields are analyzed in detail.

5.1 INTRODUCTION

The differential transformation method (DTM) was developed by Zhao (1986). It originates from the Taylor expansion of the function, but the fundamental idea is to avoid directly seeking values of higher-order derivatives or partial derivatives of an unknown function. By performing differential transformation operations on differential equations and subjected initial or boundary conditions, algebraic relations between higher-order derivatives of the adjacent order are obtained for ordinary differential equations or partial differential equations; then an iterative procedure is established to obtain the analytical solution in terms of the polynomial form.

DTM does not need to calculate derivatives or partial derivatives. Its main advantage is that not only does it require a smaller amount of computation than the traditional Taylor series method, but it can be applied directly to nonlinear differential equations of physics or mathematics. However, for differential equations in unbounded domains, results obtained by DTM are effective only in a small subregion of the problem; the result is not correct in a whole region of existing solution (Boyd, 1997). The reason is that the solution obtained by DTM is divergent when the independent variable is infinite. To overcome this problem, two new methods for solving differential equations of the infinite region have been proposed, DTM-Padé and DTM-basic function (BF), proposed by Professor Zheng's doctoral student, Xiaohong Su, while pursing a

Mathematics in Science and Engineering. http://dx.doi.org/10.1016/B978-0-12-811753-8.00005-0

doctorate at the University of Science and Technology Beijing. The basic ideas of the two methods are introduced as follows.

5.1.1 Ideas of Differential Transform—Padé and Differential Transform—Basic Function

Many scholars have applied DTM to solve differential equations arising in different fields (Abazari and Borhanifar, 2010; Arikoglu, 2005; Arikoglu and Özkol, 2009; Ayaz, 2004; Jang et al., 2010).

To solve BVPs of differential equations in unbounded domains by DTM, we usually need to introduce some assumed initial parameters such that the initial value problems (IVPs) of original differential equations with introduced initial parameters can be solved first. Then we need to determine the initial parameters by pertinent boundary conditions at infinity. Two methods can be used to determine the introduced initial parameters: one is to use DTM coupled with Padé's approximation (DTM-Padé) (Rashidi, 2009; Su et al., 2012a,b); another is to use DTM coupled with the BF method (DTM-BF).

The DTM-Padé approximation method has some problems. The solutions to algebraic equations with undetermined parameters obtained by DTM-Padé are usually not unique; in some cases the obtained algebraic equations have multiple solutions. Thus, to determine the correct initial parameter values, continuous-order Padé approximation is necessary and the initial parameters can be determined according to the convergence trend of continuous Padé approximation. In this process, according to the complexity of the equation, it is difficult to judge the correct value of the parameter; sometimes it is impossible, and one can only determine the range of the parameters. Only when the introduced initial parameters of the problem are exactly determined can solutions be obtained with high accuracy.

To overcome the difficulties in applying DTM-Padé, a novel method, i.e., DTM-BF, was developed by Su and Zheng (2011) and Su et al. (2012a,b). The proposed method can be used to solve some complex and coupled nonlinear BVPs with an infinite boundary. A few items can obtain good accuracy and fewer initial parameters are needed to be introduced for BVPs. Another advantage is that in many cases, the algebraic equations with introduced initial parameters obtained by DTM-BF have unique solutions. DTM-BF can also be used as the basis for determining the parameters of the algebraic equations obtained by DTM-Padé. The basic idea is to use DTM to give a power series solution of the IVP, according to the characteristics of the problem, and then, by performing a linear combination of the BFs, to represent the solution of BVP. The special solutions are:

1. First, initial parameters are introduced according to the BVP, such that we can solve the IVP with introduced initial parameters using DTM. For the

introduced initial parameters, in a neighboring region of the initial point, the power series solution of the IVP is obtained and the Taylor series in this region is convergent.

2. Second, according to the physical characteristics of the differential equation and boundary conditions, the appropriate basis function is chosen and the linear combination is performed to approximate the solution of the problem.
3. Third, Taylor expansion is carried out for the solution series combination by the BF, and then the power series form of the power series is organized. Furthermore, the power series solution of the corresponding IVP is matched with the same power. The approximate solution of the problem is obtained using the BF series.

5.1.2 Definition of Differential Transformation Method and Formula

5.1.2.1 Differential Transformation for Function of One Variable

Supposing that a function $w(t)$ has k order derivatives for variable t, the differential transform of function $w(t)$ is defined as Zhao (1986)

$$W(k) = M(k)\left[\frac{d^k q(t)w(t)}{dt^k}\right]_{t=t_0}, \quad k = 0, 1, 2, 3, \cdots \tag{5.1}$$

where $w(t)$ and $W(k)$ are called the original function and differential transform function, respectively. The inverse differential transform function of $W(k)$ is defined as

$$w(t) = \frac{1}{q(t)}\sum_{0}^{\infty}\frac{W(k)}{M(k)k!}(t-t_0)^k \tag{5.2}$$

where $M(k) \neq 0$, displayed for the transformation of the known function of the independent variable to be proportional to the integer. $q(t) \neq 0$ is the kernel of the transformation of a known function. If $q(t) = 1$, the proportional function $M(k) = H^k$ or $M(k) = \frac{H^k}{k!}$ (H is a proportional constant). When $M(k) = \frac{H^k}{k!}$, the product operation of the transformation is relatively simple, so the scaling function is generally used in this form, and in the following discussion, this definition is used.

5.1.2.2 Differential Transformation for Functions of Several Variables

Supposing that the function $w(x,t)$ has a continuous partial derivative, the differential transformation of the function $w(x,t)$ is defined.

$$W(k,h) = M(k)N(h)\left[\frac{\partial^{k+h} q(t)p(x)w(x,t)}{\partial x^h \partial t^k}\right]_{x=x_0,t=t_0,}, \quad k = 0, 1, 2, 3, \cdots, \tag{5.3}$$

where $w(t,x)$ and $W(k,h)$ are called the original function and differential transform function, respectively. The inverse differential transform function of $W(k,h)$ is defined as

$$w(t,x) = \frac{1}{q(t)p(x)} \sum_{0}^{\infty} \frac{W(k,h)}{M(k)N(h)h!k!}(t - t_0)^k (x - x_0)^h. \tag{5.4}$$

It is analogous to the case of one variable function, $M(k) \neq 0$ and $N(h) \neq 0$, displayed for the transformation of the known function of the independent variable to be proportional to the integer; $q(t) \neq 0$ and $p(x) \neq 0$ are the respective kernels of the transformation of a known function. If $q(t) = 1$, $p(x) = 1$, the proportional functions $M(k) = H_1^k$ and $N(h) = H_2^h$ or $M(k) = \frac{H_1^k}{k!}$ and $N(h) = \frac{H_2^h}{h!}$ [H_i $(i = 1,\ 2)$ are called proportional constants]. When $M(k) = \frac{H_1^k}{k!}$ and $N(h) = \frac{H_2^h}{h!}$, the product operations of the transformation are relatively simple. If $q(t) = p(x) = 1$ is chosen, the differential transform of the two-variable function of $w(t,x)$ can be written as

$$W(k,h) = \frac{1}{k!h!}\left[\frac{\partial^{k+h} w(x,t)}{\partial x^h \partial t^k}\right]_{x=x_0, t=t_0,}, \quad k = 0, 1, 2, 3, \cdots. \tag{5.5}$$

The inverse differential transform of function $W(k,h)$ is written as

$$w(t,x) = \sum_{0}^{\infty} W(k,h)(t - t_0)^k (x - x_0)^h, \tag{5.6}$$

In the same way, we can define the differential transform of three variables or more.

5.1.2.3 Differential Transformation Formula

According to the definition of differential transformation, the formula for calculating differential transformation with a continuous derivative or partial derivative can be derived. Tables 5.1 and 5.2 show the basic operation principle of the differential transformation of a function.

TABLE 5.1 Basic Operation Principle of Differential Transformation of Single Variable Function

Original Function	Differential Transform
$w(t) = \alpha w_1(t) + \beta w_2(t)$	$W(k) = \alpha W_1(k) + \beta W_2(k)$, α and β are constants
$w(t) = w_1^{(n)}(t)$	$W(k) = (k+1)(k+2) \cdots (k+n) W_1(k+n)$
$w(t) = w_1(t) w_2(t)$	$W(k) = \sum_{i=0}^{k} W_1(i) W_2(k-i)$
$w(t) = (t - t_0)^m$	$W(k) = \delta(k-m) = \begin{cases} 1, & k = m, \\ 0, & k \neq m. \end{cases}$

TABLE 5.2 Basic Operation Principle of Differential Transformation of Two-Variable Function

Original Function	Differential Transform
$w(x,t) = \alpha w_1(x,t) + \beta w_2(x,t)$	$W(k) = \alpha W_1(k,h) + \beta W_2(k,h)$, α, β are constants
$w(x,t) = \dfrac{\partial^n w_1(x,t)}{\partial x^n}$	$W(k,h) = (k+1)(k+2)\cdots(k+n)W_1(k+n,h)$
$w(x,t) = \dfrac{\partial^m w_1(x,t)}{\partial t^m}$	$W(k,h) = (h+1)(h+2)\cdots(h+m)W_1(k,h+m)$
$w(x,t) = \dfrac{\partial^{i+j} w_1(x,t)}{\partial x^i \partial t^j}$	$W(k,h) = (k+1)\cdots(k+i)(h+1)\cdots(h+j)$ $W_1(k+i,h+j)$
$w(x,t) = w_1(x,t)w_2(x,t)$	$W(k,h) = \sum_{j=0}^{k}\sum_{i=0}^{h} W_1(j,h-i)W_2(k-j,i)$
$w(x,t) = \dfrac{\partial w_1(x,t)}{\partial x}\dfrac{\partial w_2(x,t)}{\partial x}$	$W(k,h) = \sum_{j=0}^{k}\sum_{i=0}^{h}(j+1)(k-j+1)$ $W_1(j+1,h-i)W_2(k-j+1,i)$,

In this chapter we first apply the DTM-Padé to study the magnetohydrodynamic (MHD) Falkner–Skan boundary layer flow present in the magnetic effects over a permeable wall. Then we apply DTM-BF to study mixed convective and radiation heat transfer of MHD fluid over the stretching wedge surface and MHD nanofluids in porous media with variable surface heat flux and chemical reactions.

The reliability and validity of the method are verified by numerical comparison for all results.

5.1.3 Magnetohydrodynamic Boundary Layer Problem

MHD fluid has received much attention by scholars owing to its wide applications in engineering and technology. Many mathematical models have been proposed to characterize MHD flow and heat transfer under different conditions (Abbasbandy and Hayat, 2009a,b; Abel and Nandeppanavar, 2009; Hayat et al., 2008; Ishak et al., 2009; Parand et al., 2011; Prasad et al., 2009; Robert and Vajravelu, 2010; Soundalgekar et al., 1981; Sutton and Sherman, 1965).

Soundalgekar et al. (1981) studied the Falkner–Skan boundary layer flow and heat transfer of MHD fluid. Suitable similarity transformation was introduced to reduce the governing partial differential equations to nonlinear ordinary differential equations and then to solve them numerically. Abbasbandy and Hayat (2009a) studied the MHD Falkner–Skan boundary layer flow using the Hankel-Padé and homotopy analytical methods (Abbasbandy and Hayat, 2009b), respectively. Parand et al. (2011) investigated MHD flow around a surface of infiltration that was stationary without a wedge; the approximate solution was obtained using the spectrum of Hermite function fitting method. Robert and Vajravelu (2010) discussed the existence and uniqueness of solutions of MHD Falkner–Skan flow. Professor Zheng and coauthors carried out a series of investigations using DTM-Padé or DTM-BF, perturbation methods, homotopy analysis methods, and numerical methods on boundary layer flow and heat transfer of MHD fluid over a continuous stretching permeable sheet embedded in viscous electrically conducting fluid (Jiao et al., 2015; Li et al., 2016; Si et al., 2016; Su and Zheng, 2011; Su et al., 2012a,b; Zhang et al., 2015; Zhu et al., 2010).

5.2 MAGNETOHYDRODYNAMICS FALKNER–SKAN BOUNDARY LAYER FLOW OVER PERMEABLE WALL

5.2.1 Mathematical Physical Description

Consider a steady two-dimensional boundary layer flow of MHD fluid over a permeable stretching wedge surface. It is assumed that the magnetic field is oriented perpendicular to the wedge surface and the magnetic Reynolds number is considered to be small so that the induced magnetic field is negligible. The velocity function of suction or blowing across the permeable wall is $v_w(x)$. We also assume the strength of the electric field due to polarization of the electric charges to be negligibly small. Under these assumptions, the governing boundary layer equations of motion are

$$\frac{\partial u}{\partial x}+\frac{\partial v}{\partial y}=0, \tag{5.7}$$

$$u\frac{\partial u}{\partial x}+v\frac{\partial u}{\partial y}=U\frac{dU}{dx}+\nu\frac{\partial^2 u}{\partial y^2}-\frac{\sigma B^2(x)}{\rho}(u-U), \tag{5.8}$$

where the x axis is taken along the surface of the wedge surface and the y axis is perpendicular to it. u and v are the velocity components in the x and y directions, respectively. $U(x)=ax^m$ is the velocity of the exterior flow over the wedge surface, in which a ($a>0$) and m is a constant, ν is a kinematic viscosity, and σ is electrical conductivity. $B(x)=B_0x^{(m-1)/2}$ is the magnetic field and ρ is the fluid density. The corresponding boundary conditions are

$$u(x,0)=0, \tag{5.9}$$

$$v(x,0)=v_w, \tag{5.10}$$

$$u(x,\infty)=U(x), \tag{5.11}$$

where $v_w=v_0\,x^{(m-1)/2}$ is the suction or blowing velocity across the surface of wedge surface, because $v_w<0$ for suction and $v_w>0$ for blowing. We assume the boundary layer flow is laminar and a local similarity solution exists; thus the following transformations are introduced:

$$\eta=y\left(\frac{(m+1)U}{2\nu x}\right)^{1/2},\quad \psi=f(\eta)\left(\frac{2\nu xU}{m+1}\right)^{1/2}, \tag{5.12}$$

where η is the similarity variable and $\psi(x,y)$ is the stream function, satisfying $(u, v) = \left(\frac{\partial \psi}{\partial y}, -\frac{\partial \psi}{\partial x}\right)$.

Consequently, the following expressions of velocity components can be obtained by using the transformations (5.12):

$$u = Uf'(\eta), v = -\left(\frac{(m+1)\nu U}{2x}\right)\left(f(\eta) + \frac{m-1}{m+1} f'(\eta)\right), \tag{5.13}$$

The conservation of mass Eq. (5.7) is therefore automatically satisfied by using Eq. (5.13). Moreover, the momentum of Eq. (5.8) and the boundary conditions (5.9–5.11) are reduced as

$$f'''(\eta) + f(\eta)f''(\eta) + \beta\left[1 - (f'(\eta))^2\right] - M^2(f'(\eta) - 1) = 0, \tag{5.14}$$

$$f(0) = -C,\ f'(0) = 0,\ f'(+\infty) = 1, \tag{5.15}$$

where $M = 2\sigma B_0^2/\rho a(1+m)$ is the magnetic parameter; $C = v_0(2/m+1)^{1/2}(\nu a)^{-\frac{m-1}{2}}$ is the suction or blowing parameter; and in $C = v_0(2/m+1)^{1/2}(\nu a)^{-\frac{m-1}{2}}$, $C < 0$ is for suction and $C > 0$ is for blowing. The expression of parameter C shows that it is used to control the strength of the mass transfer of normal flow at the boundary. $m = \beta/(2-\beta)$ is the wedge angle parameter that corresponds to $\beta = \Psi/\pi$ for angle Ψ of the wedge. It is also known that the case $\beta = 0$ corresponds to the flow over a horizontal plate, whereas $\beta = 1$ represents a vertical plate. The functions $f(\eta)$ and $f'(\eta)$ denote the dimensionless stream function and the dimensionless tangential velocity, respectively.

5.2.2 Differential Transformation Method–Padé Solutions

To solve the nonlinear differential equations of (5.14) and (5.15) using DTM-Padé, we first perform the differential transform on Eq. (5.14) according to Table 5.1, which yields the following iterated function system-iteration formula

$$F(k+3) = \frac{1}{(k+1)(k+2)(k+3)}\Big(M^2(k+1)F(k+1) - (\beta + M^2)\delta(k) - \sum_{i=0}^{k}[(k-i+1)(k-i+2)F(i)F(k-i+2) - \beta(i+1)(k-i+1)F(i+1)F(k-i+1)]\Big) \tag{5.16}$$

It is noted from the boundary conditions of (5.15) two values exist at the left boundary, $\eta = 0$, and another value is at infinity, $\eta = +\infty$, which cannot be used at first transformation. Therefore, we introduce parameter α and denote it by $f''(0) = 2\alpha$; then we consider solving Eq. (5.14) under the following initial value conditions

$$f(0) = -C, \quad f'(0) = 0, \quad f''(0) = 2\alpha, \tag{5.17}$$

where $f''(0)$ is dimensionless skin friction coefficient and parameter α must be determined as part of the solution. Then the corresponding differential transforms to initial conditions (5.17) are

$$F(0) = -C, \quad F(1) = 0, \quad F(2) = \alpha. \tag{5.18}$$

Substituting (5.18) into iteration formula (5.16) yields

$$F(3) = \frac{1}{3}\alpha C - \frac{1}{6}M^2 - \frac{1}{6}\beta,$$

$$F(4) = \frac{1}{12}M^2\alpha + \frac{1}{12}\alpha C^2 - \frac{1}{24}M^2 C - \frac{1}{24}C\beta,$$

$$F(5) = \frac{1}{30}M^2\alpha C - \frac{1}{120}M^4 - \frac{1}{120}M^2\beta + \frac{1}{60}\alpha C^3 - \frac{1}{120}C^2M^2 - \frac{1}{120}C^2\beta + \frac{1}{15}C^2\beta - \frac{1}{30}\alpha^2,$$

$$F(6) = \frac{1}{360}M^4\alpha + \frac{1}{120}M^4 aC^2 - \frac{1}{360}M^4 C - \frac{1}{360}M^2 C\beta + \frac{1}{360}\alpha C^4 - \frac{1}{720}C^3M^2 - \frac{1}{720}C^3\beta + \frac{2}{45}C\beta\alpha^2 + \frac{1}{36}\beta\alpha^2 - \frac{1}{60}\beta\alpha M^2 - \frac{1}{60}\beta^2\alpha + \frac{1}{90}M^2\alpha + \frac{1}{90}\beta\alpha.$$

Other values of $F(k)$ can be calculated analogously. Furthermore, we can obtain the power series solution of the unknown function $f(\eta)$ of Eq. (5.14) under the initial condition (5.17) by substituting all of $F(k)$ into inverse differential transform formula (5.2).

$$f(\eta) = \sum_{k=0}^{\infty} F(k)\eta^k \cong \sum_{k=0}^{n} F(k)\eta^k.$$

The diagonal Padé approximant has been used by many researchers as the most accurate approximant. By performing the [*N*/*N*] diagonal Padé approximation to the power series solution of the IVP of $f'(\eta)$, we can obtain the following formula (Su and Zheng, 2011; Li et al., 2016)

$$f'(\eta) = \sum_{i=0}^{2N}(i+1)F(i+1)x^i = \frac{\sum_{i=0}^{N} a_i\eta^i}{1 + \sum_{i=1}^{N} b_i\eta^i}. \tag{5.19}$$

It is clear that there are $N+1$ independent numerator coefficients and N independent denominator coefficients, so there are $2N+1$ unknown coefficients in total to be determined. We know that the [N/N] approximant is uniquely determined. The number of unknown coefficients are suggested so that normally the [N/N] approximant ought to fit the power series $\sum_{i=0}^{\infty}(i+1)F(i+1)x^i$ through the orders $1, \eta, \eta^2, \ldots, \eta^{2N}$. Moreover, we can promptly obtain the algebra equation of $a_N = b_N$ by the condition $f'(+\infty) = 0$. Because the value of N is selected to be equal to 8, the value of $f''(0) = 2\alpha$ can easily be obtained by solving the algebraic equation of $a_N = b_N$ for the given values of the parameters β, M, and C. Furthermore, the analytical solutions can be obtained after the values of $f''(0)$ are determined. For example, the Padé approximate results of solutions for the function of $f(\eta)$ and its derivative function $f'(\eta)$ are obtained for $\beta = 4/3$, $C = -1$, and $M = 3$, and the Padé approximation solutions $f(\eta)$ and $f'(\eta)$ to the nonlinear BVPs of (5.14) and (5.15) are

$$\begin{aligned} f(\eta)_{[8,8]} = {} & (1 + 2.589611536\eta + 4.956020291\eta^2 + 4.745379570\eta^3 + 2.733096465\eta^4 \\ & + 1.246141374\eta^5 + 0.3666351390\eta^6 + 0.06469038902\eta^7 \\ & + 0.008339831059\eta^8)/(1 + 2.589611536\eta + 3.026583735\eta^2 \\ & + 2.114256148\eta^3 + 0.980459403\eta^4 + 0.312368807\eta^5 \\ & + 0.0648834530\eta^6 + 0.00702544282\eta^7 + 0.000054139359\eta^8), \end{aligned}$$

$$\begin{aligned} f'(\eta)_{[8,8]} = {} & (3.858873112\eta + 6.315360997\eta^2 + 4.00905169\eta^3 + 2.87796510\eta^4 \\ & + 1.53190034\eta^5 + .248167750\eta^6 + 0.99367688\eta^7 + 0.0077611886\eta^8) \\ & / \, (1 + 3.47548723\eta + 5.31704200\eta^2 + 4.81919153\eta^3 + 2.91649260\eta^4 \\ & + 1.24861699\eta^5 + 0.380547917\eta^6 + 0.076388508\eta^7 + 0.0077611886\eta^8). \end{aligned}$$

5.2.3 Results and Discussion

To illustrate the reliability and efficiency of the results of the DTM-Padé technique, for some specified given values of the parameter β, M, and C, we solve the nonlinear boundary value of (5.14) and (5.15) numerically by employing the Runge–Kutta method and shooting technique. The solutions obtained by DTM-Padé and the numerical results are presented in Figs. 5.1–5.8. The calculated skin friction coefficient of $f''(0)$ is presented in Table 5.3, denoted by $f''(0)_{\text{Dp}}$ and $f''(0)_{\text{num}}$, respectively.

We can see from Figs. 5.1–5.8 and Table 5.3 that for all of the parameters used in computation, the results obtained by the DTM-Padé method and the numerical method are consistent. Table 5.3 shows that the value of the wall friction coefficient increases with the increase in magnetic field parameters but decreases with the increase in suction or injection parameters.

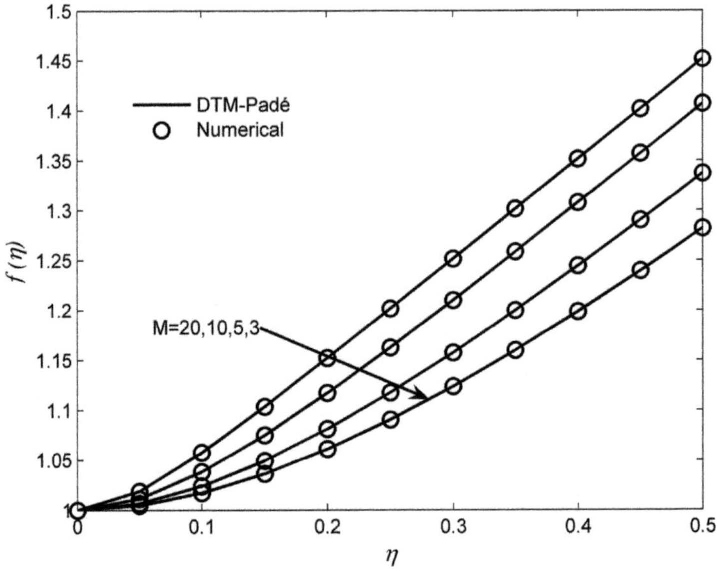

FIGURE 5.1 Differential transformation method (DTM)-Padé and numerical solutions of f for $\beta = 4/3$, $C = -1$.

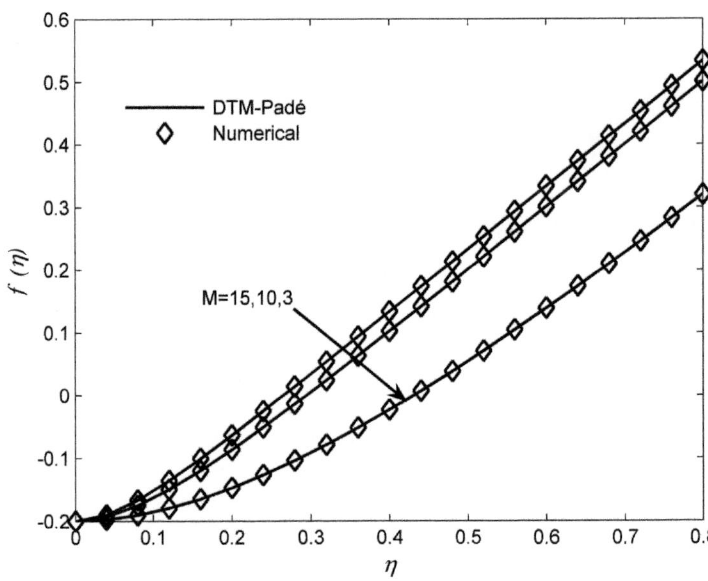

FIGURE 5.2 Differential transformation method (DTM)-Padé and numerical solutions of f for $\beta = 4/3$, $C = 1/5$.

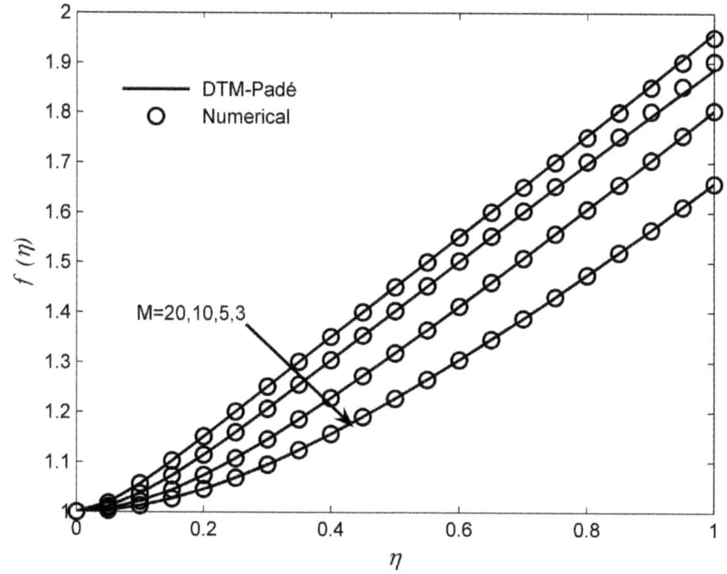

FIGURE 5.3 Differential transformation method (DTM)-Padé and numerical solutions of f for $\beta = -3$, $C = -1$.

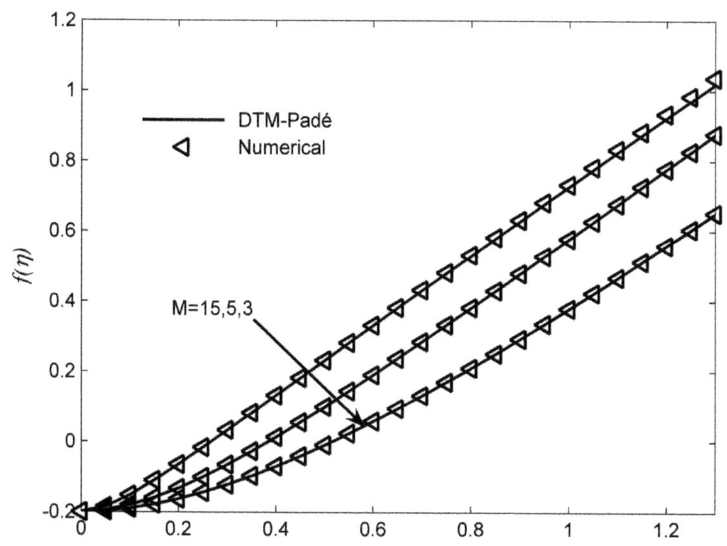

FIGURE 5.4 Differential transformation method (DTM)-Padé and numerical solutions of f for $\beta = -3$, $C = 1/5$.

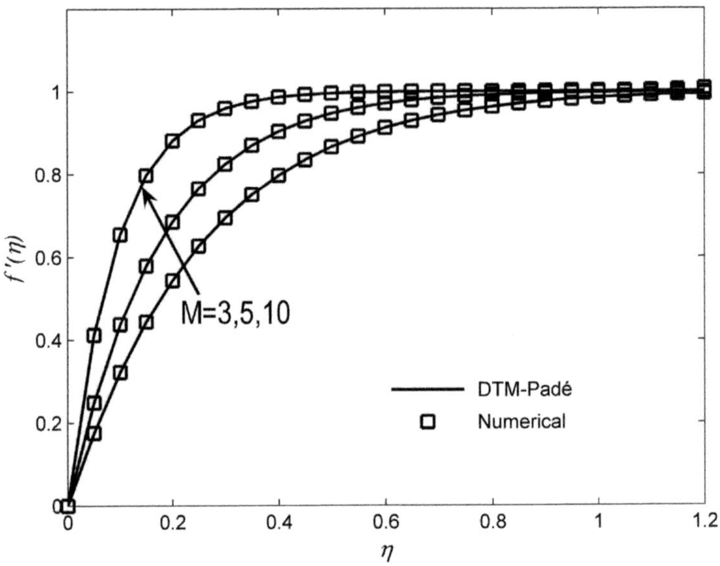

FIGURE 5.5 Differential transformation method (DTM)-Padé and numerical solutions of f' for $\beta = 4/3$, $C = -1$.

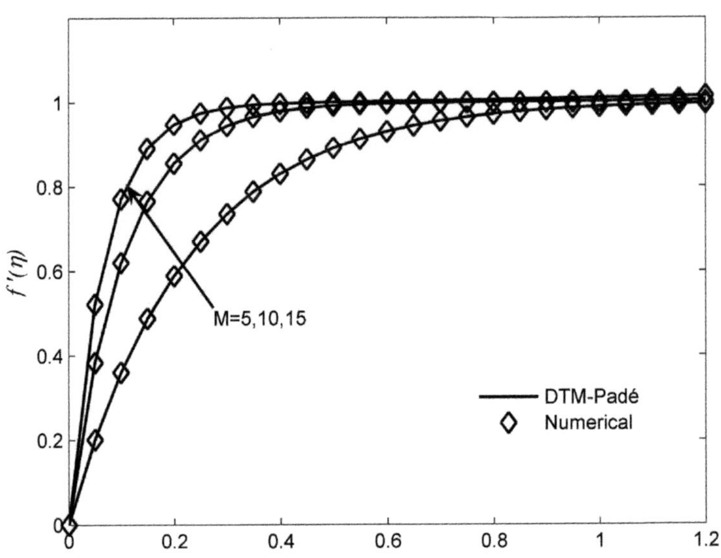

FIGURE 5.6 Differential transformation method (DTM)-Padé and numerical solutions of f' for $\beta = -3$, $C = -1$.

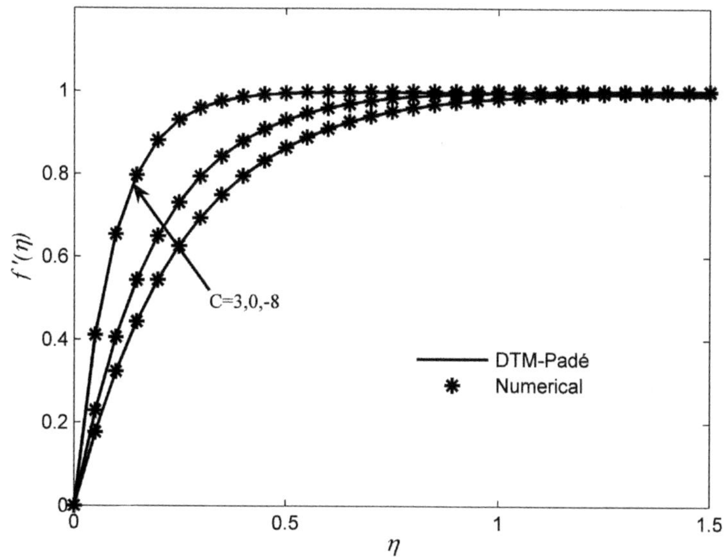

FIGURE 5.7 Differential transformation method (DTM)-Padé and numerical solutions of f' for $\beta = 4/3$, $M = 5$.

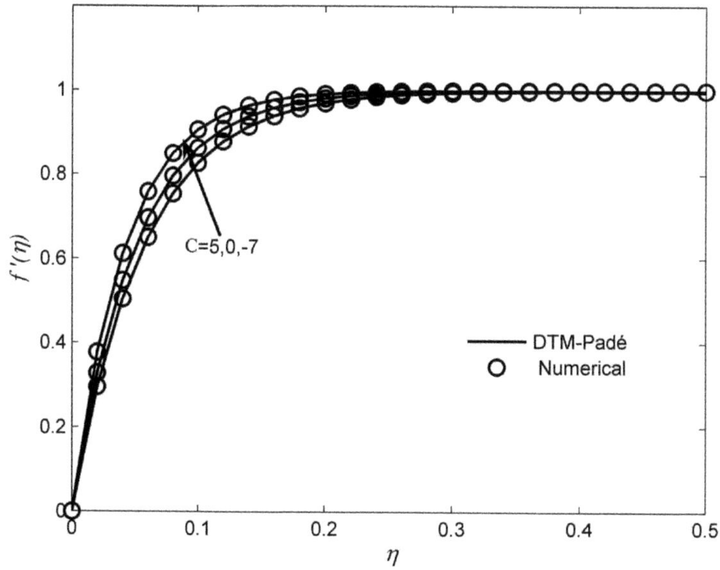

FIGURE 5.8 Differential transformation method (DTM)-Padé and numerical solutions of f' for $\beta = -3$, $M = 5$.

TABLE 5.3 Differential Transformation Method–Padé and Numerical Solutions for $C = -1, 1/5$, $\beta = 4/3, -3$, for M and $f''(0)$

		$\beta = 4/3$	$\beta = 4/3$	$\beta = -3$	$\beta = -3$
M	C	$f''(0)_{Dp}$	$f''(0)_{num}$	$f''(0)_{Dp}$	$f''(0)_{num}$
3	−1	3.858873112	3.858873707	2.813600000	2.812921191
3	1/5	3.207744568	3.207837822	2.180130322	2.180057161
5	−1	5.720170862	5.720199248	5.123657682	5.123671967
5	1/5	5.090866724	5.090869424	4.502770284	4.502773217
10	−1	10.61054613	10.61054899	10.31822832	10.31823169
10	1/5	9.996989268	9.996989738	9.707174832	9.707176594
15	−1	15.57360675	15.57360751	15.37964805	15.37964931
15	1/5	14.96488698	14.96488783	14.77209806	14.77209959
20	−1	20.55514466	20.55514515	20.40997406	20.40997483
20	1/5	19.94873263	19.94873721	19.80423623	19.80423676

Figs. 5.1–5.6 show the influence of the magnetic field parameters on the boundary layer flow characteristics. The increase in the magnetic field parameter will result in a loss in thickness of the boundary layer and an increase in the tangential velocity of the fluid in the boundary layer. It indicates that the tangential velocity of the fluid in the boundary layer tends toward 1 with an increase in the distance from the wall.

The effects of wall suction or injection parameters on the flow characteristics of the boundary layer are presented in Figs. 5.7 and 5.8. Results indicate that the thickness of the boundary layer decreases and the tangential velocity increases with suction velocity across the surface of wedge wall, and vice versa. Furthermore, the approximate solution and the numerical solution are compared; it can be seen that the approximate analytical solution obtained by the DTM-Padé method is in agreement with the numerical results.

5.3 UNSTEADY MAGNETOHYDRODYNAMICS MIXED FLOW AND HEAT TRANSFER ALONG A VERTICAL SHEET

5.3.1 Mathematical Physical Description

Consider an unsteady two-dimensional MHD boundary layer flow and heat transfer along a continuous moving vertical stretching sheet with a velocity of

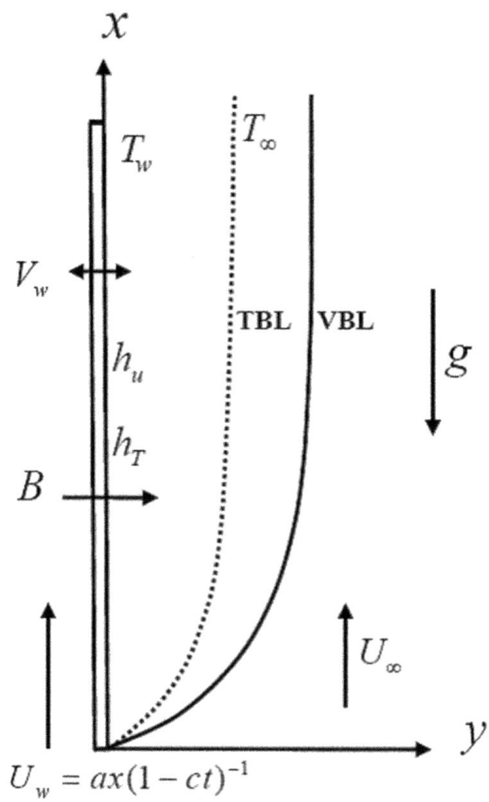

FIGURE 5.9 Magnetohydrodynamics flow and heat transfer along a vertical stretching wall. *TBL*, temperature boundary layer; *VBL*, velocity boundary layer.

$U_w = ax(1 - ct)^{-1}$ (Zhang et al., 2015; Ishak et al., 2010) in the positive x direction. The free stream velocity far from the sheet is $U_\infty = RU_w$. Here, $R \geq 0$, $a > 0$, $c > 0$, and $ct < 1$. The influence of magnetic field B exists, which acts in the direction normal to the stretching sheet (Fig. 5.9). The induced magnetic field is negligible, which is a valid assumption on a laboratory scale under the assumption of a small magnetic Reynolds number. It is also assumed that the external electric field is zero.

Under these assumptions, the boundary layer governing equations with consideration of the presence of thermal buoyancy, thermal radiation, and internal absorption/generation take the following form:

$$\frac{\partial u}{\partial x} + \frac{\partial v}{\partial y} = 0, \tag{5.20}$$

$$\frac{\partial u}{\partial t} + u\frac{\partial u}{\partial x} + v\frac{\partial u}{\partial y} = -\frac{1}{\rho}\frac{\partial p}{\partial x} + \nu\frac{\partial^2 u}{\partial y^2} - \frac{\sigma B^2}{\rho}u + g\beta_T(T - T_\infty), \tag{5.21}$$

$$\rho c_p\left(\frac{\partial T}{\partial t}+u\frac{\partial T}{\partial x}+v\frac{\partial T}{\partial y}\right)=\alpha\frac{\partial^2 T}{\partial y^2}-\frac{\partial q_r}{\partial y}+Q(T-T_\infty), \tag{5.22}$$

The boundary conditions are

$$y=0,\ \ u=U_w+D_{u0}\nu\frac{\partial u}{\partial y},\ \ v=V_w,\ \ T=T_w+D_{T0}\frac{\partial T}{\partial y}, \tag{5.23}$$

$$y=\infty,\quad u=U_\infty,\quad T=T_\infty, \tag{5.24}$$

where $\frac{\partial p}{\partial x}=-\rho\frac{dU_\infty}{dt}-\rho U_\infty\frac{dU_\infty}{dx}-\sigma B^2 U_\infty$ is the pressure gradient; x and y are the axis coordinate measures along and normal to the surface, respectively; u and v are the velocity components along the x and y directions, respectively; t is time; T is the temperature inside the boundary layer; g is the gravity field; β_T is the volumetric coefficient of thermal expansion; c_p is the specific heat at constant pressure; α is thermal conductivity; μ is fluid viscosity; $\nu=\mu/\rho$ is the kinematics viscosity of the fluid; ρ is the density of fluid; $T_w=T_\infty+ax(2\nu)^{-1}(1-ct)^{-2}$ is the temperature of the stretching sheet; T_∞ is the temperature of the fluid outside the boundary layer; $D_u=D_{u0}(1-ct)^{\frac{1}{2}}$ is the velocity slip factor, which changes with time; $D_T=D_{T0}(1-ct)^{\frac{1}{2}}$ is the temperature jump factor, which also changes with time; D_{u0} and D_{T0} are the initial values of velocity and thermal slip factors, respectively; $v_w=-C(\nu U_w)^{\frac{1}{2}}x^{-\frac{1}{2}}$ represents mass transfer on the sheet with $C<0$ for injection and $C>0$ for suction; Q is heat generation when $Q>0$ or heat absorption when $Q<0$; and the variable magnetic field B is of the form $B=B_0U_w^{\frac{1}{2}}(\nu x)^{-\frac{1}{2}}$.

The radiative heat flux q_r under the Rosseland approximation has the form $q_r=-\frac{4\sigma^1}{3k^1}\frac{\partial T^4}{\partial y}$, where σ^1 and k^1 are the Stefan–Boltzmann constant and the mean absorption coefficient, respectively. The temperature difference within the flow is assumed to be sufficiently small so that T^4 may be expressed as a linear function of temperature; thus, $T^4\approx 4T_\infty^3T-3T_\infty^4$, and yields $q_r=-\frac{16T_\infty^3\sigma^1}{3k^1}\frac{\partial T}{\partial y}$.

To obtain similarity solutions of the problem, we introduce the following dimensionless variables:

$$\eta=U_w^{\frac{1}{2}}(\nu x)^{-\frac{1}{2}}y,\ \ \psi(x,y)=(\nu xU_w)^{1/2}f(\eta),\ \ \theta(\eta)=\frac{T-T_\infty}{T_w-T_\infty} \tag{5.25}$$

where $\psi(x,y)$ is the stream function that satisfies the continuity equation. In view of $u=\frac{\partial\psi}{\partial y}$ and $v=-\frac{\partial\psi}{\partial x}$, we ascertain that $u=U_wf'(\eta)$ and $v=-(\nu U_w)^{\frac{1}{2}}x^{-\frac{1}{2}}f(\eta)$.

Thus, the problem can be expressed as

$$f''' + ff'' - f'^2 - A\left(f' + \frac{\eta}{2}f'' - R\right) - Mn(f' - R) + \gamma\theta + R^2 = 0 \quad (5.26)$$

$$\Pr^{-1}(1 + Nr)\theta'' + (\lambda - 2A)\theta - \frac{A}{2}\eta\theta' - \theta f' + f\theta' = 0 \quad (5.27)$$

with boundary conditions:

$$f(0) = C, \quad f'(0) = 1 + h_u f''(0), \quad \theta(0) = 1 + h_T\theta'(0) \quad (5.28)$$

$$f'(\infty) = R, \quad \theta(\infty) = 0 \quad (5.29)$$

In these equations, $A = c/a$ is the dimensionless measure of the unsteadiness and the prime indicates differentiation with respect to η; $\gamma = \frac{g\beta_T}{2\nu a}$ is the mixed convection parameter; $\Pr = \frac{\mu c_p}{\alpha}$ is the Prandtl number; A is a parameter that measures the unsteadiness; $Nr = \frac{16T_\infty^3\sigma^1}{3k^1\alpha}$ is the thermal radiation parameter; $Mn = \frac{\sigma B_0^2}{\rho\nu}$ is the magnetic parameter; $Re_x = \frac{U_w x}{\nu}$ is the local Reynolds number; $Re_\alpha = \frac{U_w\sqrt{\alpha}}{\nu}$; $\lambda = \frac{Q\alpha Re_x}{\mu c_p Re_\alpha^2}$ is the heat source parameter ($\lambda > 0$ corresponds to heat generation and $\lambda < 0$ corresponds to heat absorption), $h_u = D_{u0}(a\nu)^{\frac{1}{2}}$ is the dimensionless velocity slip parameter; and $h_T = D_{T0}(a\nu^{-1})^{\frac{1}{2}}$ is the dimensionless thermal slip parameter.

The physical quantities of interest are the skin friction coefficient, C_f, and the local Nusselt number, Nu_x, which are defined as

$$C_f = \frac{2\tau_w}{\rho U_w^2/2}, \quad Nu_x = \frac{xq_w}{\alpha(T_w - T_\infty)}.$$

where the wall shear stress τ_w and the wall heat flux q_w are given by

$$\tau_w = \mu\left(\frac{\partial u}{\partial y}\right)_{y=0}, \quad q_w = -\alpha\left(\frac{\partial T}{\partial y}\right)_{y=0}.$$

Using the similarity variables, we obtain

$$\frac{1}{2}C_f Re_x^{\frac{1}{2}} = f''(0), \quad Nu_x\Big/Re_x^{\frac{1}{2}} = -\theta'(0)$$

5.3.2 Differential Transformation Method—Basic Function Solutions

To obtain the approximate analytical solution of a BVP for coupled nonlinear differential equations on an unbounded domain, we propose a new analytical method named DTM-BF, which is based on the Taylor series solution to the IVP corresponding to the BVP and a set of BFs relating to the boundary conditions. The DTM-BF is employed to solve the coupled BVP on an unbounded domain in three steps. First, we need to obtain the solution in the form of the Taylor series for the IVP corresponding to the BVP by means of the DTM. Second, the solution to the BVP is expressed as a linear combination of BFs $G_i(i=0,\ 1,\ 2,\ \cdots)$ in which G_0 satisfies the boundary conditions of the BVP and $G_i(i=1,\ 2,\ \cdots)$ satisfies the corresponding homogeneous boundary conditions, respectively. Finally, we determine all unknown coefficients by solving a system of algebraic equations that are constructed by comparison of the Taylor series solutions with the IVP and the series obtained by expanding the solution expressed in the BFs.

Suppose the function $w(t)$ is analytical in the domain Ω and $t=t_0$ is in this domain. The differential transformation of function $w(t)$ is $W(k)=\frac{1}{k!}\left[\frac{d^k w(t)}{dt^k}\right]_{t=t_0}$, $k=0,1,2,3,\ldots$, where $w(t)$ and $W(k)$ are the original function and the transformed function, respectively. The inverse transformation of $W(k)$ is $w(t)=\sum_{k=0}^{\infty}W(k)(t-t_0)^k\approx\sum_{k=0}^{n}W(k)(t-t_0)^k$.

First, we solve the IVPs for Eqs. (5.26) and (5.27) under the following initial value conditions using the DTM

$$f(0)=C,\ f'(0)=1+2h_u\beta_1,\ f''(0)=2\beta_1, \tag{5.30}$$

$$\theta(0)=1+h_T\beta_2,\quad \theta'(0)=\beta_2. \tag{5.31}$$

The differential transformations of conditions (5.30) and (5.31) are:

$$F(0)=C,\quad F(1)=1+2h_u\beta_1,\quad F(2)=\beta_1, \tag{5.32}$$

$$\Theta(0)=1+h_T\beta_2,\quad \Theta(1)=\beta_2. \tag{5.33}$$

We perform the differential transformations for Eqs. (5.26) and (5.27). The following iterative formulas of the differential transformations of $f(\eta)$ and $\theta(\eta)$ can be obtained by using the fundamental operations of the DTM

$$F(k+3)=\frac{1}{(k+1)(k+2)(k+3)}\left\{\left(A+M+\frac{kA}{2}\right)(k+1)F(k+1)\right.$$

$$-\left(AR+MR+R^2\right)\delta(k)+\gamma\Theta(k)$$

$$\sum_{i=0}^{k}[(i+1)(k-i+1)F(i+1)F(k-i+1)$$

$$\left.-(k-i+1)(k-i+2)F(i)F(k-i+2)]\right\}$$

$$\Theta(k+2) = \frac{-\mathrm{Pr}}{(k+1)(k+2)(1+Nr)}\left\{(\lambda - 2A)\Theta(k) - \frac{A}{2}k\Theta(k)\right.$$
$$+\sum_{i=0}^{k}[-(k-i+1)\Theta(i)F(k-i+1)$$
$$\left.+(k-i+1)F(i)\Theta(k-i+1)]\right\}$$

All of the terms of $F(k)$ and $\Theta(k)$ can be calculated by substituting Eqs. (5.32) and (5.33) into these iterative formulas. We can obtain the solutions to the initial values problems in a power series form as

$$f(\eta) = \sum_{k=0}^{\infty} F(k)\eta^k \cong \sum_{k=0}^{n} F(k)\eta^k, \tag{5.34}$$

$$\theta(\eta) = \sum_{i=0}^{\infty} \Theta(i)\eta^i \cong \sum_{i=0}^{m} \Theta(i)\eta^i. \tag{5.35}$$

To determine the values of β_1 and β_2, we express the solutions to the coupled BVP (5.26–5.29) in the form of the BFs. From the Eqs. (5.26) and (5.27) and boundary conditions (5.28) and (5.29), it is reasonable to assume that $f(\eta)$ could be expressed by the following sets of BFs

$$\left\{f_{0,0}(\eta),\ f_{i,j}(\eta)_{(i=1,2,3,\cdots,j=2,3,\cdots,)}\right\} \text{ and } \left\{\theta_{0,0}(\eta),\ \theta_{i,j}(\eta)_{(i=1,2,\cdots,j=1,2,\cdots,)}\right\}$$

in the form

$$\begin{aligned} f(\eta) \approx f_{N_1,N_2}(\eta) &= f_{0,0}(\eta) + \sum_{j=3}^{N_1}\sum_{i=1}^{N_2} b_{i,j} f_{i,j}(\eta) \\ &= f_{0,0}(\eta) + \sum_{j=3}^{N_1}\sum_{i=1}^{N_{j,2}} b_{i,j}\eta^j e^{ia_0\eta} \end{aligned} \tag{5.36}$$

where $f_{0,0}(\eta) = C - H + R\eta + He^{a_0\eta} + b_1\eta e^{a_0\eta} + b_2\eta^2 e^{a_0\eta}$ and $\theta_{0,0}(\eta) = Le^{\gamma_0\eta} + d_1\eta e^{\gamma_0\eta}$, $H = \frac{1 + 2h_u b_0 a_0 + 2b_2 h_u - b_0 - R}{a_0 - h_u a_0^2}$, and $L = \frac{1 + h_T d_1}{1 - h_T\gamma_0}$ satisfy the inhomogeneous boundary conditions (5.30) and (5.31), respectively. $f_{i,j}(\eta) = b_{i,j}\eta^j e^{ia_0\eta}$ $(i = 1, 2, 3, \ldots, j = 3, 4, \ldots)$, and $\theta_{i,j}(\eta) = d_{i,j}\eta^j e^{i\gamma_0\eta}$ $(i = 1, 2, 3, \ldots, j = 2, 3, \ldots)$, respectively, obey the homogeneous boundary conditions:

$$f(0) = 0, \quad f'(0) = 0, \quad f'(\infty) = 0, \tag{5.37}$$

and

$$\theta(0)=0, \quad \theta'(\infty)=0. \tag{5.38}$$

where $a_0<0$ and $\gamma_0<0$ are two undetermined decaying parameters. In practical applications, it is sufficient for the truncated series $f_{N_1,N_2}(\eta)$ and $\theta_{N_3,N_4}(\eta)$ for $N_i \le 4$ ($I=1, 2, 3, 4$). In this research, we choose $N_1=4$, $N_2=3$, $N_{j,2}=2$ ($j=3, 4$), $N_{2,4}=3$, $N_{3,4}=2$. By expanding the right sides of (5.34) and (5.35) as the power series of η, we get

$$\begin{aligned} f(\eta) &= C-H+R\eta+\left(\frac{Ha_0^2}{2!}+b_1a_0+b_2\right)\eta^2 \\ &+\left(\frac{Ha_0^3}{3!}+\frac{b_1a_0^2}{2!}+b_2a_0+\sum_{i=1}^{2}b_{i,3}\right)\eta^3 \\ &+\left(\frac{Ha_0^4}{4!}+\frac{b_1a_0^3}{3!}+\frac{b_2a_0^2}{2!}+\sum_{i=1}^{2}ia_0b_{i,3}+\sum_{i=1}^{2}b_{i,4}\right)\eta^4+\cdots \end{aligned}, \tag{5.39}$$

$$\begin{aligned} \theta(\eta) &= L+(L\gamma_0+d_1)\eta^2+\left(\frac{L\gamma_0^2}{2!}+d_1\gamma_0+\sum_{i=1}^{3}d_{i,2}\right)\eta^2 \\ &+\left(\frac{L\gamma_0^3}{3!}+\frac{d_1\gamma_0^2}{2!}+\sum_{i=1}^{3}i\gamma_0d_{i,2}+\sum_{i=1}^{2}d_{i,3}\right)\eta^3 \\ &+\left(\frac{L\gamma_0^4}{4!}+\frac{d_1\gamma_0^3}{3!}+\sum_{i=1}^{3}\frac{(i\gamma_0)^2d_{i,2}}{2!}+\sum_{i=1}^{2}i\gamma_0d_{i,3}\right)\eta^4+\cdots \end{aligned}, \tag{5.40}$$

From Eqs. (5.34), (5.35), (5.39), and (5.40), the following equations are constructed

$$\begin{aligned} \sum_{i=0}^{\infty}F(i)\eta^i &= C-H+R\eta+\left(\frac{Ha_0^2}{2!}+b_1a_0+b_2\right)\eta^2 \\ &+\left(\frac{Ha_0^3}{3!}+\frac{b_1a_0^2}{2!}+b_2a_0+\sum_{i=1}^{2}b_{i,3}\right)\eta^3+\cdots \end{aligned},$$

$$\begin{aligned} \sum_{i=0}^{\infty}\Theta(i)\eta^i &= L+(L\gamma_0+d_1)\eta+\left(\frac{L\gamma_0^2}{2!}+d_1\gamma_0+\sum_{i=1}^{3}d_{i,2}\right)\eta^2 \\ &+\left(\frac{L\gamma_0^3}{3!}+\frac{d_1\gamma_0^2}{2!}+\sum_{i=1}^{3}i\gamma_0d_{i,2}+\sum_{i=1}^{2}d_{i,3}\right)\eta^3+\cdots \end{aligned}. \tag{5.41}$$

By comparing the coefficients of like powers of η on both sides of (5.41), we can obtain

$$\begin{cases} \dfrac{Ha_0^2}{2!} + b_1 a_0 + b_2 = F(2), \\ \dfrac{Ha_0^3}{3!} + \dfrac{b_1 a_0^2}{2!} + b_2 a_0 + \sum_{i=1}^{2} b_{i,3} = F(3), \\ \dfrac{Ha_0^j}{j!} + \dfrac{b_1 a_0^{j-1}}{(j-1)!} + \dfrac{b_2 a_0^{j-2}}{(j-2)!} + \sum_{i=1}^{2} \dfrac{(ia_0)^{j-3} b_{i,3}}{(j-3)!} + \sum_{i=1}^{2} \dfrac{(ia_0)^{j-4} b_{i,4}}{(j-4)!} = F(j), \\ (j = 4, 5, 6, 7, 8), \\ L\gamma_0 + d_1 = \Theta(1), \\ \dfrac{L\gamma_0^2}{2!} + d_1 \gamma_0 + \sum_{i=1}^{3} d_{i,2} = \Theta(2), \\ \dfrac{L\gamma_0^j}{j!} + \dfrac{d_1 \gamma_0^{j-1}}{(j-1)!} + \sum_{i=1}^{3} \dfrac{(i\gamma_0)^{j-2} d_{i,2}}{(j-2)!} \gamma_0 d_{i,2} + \sum_{i=1}^{2} \dfrac{(i\gamma_0)^{j-3} d_{i,3}}{(j-3)!} = \Theta(j), \\ (j = 3, 4, 5, 6, 7). \end{cases} \tag{5.42}$$

By solving the system of nonlinear algebraic equations, the 14 undetermined parameters $\beta_1 = \frac{f''(0)}{2}$, $\beta_2 = \theta'(0)$, a_0, γ_0, b_1, b_2, $b_{i,j}$ $(i = 1, 2, j = 3, 4)$, d_1, $d_{i,2}(i = 1, 2, 3)$, and $d_{i,3}(i = 1, 2)$ can be determined. Then, the approximate analytical solution of the coupled BVP (5.26–5.29) can be presented in the form of a linear combination of the basic functions $f_{i,j}(\eta)$ and $\theta_{i,j}(\eta)$. For example, the approximate analytical solutions to the coupled BVP (5.26–5.29) for $h_u = 0.1$, $h_T = 0.1$, $C = 0.5$, $A = 1.2$, $Mn = 1$, $R = 2$, $Nr = 1$, $\Pr = 1$, $\gamma = 1$, and $\lambda = -1$ are:

$$\begin{aligned} f(\eta) &= 0.2706846474 + 2\eta + 0.2293153526e^{a_0\eta} - 0.1218970384\eta e^{a_0\eta} \\ &\quad - 0.08994285063\eta^2 e^{a_0\eta} + \eta^3\left(0.05194808802e^{a_0\eta} + 0.001149181495e^{2a_0\eta}\right) \\ &\quad + \eta^4\left(0.03739482413e^{a_0\eta} + 0.0007061715022e^{2a_0\eta}\right) \end{aligned}$$

$$\theta(\eta) = 0.8475608399e^{\gamma_0\eta} - 0.1259578390\eta e^{\gamma_0\eta} + \eta^2(-0.1968616308e^{\gamma_0\eta}$$
$$+ 0.006710952631e^{2\gamma_0\eta} - 0.0007715643591e^{3\gamma_0\eta})$$
$$+ \eta^4(0.05523751050e^{\gamma_0\eta} + 0.01023535446e - 1e^{2\gamma_0\eta})$$

where $a_0 = -2.815590725$ and $\gamma_0 = -1.649950889$. The values of the skin friction coefficient and wall temperature gradient are $f''(0) = 1.162223911$ and $\theta'(0) = -1.524391601$.

5.3.3 Results and Discussion

To validate the reliability of the DTM-BF we have presented, we also solved BVP (5.26–5.29) numerically using the classical fourth-order Runge–Kutta scheme coupled with the shooting method. Computation was carried out by the DTM-BF and the numerical method for various values such as velocity slip parameter h_u, thermal slip parameter h_T, unsteadiness parameter A, suction/injection parameter C, magnetic parameter Mn, velocity ratio parameter R, radiation parameter Nr, heat absorption/generation parameter λ, mixed convection parameter γ, and Prandtl number Pr. All of the results obtained by the DTM-BF technique were compared with those presented using the numerical method. The results of the comparisons are listed in Tables 5.4 and 5.5 and are

TABLE 5.4 **Comparison of Skin Friction Coefficient and Wall Temperature Gradient Obtained Using the Differential Transformation Method–Basic Function and the Numerical Method for Various h_u When $h_T = 0.1$, $\gamma = 1$, $\lambda = -1.0$, Pr = 1.0, $Mn = 1$, $C = 0.5$, $R = 2$, $Nr = 1$, and $A = 1.2$**

	$f''(0)$		$-\theta'(0)$	
h_u	DTM-BF	Numerical	DTM-BF	Numerical
0	2.921715940	2.923224794	1.502818994	1.502860530
0.1	2.324447822	2.325302130	1.524391601	1.524131156
0.3	1.635221332	1.635465976	1.547746165	1.547332847
0.5	1.256591896	1.256609415	1.559962833	1.559456833
1.0	0.793299371	0.793157196	1.574385594	1.573754063
2.0	0.455587648	0.455490005	1.584526322	1.583835028
5.0	0.199215200	0.199796606	1.591181621	1.591383837

DTM-BF, differential transformation method–basic function.

TABLE 5.5 Comparison of Skin Friction Coefficient and Wall Temperature Gradient Obtained Using the Differential Transformation Method–Basic Function and the Numerical Method for Various h_T When $h_u = 0.1$, $\gamma = 1$, $\lambda = -1.0$, $Pr = 1.0$, $Mn = 1$, $C = 0.5$, $R = 2$, $Nr = 1$, and $A = 1.2$

	$f''(0)$		$-\theta'(0)$	
h_T	DTM-BF	Numerical	DTM-BF	Numerical
0	2.354251550	2.355000279	1.799811987	1.799427815
0.2	2.302522936	2.303447031	1.322255061	1.322067756
0.4	2.272410894	2.273423364	1.045287172	1.045173315
0.7	2.245186882	2.246265315	0.795526464	0.795462486
1.0	2.228550422	2.229550809	0.642158676	0.642117797
2.0	2.202338386	2.202112387	0.390990599	0.390976925

DTM-BF, differential transformation method–basic function.

shown in Figs. 5.10–5.22. These tables and figures demonstrate that all of the results are in excellent agreement. Thus, DTM-BF is an accurate and effective approximate analytical method for solving the coupled nonlinear BVP on an unbounded domain.

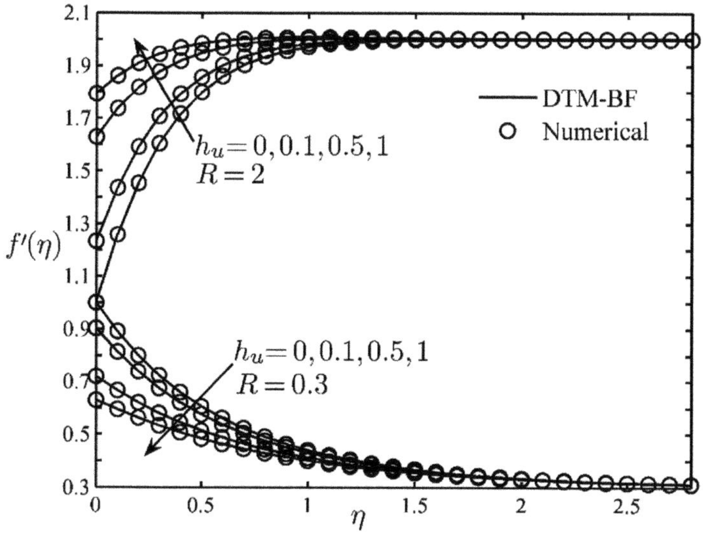

FIGURE 5.10 Velocity distributions against η obtained using the differential transformation method–basic function (DTM-BF) and the numerical method for various values of h_u and R when $h_T = 0.1$, $\gamma = 1$, $\lambda = -1.0$, $Pr = 1.0$, $Mn = 1$, $C = 0.5$, $Nr = 1$, and $A = 1.2$.

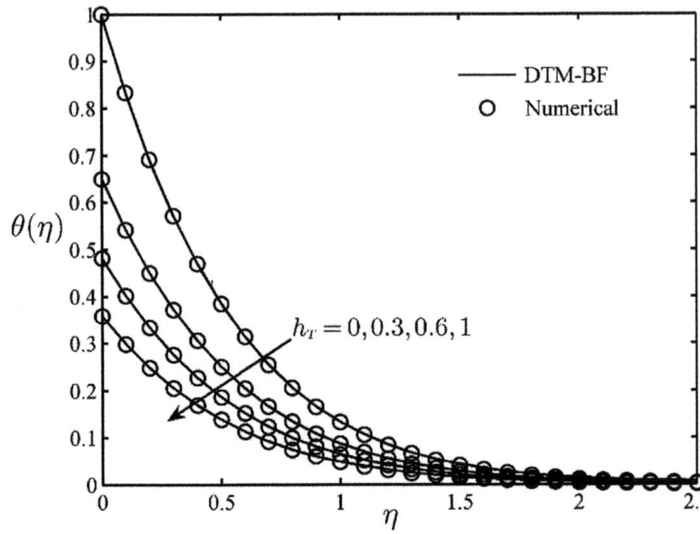

FIGURE 5.11 Temperature distributions against η obtained using the differential transformation method–basic function (DTM-BF) and the numerical method for various values of h_T when $h_u = 0.1$, $\gamma = 1$, $\lambda = -1.0$, $\Pr = 1.0$, $Mn = 1$, $C = 0.5$, $R = 2$, $Nr = 1$, and $A = 1.2$.

The effects of the velocity slip parameter h_u and thermal slip parameter h_T are presented in Tables 5.4 and 5.5 and Figs. 5.10 and 5.11. We can see from Table 5.4 and Fig. 5.10 that the values of skin friction coefficient $f''(0)$ decrease rapidly and the thickness of the momentum boundary layer reduces with the increase in the velocity slip parameter h_u because the momentum yielded by pulling of the stretching sheet can be only partly transmitted to the fluid under the partial velocity slip condition. In addition, the frictional resistance between the fluid and the surface decreases when the velocity slip starts increasing. A similar change in the surface heat transfer gradient $-\theta'(0)$ and the thermal boundary layer can be seen in Fig. 5.11 when the thermal slip parameter increases. Thus the thermal slip parameter h_T substantially decreases the heat transfer rate from the surface to the ambient fluid.

Figs. 5.12 and 5.13 illustrate the effects of velocity ratio parameter R on velocity and temperature profiles. An increase in R leads to a rise in the velocity gradient in the boundary layer in the two cases of $0 \leq R < 1$ and $R > 1$. Moreover, it shows that the fluid velocity decreases when $R \in (0,1)$ and it has an opposite trend when $R > 1$ with an increases in the values of R. On the other hand, the temperature decreases and the temperature gradient increases with increasing values of R when $R \geq 0$.

The velocity profiles for different values of mixed convection parameter γ are described in Fig. 5.14. It can be seen that the maximum peak value of velocity is obtained as $\gamma = 1.4$ and then decays to the free stream velocity. In addition, minimum peak value is attained in the absence of the thermal buoyancy force.

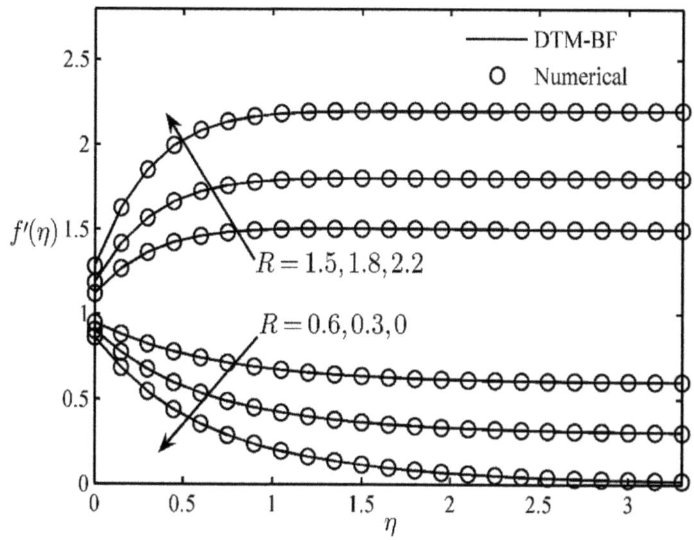

FIGURE 5.12 Velocity distributions against η obtained using the differential transformation method–basic function (DTM-BF) and the numerical method for various values of R when $h_u = 0.1$, $h_T = 0.1$, $\gamma = 1$, $\lambda = -1.0$, $\Pr = 1.0$, $Mn = 1$, $C = 0.5$, $Nr = 1$, and $A = 1.2$.

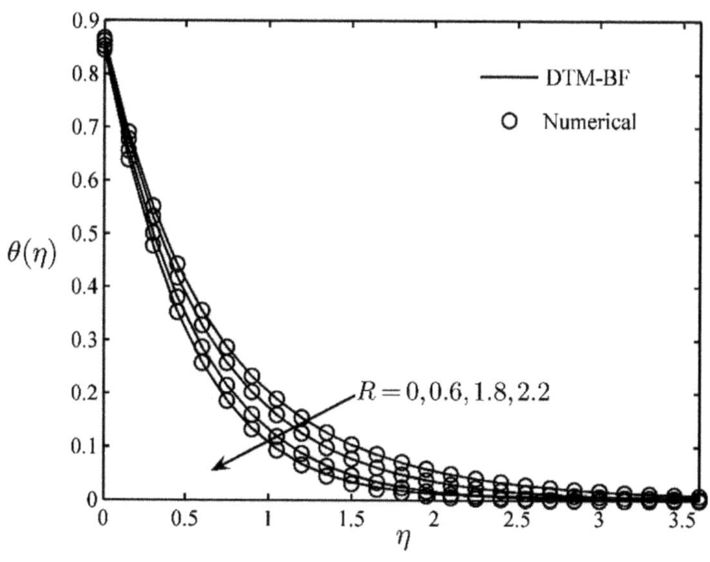

FIGURE 5.13 Temperature distributions against η obtained by using the differential transformation method–basic function (DTM-BF) and the numerical method for various values of R when $h_u = 0.1$, $h_T = 0.1$, $\gamma = 1$, $\lambda = -1.0$, $\Pr = 1.0$, $Mn = 1$, $C = 0.5$, $Nr = 1$, and $A = 1.2$.

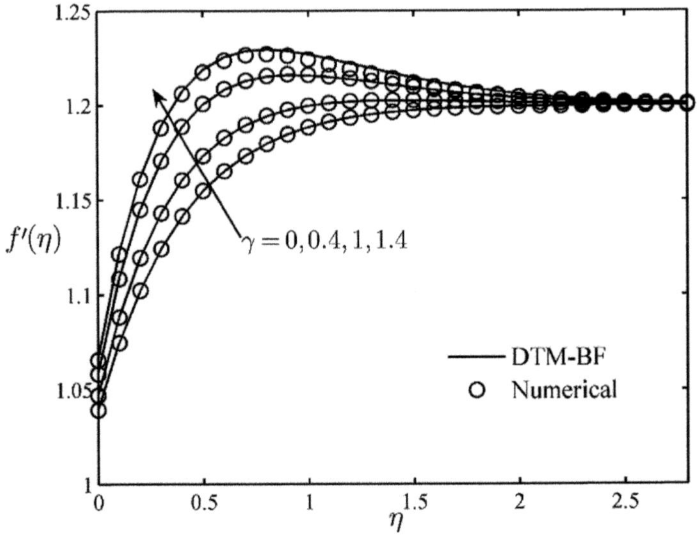

FIGURE 5.14 Velocity distributions against η obtained by using the differential transformation method—basic function (DTM-BF) and the numerical method for various values of γ when $h_u = 0.1$, $h_T = 0.1$, $\lambda = -1.0$, $\Pr = 1.0$, $Mn = 1$, $C = 0.5$, $Nr = 1$, $A = 1.2$, and $R = 1.2$.

This is because the thermal buoyancy force enhances fluid velocity and increases the velocity boundary layer (VBL) thickness with an increase in the value of γ.

Figs. 5.15 and 5.16 display the effects of suction/injection parameter C on the velocity and temperature profiles. The gradients of both fluid velocity and temperature increase with an increase in the values of C. The injection increases the momentum and thermal boundary layer thickness and decreases the velocity and temperature gradient. Nevertheless, suction has the opposite effect on the boundary layer.

Figs. 5.17 and 5.18 depict the velocity and temperature profiles for different values of unsteadiness parameter A. The gradient of the velocity increases with an increase in the unsteadiness parameter and the thickness of the boundary layer decreases with increasing values of the unsteadiness parameter. For all the cases of A considered, the increase in unsteadiness parameter A has the tendency to reduce the thermal boundary layer thickness, which results in an increase in the temperature gradient in the boundary layer.

Fig. 5.19 shows the velocity profiles for various values of magnetic parameter Mn as $R = 2$ and $R = 0.2$, respectively. The velocity curves show that an increase in magnetic parameter Mn decreases the momentum boundary layer thickness and increases the velocity gradient in the boundary layer. This is entirely because variations in Mn lead to variations in the Lorentz force owing to the magnetic field, and the Lorentz force produces more resistance with an increase in the values of Mn.

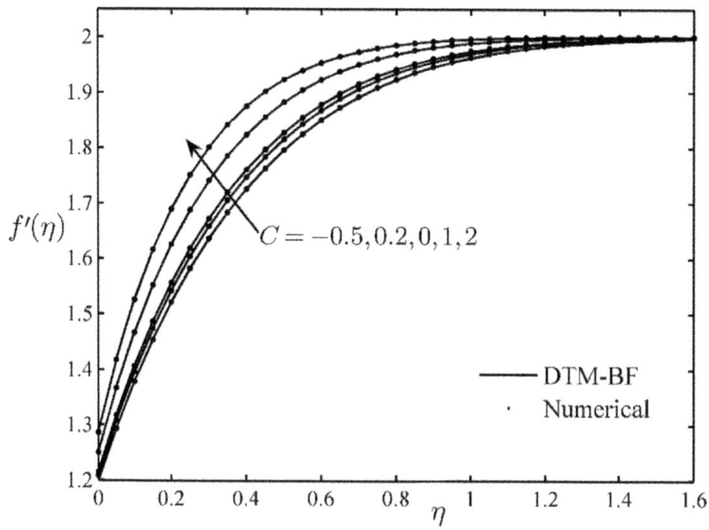

FIGURE 5.15 Velocity distributions against η obtained using the differential transformation method–basic function (DTM-BF) and the numerical method for various values of C when $h_u = 0.1$, $h_T = 0.1$, $\lambda = -1.0$, $\Pr = 1.0$, $Mn = 1$, $Nr = 1$, $\gamma = 1$, $R = 2$, and $A = 1.2$.

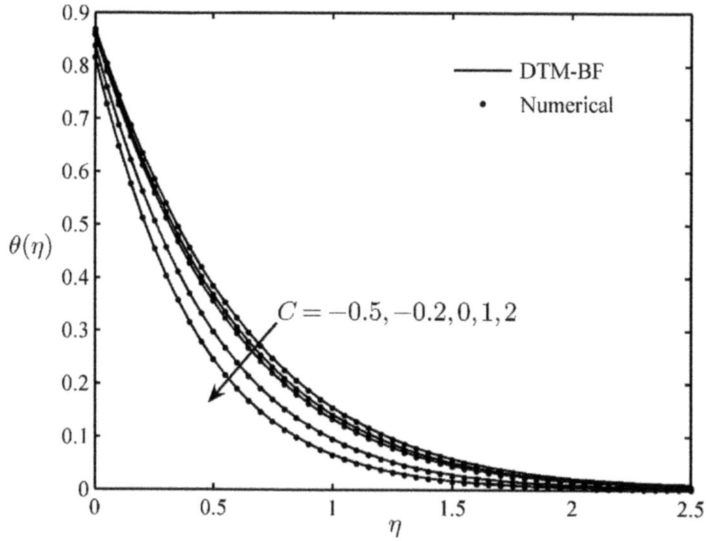

FIGURE 5.16 Temperature distributions against η obtained using the differential transformation method–basic function (DTM-BF) and the numerical method for various values of C when $h_u = 0.1$, $h_T = 0.1$, $\lambda = -1.0$, $\Pr = 1.0$, $Mn = 1$, $Nr = 1$, $\gamma = 1$, $A = 1.2$, and $R = 2$.

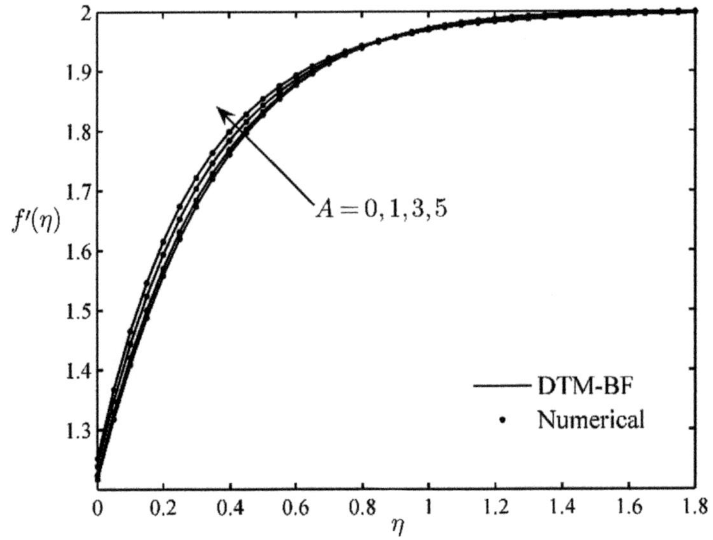

FIGURE 5.17 Velocity distributions against η obtained using the differential transformation method–basic function (DTM-BF) and the numerical method for various values of A when $h_u = 0.1$, $h_T = 0.1$, $\gamma = 1$, $\lambda = -1.0$, $\Pr = 4.0$, $Mn = 1$, $C = 0.5$, $Nr = 1$, and $R = 2$.

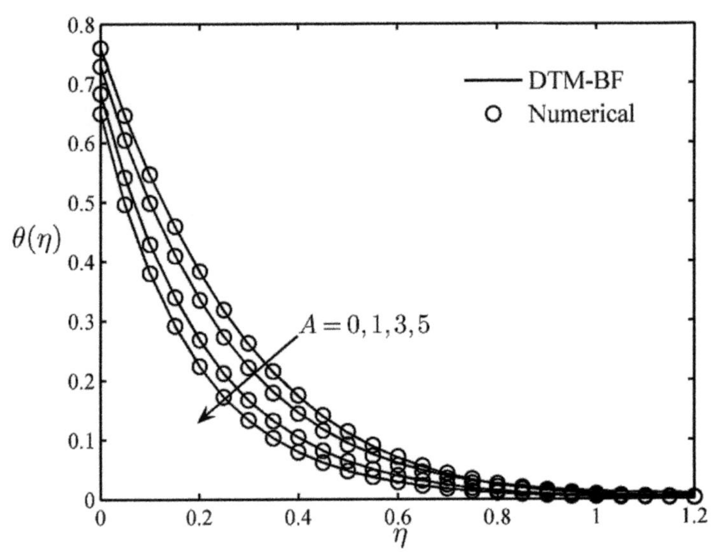

FIGURE 5.18 Temperature distributions against η obtained using the differential transformation method–basic function (DTM-BF) and the numerical method for various values of A when $h_u = 0.1$, $h_T = 0.1$, $\gamma = 1$, $\lambda = -1.0$, $\Pr = 4.0$, $Mn = 1$, $C = 0.5$, $Nr = 1$, and $R = 2$.

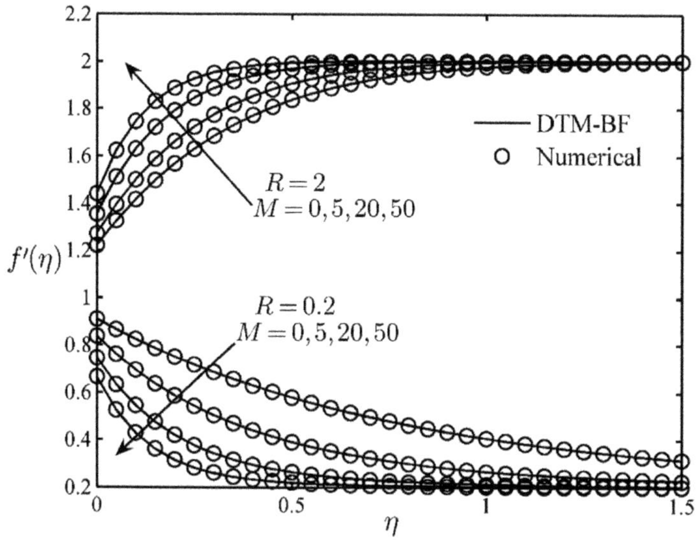

FIGURE 5.19 Velocity distributions against η obtained using the differential transformation method—basic function (DTM-BF) and the numerical method for various values of *Mn* and *R* when $h_u = 0.1$, $h_T = 0.1$, $\lambda = -1.0$, $\mathrm{Pr} = 1.0$, $Nr = 1$, $\gamma = 1$, $A = 1.2$, and $C = 0.5$.

Figs. 5.20–5.22 exhibit the effects of radiation parameter *Nr*, heat absorption/generation parameter λ, and Prandtl number Pr on the temperature profiles, respectively. From these three figures, this behavior can be understood such that an increase in Prandtl number Pr leads to a reduction in the

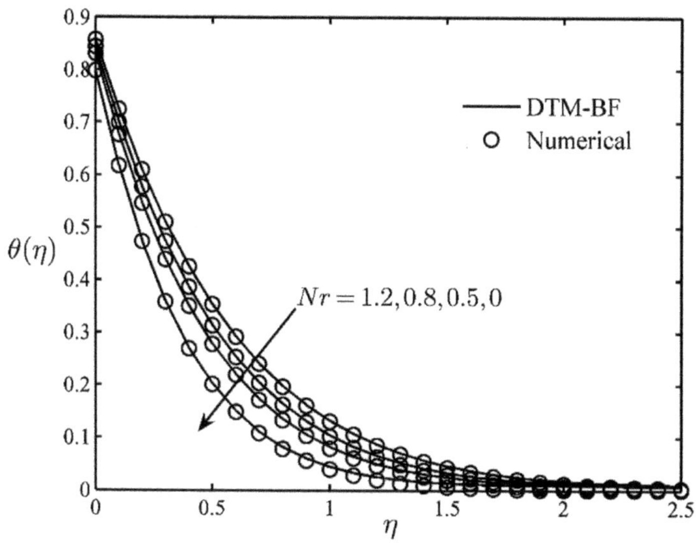

FIGURE 5.20 Temperature distributions against η obtained using the differential transformation method—basic function (DTM-BF) and the numerical method for various values of *Nr* and when $h_u = 0.1$, $h_T = 0.1$, $\lambda = -1.0$, $\mathrm{Pr} = 1.0$, $Mn = 1$, $R = 2$, $\gamma = 1$, $A = 1$, and $C = 0.5$.

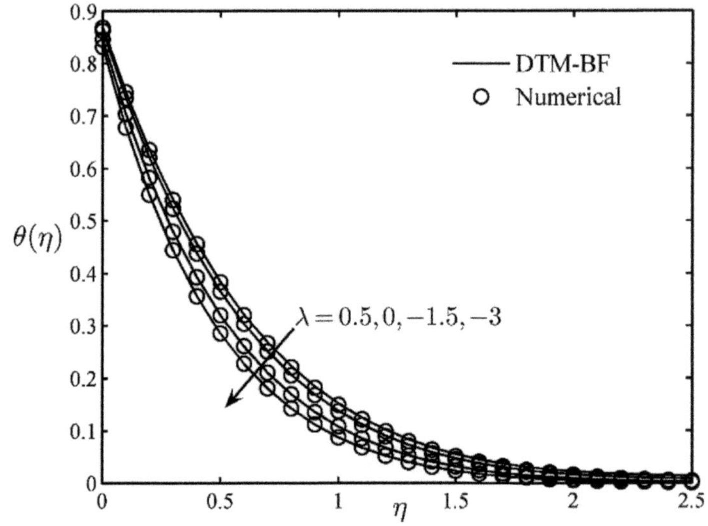

FIGURE 5.21 Temperature distributions against η obtained using the differential transformation method–basic function (DTM-BF) and the numerical method for various values of λ and when $h_u = 0.1$, $h_T = 0.1$, $Nr = 1.0$, $\Pr = 1.0$, $Mn = 1$, $R = 2$, $\gamma = 1$, $A = 1$, and $C = 0.5$.

thermal boundary layer thickness, which in turns causes a decrease in the temperature profile. On the other hand, the opposite effects are found with an increase in the values of radiation parameter Nr and heat absorption/generation parameter λ.

Tables 5.4 and 5.5 and Figs. 5.10–5.22 show that the analytical solution obtained by the DTM-BF method and numerical solution are highly in agreement with the different parameter values. This indicates that the DTM-BF is an effective method for solving the nonlinear BVP in unbounded domains.

5.4 MAGNETOHYDRODYNAMICS MIXED CONVECTIVE HEAT TRANSFER WITH THERMAL RADIATION AND OHMIC HEATING

5.4.1 Mathematical and Physical Description

Kumari and Nath (2010) and Ishak et al. (2010) investigated MHD mixed convective flow over a vertical stretching surface. Oztop et al. (2011) considered MHD mixed convection flow in a top-sided, lid-driven cavity heated by a corner heater.

Studies on mixed convective flow and heat transfer over a wedge surface are limited. Chamkha et al. (2003) explored thermal radiation effects on MHD forced convection flow adjacent to a nonisothermal wedge surface in the

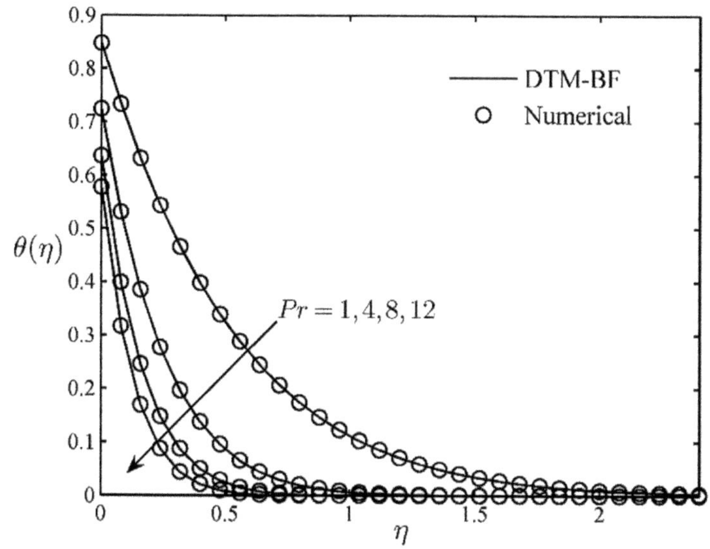

FIGURE 5.22 Temperature distributions against η obtained using the differential transformation method—basic function (DTM-BF) and the numerical method for various values of Pr when $h_u = 0.1$, $h_T = 0.1$, $Nr = 1.0$, $\lambda = -1$, $Mn = 1$, $R = 2$, $\gamma = 1$, $A = 1.2$, and $C = 0.5$.

presence of a heat source or sink when the wedge surface is fixed using the implicit finite-difference method. Hossain et al. (2006) presented numerical results for mixed convection flow over a wedge surface with a variable surface temperature. Muhaimin et al. (2009) investigated MHD mixed convective heat and mass transfer over a porous wedge surface in the presence of a chemical reaction. Kandasamy et al. (2009) studied MHD mixed convective heat and mass transfer past a porous wedge surface and considered the effects of chemical reaction, variable viscosity, and thermophoresis. Hayat et al. (2011) described the mixed convection flow of a non-Newtonian fluid past a wedge surface using homotopy analysis. In this section, DTM-BF is employed to study MHD mixed convective heat transfer over a permeable stretching wedge surface in which the effects of velocity ratio, magnetic field, suction or injection on the surface, thermal radiation, and Ohmic heating on velocity and temperature fields are taken into account. Some new characteristics of MHD mixed convective flow over a wedge surface will be addressed and analyzed.

5.4.2 Formulation of the Problem

We consider steady two-dimensional MHD mixed convective flow and heat transfer over a permeable stretching wedge surface in a moving viscous, incompressible, electrically conducting fluid, as shown in Fig. 5.23.

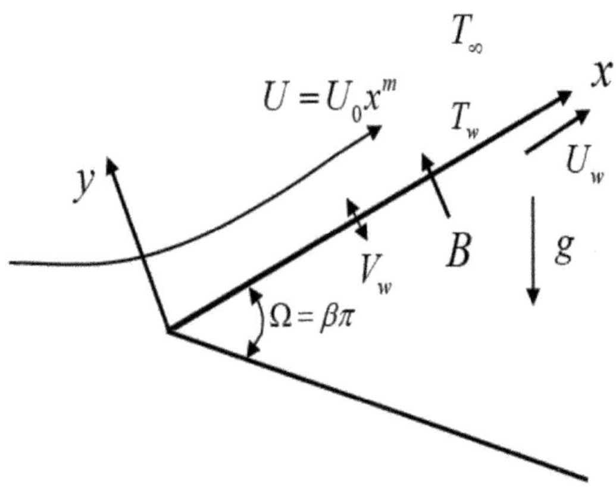

FIGURE 5.23 Magnetohydrodynamic mixed convective flow over the stretching wedge surface model and coordinate system.

Assume that the free stream velocity far from the wedge surface is $U = U_0 x^m$. The wedge surface stretches with a velocity of $U_w = RU(R \geq 0)$. The coordinate system is chosen such that the x axis is along the wedge surface and the y axis is normal to the wedge surface. A magnetic field $B(x) = B_0 x^{(m-1)/2}$ is applied in the y direction. The suction/injection velocity distribution across the wedge surface has a function form of $v_w = -C(\nu U_0)^{\frac{1}{2}} \frac{m+1}{2} x^{\frac{m+1}{2}}$. The magnetic Reynolds number is assumed to be small so that the induced magnetic field can be neglected. T_∞ is the temperature of the fluid outside the temperature boundary layer (TBL). $T_w = T_\infty + bx^{2m}$ is the temperature of the stretching wedge surface. $\beta = \frac{2m}{m+1}$ is the wedge angle parameter that corresponds to $\Omega = \beta\pi$ for a total angle of the wedge. $\beta = 0$ and $\beta = 1$ correspond to the horizontal wall case and the vertical wall case, respectively.

The boundary layer governing equations can be written as

$$\frac{\partial u}{\partial x} + \frac{\partial v}{\partial y} = 0 \tag{5.43}$$

$$u\frac{\partial u}{\partial x} + v\frac{\partial u}{\partial y} = U\frac{dU}{dx} + \nu\frac{\partial^2 u}{\partial y^2} + g\beta_0(T - T_\infty)\sin\frac{\Omega}{2} - \frac{\sigma B^2}{\rho}(u - U) \tag{5.44}$$

$$\rho c_p\left(u\frac{\partial T}{\partial x} + v\frac{\partial T}{\partial y}\right) = -\rho u\left(U\frac{dU}{dx} + \sigma B^2 U\right) + \alpha\frac{\partial^2 T}{\partial y^2} - \frac{\partial q_r}{\partial y} + \sigma B^2 u^2 \tag{5.45}$$

The boundary conditions are

$$u(x,0) = U_w, \quad v(x,0) = v_w, \quad T(x,0) = T_w \tag{5.46}$$

$$u(x,\infty) = U(x), \quad T(x,\infty) = T_\infty \tag{5.47}$$

where u and v are the velocity components along the x and y directions, respectively; T is the temperature; β_0 is the coefficient of thermal expansion; g is the gravity field; c_p is the specific heat at constant pressure; α is the thermal conductivity of the fluid; μ is the fluid viscosity; and $\nu = \mu/\rho$ is the kinematics viscosity of the fluid.

The radiative heat flux term q_r is analyzed by using the Rosseland approximation as

$$q_r = -\frac{4\sigma_0}{3k^*}\frac{\partial T^4}{\partial y} \tag{5.48}$$

where σ_0 and k^* are the Stefan–Boltzmann constant and the mean absorption coefficient, respectively. T^4 may be expressed as following approximation form

$$T^4 \approx 4T_\infty^3 T - 3T_\infty^4. \tag{5.49}$$

Approximation (5.49) is valid when the temperature difference within the flow is sufficiently small. We apply this approximation to Eq. (5.48) and thus obtain

$$q_r = -\frac{16T_\infty^3\sigma_0}{3k^*}\frac{\partial T}{\partial y}. \tag{5.50}$$

Introduce the following similarity transformations

$$\eta = U^{\frac{1}{2}}(\nu x)^{-\frac{1}{2}}y, \psi = (Ux\nu)^{\frac{1}{2}}f(\eta), \quad \theta(\eta) = \frac{T - T_\infty}{T_w - T_\infty}, \tag{5.51}$$

where $\psi(x,y)$ is the stream function. In view of $u = \frac{\partial\psi}{\partial y}$ and $v = -\frac{\partial\psi}{\partial x}$, continuity Eq. (5.43) is satisfied automatically. Functions $f' = \frac{u}{U}$ and θ are the dimensionless velocity and temperature, respectively. By substituting Eqs. (5.50) and (5.51) into (5.44) and (5.45), Eqs. (5.43–5.45) are reduced to a nonlinear coupled system of local similarity solution equations

$$f'''(\eta) + \frac{1+m}{2}f(\eta)f''(\eta) + m\left[1 - (f'(\eta))^2\right] - M(f'(\eta) - 1) + \left(\gamma\sin\frac{\Omega}{2}\right)\theta = 0, \tag{5.52}$$

$$\mathrm{Pr}^{-1}(1 + Nr)\theta'' + \frac{1+m}{2}f\theta' - 2mf'\theta + Ec\left(-(m+M)f' + Mf'^2\right) = 0. \tag{5.53}$$

The parameters that appeared in these equations are defined as

$$Gr_x = \frac{g\beta_0(T_w - T_\infty)x^3}{\nu^2}, \quad Re_x = \frac{Ux}{\nu}, \quad Ec = \frac{U^2}{c_p(T_w - T_\infty)},$$

$$\Pr = \frac{\mu c_p}{\alpha}, \quad M = \frac{\sigma B_0^2}{\rho U_0}, \quad \gamma = \frac{Gr_x}{\mathrm{Re}_x^2}, \quad Nr = \frac{16T_\infty^3 \sigma_0}{3k^* \alpha},$$

where Gr_x, Re_x, Ec, and Pr are the local Grashof number, local Reynolds number, Eckert number, and Prandtl number, respectively. M represents the magnetic parameter; Nr is the thermal radiation parameter; and γ is the mixed convection parameter, which represents the influence of the thermal buoyancy force on the flow field. Positive and negative values of γ correspond to the assisting and the opposing flow, respectively.

The corresponding boundary conditions are

$$f(0) = C, \quad f'(0) = R, \quad f'(+\infty) = 1, \tag{5.54}$$

$$\theta(0) = 1, \quad \theta(+\infty) = 0, \tag{5.55}$$

where C represents the suction or injection through the wedge surface with $C < 0$ for injection and $C > 0$ for suction; and $R = \frac{U_w}{U}$ is the velocity ratio parameter. The wedge stretches faster than that of the free stream flow when $R > 1$ and slower than that of the free stream flow when $R < 1$.

The main parameters of engineering interest for the present problem are the skin friction coefficient C_f and the local Nusselt number Nu_x, which characterize the surface shear stress and surface heat transfer rate, respectively, and are defined as:

$$C_f = \frac{\tau_w}{\rho U^2}, \quad Nu_x = \frac{x q_w}{\alpha (T_w - T_\infty)}, \tag{5.56}$$

where $\tau_w = \mu \left(\frac{\partial u}{\partial y}\right)_{y=0}$ and $q_w = -\alpha \left(\frac{\partial T}{\partial y}\right)_{y=0}$ are the wedge surface shear stress and the heat flux from the surface of the wedge, respectively. Using the similarity variables (5.51), we obtain

$$Re_x^{\frac{1}{2}} C_f = f''(0), \quad Nu_x \Big/ Re_x^{\frac{1}{2}} = -\theta'(0). \tag{5.57}$$

5.4.3 Differential Transformation Method—Basic Function Solutions

First, we solve the IVP for Eqs. (5.52) and (5.53) under the following initial value conditions using the DTM

$$f(0) = C, \quad f'(0) = R, \quad f''(0) = \alpha_1, \tag{5.58}$$

$$\theta(0) = 1, \quad \theta'(0) = \alpha_2. \tag{5.59}$$

The differential transformations of conditions (5.58) and (5.59) are

$$F(0) = C, \quad F(1) = R, \quad F(2) = \frac{\alpha_1}{2}, \tag{5.60}$$

$$\Theta(0) = 1, \quad \Theta(1) = \alpha_2 \tag{5.61}$$

where α_1 and α_2 are two priori unknown parameters that will be determined as part of the analytical solution. By performing the differential transformation for Eqs. (5.52) and (5.53), we obtain the following iterative formulas of transformed functions $F(k)$ and $\Theta(k)$

$$F(k+3) = \frac{-1}{(k+1)(k+2)(k+3)}\{-M(k+1)F(k+1) + (m+M)\delta(k)$$

$$+ \left(\gamma \sin\frac{\Omega}{2}\right)\Theta(k)\cdot\sum_{i=0}^{k}[-m(i+1)(k-i+1)F(i+1)F(k-i+1)$$

$$+ \frac{1+m}{2}(k-i+1)(k-i+2)F(i)F(k-i+2)]\}$$

$$\Theta(k+2) = \frac{-\Pr}{(k+1)(k+2)(1+Nr)}\{-E(m+M)(k+1)F(k+1)$$

$$+\sum_{i=0}^{k}\left[-2m(i+1)\Theta(i)F(i+1) + \frac{1+m}{2}(i+1)F(i)\Theta(i+1)\right.$$

$$+ EM(i+1)(k-i+1)F(i+1)F(k-i+1)]\}$$

Substituting Eqs. (5.60) and (5.61) into these iterative formulas, we can easily calculate all of the terms of $F(k)$ and $\Theta(k)$. According to the inverse transformation definition, the power series form solution of Eqs. (5.52) and (5.53) under conditions (5.58) and (5.59) are

$$f(\eta) = \sum_{k=0}^{\infty} F(k)\eta^k \approx \sum_{k=0}^{n} F(k)\eta^k. \tag{5.62}$$

$$\theta(\eta) = \sum_{i=0}^{\infty} \Theta(i)\eta^i \approx \sum_{i=0}^{m} \Theta(i)\eta^i. \tag{5.63}$$

Then we express the solution of the coupled BVP (5.52)–(5.55) in the form of a linear combination of the BFs. From Eqs. (5.52) and (5.53) and boundary conditions (5.54) and (5.55), it is reasonable to assume that $f(\eta)$ and $\theta(\eta)$ are expressed by the following BF sets

$$\left\{f_{0,0}(\eta),\ f_{i,j}(\eta)_{(i=1,2,3,\cdots,j=2,3,\cdots)}\right\} \text{ and } \left\{\theta_{0,0}(\eta),\ \theta_{i,j}(\eta)_{(i=1,2,\cdots,j=1,2,\cdots)}\right\}$$

in the form

$$f(\eta) \approx f_{N_1, N_2}(\eta) = f_{0,0}(\eta) + \sum_{j=2}^{N_1} \sum_{i=1}^{N_2} b_{i,j} f_{i,j}(\eta) = f_{0,0}(\eta) + \sum_{j=2}^{N_1} \sum_{i=1}^{N_2} b_{i,j} \eta^j e^{i a_0 \eta}, \tag{5.64}$$

$$\overline{\theta}(\eta) \approx \theta_{N_3, N_4}(\eta) = \theta_{0,0}(\eta) + \sum_{j=2}^{N_3} \sum_{i=1}^{N_4} d_{i,j} \theta_{i,j}(\eta) = \theta_{0,0}(\eta) + \sum_{j=2}^{N_3} \sum_{i=1}^{N_4} d_{i,j} \eta^j e^{i \gamma_0 \eta}, \tag{5.65}$$

where $\overline{\theta}(\eta) = \int_0^\eta \theta(\eta) d\eta$ is introduced to reduce the amount of computation. The functions

$$f_{0,0}(\eta) = C - b_0 + \eta + b_0 e^{a_0 \eta} + b_1 \eta e^{a_0 \eta}, \quad (b_1 = R - 1 - b_0 a_0)$$

and

$$\overline{\theta}_{0,0}(\eta) = -B_0 + B_0 e^{\gamma_0 \eta} + B_1 \eta e^{\gamma_0 \eta}, \quad (B_1 = 1 - B_0 \gamma_0)$$

satisfy inhomogeneous boundary condition (5.55) and the conditions

$$\overline{\theta}(0) = 0, \quad \overline{\theta}'(0) = 1, \quad \overline{\theta}'(\infty) = 0, \text{ respectively.}$$

In addition,

$$f_{i,j}(\eta) = b_{i,j} \eta^j e^{i a_0 \eta} \ (i = 1, 2, 3, \cdots, j = 3, 4, \cdots),$$

and

$$\theta_{i,j}(\eta) = d_{i,j} \eta^j e^{i \gamma_0 \eta} (i = 1, 2, 3, \cdots, j = 2, 3, \cdots)$$

respectively obey the homogeneous boundary conditions

$$f(0) = 0, \quad f'(0) = 0, \quad f'(\infty) = 0, \tag{5.66}$$

and

$$\overline{\theta}(0) = 0, \quad \overline{\theta}'(0) = 0, \quad \overline{\theta}'(\infty) = 0, \tag{5.67}$$

where $a_0 < 0$ and $\gamma_0 < 0$ are two undetermined decay parameters. In practical applications, it is sufficient to use the truncated functions $f_{N_1,N_2}(\eta)$ and $\theta_{N_3,N_4}(\eta)$ for $N_i \leq 4 (i = 1, 2, 3, 4)$. Here we choose $N_i = 3 (i = 1, 2, 3, 4)$.

The next step is to expand the right side of Eqs. (5.64) and (5.65) into power series form

$$\begin{aligned} f(\eta) = {} & C + (1 + b_0 a_0 + b_1)\eta + \left(\frac{b_0 a_0^2}{2!} + b_1 a_0 + \sum_{i=1}^{3} b_{i,2} \right) \eta^2 \\ & + \left(\frac{b_0 a_0^3}{3!} + \frac{b_1 a_0^2}{2!} + \sum_{i=1}^{3} i a_0 b_{i,2} + \sum_{i=1}^{3} b_{i,3} \right) \eta^3 \\ & + \left(\frac{b_0 a_0^4}{4!} + \frac{b_1 a_0^3}{3!} + \sum_{i=1}^{3} b_{i,2} \frac{(i a_0)^2}{2!} + \sum_{i=1}^{3} i a_0 b_{i,3} \right) \eta^4 + \cdots \end{aligned} \tag{5.68}$$

$$\overline{\theta}(\eta) = (B_0\gamma_0 + B_1)\eta + \left(\frac{B_0\gamma_0^2}{2!} + B_1\gamma_0 + \sum_{i=1}^{3} B_{i,2}\right)\eta^2$$
$$+\left(\frac{B_0\gamma_0^3}{3!} + \frac{B_1\gamma_0^2}{2!} + \sum_{i=1}^{3} i\gamma_0 B_{i,2} + \sum_{i=1}^{3} B_{i,3}\right)\eta^3 \quad , \quad (5.69)$$
$$+\left(\frac{B_0\gamma_0^4}{4!} + \frac{B_1\gamma_0^3}{3!} + \sum_{i=1}^{3} B_{i,2}\frac{(i\gamma_0)^2}{2!} + \sum_{i=1}^{3} i\gamma_0 B_{i,3}\right)\eta^4 + \cdots$$

From Eqs. (5.62), (5.63), (5.68), and (5.69), the following equations are constructed

$$\sum_{i=0}^{\infty} F(i)\eta^i = C + (1 + b_0a_0 + b_1)\eta + \left(\frac{b_0a_0^2}{2!} + b_1a_0 + \sum_{i=1}^{3} b_{i,2}\right)\eta^2$$
$$+\left(\frac{b_0a_0^3}{3!} + \frac{b_1a_0^2}{2!} + \sum_{i=1}^{3} ia_0b_{i,2} + \sum_{i=1}^{3} b_{i,3}\right)\eta^3 + \cdots \quad , \quad (5.70)$$

$$\sum_{i=0}^{\infty} \frac{\Theta(i)}{i+1}\eta^{i+1} = (B_0\gamma_0 + B_1)\eta + \left(\frac{B_0\gamma_0^2}{2!} + B_1\gamma_0 + \sum_{i=1}^{3} B_{i,2}\right)\eta^2$$
$$+\left(\frac{B_0\gamma_0^3}{3!} + \frac{B_1\gamma_0^2}{2!} + \sum_{i=1}^{3} i\gamma_0 B_{i,2} + \sum_{i=1}^{3} B_{i,3}\right)\eta^3 + \cdots, \quad (5.71)$$

Compare the coefficients of like powers of η on both sides of Eqs. (5.70) and (5.71) and we obtain the following system of algebraic equations

$$\frac{b_0a_0^2}{2!} + b_1a_0 + \sum_{i=1}^{3} b_{i,2} = F(2),$$

$$\frac{b_0a_0^3}{3!} + \frac{b_1a_0^2}{2!} + \sum_{i=1}^{3} ia_0b_{i,2} + \sum_{i=1}^{3} b_{i,3} = F(3),$$

$$\frac{b_0a_0^j}{j!} + \frac{b_1a_0^{j-1}}{(j-1)!} + \sum_{i=1}^{3} b_{i,\,2}\frac{(ia_0)^{j-2}}{(j-2)!} + \sum_{i=1}^{3} b_{i,\,3}\frac{(ia_0)^{j-3}}{(j-3)!} = F(j),\ (j = 4, 5, \cdots, 10)$$

$$\frac{B_0\gamma_0^2}{2!} + B_1\gamma_0 + \sum_{i=1}^{3} B_{i,2} = \frac{\Theta(1)}{2}$$

$$\frac{B_0\gamma_0^3}{3!}+\frac{B_1\gamma_0^2}{2!}+\sum_{i=1}^{3} i\gamma_0 B_{i,2}+\sum_{i=1}^{3} B_{i,3}=\frac{\Theta(2)}{3}$$

$$\frac{B_0\gamma_0^j}{j!}+\frac{B_1\gamma_0^{j-1}}{(j-1)!}+\sum_{i=1}^{3} B_{i,2}\frac{(i\gamma_0)^{j-2}}{(j-2)!}+\sum_{i=1}^{3} B_{i,3}\frac{(i\gamma_0)^{j-3}}{(j-3)!}=\frac{\Theta(j-1)}{j},\ (j=4,5,\cdots,10)$$

By solving this system of algebraic equations, we can determine the 18 undetermined parameters: $\alpha_1 = f''(0)$, $\alpha_2 = \theta'(0)$, a_0, γ_0, b_0, B_0, $b_{i,j}(i = 1, 2, 3, j = 2, 3)$ and $B_{i,j}(i = 1, 2, 3, j = 2, 3)$. Then the approximate analytical solution of coupled BVP (5.52–5.55) is presented in the form of a linear combination of BFs $f_{i,j}(\eta)$ and $\theta_{i,j}(\eta)$. For example, the approximate analytical solutions to coupled BVP (5.52–5.55) for $m = 0.1$, $C = 1.0$, $M = 3.0$, $\gamma = 0.8$, $Ec = 0.5$, $Nr = 1.0$, $R = 2.0$, and $\Pr = 1.0$ are:

$$\begin{aligned}f(\eta) &= 1.458643596+\eta-0.4586435961e^{a_0\eta}+0.1185697659\eta e^{a_0\eta}\\&\quad+\eta^2\left(-7.986256288\times10^{-3}e^{a_0\eta}+1.080929596\times10^{-2}e^{2a_0\eta}\right.\\&\quad\left.+5.291467881\times10^{-4}e^{3a_0\eta}\right)+\eta^3\left(-3.198870564\times10^{-2}e^{a_0\eta}\right.\\&\quad\left.+8.351466603\times10^{-3}e^{2a_0\eta}+2.229594910\times10^{-4}e^{3a_0\eta}\right)\end{aligned}$$

$$\begin{aligned}\theta(\eta) &= e^{\gamma_0\eta}+\eta-0.4586435961e^{\gamma_0\eta}+0.1185697659\eta e^{\gamma_0\eta}\\&\quad+\eta(2.066802367e^{\gamma_0\eta}+6.769525542\times10^{-2}e^{2\gamma_0\eta}\\&\quad+7.994077346\times10^{-3}e^{3\gamma_0\eta})+\eta^2(1.065918777e^{\gamma_0\eta}\\&\quad+4.00849128\times10^{-2}e^{2\gamma_0\eta}+2.188996800\times10^{-2}e^{3\gamma_0\eta})\\&\quad+\eta^3(8.687709982\times10^{-1}e^{\gamma_0\eta}+1.964316456\times10^{-1}e^{2\gamma_0\eta}\\&\quad+1.664896360\times10^{-2}e^{3\gamma_0\eta})\end{aligned}$$

where $a_0 = -1.92181956$ and $\gamma_0 = -2.40325036$. The values of the skin friction coefficient and temperature gradient on the wedge surface are $f''(0) = 2.142984882$ and $\theta'(0) = -0.4121373298$.

5.4.4 Numerical Solutions

Here, we numerically solve the coupled system of differential Eqs. (5.52) and (5.53) with the boundary condition (5.54) and (5.55) using the Runge–Kutta scheme coupled with the shooting method. In doing so, we rewrite nonlinear differential Eqs. (5.52) and (5.53) to a system of first-order differential equations in the form

$$f' = f_1,\ f_1' = f_2,\ \theta = f_3,\ f_3' = f_4,$$
$$f_2' = -\frac{1+m}{2} f f_2 + m f_1^2 + M f_1 - \left(\gamma \sin\frac{\Omega}{2}\right) f_3 - m - M \qquad .$$
$$f_4' = -\frac{\Pr}{1+Nr}\left[\frac{1+m}{2} f f_4 - 2m f_1 f_3 - Ec\left(M f_1^2 - (m+M) f_1\right)\right] \tag{5.72}$$

Corresponding boundary conditions are

$$f(0) = C,\ f_1(0) = R,\ f_2(0) = \alpha_1,\ f_3(0) = 1,\ f_4(0) = \alpha_2. \tag{5.73}$$

After suitable initial values are guessed for α_1 and α_2, the IVPs (5.72) and (5.73) are solved repeatedly using the fourth-order Runge–Kutta method with the secant method, which is used to improve the iteration for α_1 and α_2 until the boundary conditions $f'(\eta_\infty) = 1, f''(\eta_\infty) = 0$, $\theta(\eta_\infty) = 1$, and $\theta'(\eta_\infty) = 0$ are satisfied. In the problem-solving process, a step size of $\Delta\eta = 0.001$ is selected to be satisfactory with a convergence criterion of 10^{-5} in nearly all cases. The position of the edge of the boundary layer η_∞ had to be adjusted for different values of parameters to maintain accuracy.

5.4.5 Results and Discussion

Computations have been carried out using the DTM-BF and numerical methods for specified values of velocity ratio parameter R, magnetic field M, wedge angle parameter β, suction/injection parameter C, mixed convection parameter γ, thermal radiation parameter Nr, Eckert number Ec, and Prandtl number Pr. The results obtained using DTM-BF are compared with those presented using the numerical method in Tables 5.6 and 5.7 and Figs. 5.24–5.35.

Table 5.8, compares analytical and numerical results of skin friction coefficient $f''(0)$ with numerical results obtained by Ishak et al. (2007) and Yih (1998) in some simple cases. All of these comparisons are in excellent agreement, which shows that the DTM-BF method is an accurate and effective approximate analytical method for solving these kinds of nonlinear problems.

The effects of magnetic parameter M are presented in Table 5.6 and Figs. 5.24 and 5.25 for $R > 1$ and $R < 1$. Table 5.6 and Fig. 5.24 show that the absolute values of skin friction coefficient $|f''(0)|$ increase and the thickness of the momentum boundary layer decreases with an increase in magnetic parameter M for $R > 1$ or $R < 1$. Table 5.6 and Fig. 5.25 show the absolute values when the surface temperature gradient $|\theta'(0)|$ is larger and the temperature profile is lower as the magnetic parameter M becomes larger in the case of $R < 1$. However, the trends of the surface temperature gradient and the temperature profiles when $R > 1$ are opposite to those when $R < 1$ with an increase in magnetic parameter M.

TABLE 5.6 Comparison of Values of $f''(0)$ and $\theta'(0)$ for Different Values of M and R when $Pr = 1$, $C = 1$, $Nr = 1$, $\gamma = 0.8$, $R = 2$, $m = 0.1$, and $Ec = 0.5$

		$f''(0)$		$-\theta'(0)$	
M	R	DTM-BF	Numerical	DTM-BF	Numerical
0	0.1	1.028671081	1.024909003	0.623115564	0.626473534
	2.0	−1.123536646	−1.122862620	0.875723805	0.878136319
3	0.1	1.991026444	1.991260941	0.787173403	0.786144878
	2.0	−2.142984882	−2.144131581	0.412137330	0.407961083
10	0.1	3.199604583	3.200045562	0.970912005	0.969287493
	2.0	−3.501068130	−3.499789269	−0.15929260	−0.1570066807

DTM-BF, differential transformation method—basic function.

Table 5.7 shows that skin friction coefficient $f''(0)$ increases with an increase in wedge angle parameter β for $R < 1$, and first increases and then decreases with an increase in wedge angle parameter β for $R > 1$. On the other hand, the absolute values of surface temperature gradient $|\theta'(0)|$ increase with an increase in wedge angle parameter β.

Fig. 5.26 shows the velocity profiles for some values of wedge angle parameter β as $R = 0.5$ or $R = 1.1$. It can be seen from this figure that the velocity gradually rises to a peak value and then decays to the free stream velocity with an increase in β when $R = 0.5$ or $R = 1.1$. In addition, the peak

TABLE 5.7 Comparison of Values of $f''(0)$ and $\theta'(0)$ for Different Values of β and R When $Pr = 1$, $C = -0.2$, $Nr = 1$, $\gamma = 1$, and $Ec = 0.1$

		$f''(0)$		$-\theta'(0)$	
β	R	DTM-BF	Numerical	DTM-BF	Numerical
0	0.5	0.515048375	0.516899135	0.340898717	0.337387665
	1.1	−0.108271267	−0.108288744	0.373127918	0.372729011
2/3	0.5	1.058548363	1.060370852	0.750078091	0.752518431
	1.1	0.195477838	0.195539729	0.878583743	0.876601378
1	0.5	1.162627435	1.164326941	0.988602531	0.991813965
	1.1	0.136588816	0.134513028	1.173204559	1.175500685

DTM-BF, differential transformation method—basic function.

TABLE 5.8 Comparison of Values of $f''(0)$ for Different Values of C When $\gamma = 0$, $m = 1$, $M = 0$, and $R = 0$

	Numerical		Current Results	
C	Ishak et al. (2007)	Yih (1998)	Numerical	DTM-BF
−1	0.7566	0.75658	0.756575091	0.756379208
−0.5	0.9692	0.96923	0.969229466	0.969357017
0	1.2326	1.23259	1.232587947	1.233501042
0.5	1.5418	1.54175	1.541750966	1.541985282
1	1.8893	1.88931	1.889313587	1.889283054

DTM-BF, differential transformation method–basic function.

values of velocity first increase and then decrease with an increase in wedge angle parameter β when $R = 1.1$ or $R = 0.5$. A similar change in velocity profiles happens in Fig. 5.27 when the values of mixed convection parameter γ increase for assisting flow ($\gamma > 0$). However, the peak values of velocity increase with an increase in mixed convection parameter γ in the two cases of $R = 1.1$ and $R = 0.5$ for assisting flow, which differs from the effects of wedge

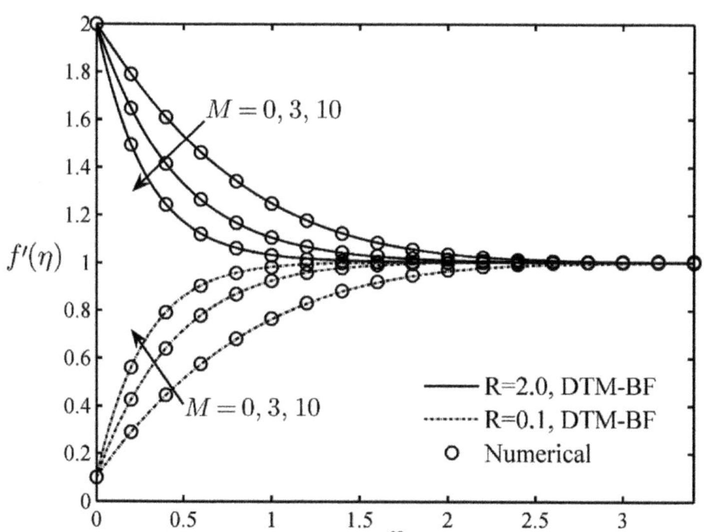

FIGURE 5.24 Velocity profiles $f'(\eta)$ for various values of M and R respectively obtained by the differential transformation method–basic function (DTM-BF) and the numerical method when $\gamma = 0.8$, $m = 0.1$, $C = 1.0$, $Ec = 0.5$, $Nr = 1.0$ and $\Pr = 1.0$.

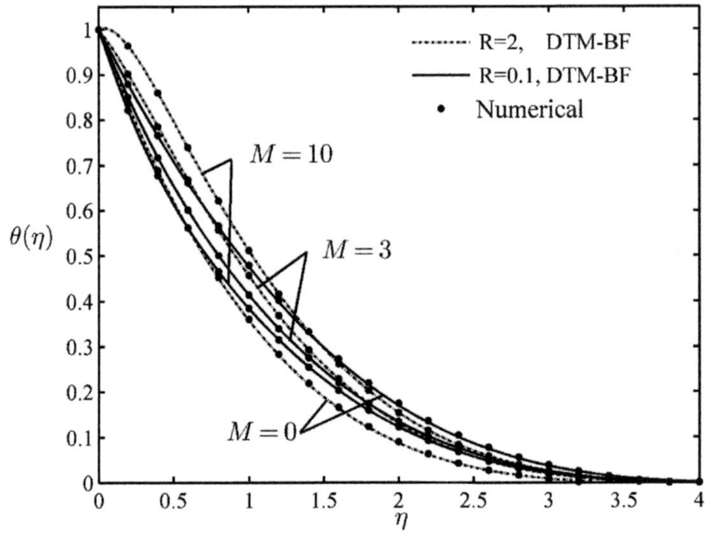

FIGURE 5.25 Temperature profiles $\theta(\eta)$ for various values of M and R respectively obtained by the differential transformation method—basic function (DTM-BF) and the numerical method when $\gamma = 0.8$, $m = 0.1$, $Ec = 0.5$, $C = 1.0$, $Nr = 1.0$, and $\Pr = 1.0$.

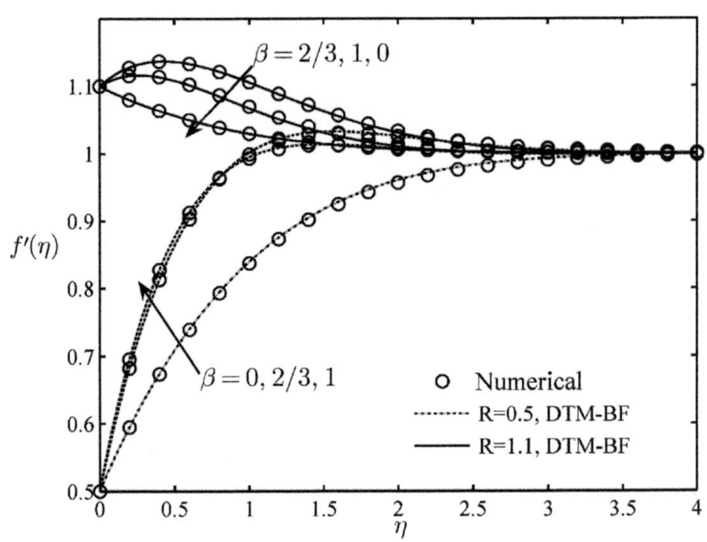

FIGURE 5.26 Velocity profiles $f'(\eta)$ for various values of β and R respectively obtained by the differential transformation method—basic function (DTM-BF) and the numerical method when $\gamma = 1.0$, $C = -0.2$, $M = 1.0$, $Ec = 0.1$, $Nr = 1.0$, and $\Pr = 1.0$.

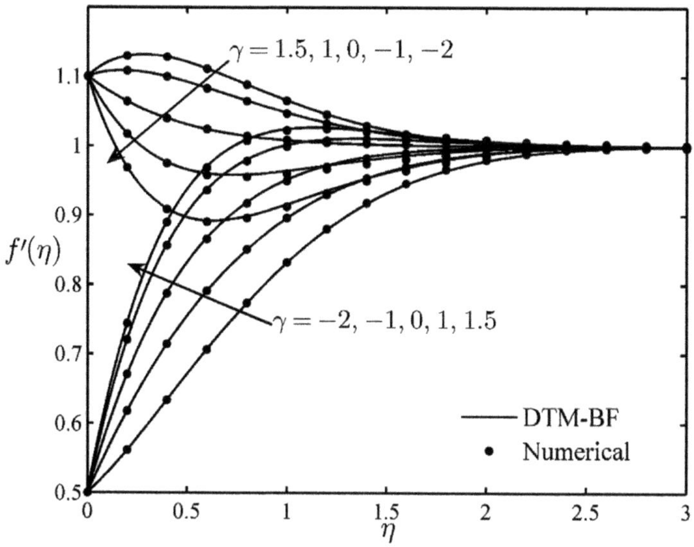

FIGURE 5.27 Velocity profiles $f'(\eta)$ for various values of γ and R respectively obtained by the differential transformation method–basic function (DTM-BF) and the numerical method when $m = 1.0$, $C = 0.5$, $M = 1.0$, $Ec = 0.1$, $Nr = 1.0$, and $\Pr = 1.0$.

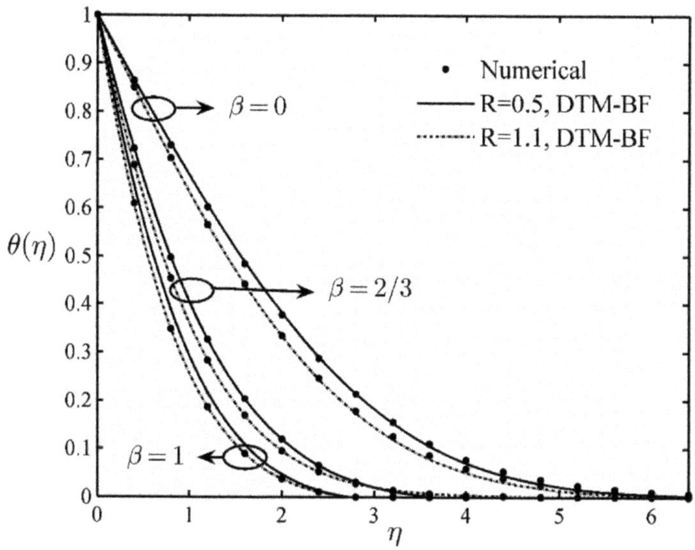

FIGURE 5.28 Temperature profiles $f'(\eta)$ for various values of β and R respectively obtained by the differential transformation method–basic function (DTM-BF) and the numerical method when $\gamma = 1.0$, $C = -0.2$, $M = 1.0$, $Ec = 0.1$, $Nr = 1.0$, and $\Pr = 1.0$.

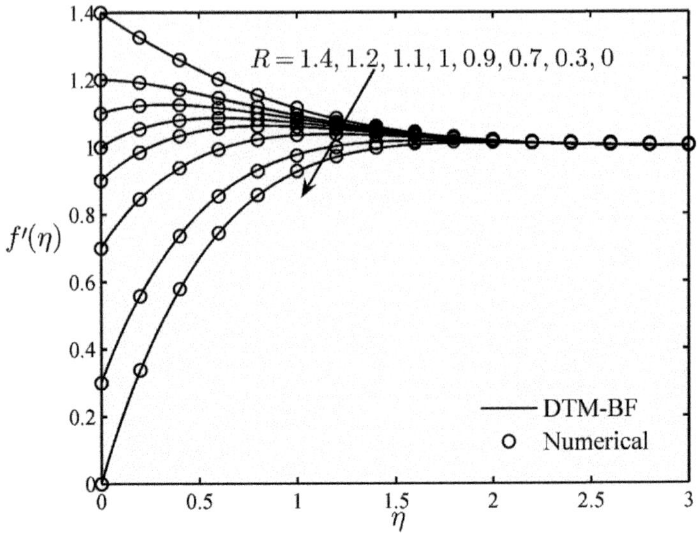

FIGURE 5.29 Velocity profiles $f'(\eta)$ for various values of R respectively obtained by the differential transformation method–basic function (DTM-BF) and the numerical method when $m = 0.5$, $C = 0.5$, $M = 1.0$, $\gamma = 1.0$, $Ec = 0.1$, $Nr = 1.0$, and $\Pr = 1.0$.

angle parameter β. On the other hand, in the case of opposing flow ($\gamma < 0$), no peak value of velocity exists when $R = 0.5$, whereas the velocity reaches a low point when $R = 1.1$. Moreover, the lowest values of velocity decrease with an increase in the absolute values $|\gamma|$ when $R = 1.1$ for opposing flow. Fig. 5.28 displays the effects of wedge angle parameter β on the temperature profile. It is indicated that an increase in β leads to a decrease in the thickness of the TBL when $R < 1$ or $R > 1$.

Fig. 5.29 illustrates the effects of velocity ratio parameter R on the velocity and temperature profiles. An increase in R leads to a rise in velocity. Moreover, the fluid velocity will reach a peak and the VBL will thin out when R approaches 1, which indicates that the effect of the mixed convection parameter γ becomes more important when R is closer to 1. Fig. 5.30 exhibits the temperature and thickness of the TBL which decreases with an increase in R.

Figs. 5.31 and 5.32 depict the effects of suction/injection parameter C on the velocity and temperature profiles. The gradients of the fluid velocity increase with an increase in the values of C. The injection increases the momentum and the TBL thickness and decreases the velocity and temperature gradient. Nevertheless, suction has the opposite effect on the boundary layer in the cases of $R < 1$ or $R > 1$.

Figs. 5.33–5.35 exhibit the effects of radiation parameter Nr, Eckert number Ec, and Prandtl number Pr on the temperature profiles, respectively.

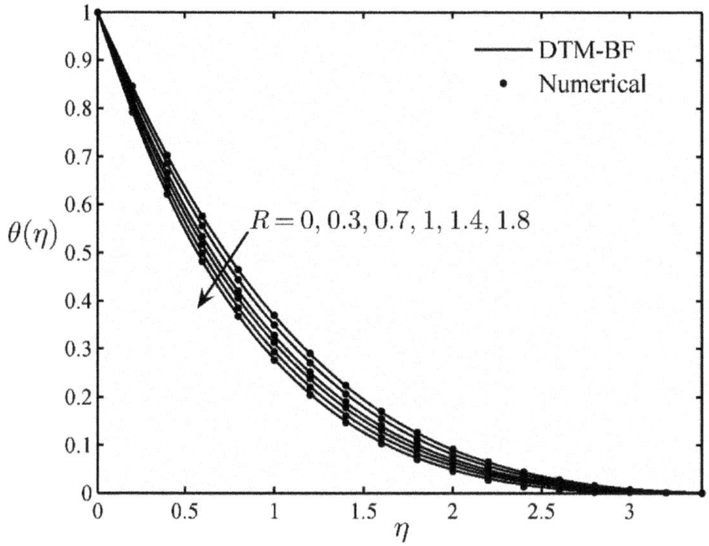

FIGURE 5.30 Temperature profiles $f'(\eta)$ for various values of R respectively obtained by the differential transformation method—basic function (DTM-BF) and the numerical method when $m = 0.5$, $C = 0.5$, $M = 1.0$, $\gamma = 1.0$, $Ec = 0.1$, $Nr = 1.0$, and $\Pr = 1.0$.

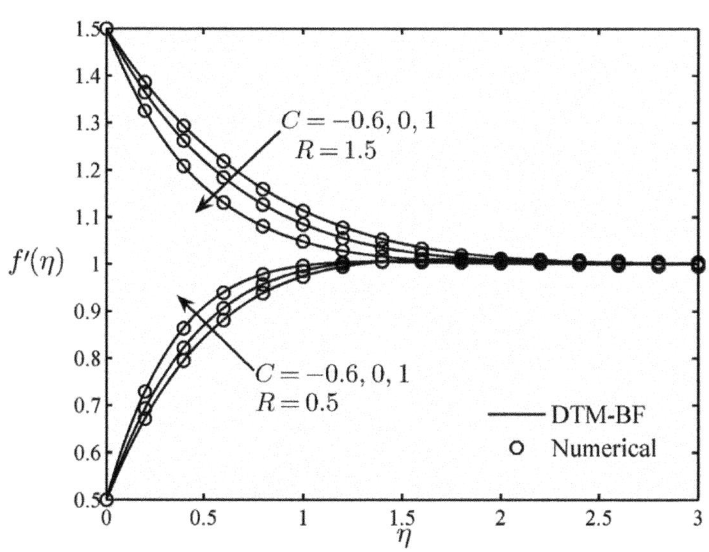

FIGURE 5.31 Velocity profiles $f'(\eta)$ for various values of C and R respectively obtained by the differential transformation method—basic function (DTM-BF) and the numerical method when $m = 1.0$, $M = 1.0$, $\gamma = 0.8$, $Ec = 0.1$, $Nr = 1.0$, and $\Pr = 1.0$.

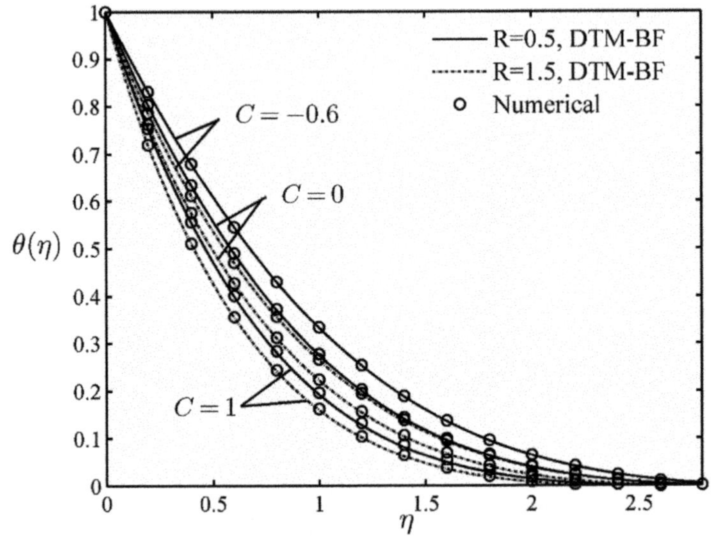

FIGURE 5.32 Temperature profiles $f'(\eta)$ for various values of C and R respectively obtained by the differential transformation method–basic function (DTM-BF) and the numerical method when $m=1.0$, $M=1.0$, $\gamma=0.8$, $Ec=0.1$, $Nr=1.0$, and $\Pr=1.0$.

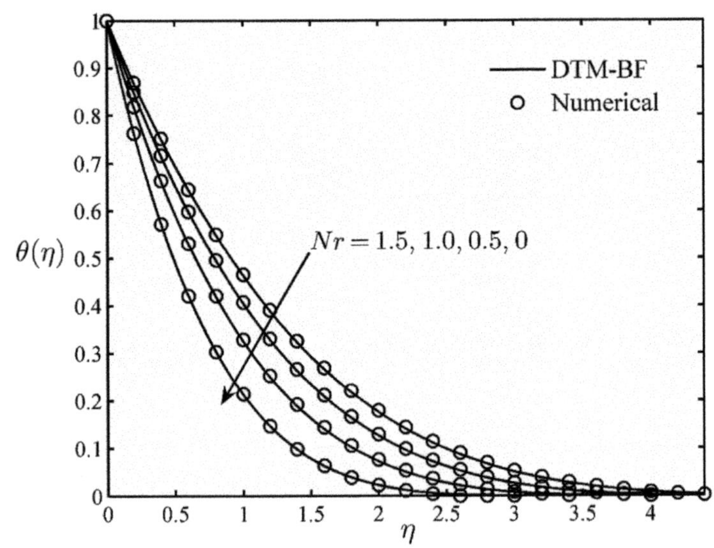

FIGURE 5.33 Temperature profiles $f'(\eta)$ for various values of Nr respectively obtained by the differential transformation method–basic function (DTM-BF) and the numerical method when $m=0.1$, $M=2.0$, $\gamma=0.8$, $R=0.5$, $Ec=0.2$, $C=1.5$, and $\Pr=1.0$.

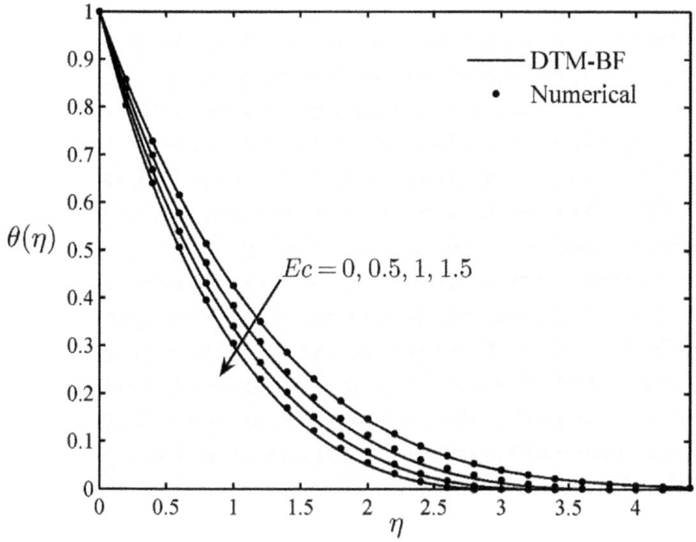

FIGURE 5.34 Temperature profiles $f'(\eta)$ for various values of Ec respectively obtained by the differential transformation method–basic function (DTM-BF) and the numerical method when $m = 0.1$, $M = 2.0$, $\gamma = 0.8$, $R = 0.5$, $Nr = 1.0$, $C = 1.5$, and $\Pr = 1.0$.

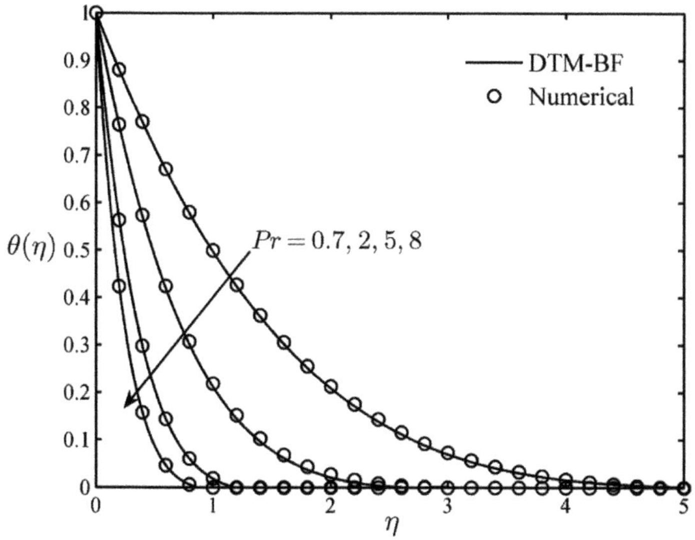

FIGURE 5.35 Temperature profiles $f'(\eta)$ for various values of Pr respectively obtained by the differential transformation method–basic function (DTM-BF) and the numerical method when $m = 0.1$, $M = 2.0$, $\gamma = 0.8$, $R = 0.5$, $Nr = 1.0$, $C = 1.5$, and $Ec = 0.2$.

An increase in the Eckert number or Prandtl number Pr leads to an increase in the temperature and a decrease in the thickness of the TBL. On the other hand, the opposite effects are found with an increase in the thermal radiation parameter, *Nr*.

5.5 MAGNETOHYDRODYNAMIC NANOFLUID RADIATION HEAT TRANSFER WITH VARIABLE HEAT FLUX AND CHEMICAL REACTION

5.5.1 Mathematical and Physical Description

The term "nanofluid" was coined by Choi (1995) in his seminal paper presented in 1995 at the American Society of Mechanical Engineers winter annual meeting. It refers to a liquid containing a dispersion of submicronic solid particles (nanoparticles) with a typical length on the order of 1−50 nm. Masuda et al. (1993) observed the phenomenon that nanofluids have an anomalous enhancement in thermal conductivity, which points to the potential of nanofluids in advanced nuclear systems. Buongiorno (2010) established an analytical model for convective transport in nanofluids by considering Brownian diffusion, and developed an explanation for abnormal convective heat transfer enhancement observed in nanofluids in which he showed that Brownian diffusion and thermophoresis are the two most important nanoparticle/base-fluid slip mechanisms. Makinde and Aziz (2011) and Makinde (2012, 2013a,b) studied the boundary layer flow of a nanofluid past a stretching sheet with a convective boundary condition, and considered buoyancy effects on MHD stagnation point flow and heat transfer of a nanofluid past a convectively heated stretching/shrinking sheet or boundary layer flow of nanofluid over a moving flat plate. Ibrahim and Makinde (2013) analyzed the effect of double stratification on boundary layer flow and heat transfer of nanofluid over a vertical plate. Motsumi and Makinde (2012) investigated the effects of thermal radiation and viscous dissipation on boundary layer flow of nanofluids over a permeable moving flat plate. Ghasemi and Aminossadati (2009), Congedo et al. (2009), and Ho et al. (2007, 2008) presented some numerical studies on the modeling of natural convection heat transfer in nanofluids using finite volume techniques or traditional finite-difference methods.

Wang (1987) studied a two-fluid stagnation flow problem when fluids with different densities are almost immiscible with constant properties. Chiam (1994) considered stagnation flow over a stretching sheet with identical stretching velocity and straining velocity. Bachok et al. (2012) investigated steady two-dimensional stagnation-point flow of a water-based nanofluid over an exponentially stretching/shrinking sheet. Using similarity transformation, the governing equations were reduced into coupled, nonlinear, ordinary differential equations and solved numerically for three types of nanoparticles.

The skin friction coefficient, local Nusselt number, and velocity and temperature profiles were presented graphically and discussed.

In this section we the study flow and radiation heat transfer of MHD nanofluids against a flat plate in a porous medium with the effects of variable surface heat flux and a first-order chemical reaction. Three types of nanoparticles, Cu, Al_2O_3, and Ag, are investigated. Similarity transformations are introduced to reduce the governing partial differential equations with boundary conditions to a system of ordinary differential equations over a semiinfinite interval. An efficient analytical approach based on DTM-BF for unbounded domains is performed to obtain approximate analytical solutions. The effects of the solid volume fraction, types of nanoparticles, permeability, magnetic field, chemical reaction, Schmidt number, radiation and suction/blowing on velocity, temperature, and concentration fields are graphically illustrated and analyzed in detail.

5.5.2 Formulation of the Problem

We consider MHD boundary layer stagnation flow and radiation heat transfer of a nanofluid with uniform suction or blowing on a flat plate embedded in a porous medium, while a variable magnetic field B is applied normally to the plate, as illustrated in Fig. 5.36.

Temperature at the surface is T_w, and the free stream velocity and ambient temperature (far from the surface) are U and T_∞, respectively. T_0 is a constant measuring the rate of temperature increase along the sheet. A first-order chemical reaction effect is assumed to exist. In the analysis of the flow in a porous medium, fluid motion is based on Darcy's law, which accounts for the drag exerted by the porous medium (Joseph et al., 1982; Pal and Hiremath, 1992).

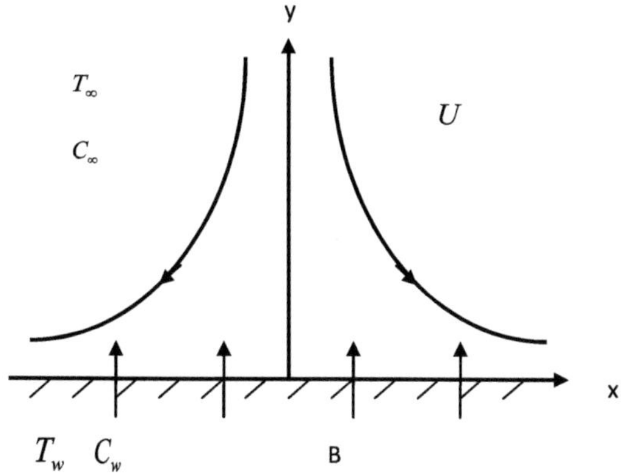

FIGURE 5.36 Sketch of saturation with a fluid with nanoparticles in suspension.

In view of the nanofluid model proposed by Kameswaran et al. (2012) with Boussinesq and boundary layer approximations, the steady boundary layer equations governing the nanofluid flow, radiation heat transfer, and concentration fields can be written in dimensional form as in Nield and Bejan (2006):

$$\frac{\partial u}{\partial x}+\frac{\partial v}{\partial y}=0, \tag{5.74}$$

$$u\frac{\partial u}{\partial x}+v\frac{\partial u}{\partial y}=-\frac{1}{\rho_{nf}}\frac{\partial p}{\partial x}+\frac{\mu_{nf}}{\rho_{nf}}\frac{\partial^2 u}{\partial y^2}-\frac{\mu_{nf}}{\rho_{nf}k}u-\frac{\sigma_{nf}B^2}{\rho_{nf}}u, \tag{5.75}$$

$$u\frac{\partial T}{\partial x}+v\frac{\partial T}{\partial y}=\alpha_{nf}\frac{\partial^2 T}{\partial y^2}-\frac{1}{(\rho c_p)_{nf}}\frac{\partial q_r}{\partial y}, \tag{5.76}$$

$$u\frac{\partial C}{\partial x}+v\frac{\partial C}{\partial y}=D\frac{\partial^2 C}{\partial y^2}-K(C-C_\infty). \tag{5.77}$$

The associated boundary conditions are:

$$\begin{aligned} &u=0,\ \ v=v_w,\ \ T=T_w=T_\infty+T_0\exp\left(\frac{x}{2L}\right),\\ &C=C_w=C_\infty+C_0\exp\left(\frac{x}{2L}\right),\ \ y=0, \end{aligned} \tag{5.78}$$

$$u\rightarrow U=a\exp\left(\frac{x}{L}\right),\ \ T\rightarrow T_\infty,\ \ C\rightarrow C_\infty,\ \ y\rightarrow\infty, \tag{5.79}$$

where u and v are velocity components in the x and y directions, respectively; T is the temperature of the nanofluid; p is the fluid pressure; and σ_s and σ_f are the electrical conductivity of the base fluid and the nanofluid. The magnetic Reynolds number is assumed to be small such that the thickness of the magnetic boundary layer is large and the induced magnetic field effect is negligible compared with the applied magnetic field. The Hall effects and viscous dissipation terms are also neglected. C is the nanoparticle concentration, C_∞ is the concentration in the free stream, and D is the mass diffusivity. $K(x)$ is the variable reaction rate given by $K(x)=K_0\exp(x/L)$ (Aminossadati and Ghasemi, 2009), L is the reference length, and K_0 is a constant. $v_w(x)=v_0\exp(x/2L)$ is a special suction or blowing velocity at the wall (Bhattacharyya, 2011), with v_0 a constant. Physically, $v_w(x)<0$ means that the fluid is injected (blowing) into the boundary layer and $v_w(x)>0$ implies that there is suction of fluid from the boundary layer. U is the free stream velocity and a is a constant number. μ_{nf} is the effective viscosity of the nanofluid, ρ_{nf} is the effective density of the nanofluid, and α_{nf} is the effective thermal diffusivity of the nanofluid (Bhattacharyya, 2012; Makinde, 2013a,b; Mutuku-Njane and Makinde, 2013; Mutuku-Njane and Makinde 2014), satisfying

$$\begin{aligned}
&\alpha_{nf}=\frac{k_{nf}}{(\rho C_p)_{nf}},\ \rho_{nf}=(1-\phi)\rho_f+\phi\rho_s,\ \mu_{nf}=\frac{\mu_f}{(1-\phi)^{2.5}},\\
&(\rho C_p)_{nf}=(1-\phi)(\rho C_p)_f+\phi(\rho C_p)_s,\\
&\frac{k_{nf}}{k_f}=\frac{(k_s+2k_f)-2\phi(k_f-k_s)}{(k_s+2k_f)+\phi(k_f-k_s)},\\
&\sigma_{nf}=(1-\phi)\sigma_f+\phi\sigma_s,
\end{aligned}\tag{5.80}$$

In the free stream, Eq. (5.75) becomes

$$U\frac{dU}{dx}=-\frac{1}{\rho_{nf}}\frac{\partial p}{\partial x}-\frac{\mu_{nf}}{\rho_{nf}K}U-\frac{\sigma_{nf}B^2}{\rho_{nf}}U.\tag{5.81}$$

Eliminating $\partial p/\partial x$ between Eqs. (5.75) and (5.81) yields

$$u\frac{\partial u}{\partial x}+v\frac{\partial u}{\partial y}=U\frac{dU}{dx}+\frac{\mu_{nf}}{\rho_{nf}}\frac{\partial^2 u}{\partial y^2}-\frac{\mu_{nf}}{\rho_{nf}k}(u-U)-\frac{\sigma_{nf}B^2}{\rho_{nf}}(u-U),\tag{5.82}$$

where $k=k_0\exp(-x/L)$ is the nonuniform permeability of the medium, k_0 is a constant that gives the initial permeability (Mandal and Mukhopadhyay, 2013), and $B=B_0\exp(x/2L)$, B_0 is the externally imposed magnetic field in the y direction (Ishak, 2011). Denote q_r as the radiative heat flux. According to Rosseland's approximation (1936), we have $q_r=-4\sigma_1\partial T^4/3k_1\partial y$, where σ_1 and k_1 are the Stefan–Boltzmann constant and absorption coefficient, respectively. We assume that temperature variation T^4 may be expanded in a Taylor's series. Neglecting higher-order terms and expanding T^4 about T_∞ yields $T^4\cong 4T_\infty^3T-3T_\infty^4$ and

$$\frac{\partial q_r}{\partial y}=-\frac{16\sigma_1T_\infty^3}{3k_1}\frac{\partial^2 T}{\partial y^2}.\tag{5.83}$$

Eq. (5.76) then reduces to

$$u\frac{\partial T}{\partial x}+v\frac{\partial T}{\partial y}=\alpha_{nf}\frac{\partial^2 T}{\partial y^2}+\frac{16\sigma_1T_\infty^3}{3(\rho c_p)_{nf}k_1}\frac{\partial^2 T}{\partial y^2}.\tag{5.84}$$

Let ψ be the stream function satisfying Eq. (5.74) with

$$u=\frac{\partial\psi}{\partial y},\ v=-\frac{\partial\psi}{\partial x}.\tag{5.85}$$

Introduce the following dimensionless functions f, θ, and g, and the similarity variable η as in (Mandal and Mukhopadhyay 2013):

$$\eta = y\sqrt{\frac{a}{2\nu_f L}}\exp\left(\frac{x}{2L}\right),\ \psi = \sqrt{2aL\nu_f}f(\eta)\exp\left(\frac{x}{2L}\right),\ \theta(\eta) = \frac{T - T_\infty}{T_w - T_\infty},$$

$$g(\eta) = \frac{C - C_\infty}{C_w - C_\infty}, \tag{5.86}$$

Substituting (5.86) into (5.77), (5.82), and (5.84) yields

$$\frac{1}{\phi_1}f''' + ff'' + 2(1 - f'^2) + \left(\frac{1}{\phi_1}P + \frac{\phi_4}{\phi_2}M\right)(1 - f') = 0, \tag{5.87}$$

$$\left[\frac{k_{nf}}{\phi_3 k_f} + \frac{R_1}{\phi_3}\right]\theta'' + \Pr(f\theta' - f'\theta) = 0, \tag{5.88}$$

$$g'' + Sc(fg' - f'g - \lambda g) = 0. \tag{5.89}$$

The boundary conditions are:

$$f = S,\ f' = 0,\ \theta = 1,\ g = 1,\ \eta = 0, \tag{5.90}$$

$$f' = 1,\ \theta = 0,\ g = 0 \text{ as } \eta \to \infty, \tag{5.91}$$

with the associated parameters:

$$P = \frac{2L\nu_f}{ak_0},\ M = \frac{2\sigma_f B_0^2 L}{a\rho_f},\ R = \frac{16\sigma_1 T_\infty^3}{3k},$$

$$S = \frac{v_0}{\sqrt{a\nu_f/2L}},\ Sc = \frac{\nu_f}{D},\ \lambda = \frac{2LK_0}{a},\ \Pr = \frac{\nu_f}{\alpha_f},$$

$$\begin{cases} \phi_1 = (1-\phi)^{2.5}[(1-\phi) + \phi\rho_s/\rho_f], \\ \phi_2 = (1-\phi) + \phi\rho_s/\rho_f, \\ \phi_3 = (1-\phi) + \phi(\rho C_p)_s/(\rho C_p)_f, \\ \phi_4 = (1-\phi) + \phi\sigma_s/\sigma_f, \end{cases} \tag{5.92}$$

where Pr is the Prandtl number, P is the permeability parameter, M is the magnetic parameter, R is the radiation parameter, S is the suction/blowing parameter, Sc is the Schmidt number, and λ is the chemical reaction parameter, $R_1 = \frac{R}{k_f}\left(= \frac{16\sigma_1 T_\infty^3}{3k_1 k_f}\right)$ with $k_f = 0.613$.

The quantities of practical interest in this study are the skin friction coefficient, C_f, the local Nusselt number, Nu_x and the local Sherwood number, Sh_x, which are defined as (Kameswaran et al., 2012)

$$C_f = \frac{\tau_w}{\rho_f U^2},\ Nu_x = \frac{xq_w}{k_f(T_w - T_\infty)},\ Sh_x = \frac{xp_m}{D(C_w - C_\infty)}, \tag{5.93}$$

where τ_w is the skin fraction, q_w is the heat flux form the sheet, and p_w is the mass flux at the wall surface, given by

$$\tau_w = \mu_{nf} \left.\frac{\partial u}{\partial y}\right|_{y=0}, q_w = -k_{nf} \left.\frac{\partial T}{\partial y}\right|_{y=0}, p_w = -D \left.\frac{\partial C}{\partial y}\right|_{y=0}, \tag{5.94}$$

Putting (5.94) into (5.93), we obtain

$$(2L/x)(1-\varphi)^{2.5} C_f \mathrm{Re}_x^{1/2} = f''(0), \;\; Nu_x \mathrm{Re}_x^{-1/2}(k_f/k_{nf}) = -\theta'(0),$$

$$Sh_x \mathrm{Re}_x^{-1/2} = -g'(0), \tag{5.95}$$

where Re_x represents the local Reynolds number, defined as $Re_x = Ux^2/(2L\nu_f)$.

5.5.3 Differential Transformation Method—Basic Function Solutions

The differential transformation of $w(t)$ is defined as:

$$W(k) = \frac{1}{k!}\left[\frac{d^k w(t)}{dt^k}\right]_{t=t_0}, \quad k = 0, 1, 2, 3, \cdots, \tag{5.96}$$

where $w(t)$ and $W(k)$ are the original function and the transformed function, respectively. The inverse transformation of $W(k)$ is defined as

$$w(t) = \sum_{k=0}^{\infty} W(k)(t-t_0)^k \approx \sum_{k=0}^{n} W(k)(t-t_0)^k. \tag{5.97}$$

Lowercase letter w and uppercase letter W are used to distinguish between the original and transformed functions according to the definitions of differential transformation (5.96) and inverse transformation (5.97).

To obtain analytical solutions to the BVP on an unbounded domain, we first need to solve the IVP. The work cannot be started at $\eta = 0$ because $f''(0)$, $\theta'(0)$, and $g'(0)$ are not known. We need to choose them as initial parameters and solve the IVP of Eqs. (5.87—5.89) with the following initial conditions:

$$f(0) = S, \;\; f'(0) = 0, \;\; f''(0) = 2\alpha_1, \tag{5.98}$$

$$\theta(0) = 1, \;\; \theta'(0) = \alpha_2, \tag{5.99}$$

$$g(0) = 1, \;\; g'(0) = \alpha_3. \tag{5.100}$$

The differential transformations for (5.98—5.100) are, respectively,

$$F(0) = S, F(1) = 0, F(2) = \alpha_1, \tag{5.101}$$

$$\Theta(0) = 1, \;\; \Theta(1) = \alpha_2, \tag{5.102}$$

$$G(0) = 1, \;\; G(1) = \alpha_3, \tag{5.103}$$

where α_1, α_2, and α_3 are three unknown parameters and will be determined as part of the solutions.

Implementing the differential transformation for Eqs. (5.87–5.89), we obtain the following iterative formulas for the transformed functions $F(k)$, $\Theta(k)$, and $G(k)$:

$$\frac{(k+1)(k+2)(k+3)F(k+3)}{\phi_1}+\sum_{i=0}^{k}\begin{bmatrix}(k-i+1)(k-i+2)F(i)F(k-i+2)-\\2(i+1)(k-i+1)F(i+1)F(k-i+1)\end{bmatrix},$$
$$+\left[\frac{1}{\phi_1}P+\frac{\phi_4}{\phi_2}M\right][\delta(k)-(k+1)F(k+1)]+2\delta(k)=0 \tag{5.104}$$

$$\left[\frac{k_{nf}/k_f}{\phi_3}+\frac{R/k_f}{\phi_3}\right](k+1)(k+2)\Theta(k+2)+\Pr\sum_{i=0}^{k}[(k-i+1)F(i)\Theta(k-i+1),$$
$$-(k-i+1)F(k-i+1)\Theta(i)]=0 \tag{5.105}$$

$$(k+1)(k+2)G(k+2)-Sc\lambda G(k)+Sc\sum_{i=0}^{k}[(k-i+1)F(i)G(k-i+1),$$
$$-(k-i+1)F(k-i+1)G(i)]=0 \tag{5.106}$$

where all of the terms of $F(k)$, $\Theta(k)$, and $G(k)$ can be calculated by substituting (5.101–5.103) into these iterative formulas. According to the inverse transformation (5.97), the power series form solutions of (5.87–5.89) under conditions (5.90) and (5.91) are:

$$f(\eta)=\sum_{k=0}^{\infty}F(k)\eta^k\approx\sum_{k=0}^{n_1}F(k)\eta^k, \tag{5.107}$$

$$\theta(\eta)=\sum_{k=0}^{\infty}\Theta(k)\eta^k\approx\sum_{k=0}^{n_2}\Theta(k)\eta^k, \tag{5.108}$$

$$g(\eta)=\sum_{k=0}^{\infty}G(k)\eta^k\approx\sum_{k=0}^{n_3}G(k)\eta^k. \tag{5.109}$$

We express the solutions of BVP (5.87–5.91) in the form of a linear combination of BFs. In view of Eqs. (5.87–5.89) and boundary conditions (5.90) and (5.91), it is reasonable to assume that $f(\eta)$, $\theta(\eta)$, and $g(\eta)$ can be expressed by the following BFs

$$\left\{f_{0,0}(\eta), f_{i,j}(\eta)_{(i=1,2,3,\ldots,j=3,4,\ldots)}\right\}, \left\{\theta_{0,0}(\eta), \theta_{i,j}(\eta)_{(i=1,2,3,\ldots,j=3,4,\ldots)}\right\},$$
$$\left\{g_{0,0}(\eta), g_{i,j}(\eta)_{(i=1,2,3,\ldots,j=3,4,\ldots)}\right\}, \tag{5.110}$$

in the form

$$f(\eta) \approx f_{N_1,N_2}(\eta) = f_{0,0}(\eta) + \sum_{j=2}^{N_1}\sum_{i=1}^{N_2} b_{i,j} f_{i,j}(\eta) = f_{0,0}(\eta) + \sum_{j=2}^{N_1}\sum_{i=1}^{N_2} b_{i,j}\eta^j e^{i a_0 \eta}, \tag{5.111}$$

$$\overline{\theta}(\eta) \approx \overline{\theta}_{N_3,N_4}(\eta) = \theta_{0,0}(\eta) + \sum_{j=2}^{N_3}\sum_{i=1}^{N_4} d_{i,j}\theta_{i,j}(\eta) = \theta_{0,0}(\eta) + \sum_{j=2}^{N_3}\sum_{i=1}^{N_4} d_{i,j}\eta^j e^{i\gamma_0\eta}, \tag{5.112}$$

$$\overline{g}(\eta) \approx \overline{g}_{N_5,N_6}(\eta) = g_{0,0}(\eta) + \sum_{j=2}^{N_5}\sum_{i=1}^{N_6} q_{i,j} g_{i,j}(\eta) = g_{0,0}(\eta) + \sum_{j=2}^{N_5}\sum_{i=1}^{N_6} q_{i,j}\eta^j e^{i\beta_0\eta}, \tag{5.113}$$

where $\overline{\theta}(\eta) = \int_0^\eta \theta(\eta)d\eta$ and $\overline{g}(\eta) = \int_0^\eta g(\eta)d\eta$ are introduced to reduce the amount of computation. The functions

$$f_{0,0}(\eta) = S - b_0 + \eta + b_0 e^{a_0\eta},$$

$$\theta_{0,0}(\eta) = -B_0 + B_0 e^{\gamma_0\eta} + B_1\eta e^{\gamma_0\eta} + B_2\eta^2 e^{\gamma_0\eta}, (B_1 = 1 - B_0\gamma_0),$$

$$g_{0,0}(\eta) = -E_0 + E_0 e^{\beta_0\eta} + E_1\eta e^{\beta_0\eta} + E_2\eta^2 e^{\beta_0\eta}, (E_1 = 1 - E_0\beta_0),$$

satisfy inhomogeneous boundary condition (5.91) and the conditions

$$\overline{\theta}(0) = 0, \quad \overline{\theta}'(0) = 1, \quad \overline{\theta}'(\infty) = 0, \quad \overline{g}(0) = 0, \quad \overline{g}'(0) = 1, \quad \overline{g}(0) = 0$$

In addition, the functions

$$f_{i,j}(\eta) = b_{i,j}\eta^j e^{i a_0\eta}, \theta_{i,j}(\eta) = d_{i,j}\eta^j e^{i\gamma_0\eta} \text{ and } g_{i,j}(\eta)$$
$$= q_{i,j}\eta^j e^{i\beta_0\eta}(i = 1, 2, 3, \cdots, j = 2, 3, 4, \cdots),$$

obey homogeneous boundary conditions

$$f(0) = 0, \quad f'(0) = 0, \quad f'(\infty) = 0, \tag{5.114}$$

$$\overline{\theta}(0) = 0, \ \overline{\theta}'(0) = 0, \ \overline{\theta}'(\infty) = 0, \tag{5.115}$$

$$\overline{g}(0) = 0, \ \overline{g}'(0) = 0, \ \overline{g}'(\infty) = 0, \tag{5.116}$$

where $a_0 < 0$, $\gamma_0 < 0$ and $\beta_0 < 0$ are three undetermined decay parameters. In applications, it is sufficient to consider truncated functions $f_{N_1,N_2}(\eta)$, $\theta_{N_3,N_4}(\eta)$ and $g_{N_5,N_6}(\eta)$ for $N_i \le 4$ ($i = 1, 2, 3, 4$). Here we choose $N_1 = N_4 = N_6 = 2$, $N_2 = N_3 = N_5 = 4$.

Expanding the right side of (5.111–5.113) into power series yields

$$f(\eta) = S - b_0 + \eta + \left(\frac{b_0 a_0^2}{2!} + \sum_{i=1}^{4} b_{i,2}\right)\eta^2 + \left(\frac{b_0 a_0^3}{3!} + \sum_{i=1}^{4} i a_0 b_{i,2}\right)\eta^3 + \left(\frac{b_0 a_0^4}{4!} + \sum_{i=1}^{4} b_{i,2}\frac{(i a_0)^2}{2!}\right)\eta^4 + \cdots, \tag{5.117}$$

$$\overline{\theta}(\eta) = (B_0\gamma_0 + B_1)\eta + \left(\frac{B_0\gamma_0^2}{2!} + B_1\gamma_0 + B_2\right)\eta^2 + \left(\frac{B_0\gamma_0^3}{3!} + \frac{B_1\gamma_0^2}{2!} + B_2\gamma_0 + \sum_{i=1}^{2} B_{i,3} i\gamma_0 + \sum_{i=1}^{2} B_{i,4}\right)\eta^3 + \left(\frac{B_0\gamma_0^4}{4!} + \frac{B_1\gamma_0^3}{3!} + \frac{B_2\gamma_0^2}{2!} + \sum_{i=1}^{2} B_{i,3}\frac{(i\gamma_0)^2}{2!} + \sum_{i=1}^{2} B_{i,4} i\gamma_0\right)\eta^4 + \cdots, \tag{5.118}$$

$$\overline{g}(\eta) = (E_0\beta_0 + E_1)\eta + \left(\frac{E_0\beta_0^2}{2!} + E_1\beta_0 + E_2\right)\eta^2 + \left(\frac{E_0\beta_0^3}{3!} + \frac{E_1\beta_0^2}{2!} + E_2\beta_0 + \sum_{i=1}^{2} E_{i,3} i\beta_0 + \sum_{i=1}^{2} E_{i,4}\right)\eta^3 + \left(\frac{E_0\beta_0^4}{4!} + \frac{E_1\beta_0^3}{3!} + \frac{E_2\beta_0^2}{2!} + \sum_{i=1}^{2} E_{i,3}\frac{(i\beta_0)^2}{2!} + \sum_{i=1}^{2} E_{i,4} i\beta_0\right)\eta^4 + \cdots. \tag{5.119}$$

In view of (5.107–5.109) and (5.117–5.119), the following equations are constructed and obtained:

$$\sum_{i=0}^{\infty} F(i)\eta^i = S - b_0 + \eta + \left(\frac{b_0 a_0^2}{2!} + \sum_{i=1}^{4} b_{i,2}\right)\eta^2 + \left(\frac{b_0 a_0^3}{3!} + \sum_{i=1}^{4} i a_0 b_{i,2}\right)\eta^3 + \left(\frac{b_0 a_0^4}{4!} + \sum_{i=1}^{4} b_{i,2}\frac{(i a_0)^2}{2!}\right)\eta^4 + \cdots, \tag{5.120}$$

$$\sum_{i=0}^{\infty}\Theta(i)\eta^{i}=(B_0\gamma_0+B_1)\eta+\left(\frac{B_0\gamma_0^2}{2!}+B_1\gamma_0+B_2\right)\eta^2$$
$$+\left(\frac{B_0\gamma_0^3}{3!}+\frac{B_1\gamma_0^2}{2!}+B_2\gamma_0+\sum_{i=1}^{2}B_{i,3}i\gamma_0+\sum_{i=1}^{2}B_{i,4}\right)\eta^3$$
$$+\left(\frac{B_0\gamma_0^4}{4!}+\frac{B_1\gamma_0^3}{3!}+\frac{B_2\gamma_0^2}{2!}+\sum_{i=1}^{2}B_{i,3}\frac{(i\gamma_0)^2}{2!}+\sum_{i=1}^{2}B_{i,4}i\gamma_0\right)\eta^4+\cdots, \tag{5.121}$$

$$\sum_{i=0}^{\infty}G(i)\eta^{i}=(E_0\beta_0+E_1)\eta+\left(\frac{E_0\beta_0^2}{2!}+E_1\beta_0+E_2\right)\eta^2$$
$$+\left(\frac{E_0\beta_0^3}{3!}+\frac{E_1\beta_0^2}{2!}+E_2\beta_0+\sum_{i=1}^{2}E_{i,3}i\beta_0+\sum_{i=1}^{2}E_{i,4}\right)\eta^3$$
$$+\left(\frac{E_0\beta_0^4}{4!}+\frac{E_1\beta_0^3}{3!}+\frac{E_2\beta_0^2}{2!}+\sum_{i=1}^{2}E_{i,3}\frac{(i\beta_0)^2}{2!}+\sum_{i=1}^{2}E_{i,4}i\beta_0\right)\eta^4+\cdots. \tag{5.122}$$

Comparing the coefficients of powers of η on both sides of (5.120–5.122), we obtain a system of algebraic equations

$$1+b_0a_0=F(1),$$
$$\frac{b_0a_0^2}{2!}+\sum_{i=0}^{4}b_{i,2}=F(2),$$
$$\frac{b_0a_0^3}{3!}+\sum_{i=1}^{4}ia_0b_{i,2}=F(3),$$
$$\frac{b_0{a_0}^j}{j!}+\sum_{i=1}^{3}b_{i,2}\frac{(ia_0)^{j-2}}{(j-2)!}=F(j)j,\quad (j=4,5,6,7),$$
$$\frac{B_0\gamma_0^2}{2!}+B_1\gamma_0+B_2=\frac{\Theta(2)}{2},$$
$$\frac{B_0\gamma_0^3}{3!}+\frac{B_1\gamma_0^2}{2!}+B_2\gamma_0+\sum_{i=1}^{2}B_{i,3}=\frac{\Theta(3)}{3},$$
$$\frac{B_0{\gamma_0}^j}{j!}+\frac{B_1{\gamma_0}^{j-1}}{(j-1)!}+\frac{B_2{\gamma_0}^{j-2}}{(j-2)!}+\sum_{i=1}^{2}B_{i,3}\frac{(i\gamma_0)^{j-3}}{(j-3)!}$$
$$+\sum_{i=1}^{2}B_{i,4}\frac{(i\gamma_0)^{j-4}}{(j-4)!}=\frac{\Theta(j)}{j},\quad (j=4,5,\cdots,9),$$

$$\frac{E_0\beta_0^2}{2!}+E_1\beta_0+E_2=\frac{G(2)}{2},$$

$$\frac{E_0\beta_0^3}{3!}+\frac{E_1\beta_0^2}{2!}+E_2\beta_0+\sum_{i=1}^{2}E_{i,3}=\frac{G(3)}{3},$$

$$\frac{E_0\beta_0^j}{j!}+\frac{E_1\beta_0^{j-1}}{(j-1)!}+\frac{E_2\beta_0^{j-2}}{(j-2)!}+\sum_{i=1}^{2}E_{i,3}\frac{(i\beta_0)^{j-3}}{(j-3)!}$$
$$+\sum_{i=1}^{2}E_{i,4}\frac{(i\beta_0)^{j-4}}{(j-4)!}=\frac{G(j)}{j},\quad (j=4,5,\cdots,9).$$

By solving these algebraic equations, we obtain a total of 25 undetermined parameters: $\alpha_1=f''(0)$, $\alpha_2=\theta'(0)$, $\alpha_3=g'(0)$, a_0, γ_0, β_0, b_0, B_0, B_1, B_2, E_0, E_1, E_2, $b_{i,j}(i=1,2,3,4,j=2)$, $B_{i,j}(i=1,2,j=3,4)$, and $E_{i,j}(i=1,2,j=3,4)$.

The solutions of BVP (5.86–5.89) are presented in the form of a linear combination of BFs $f_{i,j}(\eta)$, $\theta_{i,j}(\eta)$, and $g_{i,j}(\eta)$. For example, the approximate analytical solutions to BVP (5.86–5.89) for $\mathrm{Pr}=6.2$, $\phi=0.1$, $M=10^{-12}$, $P=S=R=Sc=\lambda=1$ are:

$$\begin{aligned}f(\eta)=&0.7223670337+\eta-e^{a_0\eta}+\big(-0.3575790524e^{a_0\eta}\\&-0.003391048884e^{2a_0\eta}+0.001113264561e^{3a_0\eta}\\&-0.001343054338e^{4a_0\eta}\big)\eta+\big(0.6439780138e^{a_0\eta}+0.01221414384e^{2a_0\eta}\\&-0.006014764254e^{3a_0\eta}+0.0009675035040e^{4a_0\eta}\big)\eta^2+\ldots,\end{aligned}$$

$$\begin{aligned}\theta(\eta)=&1+(3.645149747e^{\gamma_0\eta})\eta+\big(5.786414211e^{\gamma_0\eta}-0.004722807291e^{2\gamma_0\eta}\big)\eta^2\\&+\big(11.69383987e^{\gamma_0\eta}+0.01162450104e^{2\gamma_0\eta}\big)\eta^3\\&+\big(2.552407116e^{\gamma_0\eta}+0.01321880658e^{2\gamma_0\eta}\big)\eta^4+\ldots,\end{aligned}$$

$$\begin{aligned}g(\eta)=&1+\big(2.476752274e^{\beta_0\eta}\big)\eta+\big(2.632823681e^{\beta_0\eta}-0.002027833767e^{2\beta_0\eta}\big)\eta^2\\&+\big(1.555431850e^{\beta_0\eta}+0.003507021005e^{2\beta_0\eta}\big)\eta^3\\&+\big(-4.433704611e^{\beta_0\eta}-8.867409222^{2\beta_0\eta}\big)\eta^4+\ldots,\end{aligned}$$

where $a_0=-3.601877736$, $\gamma_0=-6.340607087$, and $\beta_0=-4.43370461$.

The skin friction coefficient, temperature, and concentration gradients are, $f''(0)=1.536927800$, $\theta'(0)=-2.695457337$, and $g'(0)=-1.956952340$, respectively.

5.5.4 Numerical Solutions

To validate the effectiveness and reliability of analytical results, the nonlinear differential Eqs. (5.87–5.89) with boundary conditions (5.90) and (5.91) are solved using a shooting technique coupled with the Runge–Kutta scheme and

Newton's method. In doing so, we convert the problems into a system of first-order differential equations:

$$f' = f_1,\ f_1' = f_2,\ f_2' = \varphi_1\left(2f_1^2 - ff_2 - 2\right) + \left[P + \frac{\varphi_1\varphi_4}{\varphi_2}M\right](f_1 - 1), \quad (5.123)$$

$$\theta' = h, \quad h' = -\frac{\Pr\varphi_3}{k_{nf}/k_f + R/k_f}(fh - f_1\theta), \quad (5.124)$$

$$g' = p, \quad p' = -Sc(fp - f_1 g - \lambda g), \quad (5.125)$$

with boundary conditions

$$f(0) = S, \quad f_1(0) = 0, \quad \theta(0) = 1, \quad g(0) = 1. \quad (5.126)$$

Let the range of numerical integration be finite dimensions (such that $\eta_{max} = 10$ is considered). Nevertheless numerical integration cannot be started at $\eta = 0$ because $f''(0)$, $\theta'(0)$, and $g'(0)$ are not known here. Thus we choose initial values for them as initial solution parameters. Then we compare the calculated value of $f''(0)$, $\theta'(0)$, $g'(0)$ at $\eta_\infty(= 10)$ with given boundary conditions $f'(\eta_\infty) = 1$, $\theta(\eta_\infty) = 0$, and $g(\eta_\infty) = 0$ and adjust the values of $f''(0)$, $\theta'(0)$, $g'(0)$ using a "secant method" to give a better approximation for the solutions. The process is repeated until we obtain results correct to the desired accuracy, say $\varepsilon = 10^{-6}$.

5.5.5 Results and Discussion

Calculations are carried out by DTM-BF and the numerical method for various specific values: admissible permeability parameter P, magnetic parameter M, solid volume fraction of nanoparticles φ, chemical reaction coefficient λ, radiation parameter R, and suction/blowing parameter S, and the type of nanoparticles. Table 5.9 lists data for the thermophysical properties of the fluid and the nanoparticles. Based on Masuda et al. (1993), we consider the nanoparticle volume fraction in the range of $0 \leq \varphi \leq 0.2$. The Prandtl number Pr of the base

TABLE 5.9 Thermophysical Properties of Fluid and Nanoparticles (Mutuku-Njane and Makinde, 2013; Hamad, 2010)

Physical Properties	Fluid Phase (Water)	Cu	Al_2O_3	Ag
$C_p(J/kgK)$	4179	385	765	235
$\rho(kg/m^3)$	997.1	8933	3970	10.50
$k(W/mK)$	0.613	400	40	429
$\sigma(S/m)$	5.5×10^{-6}	59.6×10^6	35×10^6	63.01×10^6

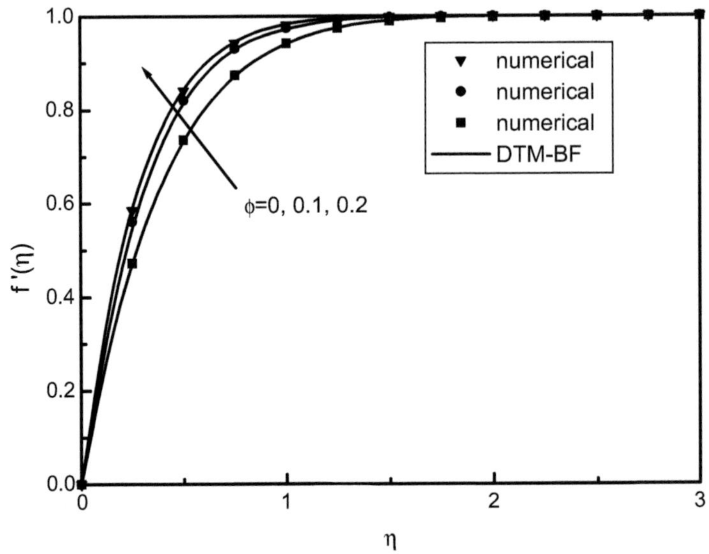

FIGURE 5.37 Velocity profiles for different values of $\varphi(0 \leq \varphi \leq 0.2)$ for Cu–water working nanofluid with $P = 0.5$, $R = S = 1$, and $M = 10^{-12}$. *DTM-BF*, differential transformation method–basic function.

fluid is kept constant. All of the solutions obtained by the DTM-BF method compared with the numerical ones in Figs. 5.37–5.48 are in close agreement. The results demonstrate that DTM-BF is an accurate and effective approximate analytical method for solving this type of nonlinear problem.

Figs. 5.37 and 5.38 show the effects of the solid volume fraction of nanoparticles φ when $M = 10^{-12}$, $P = 0.5$, and $R = S = 1$ for velocity and temperature profiles for a Cu–water nanofluid. The profiles of velocity and temperature distribution increase with increases in the value of φ. The increase in volume fraction of nanoparticles results in an increase in thermal conductivity of nanofluid and in the thickness of the thermal boundary layer as well.

Figs. 5.39 and 5.40 display the profiles of velocity and temperature distributions for different types of nanoparticles with $M = 10^{-12}$, $P = S = R = 1$, and $\varphi = 0.1$. The results indicate that the fluid velocity and temperature fields are strongly affected by the types of nanoparticles.

Figs. 5.41–5.43 display the effects of suction/blowing parameter S on the velocity, temperature, and concentration profiles, respectively. The velocity profiles significantly increases with increasing suction whereas they decreases with blowing in Fig. 5.41. When wall suction ($S > 0$) is considered, the results suggest a decrease in boundary layer thickness and an increase in fluid velocity. $S = 0$ represents the case of a nonporous plate. A contrary behavior is noted for blowing ($S < 0$). Fig. 5.42 displays the temperature distribution for variable suction/blowing parameter S with variable surface heat flux. The

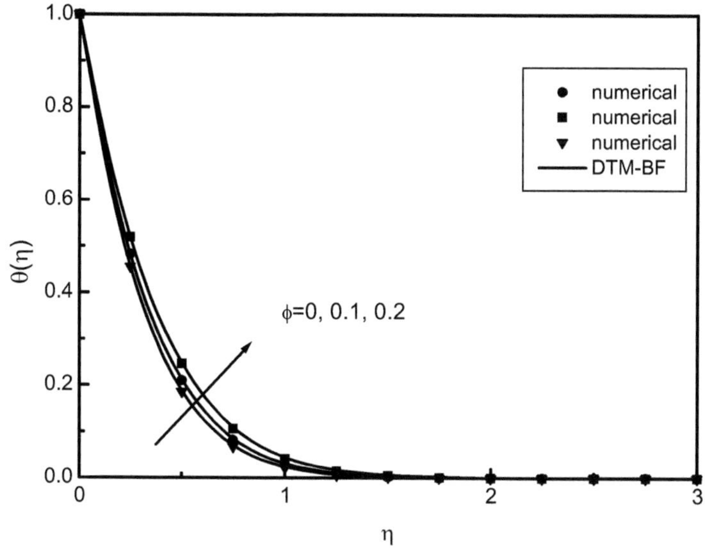

FIGURE 5.38 Temperature profiles for different values of $\varphi(0 \leq \varphi \leq 0.2)$ for Cu–water working nanofluid with $P = 0.5$, $R = S = 1$, and $M = 10^{-12}$. *DTM-BF*, differential transformation method–basic function.

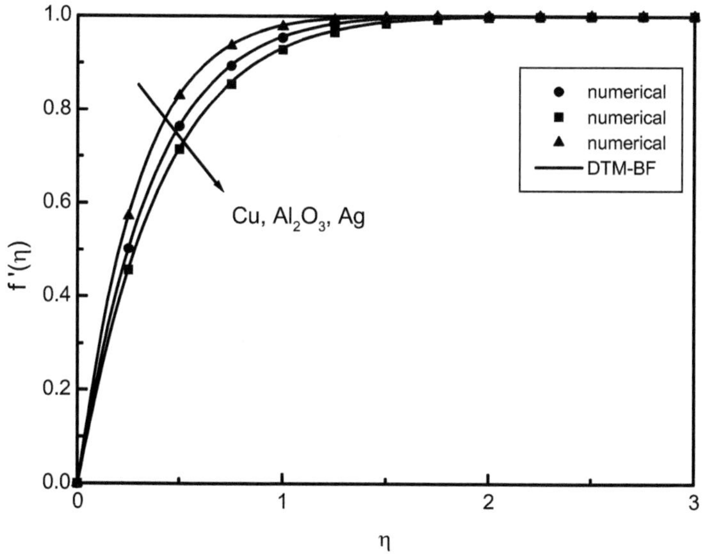

FIGURE 5.39 Velocity profiles for different types of nanofluids with $P = S = 1$, and $M = 10^{-12}$ when $R = 1$ and $\phi = 0.1$. *DTM-BF*, differential transformation method–basic function.

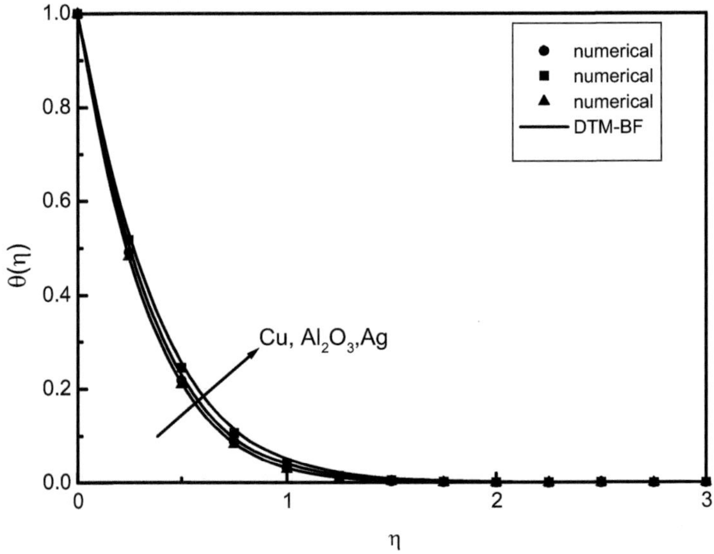

FIGURE 5.40 Temperature profiles for different types of nanofluids with $P = S = 1, M = 10^{-12}$ when $R = 1$ and $\phi = 0.1$. *DTM-BF*, differential transformation method—basic function.

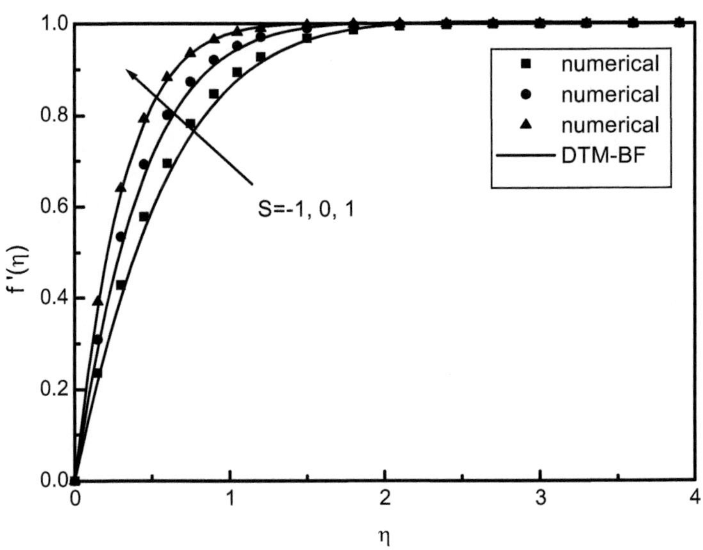

FIGURE 5.41 Velocity profiles for different values of suction/blowing parameter S for Cu—water working nanofluid with $M = 10^{-12}$, $P = 0.2$, $Sc = \lambda = 1$, $\varphi = 0.1$, and $R = 1$. *DTM-BF*, differential transformation method—basic function.

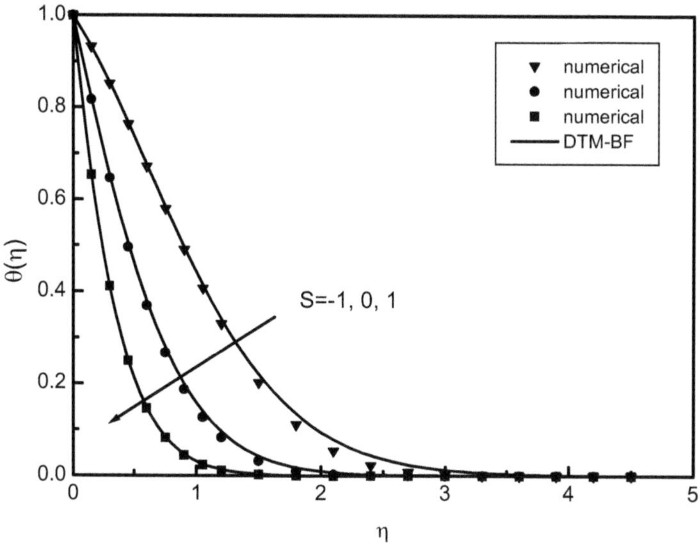

FIGURE 5.42 Temperature profiles for different values of the suction/blowing parameter S for Cu−water working nanofluid with $M = 10^{-12}$, $P = 0.2$, $Sc = \lambda = 1$, $\varphi = 0.1$, and $R = 1$. *DTM-BF*, differential transformation method−basic function.

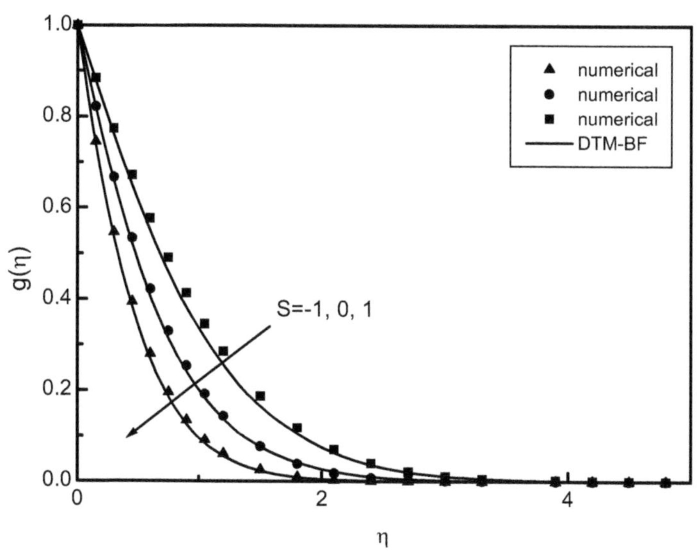

FIGURE 5.43 Concentration profiles for different values of suction/blowing parameter S for Cu−water working nanofluid with $M = 10^{-12}$, $P = 0.2$, $Sc = \lambda = 1$, $\varphi = 0.1$, and $R = 1$. *DTM-BF*, differential transformation method−basic function.

results indicate that the temperature profiles become lower, which implies a loss in the thickness of the TBL with increasing suction, whereas it increases as a result of increasing blowing, suggesting an increase in the thickness of the TBL. From Fig. 5.43, we can see that the concentration and thickness of the concentration boundary layer are reduced by an increase in suction/blowing parameter S.

Fig. 5.44 depicts the effects of a magnetic field on velocity distribution for a Cu–water nanofluid. It shows that an increase in M leads the velocity to increase whereas the thickness of the VBL decreases. Fig. 5.45 shows the effects of permeability parameter P on the fluid velocity for a Cu–water working nanofluid when $M = 10^{-12}$, $S = R = 1$, and $\varphi = 0.1$; the results indicate that the fluid velocity increases with an increase in permeability parameter P.

From Fig. 5.46, we can see that the thermal boundary layer thickness increases with an increase in thermal radiation parameter R for a Cu–water nanofluid when $M = 10^{-12}$, $P = 0.5$, $S = 1$, and $\varphi = 0.1$. Fig. 5.47 indicates that the concentration and thickness of the boundary layer decrease as the Schmidt number increases. Fig. 5.48 shows the effects of different values of the chemical reaction coefficient on the concentration profiles. It indicates that the concentration profiles increase with the chemical reaction coefficient. It also implies an increase in the thickness of the concentration boundary layer.

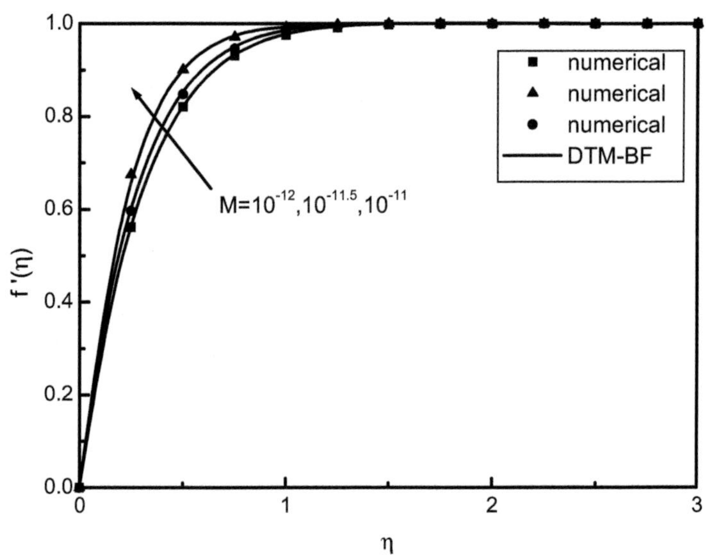

FIGURE 5.44 Velocity profiles for different values of magnetic parameter M for Cu–water working nanofluid with $P = 0.5$ and $S = 1$ when $R = 1$ and $\phi = 0.1$. *DTM-BF*, differential transformation method–basic function.

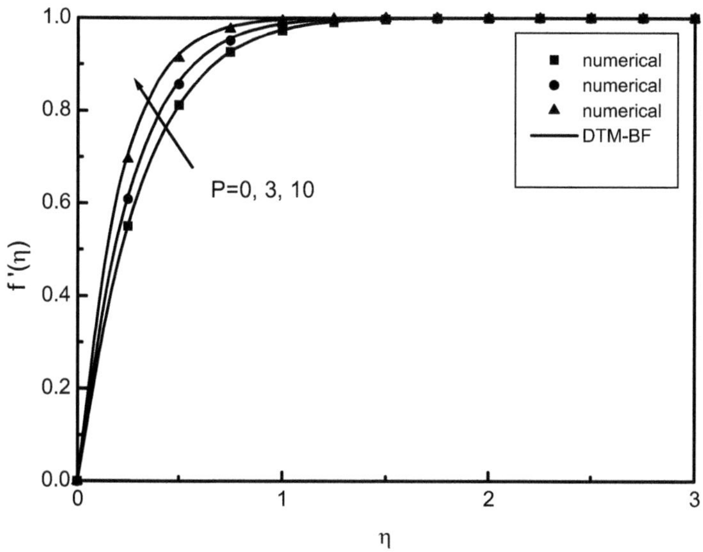

FIGURE 5.45 Velocity profiles for different values of permeability parameter P for Cu–water working nanofluid with $M = 10^{-12}$ and $S = 1$ when $R = 1$ and $\phi = 0.1$. *DTM-BF*, differential transformation method–basic function.

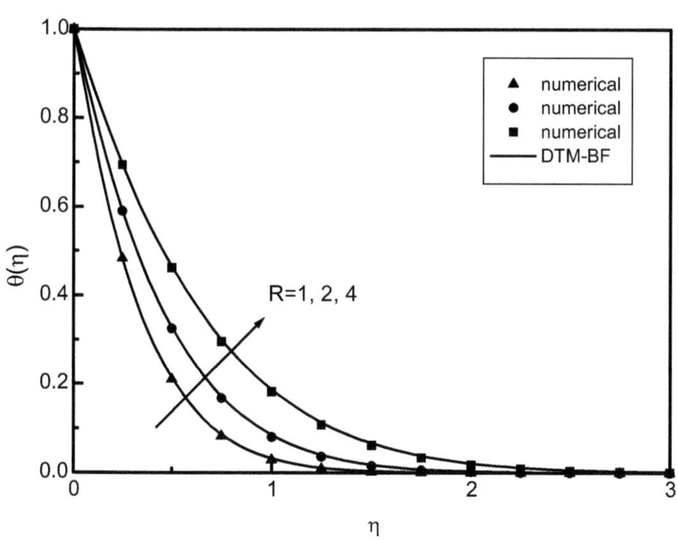

FIGURE 5.46 Effects of radiation parameter R on the temperature distribution $\theta(\eta)$ of a Cu–water working nanofluid with $M = 10^{-12}$ and $S = 1$ when $P = 0.5$ and $\phi = 0.1$. *DTM-BF*, differential transformation method–basic function.

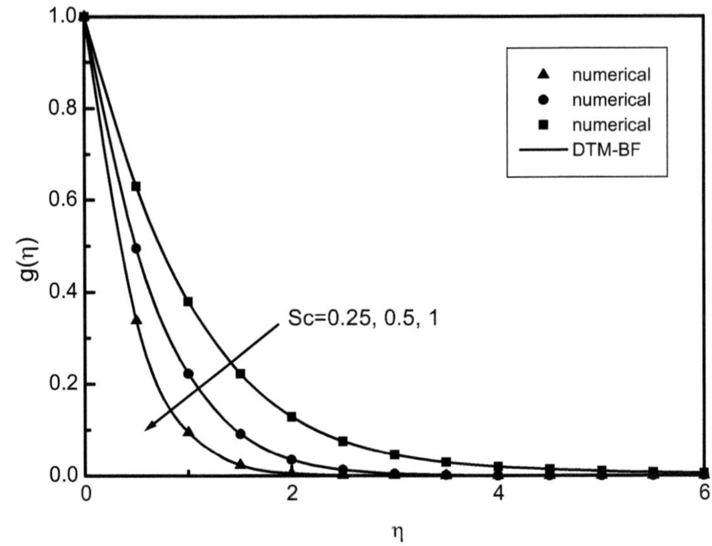

FIGURE 5.47 Concentration profiles for different values of Schmidt number Sc for a Cu–water working nanofluid with $M = 10^{-12}$, $P = S = \lambda = R = 1$, and $\varphi = 0.1$. *DTM-BF*, differential transformation method–basic function.

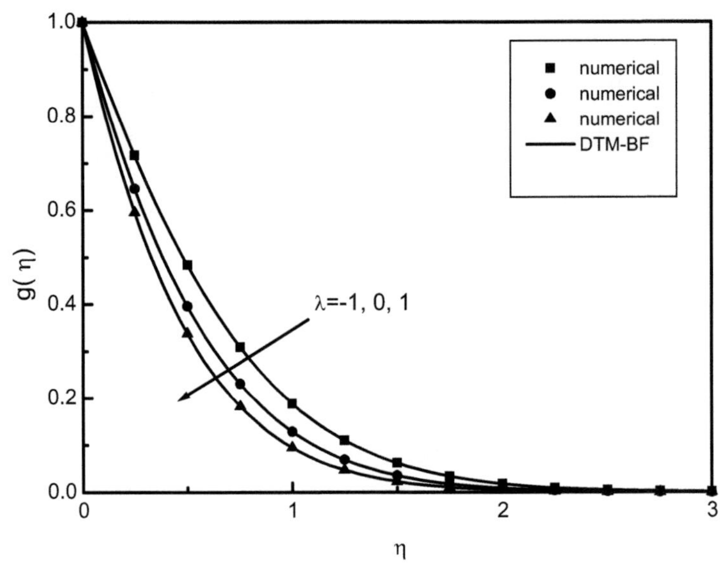

FIGURE 5.48 Concentration profiles for different values of the chemical reaction coefficient λ for Cu–water working nanofluid with $M = 10^{-12}$, $P = S = Sc = R = 1$ and $\varphi = 0.1$. *DTM-BF*, differential transformation method–basic function.

TABLE 5.10 $-\theta'(0)$ for Various Values of φ and R When $Pr = 6.2$

φ	R	Cu–Water	Al_2O_3–Water	Ag–Water
0	1	2.9174	2.9174	2.9174
	2	2.0294	2.0294	2.0294
	4	1.3733	1.3733	1.3733
0.1	1	2.6876	2.6269	2.4348
	2	1.9518	1.9007	1.7688
	4	1.3597	1.3211	1.2351
0.2	1	2.4201	2.3358	2.0037
	2	1.8277	1.7551	1.5191
	4	1.3101	1.2542	1.0966

Tables 5.10 and 5.11 and Figs. 5.49–5.51 demonstrate the effects of various pertinent parameters at the surface of the plate for the skin friction coefficient, heat transfer coefficients, and mass transfer coefficients.

As seen in Table 5.10, the heat transfer coefficient decreases considerably with an increase in R. The increase in the value of φ results in a decrease in heat transfer coefficients. The mass transfer coefficients are shown in Table 5.11 for different values of nanoparticle volume fraction φ and Schmidt

TABLE 5.11 $-g'(0)$ for Various Values of Sc and φ When $\lambda = 1$

φ	Sc	Cu–Water	Al_2O_3–Water	Ag–Water
0	0.25	0.8438	0.8438	0.8438
	0.5	1.2530	1.2530	1.2530
	1	1.9172	1.9172	1.9172
0.1	0.25	0.8428	0.8472	0.8397
	0.5	1.2776	1.2580	1.2453
	1	1.9525	1.9241	1.9055
0.2	0.25	0.8525	0.8464	0.8340
	0.5	1.2838	1.2561	1.2359
	1	1.9616	1.9211	1.8920

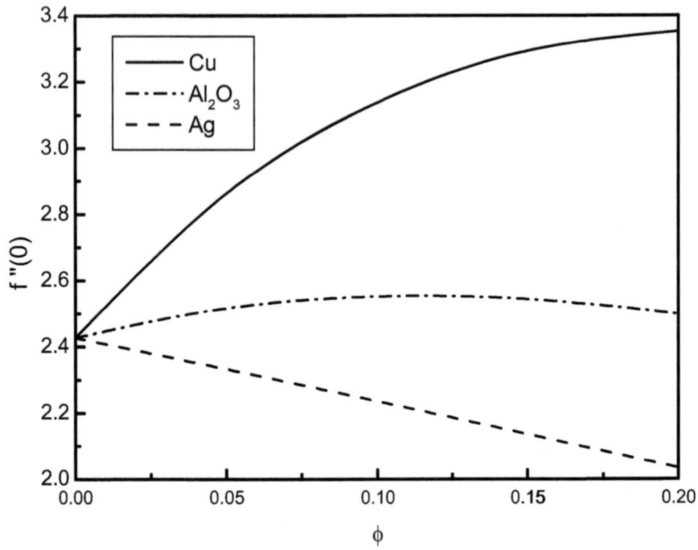

FIGURE 5.49 Local skin friction profiles for different nanofluids with $M = 10^{-12}$, $S = Sc = R = 1$, and $P = 0.5$.

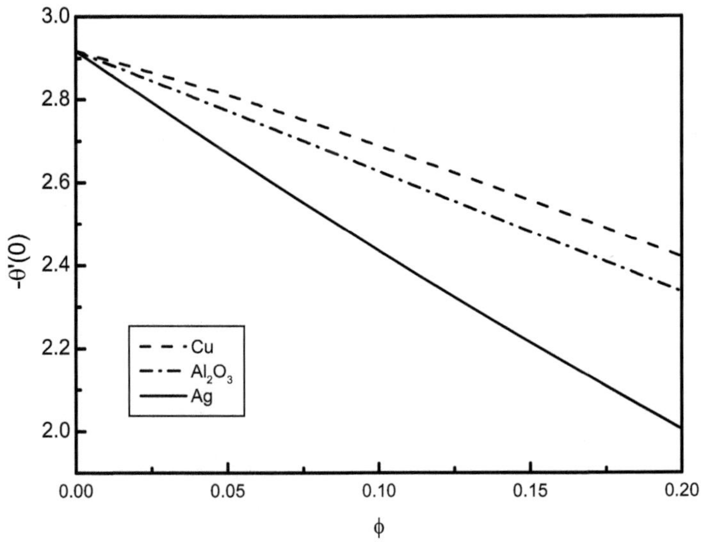

FIGURE 5.50 Local Nusselt number for different nanofluids with $M = 10^{-12}$, $S = Sc = R = 1$, and $P = 0.5$.

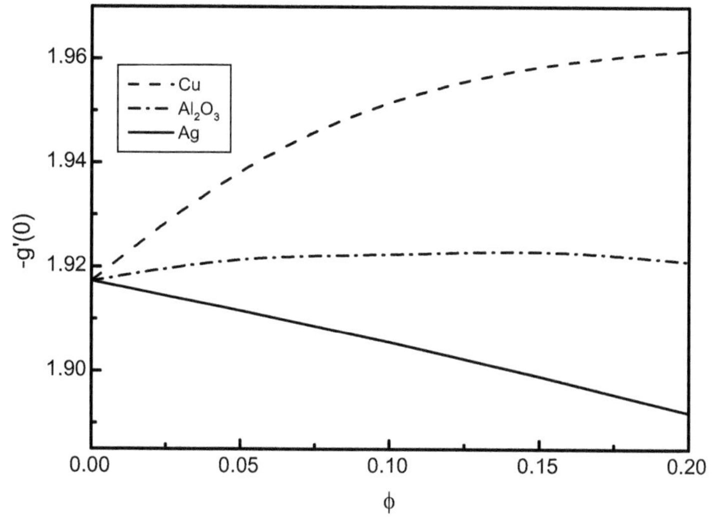

FIGURE 5.51 Local Sherwood number for different nanofluids with $M = 10^{-12}$, $S = Sc = R = 1$, and $P = 0.5$.

number Sc. Increasing the values of Sc results in an increase in the mass transfer coefficients whereas different values of nanoparticle volume fraction φ have a slight effect on the mass transfer coefficients.

Fig. 5.49 shows that different nanoparticles in the fluid have different effects on skin friction. For Cu—water, increasing the nanoparticle volume fraction increases skin friction. For Al_2O_3—water, skin friction changes slightly with an increase in the nanoparticle volume fraction. Skin friction decreases distinctly as the nanoparticle volume fraction increases for Ag—water. As seen in Fig. 5.50, the three nanoparticles (Cu, Al_2O_3, and Ag) in the fluid result in a decrease in the local Nusselt number. Fig. 5.51 shows that for Cu—water, the nanoparticle volume fraction increases the local Sherwood number, whereas there is a decrease in the rate of mass transfer with an increase in volume fraction φ. For Al_2O_3—water, the local Sherwood number changes slightly with an increase in volume fraction φ.

5.6 SUMMARY

In this chapter, the author used the DTM-Padé method to solve the problem of Falkner—Skan boundary layer flow in the permeable wall under the action of a magnetic field, and to obtain numerical and approximate analytical solutions for the wall friction system. Moreover, numerical solutions are presented that verify the reliability and validity of the results obtained by DTM-Padé.

For the problem of BVPs in unbounded domains, we propose a method to solve the problem of BVPs, which is based on DTM-BF. In this method, the boundary layer flow and heat transfer problem of a class of nonconstant magnetic fluid is solved. The analytical and numerical results are compared with the results of DTM-BF. The results show that the analytical and numerical solutions are highly consistent with all of the parameters in different values, which verifies that DTM-BF is an effective method for solving nonlinear problems in unbounded domains.

REFERENCES

Abazari, R., Borhanifar, A., 2010. Numerical study of the solution of the Burgers and coupled Burgers equations by a differential transformation method. Computers & Mathematics with Applications 59 (8), 2711–2722.

Abbasbandy, S., Hayat, T., 2009a. Solution of the MHD Falkner–Skan flow by Hankel–Padé method. Physics Letters A 373 (7), 731–734.

Abbasbandy, S., Hayat, T., 2009b. Solution of the MHD Falkner-Skan flow by homotopy analysis method. Communications in Nonlinear Science and Numerical Simulation 14, 3591–3598.

Abel, M.S., Nandeppanavar, M.M., 2009. Heat transfer in MHD viscoelastic boundary layer flow over a stretching sheet with non-uniform heat source/sink. Communications in Nonlinear Science and Numerical Simulation 14 (5), 2120–2131.

Aminossadati, S.M., Ghasemi, B., 2009. Natural convection cooling of a localized heat source at the bottom of a nanofluid- filled enclosure. European Journal of Mechanics-B/Fluids 28, 630–640.

Arikoglu, A.I., 2005. Solution of boundary value problems for integro-differential equations by using differential transform method. Applied Mathematics and Computation 168 (2), 1145–1158.

Arikoglu, A., Özkol, I., 2009. Solution of fractional integro-differential equations by using fractional differential transform method. Chaos, Solitons Fractals 40, 521–529.

Ayaz, F., 2004. Solutions of the systems of differential equations by differential transform method. Applied Mathematics and Computation 147 (2), 547–567.

Bachok, N., Ishak, A., Pop, I., 2012. Boundary layer stagnation-point flow and heat transfer over an exponentially stretching/shrinking sheet in a nanofluid. International Journal of Heat and Mass Transfer 55, 8122–8128.

Bhattacharyya, K., 2011. Boundary layer flow and heat transfer over an exponentially shrinking sheet. Chinese Physics Letters 28, 27–29.

Bhattacharyya, K., 2012. Steady boundary layer flow and reactive mass transfer past an exponentially stretching surface in an exponentially moving free stream. Journal of the Egyptian Mathematical Society 20, 223–228.

Boyd, J., 1997. Padé approximant algorithm for solving nonlinear ordinary differential equation boundary value problems on an unbounded domain. Computers in Physics 11 (3), 299–303.

Buongiorno, J., 2010. Convective transport in nanofluids. Journal of Heat Transfer 128, 240–250.

Chamkha, A.J., Mujtaba, M., Quadri, A., Issa, C., 2003. Thermal radiation effects on MHD forced convection flow adjacent to a non-isothermal wedge in the presence of heat source or sink. Heat Mass Transfer 39, 305–312.

Chiam, T.C., 1994. Stagnation-point flow towards a stretching plate. Journal of the Physical Society of Japan 63, 2443–2444.

Choi, S., 1995. Enhancing thermal conductivity of fluids with nanoparticle. In: Siginer, D.A., Wang, H.P. (Eds.), Developments and Applications of Non-Newtonian Flows, American Society of Mechanical Engineers, vol. 231, pp. 99–105. FED, vol. 66.

Congedo, P.M., Collura, S., Congedo, P.M., 2009. Modeling and analysis of natural convection heat transfer in nanofluids. In: Proceeding ASME Summer Heat Transfer Conference, vol. 3, pp. 569–579.

Ghasemi, B., Aminossadati, S.M., 2009. Natural convection heat transfer in an inclined enclosure filled with a water-Cu nanofluid. Numerical Heat Transfer Heat Transfer; Part A: Applications 55, 807–823.

Hamad, M.A.A., 2010. Analytical solution of natural convection flow of a nanofluid over a linearly stretching sheet in the presence of magnetic field. International Journal of Heat and Mass Transfer 38, 487–492.

Hayat, T., Hussain, M., Nadeem, S., Mesloub, S., 2011. Falkner-Skan wedge flow of a power-law fluid with mixed convection and porous medium. Computers & Fluids 49, 22–28.

Hayat, T., Javedb, T., Sajid, M., 2008. Analytic solution for MHD rotating flow of a second grade fluid over a shrinking surface. Physics Letters A 372, 3264–3273.

Ho, C.J., Chen, M.W., Li, Z.W., 2007. Effect of natural convection heat transfer of nanofluid in an enclosure due to uncertainties of viscosity and thermal conductivity. In: Asme/Jsme Thermal Engineering Heat Transfer Summer Conference Collocated with the Asme Interpack Conference, vol. 1, pp. 833–841.

Ho, C.J., Chen, M.W., Li, Z.W., 2008. Numerical simulation of natural convection of nanofluid in a square enclosure: effects due to uncertainties of viscosity and thermal conductivity. International Journal of Heat and Mass Transfer 51, 4506–4516.

Hossain, M.A., Bhowmik, S., Gorla, R.S.R., 2006. Unsteady mixed-convection boundary layer flow along a symmetric wedge with variable surface temperature. International Journal of Engineering Science 44, 607–620.

Ibrahim, W., Makinde, O.D., 2013. The effect of double stratification on boundary-layer flow and heat transfer of nanofluid over a vertical plate. Computers & Fluids 86, 433–441.

Ishak, A., 2011. MHD boundary layer flow due to an exponentially stretching sheet with radiation effect. Sains Malaysiana 40, 391–395.

Ishak, A., Nazar, R., Pop, I., 2007. Falkner–Skan equation for flow past a moving wedge with suction or injection. Journal of Applied Mathematics and Computing 25, 67–89.

Ishak, A., Nazar, R., Pop, I., 2009. MHD boundary-layer flow of a micropolar fluid past a wedge with constant wall heat flux. Communications in Nonlinear Science and Numerical Simulation 14 (1), 109–118.

Ishak, A., Nazar, R., Pop, I., 2010. MHD mixed convection boundary layer flow towards a stretching vertical surface with constant wall temperature. International Journal of Heat and Mass Transfer 53, 5330–5334.

Jang, M.J., Yeh, Y.L., Chen, C.L., Yeh, W.C., 2010. Differential transformation approach to thermal conductive problems with discontinuous boundary condition. Applied Mathematics and Computation 216, 2339–2350.

Jiao, C.R., Zheng, L.C., Ma, L.X., 2015. MHD thermosolutal marangoni convection heat and mass transport of power law fluid driven by temperature and concentration Gradient. AIP Advances 5, 49–317.

Joseph, D.D., Nield, D.A., Papanicolaou, G., 1982. Nonlinear equation governing flow in a saturated porous media. Water Resources Research 18, 1049–1052.

Kameswaran, P.K., Narayana, M., Sibanda, P., Murthy, P.V.S.N., 2012. Hydromagnetic nanofluid flow due to a stretching or shrinking sheet with viscous dissipation and chemical reaction effects. International Journal of Heat and Mass Transfer 55, 7587–7595.

Kandasamy, R., Muhaimin, I., Khamis, A.B., 2009. Thermophoresis and variable viscosity effects on MHD mixed convective heat and mass transfer past a porous wedge in the presence of chemical reaction. Heat Mass Transfer 45, 703–712.

Kumari, M., Nath, G., 2010. Unsteady MHD mixed convection flow over an impulsively stretched permeable vertical surface in a quiescent fluid. International Journal of Non-Linear Mechanics 45, 310–319.

Li, J., Zheng, L.C., Liu, L., 2016. MHD viscoelastic flow and heat transfer over a vertical stretching sheet with Cattaneo-Christov heat flux effects. Journal of Molecular Liquids 221, 19–25.

Makinde, O.D., 2012. Analysis of Sakiadis flow of nanofluids with viscous dissipation and Newtonian heating. Applied Mathematics and Mechanics (English Edition) 12, 1545–1554.

Makinde, O.D., 2013a. Computational modelling of nanofluids flow over a convectively heated unsteady stretching sheet. Current Nanoscience 9, 673–678.

Makinde, O.D., 2013b. Effects of viscous dissipation and Newtonian heating on boundary layer flow of nanofluids over a flat plate. International Journal of Numerical Methods for Heat and Fluid Flow 23, 1291–1303.

Makinde, O.D., Aziz, A., 2011. Boundary layer flow of a nanofluid past a stretching sheet with a convective boundary condition. International Journal of Thermal Sciences 50, 1326–1332.

Makinde, O.D., Khan, W.A., Khan, Z.H., 2013. Buoyancy effects on MHD stagnation point flow and heat transfer of a nanofluid past a convectively heated stretching/shrinking sheet. International Journal of Heat and Mass Transfer 62, 526–533.

Mandal, I.C., Mukhopadhyay, S., 2013. Heat transfer analysis for fluid flow over an exponentially stretching porous sheet with surface heat flux in porous medium. Ain Shams Engineering Journal 4, 103–110.

Masuda, H., Ebata, A., Teramae, K., Hishinuma, N., 1993. Alteration of thermal conductivity and viscosity of liquid by dispersing ultra-fine particles. Scientific Information Database 227–233.

Motsumi, T.G., Makinde, O.D., 2012. Effects of thermal radiation and viscous dissipation on boundary layer flow of nanofluids over a permeable moving flat plate. Physical Scripta 86, 2210–2219.

Muhaimin, I., Kandasamy, R., Khamis, A.B., 2009. Numerical investigation of variable viscosities and thermal stratification effects on MHD mixed convective heat and mass transfer past a porous wedge in the presence of a chemical reaction. Applied Mathematics and Mechanics 5 (103), 1353–1364.

Mutuku-Njane, W.N., Makinde, O.D., 2013. Combined effect of buoyancy force and Navier Slip on MHD Flow of a nanofluid over a convectively heated vertical porous plate. The Scientific World Journal 2013, 160–169.

Mutuku-Njane, W.N., Makinde, O.D., 2014. MHD nanofluid flow over a permeable vertical plate with convective heating. Journal of Computational and Theoretical Nanoscience 11, 667–675.

Nield, D.A., Bejan, A., 2006. Convection in Porous Media, third ed. Springer, New York, NY.

Oztop, H.F., Al-Salem, K., Pop, I., 2011. MHD mixed convection in a lid-driven cavity with corner heater. International Journal of Heat and Mass Transfer 54, 3494–3504.

Pal, D., Hiremath, P.S., 1992. Computational modeling of heat transfer over an unsteady stretching surface embedded in a porous medium. Meccanica 45, 415–424.

Parand, K., Rezaei, A.R., Ghaderi, S.M., 2011. An approximate solution of the MHD Falkner-Skan flow by Hermite functions pseudospectral method. Communications in Nonlinear Science and Numerical Simulation 16 (1), 274–283.

Prasad, K.V., Pal, D., Datti, P.S., 2009. MHD power-law fluid flow and heat transfer over a non-isothermal stretching sheet. Communications in Nonlinear Science and Numerical Simulation 14 (5), 2178–2189.

Rashidi, M.M., 2009. The modified differential transform method for solving MHD boundary-layer Equations. Computer Physics Communications 180 (11), 2210–2217.
Robert, A.V.G., Vajravelu, K., 2010. Existence and uniqueness results for a nonlinear differential equation arising in MHD Falkner–Skan flow. Communications in Nonlinear Science and Numerical Simulation 15 (9), 2272–2277.
Rosseland, S., 1936. Theoretical Astrophysics. Oxford University, New York, NY.
Si, X.H., Yuan, L.L., Zheng, L.C., Shen, Y.N., Cao, L., 2016. Lie group method for the modified model of MHD flow and heat transfer of a non-Newtonian fluid with prescribed heat flux over a moving porous plate. Journal of Molecular Liquids 220, 768–777.
Soundalgekar, V.M., Takhar, H.S., Singh, M., 1981. Velocity and temperature field in MHD Falkner-Skan flow. Journal of the Physical Society of Japan 50 (9), 3139–3143.
Sutton, G.W., Sherman, A., 1965. Engineering Magnetohydrodynamics. McGraw-Hill, New York, NY.
Su, X.H., Zheng, L.C., 2011. Approximate solutions to MHD Falkner-Skan flow over permeable wall. Applied Mathematics and Mechanics 32 (4), 401–408.
Su, X.H., Zheng, L.C., Zhang, X.X., 2012a. On DTM-BF method and dual solutions for an unsteady MHD flow over a permeable shrinking sheet with velocity slip. Applied Mathematics and Mechanics 33 (12), 1555–1568.
Su, X.H., Zheng, L.C., Zhang, X.X., Zhang, J.H., 2012b. MHD mixed convective heat transfer over a permeable stretching wedgewith thermal radiation and ohmic heating. Chemical Engineering Science 78, 1–8.
Wang, C.Y., 1987. Impinging stagnation flows. Physics of Fluids (1958-1988) 30, 915–917.
Yih, K.A., 1998. Uniform suction/blowing effect on the forced convection about a wedge: uniform heat flux. Acta Mechanica 128, 173–181.
Zhang, C.L., Zheng, L.C., Zhang, X.X., Chen, G., 2015. MHD flow and radiation heat transfer of nanofluids in porous media with variable surface heat flux and chemical reaction. Applied Mathematical Modelling 39, 165–181.
Zhao, J.K., 1986. Differential Transformation and its Applications for Electrical Circuits. Huazhong University Press, Wuhan, China (in Chinese).
Zhu, J., Zheng, L.C., Zhang, Z.G., 2010. Effects of slip condition on MHD stagnation-point flow over a power-law stretching sheet. Applied Mathematics and Mechanics 31 (4), 439–448.

Chapter 6

Variational Iteration Method and Homotopy Perturbation Method

ABSTRACT

In this chapter, two approximate analytical methods, i.e., the variational iteration method and homotopy perturbation method, are introduced to solve the time–space fractional derivative anomalous diffusion equations and the two-dimensional unsteady flow and heat transfer of power law non-Newtonian fluid over a nonisothermal and horizontally stretching surface with a time-dependent stretching velocity and modified Fourier's heat conduction law. Approximate analytical solutions are obtained and the effects of pertinent parameters on anomalous diffusion behavior, velocity, and temperature fields transport characteristics are analyzed and discussed.

6.1 REVIEW OF VARIATIONAL ITERATION METHOD

The variational iteration method (VIM), developed by Professor Jihuan He (1999a), is an improvement on Lagrange's multiplication. This method does not require the presence of small parameters in the differential equation and provides the solution (or an approximation to it) as a sequence of iterations, which has been widely applied to solve nonlinear ordinary and partial and fractional differential equations.

The variational method is an ancient branch of mathematics. Its object is to establish a functional extremum. The earliest variational problem was put forward by the famous scientist Pierre de Fermat (1601–1665). At the beginning of the calculus formation, Newton, Bernoulli, and many scientists proposed questions about the functional extremum, called the variational problem. Methods to deal with these problems are called the variational method (He, 1997, 1998).

Mathematics in Science and Engineering. http://dx.doi.org/10.1016/B978-0-12-811753-8.00006-2

In 1687, Newton proposed the least resistance problem of a rotating body, i.e., to choose an appropriate function of $y(x)$, such that

$$J = \min \int_{x_1}^{x_2} \frac{y(x)y'^2(x)}{1+y'^2(x)} \mathrm{d}x. \tag{6.1}$$

The variational method was developed rapidly in the 20th century, mainly owing to the development of mechanics. In 1950, Lingxi Qian proposed the principle of residual energy. The weighted residual method was also applied to modify and complete the principle of least dissipation of energy to represent the transport process of available energy. In 1983, Qian Wei-Chang discovered the critical variational phenomenon.

He (1999a) proposed a correction functional that can be easily constructed by a general Lagrange multiplier, which can be optimally identified by variational theory. The application of restricted variations in a correction functional makes it much easier to determine the multiplier. The initial approximation can be freely selected with unknown constants, which can be determined by various methods. The approximations obtained are valid not only for small parameters but also for very large parameters; their first-order approximations are of extreme accuracy. Many scholars have proved that the variational iteration method is a good mathematical tool for solving nonlinear problems. It has been applied successfully to solving the Burgers equation, Hyperbolic partial differential equation, Delay differential equation, nonlinear differential equation, generalized nonlinear Boussinesq equation, etc. (He, 2000a, 2003a, 2006; He et al., 2014; He and Lee, 2009; He and Wu, 2007; Wang and He, 2007).

The basic idea of the variational iteration method is illustrated as follows. For a given nonlinear differential equation of the form

$$Ly(t) + Ny(t) = g(t), \tag{6.2}$$

where L is the linear operator, N is a nonlinear operator, and g is a known differential function, in terms of the variational iteration method, a correction functional is constructed

$$y_{n+1}(t) = y_n(t) + \int_0^t \lambda\{Ly_n(s) + N\widetilde{y}_n(s) - g(s)\}\mathrm{d}s, \tag{6.3}$$

where λ is a general Lagrange multiplier that can be identified optimally via variational theory. The second term on the right is called the correction. y_n is the nth approximate solution and $\widetilde{y}_n$ is considered a restricted variation, i.e., $\delta\widetilde{y}_n = 0$. Making this correction functional stationary and noticing that

$$\begin{aligned} \delta y_{n+1}(t) &= \delta y_n(t) + \delta \int_0^t \lambda\{Ly_n(s) + N\widetilde{y}_n(s) - g(s)\}\mathrm{d}s \\ &= \delta y_n(t) + \int_0^t \lambda \delta L y_n(s)\mathrm{d}s \\ &= 0 \end{aligned} . \tag{6.4}$$

One can obtain the stationary condition equations and the value of λ. As a result, we obtain the iteration formula. Thus, for the initial value y_0 we can establish the approximate solution y_{n+1} and the exact solution of $y = \lim_{n \to \infty} y_n$.

For a linear differential equation, its exact solution can be obtained by only one iteration step owing to the fact that the Lagrange multiplier can be identified exactly. For example, consider the following problem

$$\begin{cases} y'' + \omega^2 y = A \sin \omega t + B \sin t \\ y(0) = 1, \;\; y'(0) = 0 \end{cases}. \tag{6.5}$$

The correction functional can be written as

$$y_{n+1}(t) = y_n(t) + \int_0^t \lambda\{y''_n(s) + \omega^2 y_n(s) - A \sin \omega s - B \sin s\} \mathrm{d}s, \tag{6.6}$$

making this correction functional stationary and noticing that $\delta y_n(0) = 0$,

$$\begin{aligned} \delta y_{n+1}(t) &= \delta y_n(t) + \delta \int_0^t \lambda\{y''_n(s) + \omega^2 y_n(s) - A \sin \omega s + B \sin t\} \mathrm{d}s \\ &= \delta y_n(t) + \lambda'(s)\delta y'_n(s)\big|_{s=t} - \lambda'(s)\delta y_n(s)\big|_{s=t} + \int_0^t (\lambda'' + \omega^2\lambda)\delta y_n(s)\mathrm{d}s, \\ &= (1 - \lambda'(t))\delta y_n(t) + \lambda(t)\delta y'_n(t) + \int_0^t (\lambda'' + \omega^2\lambda)\delta y_n(s)\mathrm{d}s \\ &= 0 \end{aligned}$$

yields the stationary equations,

$$\begin{cases} \delta y_n : \lambda'(s) + \omega^2 \lambda(s) = 0 \\ \delta y'_n : \lambda(s)\big|_{s=t} = 0 \\ \delta y_n : 1 - \lambda'(s)\big|_{s=t} = 0 \end{cases}. \tag{6.7}$$

The Lagrange multiplier can be readily identified

$$\lambda = \frac{1}{\omega} \sin \omega(s - t). \tag{6.8}$$

Therefore, we obtain the iteration formula

$$y_{n+1}(t) = y_n(t) + \frac{1}{\omega} \int_0^t \sin \omega(s - t)\{y''_n(s) + \omega^2 y_n(s) - A \sin \omega s - B \sin s\} \mathrm{d}s. \tag{6.9}$$

If we choose $y_0 = \cos \omega t$ as the initial approximate solution, in terms of Eq. (6.9), we obtain

$$
\begin{aligned}
y_1(t) &= \cos \omega t + \frac{1}{\omega}\int_0^t \sin \omega(\tau - t)\{-A \sin \omega t - B \sin \tau\}d\tau \\
&= \cos \omega t - \frac{A}{2\omega}\int_0^t \{\cos(-\omega t) - \cos(2\omega\tau - \omega t)\}d\tau \\
&\quad -\frac{B}{2\omega}\int_0^t \{\cos[(\omega-1)\tau - \omega t] - \cos[(\omega+1)\tau - \omega t]\}d\tau \\
&= \cos \omega t - \frac{A}{2\omega} t \cos \omega t - \frac{B}{2\omega}\left[\frac{1}{\omega - 1}[\sin(-t) - \sin(-\omega t)]\right. \\
&\quad \left. -\frac{1}{\omega+1}[\sin t - \sin(-\omega t)]\right] \\
&= \cos \omega t - \frac{A}{2\omega} t \cos \omega t + \frac{B}{\omega^2 - 1}(\sin t + \sin \omega t)
\end{aligned}
$$

This is the exact solution of problem Eq. (6.5).

For a nonlinear differential equation, to determine the Lagrange multiplier, the nonlinear terms have to be considered restricted variables. The Duffing equation is an example of a dynamical system that exhibits chaotic behavior. Moreover, in the frequency response the Duffing system presents a jump resonance phenomenon that is a sort of frequency hysteresis behavior (Finlayson, 1972).

$$
\begin{cases}
\dfrac{d^2u}{dt^2} + u + \varepsilon u^3 = 0 \\
u(0) = 1, u'(0) = 0
\end{cases}, \tag{6.10}
$$

Its correction functional can be written as

$$
u_{n+1}(t) = u_n(t) + \int_0^t \lambda\left\{\frac{d^2u_n(\tau)}{d\tau^2} + u_n(\tau) + \varepsilon\tilde{u}^3(\tau)\right\}d\tau, \tag{6.11}
$$

where $\tilde{u}^3(\tau)$ is considered a restricted variation. The stationary conditions of equation are

$$
\begin{cases}
\lambda''(\tau) + \lambda(\tau) = 0 \\
\lambda(\tau)|_{\tau=t} = 0 \\
1 - \lambda'(\tau)|_{\tau=t} = 0
\end{cases}, \tag{6.12}
$$

The Lagrange multiplier can be readily identified as

$$
\lambda = \sin(\tau - t). \tag{6.13}
$$

Substituting Eq. (6.13) into Eq. (6.11), we obtain

$$
u_{n+1}(t) = u_n(t) + \int_0^t \sin(\tau - t)\left\{\frac{d^2u_n(\tau)}{d\tau^2} + u_n(\tau) + \varepsilon\tilde{u}_n^3(\tau)\right\}d\tau. \tag{6.14}
$$

The initial approximate solution may be chosen as

$$u_0(t) = \cos \alpha t.$$

In view of Eq. (6.14), this yields

$$u_1(t) = \cos \alpha t + \int_0^t \sin(\tau - t)\left\{\left(-\alpha^2 + 1 + \frac{3}{4}\,\varepsilon\right)\cos \alpha\tau + \frac{\varepsilon}{4}\cos 3\alpha\tau\right\}d\tau, \tag{6.15}$$

parameter α can be identified by various methods, such as the method of weighted residuals (least-square method, collocation method, Galerkin method, and so on) (Finlayson, 1972). For example, let $-\alpha^2 + 1 + \frac{3}{4}\varepsilon = 0$, which yields $\alpha = \sqrt{1 + \frac{3}{4}\varepsilon}$. Therefore, the first-order approximate solution is obtained as

$$\begin{aligned} u_1(t) &= \cos \alpha t + \frac{\varepsilon}{4}\int_0^t (\tau - t)\cos 3\alpha\tau \, d\tau \\ &= \cos \alpha t + \frac{\varepsilon}{8}\int_0^t \{\sin[(1+3\alpha)\tau - t] + \sin[(1-3\alpha)\tau - t]\}d\tau \\ &= \cos \alpha t - \frac{\varepsilon}{8}\left(\frac{1}{1+3\alpha} + \frac{1}{1-3\alpha}\right)(\cos 3\alpha t - \cos t) \\ &= \cos \alpha t - \frac{\varepsilon}{4(1-9\alpha^2)}(\cos 3\alpha t - \cos t) \end{aligned}, \tag{6.16}$$

In the next example, the Riemann-Liouville fractional differential equation.

$$\frac{du(t)}{dt} + D^{\frac{1}{2}}u(t) - 2u^2(t) = 0, \tag{6.17}$$

is subject to the initial condition $v(0) = c$ (c is a constant).

Its correction functional can be written as

$$u_{n+1}(t) = u_n(t) + \int_0^t \lambda(s)\left\{u_n'(s) + D^{\frac{1}{2}}\tilde{u}_n^2(s)\right\}ds, \tag{6.18}$$

where $\tilde{u}(s)$ is considered a restricted variation. The Lagrange multiplier may be readily identified as $\lambda(s) = -1$. Therefore, the iteration formula can be obtained

$$u_{n+1}(t) = u_n(t) - \int_0^t \left\{u_n'(s) + D^{\frac{1}{2}}u_n(s) - 2u_n^2(s)\right\}ds. \tag{6.19}$$

Choosing $u_0(t) = u(0) = c$, we can obtain iteration approximate solutions as

$$u_0(t) = c$$

$$u_1(t) = c - \frac{2c\sqrt{t}}{\sqrt{\pi}} + 2c^2 t$$

$$u_1(t) = c - \frac{2c\sqrt{t}}{\sqrt{\pi}} + (c + 2c^2 t) - \frac{8c^2 t^{\frac{3}{2}}}{\sqrt{\pi}} + \left(4c^3 + \frac{4c^2}{\pi}\right)t^2 - \frac{32c^2 t^{\frac{5}{2}}}{\sqrt{\pi}} + \frac{8c^4 t^3}{3}$$

$, \ldots, .$

6.2 FRACTIONAL DIFFUSION PROBLEM

Fractional differential equations have attracted much attention and have been widely used in engineering, physics, chemistry, biology, and other fields (Podlubny, 1999; Xuan et al., 2005).

Most nonlinear fractional diffusion equations have no exact solution, the approximate solution or numerical solution may be a good approach. Many numerical methods (Dehghan, 2005; Liu et al., 2007; Meerschaert and Tadjeran, 2004; Odibat and Momani, 2008; Tadjeran and Meerschaert, 2007) have been proposed for fractional differential equations. And various approximate analytical methods are also developed, such as the VIM (Abbasbandy, 2007a; Abuasad and Odibat, 2006; Ghorbani and Nadjafi, 2009; Hemeda, 2008; Inc, 2008; Lu, 2007; Odibat and Momani, 2006; Tatari and Dehghan 2007), Adomian decomposition method (Momani, 2005; Momani and Odibat, 2006a), rectangular decomposition method (Odibat, 2006), homotopy analysis method (Jafari and Seifi, 2009; Xu and Cang, 2008), homotopy perturbation method (HPM) (Wang, 2007; Momani and Odibat, 2007; Momani and Odibat, 2006b), and differential transform method (Momani and Odibat, 2008; Tripathi et al., 2015).

In this section, we investigate applications of the VIM for solving time−space fractional derivative diffusion equations. Special emphasis is given to the formulation of fractional diffusion equation, which provides approximation analytic solutions.

6.3 FRACTIONAL ADVECTION-DIFFUSION EQUATION

6.3.1 Formulation of the Problem

The classical advection-diffusion equation is based on the assumption of continuum traffic flow, but the actual medium has a large number of fractal structures that can induce anomalous diffusion. To eliminate the deficiency, Xuan et al. (2005) extended the classical Fick's law for standard diffusion to a general fractional Fick's law, and developed a nonlocal method with temporal and spatial correlations to introduce a fractional order advection-diffusion equation.

Consider the following two-dimensional fractional diffusion equation

$$\frac{\partial^{\alpha} u(x,y,t)}{\partial t^{\alpha}} = f(x,y)\frac{\partial^{\beta} u(x,y,t)}{\partial x^{\beta}} + g(x,y)\frac{\partial^{\gamma} u(x,y,t)}{\partial y^{\gamma}} + q(x,y,t), \tag{6.20}$$

where $0 < x < a$, $0 < y < b$, $0 < \alpha \leq 1$, $1 < \beta, \gamma \leq 2$, $u(x, y, t)$ is the solute concentration, $f(x, y) > 0$ and $g(x, y) > 0$ represent the diffusion coefficients, and the forcing function $q(x, y, t)$ can be used to represent sources and sinks.

The initial condition:

$$u(x,y,0) = \psi(x,y). \tag{6.21}$$

The boundary conditions:

$$\begin{matrix} u(0,y,t)=f_1(y,t), & u(a,y,t)=f_2(y,t) \\ u(x,0,t)=g_1(x,t), & u(x,b,t)=g_2(x,t) \end{matrix}. \tag{6.22}$$

The time fractional derivative ${}_0D_t^{\alpha}=\frac{\partial^{\alpha}c(x,t)}{\partial t^{\alpha}}$ is the Caputo fractional derivative (Podlubny, 1999):

$$\frac{\partial^{\alpha}c(x,y,t)}{\partial t^{\alpha}}=\frac{1}{\Gamma(1-\alpha)}\int_0^t (t-\tau)^{-\alpha}\frac{\partial c(x,y,\tau)}{\partial \tau}\mathrm{d}\tau. \tag{6.23}$$

whereas the space fractional derivatives $\frac{\partial^{\beta}u(x,y,t)}{\partial x^{\beta}}$ and $\frac{\partial^{\gamma}u(x,y,t)}{\partial y^{\gamma}}$ are the Riemann–Liouville fractional derivatives

$$\frac{\partial^{\beta}u(x,y,t)}{\partial x^{\beta}}=D_x^{\beta}u(x,y,t)=\frac{1}{\Gamma(m-\beta)}\frac{\mathrm{d}^m}{\mathrm{d}\xi^m}\int_0^t (x-\xi)^{m-\beta-1}u(\xi,y,t)\mathrm{d}\xi, \tag{6.24}$$

$$\frac{\partial^{\gamma}u(x,y,t)}{\partial y^{\gamma}}=D_y^{\gamma}u(x,y,t)=\frac{1}{\Gamma(n-\gamma)}\frac{\mathrm{d}^n}{\mathrm{d}\eta^n}\int_0^t (y-\eta)^{n-\gamma-1}u(x,\eta,t)\mathrm{d}\eta. \tag{6.25}$$

6.3.2 Variational Iteration Method Solutions

According to the VIM, the correction functional for Eq. (6.20) can be approximately expressed as follows:

$$u_{k+1}(x,y,t)=u_k(x,y,t)+\int_0^t \lambda(\xi)\left(\begin{matrix} \frac{\partial^m}{\partial \xi^m}u_k(x,y,\xi)-f(x,y)\frac{\partial^{\beta}}{\partial x^{\beta}}\widetilde{u}_k(x,y,\xi) \\ -g(x,y)\frac{\partial^{\gamma}}{\partial y^{\gamma}}\widetilde{u}_k(x,y,\xi)-q(x,y,\xi) \end{matrix}\right)\mathrm{d}\xi, \tag{6.26}$$

where λ is a general Lagrange multiplier that can be identified optimally by variational theory and u_n is the nth-order approximate. $\frac{\partial^{\beta}\widetilde{u_k}}{\partial x^{\beta}}$ and $\frac{\partial^{\gamma}\widetilde{u_k}}{\partial y^{\gamma}}$ are considered restricted variations, i.e., $\delta\widetilde{u}_n=0$. Making this functional stationary,

$$\delta u_{k+1}(x,y,t)=\delta u_k(x,y,t)+\delta\int_0^t \lambda(\xi)\left(\frac{\partial^m}{\partial \xi^m}u_k(x,y,\xi)-q(x,y,\xi)\right)\mathrm{d}\xi, \tag{6.27}$$

For $m=1$, we obtain for Eq. (6.27) the following stationary conditions

$$1+\lambda(t)|_{\xi=t}=0, \tag{6.28}$$

$$\lambda'(\xi)=0. \tag{6.29}$$

Therefore, the general Lagrange multiplier can be identified as:

$$\lambda(\xi) = -1. \tag{6.30}$$

We obtain the following iteration formula:

$$u_{k+1}(x,y,t) = u_k(x,y,t) - \int_0^t \left(\begin{array}{c} \frac{\partial^\alpha}{\partial \xi^\alpha} u_k(x,y,\xi) - f(x,y) \frac{\partial^\beta}{\partial x^\beta} \widetilde{u}_k(x,y,\xi) \\ -g(x,y) \frac{\partial^\gamma}{\partial y^\gamma} \widetilde{u}_k(x,y,\xi) - q(x,y,\xi) \end{array} \right) d\xi. \tag{6.31}$$

There, we begin with the initial approximation

$$u_0(x,y,t) = \psi(x,y). \tag{6.32}$$

Consequently, the exact solution is obtained as

$$u(x,y,t) = \lim_{k \to \infty} u_k(x,y,t). \tag{6.33}$$

The convergence of the variation iteration method has been investigated in He and Lee (2009). In the next sections, we present two examples to demonstrate that the variation iteration method is effective and the results are in good agreement with the exact solution.

6.3.3 Examples

Example 1:

We consider the time−space fractional diffusion equation with variable coefficients on a finite domain

$$\frac{\partial^{0.8} u(x,y,t)}{\partial t^{0.8}} = f(x,y) \frac{\partial^{1.8} u(x,y,t)}{\partial x^{1.8}} + g(x,y) \frac{\partial^{1.5} u(x,y,t)}{\partial y^{1.5}} + q(x,y,t), \tag{6.34}$$

$$0 < x < 1, \; 0 < y < 1, \; 0 < \alpha \le 1, \; 0 \le t \le T_{end}.$$

The diffusion coefficients are

$$f(x,y) = x^{1.8}/\Gamma(3.8), \quad g(x,y) = \Gamma(2.5) y^{1.5}/6. \tag{6.35}$$

The source term is

$$q(x,y,t) = -2\left(t^2 - t^{1.2}/\Gamma(2.2) + 1\right) x^{2.8} y^3. \tag{6.36}$$

subject to the initial condition:

$$u(x,y,0) = x^{2.8} y^3, \tag{6.37}$$

and the boundary conditions:

$$u(0,y,t) = u(x,0,t) = 0, \; u(1,y,t) = y^3\left(1+t^2\right), \; u(x,1,t) = x^{2.8}\left(1+t^2\right). \tag{6.38}$$

When $\alpha = 0.8$, one can see that the problem has an exact solution as

$$u(x,y,t) = x^{2.8}y^3\left(1+t^2\right). \tag{6.39}$$

To solve the problem by the VIM, we use Eq. (6.31) to construct the iteration formula for Eq. (6.39) as follows

$$u_{k+1}(x,y,t) = u_k(x,y,t) - \int_0^t \left(\begin{array}{l} \frac{\partial^{0.8}}{\partial \xi^{0.8}} u_k(x,y,\xi) - f(x,y)\frac{\partial^{1.8}}{\partial x^{1.8}}\widetilde{u}_k(x,y,\xi) \\ -g(x,y)\frac{\partial^{1.5}}{\partial y^{1.5}}\widetilde{u}_k(x,y,\xi) - q(x,y,\xi) \end{array} \right) d\xi. \tag{6.40}$$

Using this variational iteration formula with $u_0(x,\ y,\ t) = x^{2.8}y^3$, we can obtain the following approximations

$$u_0(x,y,t) = x^{2.8}y^3, \tag{6.41}$$

$$u_1(x,y,t) = x^{2.8}y^3\left(1+\frac{2}{\Gamma(3.2)}t^{2.2}\right), \tag{6.42}$$

$$u_2(x,y,t) = x^{2.8}y^3\left(1+\frac{4}{\Gamma(3.2)}t^{2.2} - \frac{2}{\Gamma(3.4)}t^{2.4} - \frac{2}{3}t^3 + \frac{4}{\Gamma(4.2)}t^{3.2}\right), \tag{6.43}$$

$$\begin{aligned} u_3(x,y,t) = x^{2.8}y^3\Bigg(&1+\frac{6}{\Gamma(3.2)}t^{2.2} - \frac{6}{\Gamma(3.4)}t^{2.4} + \frac{2}{\Gamma(3.6)}t^{2.6} \\ &-\frac{4}{3}t^3 + \frac{16}{\Gamma(4.2)}t^{3.2} - \frac{8}{\Gamma(4.4)}t^{3.4} - \frac{1}{3}t^4 + \frac{8}{\Gamma(5.2)}t^{4.2}\Bigg), \end{aligned} \tag{6.44}$$

$$\begin{aligned} u_4(x,y,t) = x^{2.8}y^3\Bigg(&1+\frac{8}{\Gamma(3.2)}t^{2.2} - \frac{12}{\Gamma(3.4)}t^{2.4} + \frac{8}{\Gamma(3.6)}t^{2.6} \\ &-\frac{2}{\Gamma(3.8)}t^{2.8} - 2t^3 + \frac{36}{\Gamma(4.2)}t^{3.2} - \frac{36}{\Gamma(4.4)}t^{3.4} \\ &+\frac{12}{\Gamma(4.6)}t^{3.6} - t^4 + \frac{48}{\Gamma(5.2)}t^{4.2} - \frac{24}{\Gamma(5.4)}t^{4.4} \\ &-\frac{2}{15}t^5 + \frac{16}{\Gamma(6.2)}t^{5.2}\Bigg), \\ &\ldots, \end{aligned} \tag{6.45}$$

In this manner, the rest of the components can be obtained using the Mathematica package.

Fig. 6.1 compares the approximate solution and the exact solution when $x = y = 0.5$. It is obvious that the approximate results are in good agreement with the exact ones.

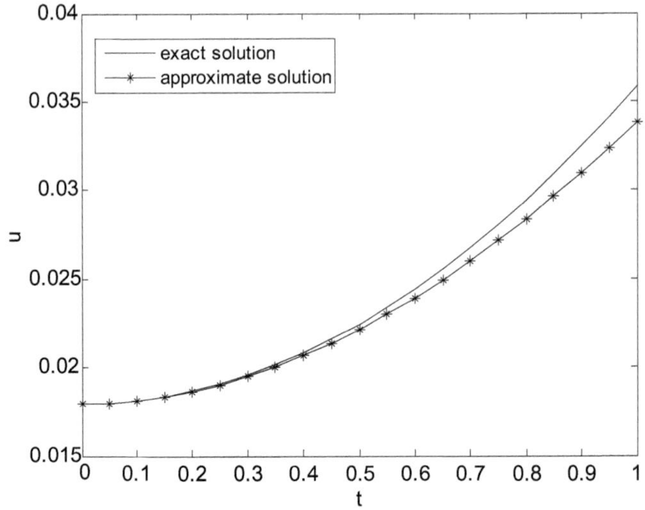

FIGURE 6.1 For $x = y = 0.5$, a comparison of the exact solution with the approximate solution.

Fig. 6.2 indicates the approximation of $u(x, y, t)$ with the variation of k. It is seen that the efficiency of this approach can be dramatically enhanced by computing further terms when the VIM is used.

Example 2:

Here, we consider the following two-dimensional fractional heat-like problem

$$D_t^{\alpha} u = \frac{1}{2}\left(x^2 u_{xx} + y^2 u_{yy}\right) + x^2 + y^2, \quad 0 < x, y < 1, \quad 0 < \alpha \le 1, \quad t > 0, \tag{6.46}$$

subject to the following initial condition

$$u(x, y, 0) = x^2 + y^2, \tag{6.47}$$

and boundary conditions

$$\begin{aligned} u(0, y, t) &= y^2(2e^t - 1), & u(x, 0, t) &= x^2(2e^t - 1) \\ u(1, y, t) &= (1 + y^2)(2e^t - 1), & u(x, 1, t) &= (1 + x^2)(2e^t - 1) \end{aligned}, \tag{6.48}$$

When $\alpha = 1$, it is easily seen that the problem has an exact solution as

$$u(x, y, t) = -(x^2 + y^2)(1 - 2e^t). \tag{6.49}$$

To solve the problem using the VIM, we use Eq. (6.31) to construct the iteration formula for Eq. (6.46)

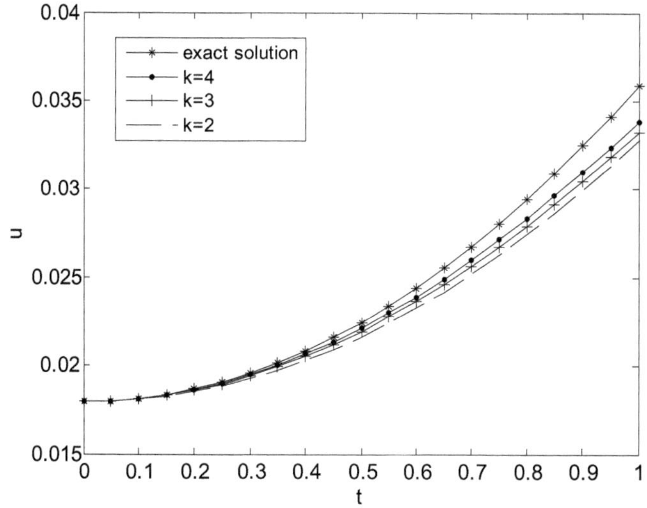

FIGURE 6.2 Comparing the exact and approximate solutions with the value of k.

$$u_{k+1}(x,y,t) = u_k(x,y,t) - \int_0^t \begin{pmatrix} \dfrac{\partial^\alpha}{\partial \xi^\alpha} u_k(x,y,\xi) - \dfrac{1}{2}x^2 \dfrac{\partial^2}{\partial x^2}\widetilde{u}_k(x,y,\xi) \\ -\dfrac{1}{2}y^2 \dfrac{\partial^2}{\partial y^2}\widetilde{u}_k(x,y,\xi) - x^2 - y^2 \end{pmatrix} d\xi. \tag{6.50}$$

Using the variational iteration formula with $u_0(x, y, t) = x^2 + y^2$, we can obtain the following approximations

$$u_0(x,y,t) = x^2 + y^2, \tag{6.51}$$

$$u_1(x,y,t) = \left(x^2 + y^2\right)(1 + 2t), \tag{6.52}$$

$$u_2(x,y,t) = \left(x^2 + y^2\right)\left[1 + 4t + t^2 - \frac{2}{\Gamma(3-\alpha)}t^{2-\alpha}\right], \tag{6.53}$$

$$u_3(x,y,t) = \left(x^2 + y^2\right)\begin{pmatrix} 1 + 6t + 3t^2 + \dfrac{1}{3}t^3 - \dfrac{6}{\Gamma(3-\alpha)}t^{2-\alpha} \\ -\dfrac{4}{\Gamma(4-\alpha)}t^{3-\alpha} + \dfrac{2}{\Gamma(4-2\alpha)}t^{3-2\alpha} \end{pmatrix}, \tag{6.54}$$

$$u_4(x,y,t) = (x^2+y^2)\begin{pmatrix} 1+8t+6t^2+\frac{4}{3}t^3+\frac{1}{12}t^4-\frac{12}{\Gamma(3-\alpha)}t^{2-\alpha} \\ -\frac{16}{\Gamma(4-\alpha)}t^{3-\alpha}-\frac{6}{\Gamma(5-\alpha)}t^{4-\alpha}+\frac{8}{\Gamma(4-2\alpha)}t^{3-2\alpha} \\ +\frac{6}{\Gamma(5-2\alpha)}t^{4-2\alpha}-\frac{2}{\Gamma(5-3\alpha)}t^{4-3\alpha} \end{pmatrix},$$

$\cdots.$

(6.55)

In the same manner, the rest of components can be obtained using the Mathematica package.

We can see that the solution for the standard heat-like equation ($\alpha = 1$) is given by

$$u(x,y,t) = (x^2+y^2)\left(1+2t+t^2+\frac{1}{3}t^3+\frac{1}{12}t^4\right). \quad (6.56)$$

The solution in a closed form

$$u(x,y,t) = -(x^2+y^2)(1-2e^t), \quad (6.57)$$

is readily obtained.

Fig. 6.3 compares the approximate and exact solutions when $x^2+y^2=1$. It is obvious that the approximate results are in good agreement with the exact solutions. Fig. 6.4 indicates the approximation of $u(x, y, t)$ with the variation of k.

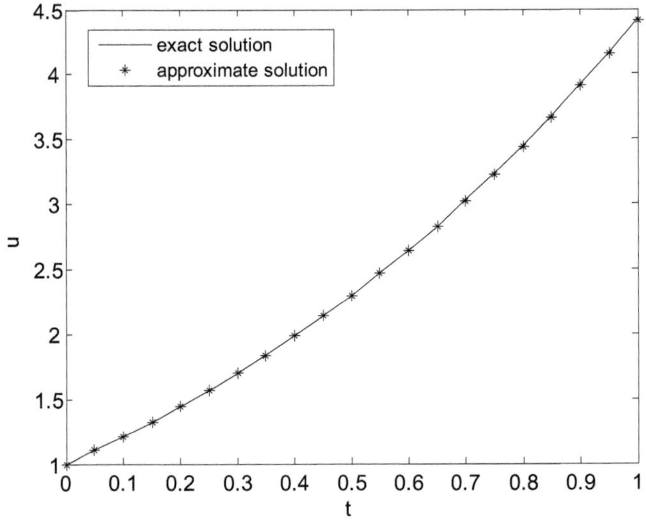

FIGURE 6.3 For $x^2+y^2=1$, comparing the exact and approximate solutions.

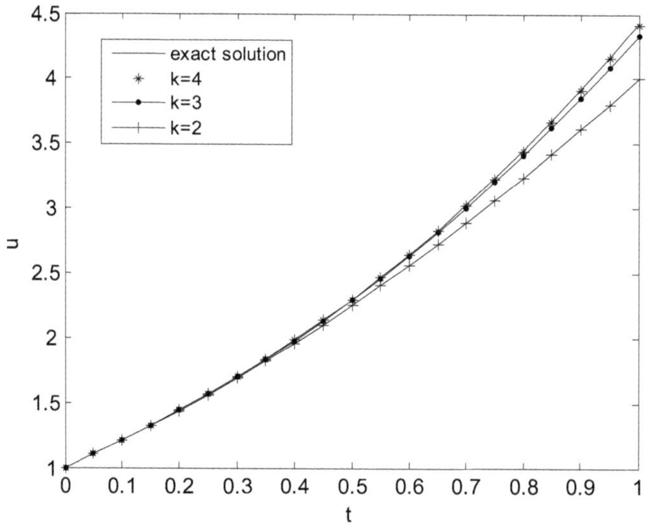

FIGURE 6.4 Comparing the exact and approximate solutions $u(x, y, t)$ with the variation of k.

It is seen that the efficiency of this approach can be dramatically enhanced by computing further terms when the VIM is used. Fig. 6.5 describes the characteristics of $u(x, y, t)$ with different values of α.

For the case of $\alpha = 1$, we know that the exact solution and the approximate solution by VIM are efficient. The approximate solution has the better

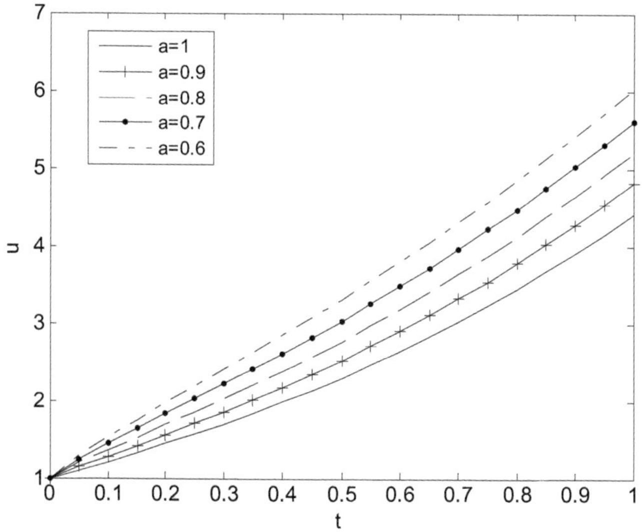

FIGURE 6.5 The characteristics of $u(x, y, t)$ and the variation of α.

characteristics: the closer value of α is to 1, the better the approach of approximate solution to the exact solution will be obtained.

6.4 REVIEW OF HOMOTOPY PERTURBATION METHOD

HPM is a nonlinear analytical technique that was proposed by He (1999b). In contrast to traditional perturbation methods, this technique does not require a small parameter in an equation. According to homotopy theory, a homotopy embedding parameter $p \in [0, 1]$ is introduced as a perturbation small parameter, so the method is called HPM, which can take full advantage of traditional perturbation methods and homotopy techniques.

To illustrate the basic ideas of this method, we consider the following nonlinear differential equation (Ganji et al., 2009)

$$A(u) - f(r) = 0, \quad r \in \Omega, \tag{6.58}$$

subject to boundary conditions

$$B(u, \partial u/\partial n) = 0, \quad r \in \Gamma, \tag{6.59}$$

where A is a general differential operator, B is boundary operator, $f(r)$ is a known analytical function, and Γ is the boundary of the domain Ω.

Operator A can be divided into two parts of L and N, where L is the linear part and N is a nonlinear part. Eq. (6.58) can therefore be rewritten as

$$L(u) + N(u) - f(r) = 0, \quad r \in \Omega. \tag{6.60}$$

With the homotopy technique, we construct a homotopy as $v(r, p) = 0$: $\Omega \times [0, 1] \rightarrow R$, satisfies

$$H(v,p) = (1-p)[L(v) - L(u_0)] + p[A(v) - f(r)] = 0, \quad p \in [0,1], \; r \in \Omega, \tag{6.61}$$

where $p \in [0, 1]$ is an embedding parameter and $u_0(r)$ is an initial approximation that satisfies the boundary conditions. Obviously, in terms of Eq. (6.60), when $p = 0, 1$, we have, respectively,

$$\begin{aligned} H(v,0) &= L(v) - L(u_0) = 0 \\ H(v,1) &= A(v) - f(r) = 0 \end{aligned}. \tag{6.62}$$

The changing process of p from zero to unity is just that of $v(r, p)$ from initial approximation solution $u_0(r)$ to the final solution $u(r)$. In topology, this is called deformation; $L(v) - L(u_0)$ and $A(v) - f(r)$ are called homotopy. According to HPM, we can first use the embedding parameter p as the "small parameter." Assume that the solution of Eq. (6.61) can be written as a power series in p

$$v = v_0 + p \cdot v_1 + p^2 \cdot v_2 + \cdots \tag{6.63}$$

Setting $p = 1$ results in the approximate solution of Eq. (6.61)

$$u = \lim_{p \to 1} v = v_0 + v_1 + v_2 + \cdots \tag{6.64}$$

The combination of the perturbation method and the homotopy method is called the HPM, which lacks the limitations of traditional perturbation methods, although this technique can take full advantage of traditional perturbation techniques. The series Eq. (6.64) is convergent for most cases. However, the convergence rate depends on the nonlinear operator $A(v)$. The following opinions are suggested by He (1999b, 2000b):

1. The second derivative of $N(v)$ with respect to v must be small because the parameter p may be relatively large, i.e., $p \to 1$.
2. The norm of $L^{-1}\partial N/\partial v$ must be smaller than 1 so that the series converges.

The applications and developments of HPM in linear and nonlinear problems have been studied by many scholars (Abbasbandy, 2007b; Cuce and Cuce, 2015; Ganji et al., 2009; Golbabai and Javidi, 2007; Grysa and Maciag, 2016; He, 2003b, 2004, 2005a,b; Ishak et al., 2009; Rashidi and Ganji, 2009; Roy et al., 2015; Sushila et al., 2013; Yun and Temuer, 2015).

6.5 UNSTEADY FLOW AND HEAT TRANSFER OF A POWER LAW FLUID OVER A STRETCHING SURFACE

Considerable effort has been devoted to studying non-Newtonian fluids with the hope of understanding and predicting the flow behavior of these fluids in various physical regimes. Fluids such as molten plastic, pulp, slurries, emulsions, etc., are found in many engineering applications where they exhibit some significantly different dynamic behavior from Newtonian fluids whose stress tensors are directly proportional to their deformation tensors. Many constitutive equations have been proposed to describe the flow and heat transfer mechanisms of non-Newtonian fluids, among them is the Ostwaald-de Waele model (also called the power law model), which applies the boundary layer assumption that has been the most successful.

In following discussion, we apply HPM to study unsteady flow and heat transfer of a power law fluid over a stretching surface. Approximate analytical solutions are obtained and the effects of pertinent parameters on velocity and temperature fields are analyzed.

6.5.1 Boundary Layer Governing Equations

We consider in part the two-dimensional unsteady boundary layer flow and heat transfer of power law fluid over a nonisothermal and horizontally stretching surface with a time-dependent stretching velocity $u_w(x, t)$. For simplicity, we assume that there is no shear stress at the free boundary, that the temperature of the uniform main stream is constant and equals T_∞, and that the surface is maintained at a uniform temperature T_w with $T_w > T_\infty$. Considering the effects of power law kinematic viscosity on heat transfer, the boundary layer governing equations for mass, momentum, and energy conservation are written, respectively, as

$$\frac{\partial u}{\partial x} + \frac{\partial v}{\partial y} = 0, \tag{6.65}$$

$$\frac{\partial u}{\partial t} + u\frac{\partial u}{\partial x} + v\frac{\partial u}{\partial y} = \frac{1}{\rho}\left(\frac{\partial \tau_{xy}}{\partial y}\right), \tag{6.66}$$

$$\frac{\partial T}{\partial t} + u\frac{\partial T}{\partial x} + v\frac{\partial T}{\partial y} = a_0\frac{\partial}{\partial y}\left[\left|\frac{\partial u}{\partial y}\right|^{n-1}\frac{\partial T}{\partial y}\right], \tag{6.67}$$

where x and y are the Cartesian coordinates along and normal to the surface of the body, u and v are the velocity components along the x and y axes, τ_{xy} is the shear stress tensor, ρ is the density of the fluid, and a_0 is a constant related to the thermal diffusivity. In the current problem, we have $\partial u/\partial y \le 0$ and so the shear stress is

$$\tau_{xy} = K\left(-\frac{\partial u}{\partial y}\right)^n, \tag{6.68}$$

where K and n are called the consistency coefficient index and the power law exponent, respectively. Parameter n is an important index that categorizes all fluids into two main types: the pseudoplastic fluids ($0 < n < 1$) and the dilatant fluids ($n > 1$). A fluid is Newtonian type when $n = 1$ with a dynamic coefficient of viscosity K.

The associated boundary conditions are

$$u_w = \frac{cx}{1 - \alpha t}, \quad v = 0, \quad \text{at} \quad y = 0; \quad u \to 0 \text{ as } y \to \infty, \tag{6.69}$$

and

$$T = T_w \quad \text{at } y = 0, \quad T \to T_\infty \text{ as } y \to \infty, \tag{6.70}$$

for some positive constants c and α, with the analysis being valid only for time $t < (1/\alpha)$.

Introduce the similarity transformation as

$$u = cx(1 - \alpha t)^{-1} f'(\eta), \tag{6.71}$$

and

$$v = -\left(\frac{c^{1-2n}}{K/\rho}\right)^{-1/(1+n)} x^{(n-1)/(1+n)}(1-\alpha t)^{(1-2n)/(1+n)} \times \left[\frac{2n}{n+1}f(\eta) + \frac{1-n}{1+n}\eta f'(\eta)\right],$$

where the similarity variable η is given by

$$\eta = \left(\frac{c^{2-n}}{K/\rho}\right)^{1/(n+1)} x^{(1-n)/(1+n)}(1-\alpha t)^{(n-2)/(1+n)} y, \tag{6.72}$$

The mass conservation (Eq. 6.65) is therefore automatically satisfied and the momentum (Eq. 6.66) is

$$n(-f'')^{n-1}f''' - (f')^2 + \left(\frac{2n}{n+1}\right)ff'' - S\left[f' + \frac{2-n}{n+1}\eta f''\right] = 0, \tag{6.73}$$

subject to the boundary conditions

$$f(0) = 0,\ f'(0) = 1,\ f'(\infty) = 0. \tag{6.74}$$

To facilitate the heat transfer analysis, we define a dimensionless temperature

$$\theta(\eta) = \frac{T - T_\infty}{T_w - T_\infty}, \tag{6.75}$$

Substituting Eqs. (6.71), (6.72), and (6.75) into energy Eqs. (6.67) and (6.70) yields,

$$\frac{1}{\Pr}\left[-(n-1)(-f'')^{n-2}f'''\theta' + (-f'')^{n-1}\theta''\right] - S\frac{2-n}{1+n}\eta\theta' + \frac{2n}{n+1}f\theta' = 0, \tag{6.76}$$

$$\theta(0) = 1, \quad \theta(\infty) = 0, \tag{6.77}$$

where $S(\equiv \alpha/c)$ is the unsteadiness parameter, $\Pr = \nu/a_0$ is the Prandtl number, and the prime denotes differentiation with respect to η.

6.5.2 Modified Homotopy Perturbation Method Solutions

One expects from the boundary conditions in Eqs. (6.74) and (6.77) that the solution series will take the forms

$$f(\eta) = \sum_{n=0}^{+\infty}\sum_{k=0}^{+\infty} a_{n,k}\eta^k e^{-n\eta}, \tag{6.78}$$

and

$$\theta(\eta) = \sum_{n=1}^{+\infty}\sum_{k=0}^{+\infty} b_{n,k}\eta^k e^{-n\eta}, \tag{6.79}$$

where $a_{n,k}$ and $b_{n,k}$ are coefficients that have to be determined. Again, from the boundary conditions Eqs. (6.74) and (6.77), we choose the initial solutions as

$$f_0(\eta) = -\exp(-\eta) + 1, \tag{6.80}$$

and

$$\theta_0(\eta) = \exp(-\eta). \tag{6.81}$$

and the auxiliary linear operators are chosen as

$$L_1[\Phi(\eta,p)] = \frac{d^3\Phi}{d\eta^3} - \frac{d\Phi}{d\eta}, \tag{6.82}$$

and

$$L_2[\Omega(\eta,p)] = \frac{d^2\Omega}{d\eta^2} + \frac{d\Omega}{d\eta}, \tag{6.83}$$

that possess the properties

$$L[C_1 + C_2\exp(-\eta) + C_3\exp(\eta)] = 0, \tag{6.84}$$

and

$$L[C_4 + C_5\exp(-\eta)] = 0, \tag{6.85}$$

for arbitrary constants C_i ($i = 1, 2, \ldots, 5$) that are independent of η. In view of Eqs. (6.73) and (6.76), we define the nonlinear operators

$$N_1[\Phi(\eta;p)] = n(-\Phi'')^{n-1}\Phi''' - (\Phi')^2 + \left(\frac{2n}{n+1}\right)\Phi\Phi'' - S\left[\Phi' + \frac{2-n}{n+1}\eta\Phi''\right], \tag{6.86}$$

and

$$\begin{aligned} N_2[\Omega(\eta;p)] = {} & \frac{1}{\Pr}\left[-(n-1)(-\Phi'')^{n-2}\Phi'''\Omega' + (-\Phi'')^{n-1}\Omega''\right] \\ & - S\frac{2-n}{1+n}\eta\Omega' + \frac{2n}{n+1}\Phi\Omega', \end{aligned} \tag{6.87}$$

and construct the homotopy equations

$$(1-p)L_1[\Phi(\eta;p) - f_0(\eta)] = pN_1[(\Phi(\eta;p)], \tag{6.88}$$

and

$$(1-p)L_2[\Omega(\eta;p)-\theta_0(\eta)] = pN_2[\Omega(\eta;p)], \tag{6.89}$$

subject to the boundary conditions

$$\Phi(0;p)=0, \Phi'(0;p)=1, \Phi'(\infty;p)=0, \tag{6.90}$$

and

$$\Omega(0;p)=1, \Omega(\infty;p)=0, \tag{6.91}$$

for some homotopy parameter $p \in [0, 1]$. Obviously, we have

$$\Phi(\eta;0)=f_0(\eta), \Omega(\eta;0)=\theta_0(\eta), \tag{6.92}$$

and

$$\Phi(\eta;1)=f(\eta), \Omega(\eta;1)=\theta(\eta), \tag{6.93}$$

for $p=0$ and $p=1$, and the basic assumption is that the solutions of Eqs. (6.88) and (6.89) have a power series expansion in p

$$\Phi(\eta;p)=f_0(\eta)+\sum_{m=1}^{+\infty} f_m(\eta)p^m, \tag{6.94}$$

and

$$\Omega(\eta;p)=\theta_0(\eta)+\sum_{m=1}^{+\infty} \theta_m(\eta)p^m. \tag{6.95}$$

Taking $p=1$ thus yields the approximate analytical solutions

$$\Phi(\eta;p)=f_0(\eta)+\sum_{m=1}^{+\infty} f_m(\eta), \quad \Omega(\eta;p)=\theta_0(\eta)+\sum_{m=1}^{+\infty} \theta_m(\eta), \tag{6.96}$$

which upon substituting Eqs. (6.94) and (6.95) into Eqs. (6.88) and (6.89) and comparing the corresponding terms with order p^3 gives

$$p^1: f_1' + \left[-Sf_0' - (f_0')^2 - \frac{2S\eta f_0''}{1+n} + \frac{nS\eta f_0''}{1+n} + \frac{2nf_0f_0''}{1+n} + n\left(-f_0''\right)^{n-1} f_0'''\right] - f_1''' = 0,$$

$$-\theta_1' - \theta_1'' + \left\{-\frac{(2-n)S\eta\theta_0'}{1+n} + \frac{2nf_0\theta_0'}{1+n} + \frac{1}{\Pr}\left[\left(-f_0''\right)^{n-1}\theta_0'' + (1-n)\theta_0' \times \left(-f_0''\right)^{n-2} f_0'''\right]\right\} = 0;$$

$$p^2: -f_1' + f_2' + f_1''' + \Bigg\{ -Sf_1' - 2f_0'f_1' + \frac{2nf_1f_0''}{1+n} - \frac{2S\eta f_1''}{1+n} + \frac{nS\eta f_1''}{1+n} + \frac{2nf_0f_1''}{1+n}$$

$$+ n\Big[-(n-1)\left(-f_0''\right)^{n-2} f_1'' f_0''' + \left(-f_0''\right)^{n-1} f_1''' \Big] \Bigg\} - f_2''' = 0,$$

$$-\theta_1' - \theta_2' + \theta_1'' - \theta_2'' + \Bigg\{ - \frac{(2-n)S\eta\theta_1'}{1+n} + \frac{2n\left(f_1\theta_0' + f_0\theta_1'\right)}{1+n}$$

$$+ \frac{1}{\Pr}\Big[(n-1)\left(-f_0''\right)^{n-2} f_1''\theta_0'' + \left(-f_0''\right)^{n-1}\theta_1'' + (1-n)\Big(\Big(\theta_1'\Big(\left(-f_0''\right)^{n-2}$$

$$-(n-2)\theta_0'(-f''_0)^{n-2} f_1''\Big) f_0''' + \theta_0'\left(-f_0''\right)^{n-2} f_1'''\Big)\Big)\Big] \Bigg\} = 0;$$

$$p^3: \quad -f_2' + f_3' + f_2''' + \Bigg\{ -\left(f_1'\right)^2 - Sf_2' - 2f_0'f_2' + \frac{2nf_2f_0''}{1+n} + \frac{2nf_1f_1''}{1+n} - \frac{2S\eta f_2''}{1+n} + \frac{nS\eta f_2''}{1+n}$$

$$+ \frac{2nf_0f_2''}{1+n} + n\Bigg[\left(\frac{1}{2}(n-2)(n-1)\left(-f_0''\right)^{n-3}\left(f_1''\right)^2 - (n-1)\left(-f_0''\right)^{n-2} f_2'' \right) f_0'''$$

$$- (n-1)\left(-f_0''\right)^{n-2} f_1'' f_1''' + \left(-f_0''\right)^{n-1} f_2''' \Bigg] \Bigg\} - f_3''' = 0,$$

$$-\theta_2' - \theta_3' + \theta_2'' - \theta_3''' + \Bigg\{ - \frac{(2-n)S\eta\theta_2'}{1+n} + \frac{2n\left(f_2\theta_0' + f_1\theta_1' + f_0\theta_2'\right)}{1+n}$$

$$+ \frac{1}{\Pr}\Bigg[\frac{1}{2}(n-2)(n-1)\left(-f_0''\right)^{n-3}\left(f_1''\right)^2 - (n-1)\left(-f_0''\right)^{n-2} f_2'' \Big) \theta_0''$$

$$- (n-1)\left(-f_0''\right)^{n-2} f_1'' \theta_2'' + \left(-f_0''\right)^{n-1}\theta_2'' + (1-n)\Big(\Big(\left(\left(-f_0''\right)^{n-1}\theta_2''\right.$$

$$- (n-2)\left(-f_0''\right)^{n-2} f_1''\theta_1' \quad + \theta_0' \left(\frac{1}{2}(n-3)(n-2)\left(-f_0''\right)^{n-4}\left(f_1''\right)^2\right.$$

$$\left. - (n-2)\left(-f_0''\right)^{n-3} f_2'' \right)\Big) f_0''' + \left(-f_0''\right)^{n-1}\theta_1' - (n-2)\theta_0'\left(-f_0''\right)^{n-3} f_1''\Big) f_1'''$$

$$+ \theta_0'\left(-f_0''\right)^{n-2} f_2'''\Big)\Big)\Bigg] \Bigg\} = 0;$$

with the boundary conditions

$$f_i(0) = 0, \; f_i'(0) = 0, \; f_i(\infty) = 0 (i = 1, 2, 3). \tag{6.97}$$

6.5.3 Results and Discussion

We now present some graphs of the velocity, shear, and temperature profiles that show the effects of the parameters on the characteristics of flow and heat transfer. For simplicity, the parameter n, S, and Pr are all set in the ranges $0.8 \le n \le 3$, $0 \le S \le 2$, and $0.7 \le \Pr \le 7$.

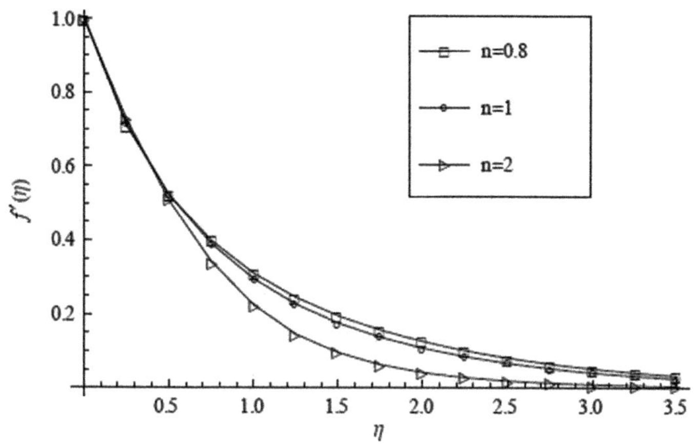

FIGURE 6.6 Velocity profiles $f'(\eta)$ for $n = 0.8$, 1, 2 and $S = 1.2$.

Fig. 6.6 shows the nondimensional velocity profiles $f'(\eta)$ for $n = 0.8$, 1, 2 and $S = 1.2$, where it is seen that the dimensionless velocity $f'(\eta)$ is monotonically decreasing with η for every fixed n. Indeed, this functional relationship is consistent with the behavior of dilatant fluids, which generally exhibit greater surface velocities than Newtonian ones. Fig. 6.7 shows the nondimensional shear stress profiles$[-f''(\eta)]^n$ for $n = 0.8$, 1, 2 and $S = 0.5$, where an identical monotonically decreasing relationship exists for every given n. $f'(\eta)$ and $-f''(\eta)$ are first and second derivatives, respectively.

We can see from these results that pseudoplastic fluids have a viscosity that decreases with increasing shear rates and dilatant fluids become progressively

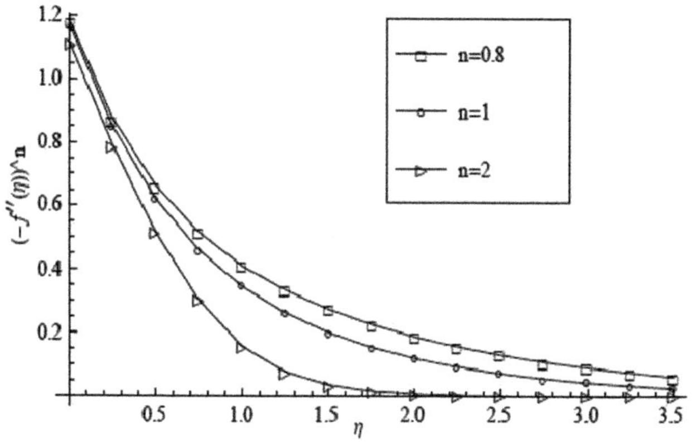

FIGURE 6.7 Shear stress profiles $(-f''(\eta))^n$ for $n = 0.8$, 1, 2 and $S = 0.5$.

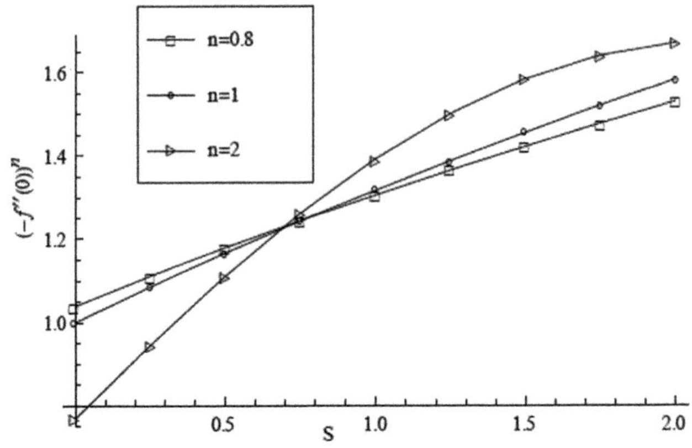

FIGURE 6.8 Skin friction $(-f''(0))^n$ versus unsteadiness parameter S for $n = 0.8$, 1, 2.

more viscous (in other words, thickens) with increasing rates of shear. Hence, the overall effect is to produce a thicker and more viscous boundary layer as n increases. Fig. 6.8 shows the dependence of the surface skin friction $[-f''(0)]^n$ of the stretching surface on S for $n = 0.8$, 1, 2, and it can be seen that although $[-f''(0)]^n$ monotonically increases with S for every fixed n, the dependence of $[-f''(0)]^n$ on n is completely reversed for S values on opposite sides of $S = 0.7$.

Figs. 6.9–6.11 show the temperature profiles of the parameters n, S, and Pr where the temperature distributions all become broader with increasing values of n. The heat transfer rate $\theta(\eta)$, on the other hand, increases with S and decreases with Pr, in full agreement with theory.

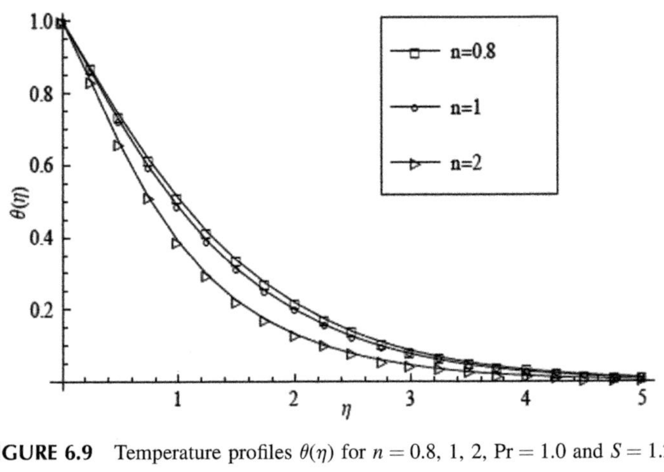

FIGURE 6.9 Temperature profiles $\theta(\eta)$ for $n = 0.8$, 1, 2, $\Pr = 1.0$ and $S = 1.2$.

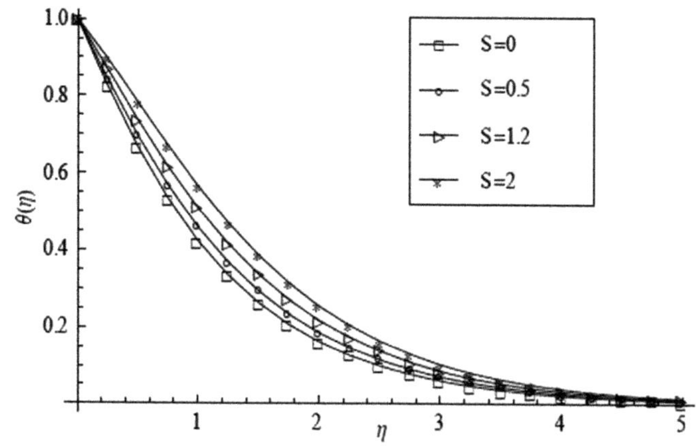

FIGURE 6.10 Temperature profiles $\theta(\eta)$ for different values of S for $\mathrm{Pr} = 1.0$ and $n = 0.8$.

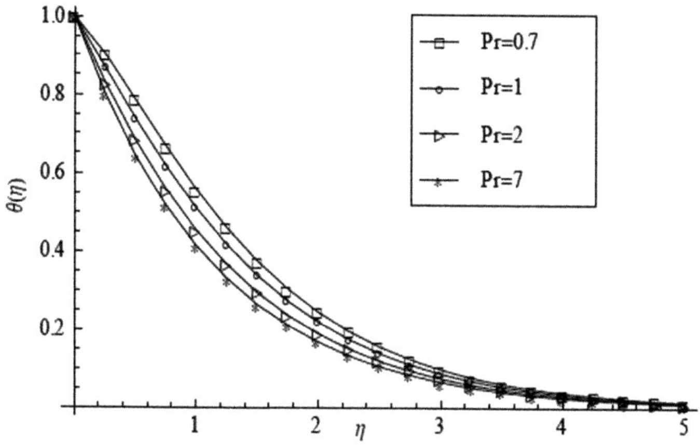

FIGURE 6.11 Temperature profiles $\theta(\eta)$ for different values of Pr for $S = 1.2$ and $n = 0.8$.

6.6 SUMMARY

In this chapter, the VIM and modified HPM are introduced to solve the time—space fractional derivative diffusion equations and the two-dimensional unsteady flow and heat transfer of an incompressible power law fluid over a nonisothermal and horizontally stretching surface with a time-dependent stretching velocity and modified Fourier's heat conduction law. Approximate analytical solutions are obtained and the effects of pertinent parameters on velocity and temperature fields are analyzed and discussed in detail.

Remark. The author's work in the applications of VIM and HPM is limited. Readers may see pertinent references in these fields.

REFERENCES

Abbasbandy, S., 2007a. An approximation solution of a nonlinear equation with Riemann–Liouville's fractional derivatives by He's variational iteration method. Journal of Computational and Applied Mathematics 207, 53–58.

Abbasbandy, S., 2007b. Application of He's homotopy perturbation method to functional integral equations. Chaos, Solitons and Fractals 31, 1243–1247.

Cuce, E., Cuce, P.M., 2015. A successful application of homotopy perturbation method for efficiency and effectiveness assessment of longitudinal porous fins. Energy Conversion and Management 93, 92–99.

Dehghan, M., 2005. Numerical approximations for solving a time-dependent partial differential equation with non-classical specification on four boundaries. Applied Mathematics and Computation 167, 28–45.

Finlayson, B.A., 1972. The Method of Weighted Residuals and Variational Principles. Academic Press, New York.

Ganji, D.D., Sahouli, A.R., Famouri, M., 2009. A new modification of He's homotopy perturbation method for rapid convergence of nonlinear undamped oscillators. Journal of Applied Mathematics and Computing 30, 181–192.

Ghorbani, A., Nadjafi, J.S., 2009. An effective modification of He's variational iteration method. Nonlinear Analysis: Real World Applications 10, 2828–2833.

Golbabai, A., Javidi, M., 2007. New iterative methods for nonlinear equations by modified HPM. Applied Mathematics and Computation 191, 122–127.

Grysa, K., Maciag, A., 2016. Temperature dependent thermal conductivity determination and source identification for nonlinear heat conduction by means of the Trefftz and homotopy perturbation methods. International Journal of Heat and Mass Transfer 100, 627–633.

He, J.H., 1997. A new approach to non-linear partial differential equations. Communications in Nonlinear Science and Numerical Simulation 2, 230–235.

He, J.H., Lee, E.W.M., 2009. Variational principle for the differential–difference system arising in stratified hydrostatic flows. Physics Letters A 373, 1644–1645.

He, J.H., Wu, X.H., 2007. Variational iteration method: new development and applications. Computers and mathematics with Applications 54, 881–894.

He, J.H., Kong, H.Y., et al., 2014. Variational iteration method for Bratu-like equation arising in electrospinning. Carbohydrate Polymers 105, 229–230.

He, J.H., 1998. Approximate analytical solution for seepage flow with fractional derivatives in porous media. Computer Methods in Applied Mechanics Engineering 167, 57–68.

He, J.H., 1999a. Variational iteration method-a kind of non-linear analytical technique: some examples. International Journal of Non-Linear Mechanics 34, 699–708.

He, J.H., 1999b. Homotopy perturbation technique. Computer Methods in Applied Mechanics and Engineering 178, 257–262.

He, J.H., 2000a. A coupling method of homotopy technique and perturbation technique for nonlinear problems. International Journal of Non-linear Mechanics 35, 37–43.

He, J.H., 2000b. Variational iteration method for autonomous ordinary differential systems. Applied Mathematics and Computation 114, 115–123.

He, J.H., 2003a. A Lagrangian for von Karman equations of large deflection problem of thin circular plate. Applied Mathematics and Computation 143, 543–549.

He, J.H., 2003b. Homotopy perturbation method: a new nonlinear analytical technique. Applied Mathematics and Computation 135, 73–79.

He, J.H., 2004. Comparison of homotopy perturbation method and homotopy analysis method. Applied Mathematics and Computation 156, 527–539.

He, J.H., 2005a. Homotopy perturbation method for bifurcation of nonlinear problems. International Journal of Nonlinear Sciences and Numerical Simulation 6, 207–208.

He, J.H., 2005b. Limit cycle and bifurcation of nonlinear problems. Chaos, Solitons and Fractals 26, 827–833.

He, J.H., 2006. Non-perturbative Methods for Strongly Nonlinear Problems. Dissertation, de-Verlag in GmbH, Berlin.

Hemeda, A.A., 2008. Variational iteration method for solving non-linear partial differential equations. Computers and Mathematics with Applications 56, 1948–1953.

Inc, M., 2008. The approximate and exact solutions of the space- and time-fractional Burgers equations with initial conditions by variational iteration method. Journal of Mathematical Analysis and Applications 345, 476–484.

Ishak, A., Nazar, R., Pop, I., 2009. Heat transfer over an unsteady stretching permeable surface with prescribed wall temperature. Nonlinear Analysis: Real World Applications 10, 2909–2913.

Jafari, H., Seifi, S., 2009. Homotopy analysis method for solving linear and nonlinear fractional diffusion-wave equation. Communications in Nonlinear Science and Numerical Simulation 14, 2006–2012.

Liu, F.W., Zhuang, P., Anh, V., Turner, I., Burrage, K., 2007. Stability and convergence of the difference methods for the space–time fractional advection–diffusion equation. Applied Mathematics and Computation 191, 12–20.

Lu, J.F., 2007. Variational iteration method for solving a nonlinear system of second-order boundary value problems. Computers and Mathematics with Applications 54, 1133–1138.

Meerschaert, M.M., Tadjeran, C., 2004. Finite difference approximation for fractional advection-dispersion flow equations. Journal of Computational and Applied Mathematics 172, 65–77.

Momani, S., Odibat, Z., 2006a. Analytical solution of a time-fractional Navier–Stokes equation by Adomian decomposition method. Applied Mathematics and Computation 177, 488–494.

Momani, S., Odibat, Z., 2006b. Analytical approach to linear fractional partial differential equations arising in fluid mechanics. Physics Letters A 355, 271–279.

Momani, S., Odibat, Z., 2007. Comparison between the homotopy perturbation method and the variational iteration method for liner fractional partial differential equations. Computer and Mathematics with Application 54, 910–919.

Momani, S., Odibat, Z., 2008. A novel method for nonlinear fractional partial differential equations: combination of DTM and generalized Taylor's formula. Journal of Computational and Applied Mathematics 220, 85–95.

Momani, S., Abuasad, S., Odibat, Z., 2006. Variational iteration method for solving nonlinear boundary value problems. Applied Mathematics and Computation 183, 1351–1358.

Momani, S., 2005. Approximate analytical solution for fractional heat-like and wave-like equations with variable coefficients using the decomposition method. Applied Mathematics and Computation 165, 459–472.

Odibat, Z., Momani, S., 2006. Application of variational iteration method to nonlinear differential equations of fractal order. International Nonlinear Science and Numerical Simulation 7, 27–34.

Odibat, Z., Momani, S., 2008. Numerical methods for nonlinear partial differential equations of fractional order. Applied Mathematical Modeling 32, 28–39.

Odibat, Z.M., 2006. Rectangular decomposition method for fractional diffusion-wave equations. Applied Mathematics and Computation 179, 92–97.

Podlubny, I., 1999. Fractional Differential Equations. Academic Press, New York.

Rashidi, M.M., Ganji, D.D., 2009. Homotopy perturbation combined with Padé approximation for solving two dimensional viscous flow in the extrusion process. International Journal of Nonlinear Science 7, 387–394.

Roy, P.K., Das, A., Mondal, H., Mallick, A., 2015. Application of homotopy perturbation method for a conductive–radiative fin with temperature dependent thermal conductivity and surface emissivity. Ain Shams Engineering Journal 6, 1001–1008.

Sushila, Singh, J., Shishodia, Y.S., 2013. An efficient analytical approach for MHD viscous flow over a stretching sheet via homotopy perturbation sumudu transform method. Aim Shams Engineering Journal 4, 549–555.

Tadjeran, C., Meerschaert, M.M., 2007. A second-order accurate numerical method for the two-dimensional fractional diffusion equation. Journal of Computational Physics 220, 813–823.

Tatari, M., Dehghan, M., 2007. On the convergence of He's variational iteration method. Journal of Computational and Applied Mathematics 207, 121–128.

Tripathi, D., Bég, O.A., Gupta, P.K., Radhakrishnamacharya, G., Mazumdar, J., 2015. DTM simulation of peristaltic viscoelastic biofluid flow in asymmetric porous media: a digestive transport model. Journal of Bionic Engineering 12, 643–655.

Wang, S., He, J.H., 2007. Variational iteration method for solving integro-differential equations variational iteration method for solving integro-differential equations. Physics Letters A 36, 188–191.

Wang, Q., 2007. Homotopy perturbation method for fractional KdV equation. Applied Mathematics and Computation 190, 1795–1802.

Xu, H., Cang, J., 2008. Analysis of a time fractional wave-like equation with the homotopy analysis method. Physics Letters A 372, 1250–1255.

Xuan, C.F., Jin, C., Wei, H., 2005. Anomalous diffusion and fractional advection- diffusion equation. Acta Physica Sinica 53, 1113–1117.

Yun, Y., Temuer, C., 2015. Application of the homotopy perturbation method for the large deflection problem of a circular plate. Applied Mathematical Modelling 39, 1308–1316.

Chapter 7

Exact Analytical Solutions for Fractional Viscoelastic Fluids

ABSTRACT

Fractional derivatives were to be quite flexible for describing the behaviors of viscoelastic fluids, which have been successfully applied to the characterization of the constitutive relationship of viscoelastic non-Newtonian fluids. In this chapter, we study flow and heat transfer problems for fractional viscoelastic fluid. Fractional derivatives governing differential equations are formulated for several classical types of viscoelastic fluids and specified physical regimes. Exact closed-form analytical solutions are obtained by using the fractional integral transformation techniques. The solutions are presented in series forms and generalized Gamma function, Beta function, *E*-function, *G*-function, *R*-function, *H*-function, Mittag–Leffler function, Wright function, etc. The effects of pertinent parameters on the velocity and temperature fields and the shear stress distributions are shown graphically and analyzed in detail.

7.1 INTRODUCTION

7.1.1 The Viscoelastic Non-Newtonian Fluids

The viscoelastic non-Newtonian fluids typically include polymer solutions, thermoplastics, granular materials, resin, paints, asphalt, gel, biological fluids such as blood, cerebrospinal fluid, endobronchial secretions, etc. They are the subject of intensive research and development in medicine and many industries including polymer, metallurgy, chemical, plastics, oil, and food industries (Deville and Gatski, 2012; Miller and Ross, 1993; Oldham and Spanier, 1974; Podlubny, 1999; Samko et al., 1993).

In industries, an increasing number of metallic materials has been replaced by polymers and composites. The motivation for this replacement is weight reduction, faster production, and additional functionality. Polymers involve a broad range of scales from the nanometer scale (carbon nanotube, graphene, fullerene, etc.) to the millimeter scale (particles, fibers, etc.) to the macroscopic scale (fibrous reinforcements made of continuous fibers in bundles, etc.).

Mathematics in Science and Engineering. http://dx.doi.org/10.1016/B978-0-12-811753-8.00007-4

In continuous fiber-reinforced polymers, the impregnation of the reinforcement with a low-viscosity polymer involves the flow of a Newtonian or non-Newtonian fluid in the complex multiscale microstructure related to the fiber. In the development of this type of composite materials, it is necessary to understand the composite manufacturing processes to ensure complete and uniform solidification, and accurate fiber placement and control.

In general, polymer solutions exhibit a variety of non-Newtonian fluid properties—the shear viscosity of most polymeric fluids is not constant. As a result, the traditional models of Newtonian fluids must be modified. In addition, polymeric fluids are viscoelastic materials, meaning that the stress endured by a fluid element depends on the history of the deformation experienced by that element. This is the memory effect of polymeric fluids. These two effects of nonlinearity and memory are responsible for numerous flow phenomena of industrial relevance that need to be modeled, predicted, and possibly controlled by means of a combination of mathematical and physical models, and computational and statistical methods (Bellout and Bloom, 2014; Weigand, 2015).

The class of viscoelastic fluid models typically contains the Maxwell, Kelvin–Voigt, Jeffreys, Oldroyd-B, and Burgers models. These mechanical models combine the discrete elastic element (the spring) and the viscidity element (the damper) in different ways (Miller and Ross, 1993; Oldham and Spanier, 1974; Podlubny, 1999; Samko et al., 1993). These models involve integer-order derivatives, which have the locality property (Havlin and Ben-Avraham, 2002; Khan et al., 2009a,b,c,d; Mahmood et al., 2009).

However, the classical integer-order differential theory was not flexible enough to fit experimental data on the transport process of viscoelastic fluids. Fractional Maxwell model (Fetecau et al., 2009; Fetecau et al., 2010; Li et al., 2016; Vieru et al., 2008; Zheng et al., 2011c), fractional Oldroyd-B model (Khan et al., 2012; Qi and Xu, 2007; Zheng et al., 2011a,b), and fractional Burgers model (Liu, 2011; Lorenzo and Hartley, 2008) have been proposed. These fractional dynamical models can be reformulated as multiterm time-fractional dynamical systems with fractional differential operators. A list of the key literature is given in (Li, 2014; Liu, 2011; Zheng and Zhang, 2011).

Maxwell model, for example, describes the physical characteristics that are equal to a spring and a damper in series, which can accurately describe the stress relaxation phenomenon (Fig. 7.1).

Assume that γ denotes the length of the total elongation, γ_1 refers to the spring (elasticity), and γ_2 refers to the damper (viscosity). The Maxwell model is, according to connection in series,

$$\gamma = \gamma_1 + \gamma_2, \quad \text{i.e,} \quad \dot{\gamma} = \dot{\gamma}_1 + \dot{\gamma}_2 = \frac{\dot{\tau}}{G} + \frac{\tau}{\mu} \quad \text{or} \quad \tau + \frac{\mu}{G}\dot{\tau} = \mu\dot{\gamma}. \tag{7.1}$$

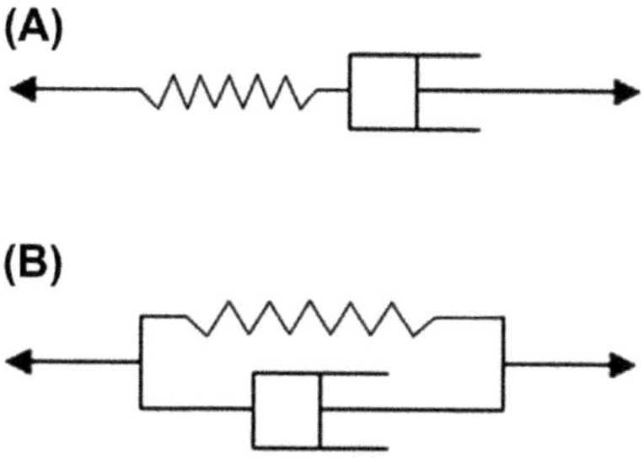

FIGURE 7.1 (A) Maxwell model (B) Kelvin–Voigt model.

The Kelvin–Voigt model is equal to a spring and a damper, according to connection in parallel,

$$\tau = \tau_1 + \tau_2, \quad \tau = \mu k + G\gamma \quad \text{or} \quad \tau = G\gamma + \mu\dot{\gamma}. \tag{7.2}$$

The Jeffreys model describes a physical model of a spring and a damper, according to connection in series, and then connection in parallel with a damper:

$$\tau + \lambda_1 \dot{\tau} = \mu_0(\gamma + \lambda_2 \dot{\gamma}), \tag{7.3}$$

where λ_1, λ_2, μ_0 all refer to the material parameters.

The constitutive equation of the Burgers model is:

$$\tau + p_1 \dot{\tau} + p_2 \ddot{\tau} = q_1 \gamma + q_2 \dot{\gamma}. \tag{7.4}$$

where p_1, p_2, q_1, q_2 all refer to the material parameters.

7.1.2 The Fractional Calculus

The fractional derivatives have been successfully applied to characterize the constitutive relationship of viscoelastic non-Newtonian fluid. The scalar form of fractional calculus operator theory is applied to the viscoelastic fluid constitutive equations, by using the fractional derivatives to replace the integer order derivatives, so that the characterizations of the problems are more extensive. For some special geometry boundary conditions, some special functions are applied to the theory of fractional calculus operators, such as generalized Gamma function, Beta function, *E*-function, *G*-functions, *R*-functions, *H*-function, Mittag–Leffler function, Wright function, etc. One can often get the closed form exact analytical solutions to the problems, so as to reveal the flow and heat transfer characteristic of viscoelastic fluid, and when the fractional order derivative of $\alpha \to 1$ the obtained solutions tended to integer order derivative of Newton fluid solutions (Fetecau et al., 2009;

Khan et al., 2012; Li, 2014; Li et al., 2016; Liu, 2011; Oldham and Spanier, 1974; Weigand, 2015; Zheng et al., 2011a,b,c; Zheng and Zhang, 2011). In the following, we show the applications of fractional calculus in studying flow and heat transfer behavior for several classical types models of fractional viscoelastic fluids.

7.2 FRACTIONAL MAXWELL FLUID FLOW DUE TO ACCELERATING PLATE

7.2.1 Governing Equations

Unsteady incompressible flow is governed by the following equations:

$$\operatorname{div} \mathbf{V} = 0, \quad \rho \frac{\mathrm{d}\mathbf{V}}{\mathrm{d}t} = \operatorname{div} \mathbf{T}, \tag{7.5}$$

where **T** is the Cauchy stress tensor, **V** is the velocity vector.

We select the velocity field vector in the following form:

$$\mathbf{V} = \left(u(y,t), 0, 0\right), \tag{7.6}$$

where $u(y, t)$ is the velocity in the x-coordinate direction. Together with the initial condition (the fluid being at rest up to the moment $t = 0$),

$$\mathbf{S}(y,0) = \mathbf{0}, \quad y > 0. \tag{7.7}$$

The generalized Maxwell fluid being at rest up to the time $t = 0$, we get

$$\left(1 + \lambda \mathrm{D}_t^{\alpha}\right) \frac{\partial u(y,t)}{\partial t} = \nu \frac{\partial^2 u(y,t)}{\partial y^2}, \tag{7.8}$$

where $\nu = \mu/\rho$ is the kinematic viscosity of the fluid, λ is the material constant, and D_t^{α} is the Riemann–Liouville fractional differentiation operator of order α with respect to t, defined as:

$$\mathrm{D}_t^{\alpha} f(t) = \frac{1}{\Gamma(1-\alpha)} \frac{\mathrm{d}}{\mathrm{d}t} \int_0^t \frac{f(\tau)}{(t-\tau)^{\alpha}} \mathrm{d}\tau, \quad 0 \le \alpha < 1, \tag{7.9}$$

where $\Gamma(\cdot)$ is the Gamma function.

7.2.2 Statement of the Problem

Consider an incompressible generalized Maxwell fluid over an infinitely extended flat plate. Initially the fluid as well as the plate are at rest. After time $t = 0$, the plate begins to oscillate and slide with constant acceleration in the x-direction and induces the motion into the fluid. Owing to the shear the fluid is gradually moved, its velocity being of the form Eq. (7.6) and the governing equation is given by Eq. (7.8). Accordingly, the initial and boundary conditions are:

$$u(0,t) = At + U\cos(\omega t) \text{ or } u(0,t) = U\sin(\omega t);\ t > 0, \tag{7.10}$$

where A is the constant acceleration parameter, U the amplitude, and ω the frequency of oscillation of the plate, and

$$u(y,t) = \frac{\partial u(y,t)}{\partial t} = 0 \quad \text{when } t = 0,\ y > 0, \tag{7.11}$$

Moreover, the natural conditions are:

$$u(y,t),\ \frac{\partial u(y,t)}{\partial y} \to 0 \quad \text{as } y \to \infty,\ t > 0. \tag{7.12}$$

7.2.3 Calculation of the Velocity Field

1. **The case**. $u(0, t) = At + U\cos(\omega t)$

   In terms of Fourier transformation formula, multiplying both sides of Eq. (7.8) by $\sqrt{2/\pi}\sin(\xi y)$ and integrating the result with respect to y from 0 to ∞ with the initial and boundary conditions Eqs. (7.10)–(7.12), yields:

$$\left(1 + \lambda D_t^{\alpha}\right)\frac{\partial u_s(\xi,t)}{\partial t} = \sqrt{\frac{2}{\pi}}\nu\xi(At + U\cos\omega t) - \nu\xi^2 u_s(\xi,t), \tag{7.13}$$

where $u_s(\xi,t)$ is the Fourier sine transform function of $u(y,t)$, which satisfies the conditions

$$u_s(\xi,0) = \frac{\partial u_s(\xi,0)}{\partial t} = 0 \quad \text{for } \xi > 0. \tag{7.14}$$

Applying the Laplace transform to Eq. (7.13) and using the Laplace transform formula for sequential fractional derivatives (Samko et al., 1993), we obtain the image function $\overline{u}_s(\xi,s)$ of $u_s(\xi,t)$ with the form

$$\overline{u}_s(\xi,s) = \frac{1}{s + \lambda s^{\alpha+1} + \nu\xi^2}\left(\frac{1}{s^2}\nu\xi\sqrt{\frac{2}{\pi}}A + \nu\xi\sqrt{\frac{2}{\pi}}U\frac{s}{s^2+\omega^2}\right), \tag{7.15}$$

In order to obtain $u_s(\xi,t) = L^{-1}\overline{u}_s(\xi,s)$ and to avoid the lengthy calculations of residues and contour integrals, we apply the discrete inverse Laplace transformation. In doing so, for a more suitable presentation of the final results, we rewrite Eq. (7.15) in the equivalent forms:

$$\begin{aligned}
\overline{u}_s(\xi,s) &= A\sqrt{\frac{2}{\pi}}\frac{1}{\xi s^2}+\sqrt{\frac{2}{\pi}}\frac{U}{\xi}\cdot\frac{s}{s^2+\omega^2}-\frac{A}{\nu}\sqrt{\frac{2}{\pi}}\left(\frac{1}{s}-\frac{1}{s+\nu\xi^2}\right)\frac{1}{\xi^3}\\
&-A\nu\sqrt{\frac{2}{\pi}}\cdot\frac{\xi}{s+\nu\xi^2}\sum_{k=0}^{\infty}\left(\frac{-1}{\lambda}\right)^k\frac{s^{k+\alpha-1}}{\left(s^{\alpha+1}+\frac{\nu\xi^2}{\lambda}\right)^{k+1}}\\
&-A\nu\sqrt{\frac{2}{\pi}}\cdot\frac{\xi}{s+\nu\xi^2}\sum_{k=0}^{\infty}\left(\frac{-1}{\lambda}\right)^k\frac{s^{k+\alpha-1}}{\left(s^{\alpha+1}+\frac{\nu\xi^2}{\lambda}\right)^{k+1}}\\
&+\sqrt{\frac{2}{\pi}}\cdot\frac{U}{\xi}\sum_{k=0}^{\infty}\left(\frac{-1}{\lambda}\right)^{k+1}\times\frac{s^{k+1}}{\left(s^{\alpha+1}+\frac{\nu\xi^2}{\lambda}\right)^{k+1}}\cdot\frac{s}{s^2+\omega^2}\\
&-\sqrt{\frac{2}{\pi}}\cdot\frac{U}{\xi}\sum_{k=0}^{\infty}\left(\frac{-1}{\lambda}\right)^k\frac{s^{\alpha+k+1}}{\left(s^{\alpha+1}+\frac{\nu\xi^2}{\lambda}\right)^{k+1}}\cdot\frac{s}{s^2+\omega^2}.
\end{aligned}\tag{7.16}$$

Inverting this result by means of the Fourier sine formula (Li et al., 2016), we find that

$$\begin{aligned}
\overline{u}(y,s) &= \frac{A}{s^2}+U\cdot\frac{s}{s^2+\omega^2}-\frac{2A}{\nu\pi}\int_0^{\infty}\left(\frac{1}{s}-\frac{1}{s+\nu\xi^2}\right)\cdot\frac{\sin(y\xi)}{\xi^3}d\xi\\
&-\frac{2A\nu}{\pi}\int_0^{\infty}\frac{\xi\sin(y\xi)}{s+\nu\xi^2}\sum_{k=0}^{\infty}\frac{(-1)^k}{\lambda^k}\times\frac{s^{k+\alpha-1}}{\left(s^{\alpha+1}+\frac{\nu\xi^2}{\lambda}\right)^{k+1}}d\xi\\
&+\frac{2U}{\pi}\int_0^{\infty}\frac{\sin(y\xi)}{\xi}\sum_{k=0}^{\infty}\left(\frac{-1}{\lambda}\right)^{k+1}\frac{s^{k+1}}{\left(s^{\alpha+1}+\frac{\nu\xi^2}{\lambda}\right)^{k+1}}\cdot\frac{s}{s^2+\omega^2}d\xi\,.\\
&-\frac{2U}{\pi}\int_0^{\infty}\frac{\sin(y\xi)}{\xi}\sum_{k=0}^{\infty}\left(\frac{-1}{\lambda}\right)^k\frac{s^{\alpha+k+1}}{\left(s^{\alpha+1}+\frac{\nu\xi^2}{\lambda}\right)^{k+1}}\cdot\frac{s}{s^2+\omega^2}d\xi
\end{aligned}\tag{7.17}$$

Finally, applying the inverse Laplace transform to Eq. (7.17) and taking into account all of the previous results, it is not difficult to show that

$$u(y,t) = At + U\cos(\omega t) - \frac{2A}{\nu\pi}\int_0^\infty \left(1 - e^{-\nu\xi^2 t}\right)\cdot\frac{\sin(y\xi)}{\xi^3}\mathrm{d}\xi$$
$$-\frac{2A\nu}{\pi}\int_0^\infty\int_0^t \xi\sin(y\xi)\cdot e^{-\nu\xi^2(t-s)}\sum_{k=0}^\infty \frac{(-1)^k}{\lambda^k}\cdot G_{\alpha+1,k+\alpha-1,k+1}\left(-\frac{\nu\xi^2}{\lambda}, s\right)\mathrm{d}s\mathrm{d}\xi$$
$$+\frac{2U}{\pi}\int_0^\infty\int_0^t \frac{\sin(y\xi)}{\xi}\cdot\sum_{k=0}^\infty\frac{(-1)^{k+1}}{\lambda^{k+1}}\cdot G_{\alpha+1,k+1,k+1}\left(-\frac{\nu\xi^2}{\lambda}, s\right)\cdot\cos(\omega(t-s))\mathrm{d}s\mathrm{d}\xi$$
$$-\frac{2U}{\pi}\int_0^\infty\int_0^t \frac{\sin(y\xi)}{\xi}\cdot\sum_{k=0}^\infty\frac{(-1)^{k}}{\lambda^{k}}\cdot G_{\alpha+1,\alpha+k+1,k+1}\left(-\frac{\nu\xi^2}{\lambda}, s\right)\cdot\cos(\omega(t-s))\mathrm{d}s\mathrm{d}\xi, \tag{7.18}$$

where $G_{a,b,c}(p,\, t)$ (Lorenzo and Hartley, 2008) is defined as:

$$G_{a,b,c}(p,t) = L^{-1}\left\{\frac{s^b}{(s^a - p)^c}\right\}, \tag{7.19}$$

or equivalently,

$$u(y,t) = At + U\cos(\omega t) - \frac{2A}{\nu\pi}\int_0^\infty \left(1 - e^{-\nu\xi^2 t}\right)\cdot\frac{\sin(y\xi)}{\xi^3}\mathrm{d}\xi$$
$$-\frac{2A\nu}{\pi}\int_0^\infty\int_0^t \xi\sin(y\xi)\times e^{-\nu\xi^2(t-s)}\sum_{k=0}^\infty\sum_{j=0}^\infty\left(\frac{-1}{\lambda}\right)^k\left(-\frac{\nu\xi^2}{\lambda}\right)^j$$
$$\times\frac{(k+1)_j s^{k\alpha+j\alpha+j+1}}{\Gamma(1+j)\Gamma(k\alpha+j\alpha+j+2)}\mathrm{d}s\mathrm{d}\xi$$
$$+\frac{2U}{\pi}\int_0^\infty\int_0^t\frac{\sin(y\xi)}{\xi}\cdot\sum_{k=0}^\infty\sum_{j=0}^\infty\left(\frac{-1}{\lambda}\right)^{k+1}\left(-\frac{\nu\xi^2}{\lambda}\right)^j\frac{(k+1)_j}{\Gamma(1+j)} \tag{7.20}$$
$$\times\frac{s^{k\alpha+j\alpha+\alpha+j-1}}{\Gamma(k\alpha+j\alpha+\alpha+j)}\times\cos(\omega(t-s))\mathrm{d}s\mathrm{d}\xi$$
$$-\frac{2U}{\pi}\int_0^\infty\int_0^t\frac{\sin(y\xi)}{\xi}\sum_{k=0}^\infty\sum_{j=0}^\infty\left(\frac{-1}{\lambda}\right)^k\left(-\frac{\nu\xi^2}{\lambda}\right)^j$$
$$\times\frac{(k+1)_j s^{k\alpha+j\alpha+j-1}}{\Gamma(1+j)\Gamma(k\alpha+j\alpha+j)}\cos(\omega(t-s))\mathrm{d}s\mathrm{d}\xi.$$

2. The case. $u(0, t) = U\sin(\omega t)$

In the same way, we can obtain the solution

$$\begin{aligned} u(y,t) &= U\sin(\omega t) \\ &+ \frac{2U}{\pi}\int_0^\infty \int_0^t \frac{\sin(y\xi)}{\xi}\cdot\sum_{k=0}^{\infty}\frac{(-1)^{k+1}}{\lambda^{k+1}}\cdot G_{\alpha+1,k+1,k+1}\left(-\frac{v\xi^2}{\lambda},s\right) \\ &\times \sin(\omega(t-s))\mathrm{d}s\mathrm{d}\xi \\ &- \frac{2U}{\pi}\int_0^\infty \int_0^t \frac{\sin(y\xi)}{\xi}\cdot\sum_{k=0}^{\infty}\frac{(-1)^{k}}{\lambda^{k}}\cdot G_{\alpha+1,\alpha+k+1,k+1}\left(-\frac{v\xi^2}{\lambda},s\right) \\ &\times \sin(\omega(t-s))\mathrm{d}s\mathrm{d}\xi. \end{aligned} \tag{7.21}$$

The similar solutions corresponding to generalized Maxwell fluids have been obtained as limiting cases of general solutions. The present solutions are written in terms of the generalized G-functions.

7.2.4 Calculation of the Shear Stress

1. The case. $u(0, t) = At + U\cos(\omega t)$

To determine the shear stress $\tau(y, t)$, we use the relation

$$\bar{\tau}(y,s) = \frac{\mu}{\lambda s^\alpha + 1}\frac{\partial \bar{u}(y,s)}{\partial y}, \tag{7.22}$$

in view of Eq. (7.17),

$$\begin{aligned} \frac{\partial \bar{u}(y,s)}{\partial y} &= -\frac{2A}{v\pi}\int_0^\infty \left(\frac{1}{s}-\frac{1}{s+v\xi^2}\right)\cdot\frac{\cos(y\xi)}{\xi^2}\mathrm{d}\xi \\ &- \frac{2Av}{\pi}\int_0^\infty \frac{\xi^2\cos(y\xi)}{s+v\xi^2}\sum_{k=0}^{\infty}\frac{(-1)^k}{\lambda^k}\times\frac{s^{k+\alpha-1}}{\left(s^{\alpha+1}+\frac{v\xi^2}{\lambda}\right)^{k+1}}\mathrm{d}\xi \\ &+ \frac{2U}{\pi}\int_0^\infty \cos(y\xi)\sum_{k=0}^{\infty}\left(\frac{-1}{\lambda}\right)^{k+1}\frac{s^{k+1}}{\left(s^{\alpha+1}+\frac{v\xi^2}{\lambda}\right)^{k+1}}\cdot\frac{s}{s^2+\omega^2}\mathrm{d}\xi \\ &- \frac{2U}{\pi}\int_0^\infty \cos(y\xi)\sum_{k=0}^{\infty}\left(\frac{-1}{\lambda}\right)^{k}\frac{s^{\alpha+k+1}}{\left(s^{\alpha+1}+\frac{v\xi^2}{\lambda}\right)^{k+1}}\cdot\frac{s}{s^2+\omega^2}\mathrm{d}\xi \end{aligned} \tag{7.23}$$

Substituting Eq. (7.23) into (7.22) yields:

$$\overline{\tau}(y,s) = -\frac{2\rho A}{\pi}\int_0^{\infty}\left(\frac{1}{s}-\frac{1}{s+\nu\xi^2}\right)\cdot\frac{\cos(y\xi)}{\xi^2}\mathrm{d}\xi$$
$$+\frac{2\mu A}{\pi}\int_0^{\infty}\frac{s^{\alpha-1}}{\left(s+\nu\xi^2\right)\left(s^{\alpha}+\frac{1}{\lambda}\right)}\cos(y\xi)\mathrm{d}\xi$$
$$-\frac{2A\nu}{\pi}\frac{\mu}{\lambda}\int_0^{\infty}\frac{\xi^2\cos(y\xi)}{s+\nu\xi^2}\sum_{k=0}^{\infty}\frac{(-1)^k}{\lambda^k}\cdot\frac{s^{k+\alpha-1}}{\left(s^{\alpha+1}+\frac{\nu\xi^2}{\lambda}\right)^{k+1}}\frac{1}{s^{\alpha}+\frac{1}{\lambda}}\mathrm{d}\xi$$
$$+\frac{2U}{\pi}\frac{\mu}{\lambda}\int_0^{\infty}\cos(y\xi)\sum_{k=0}^{\infty}\left(\frac{-1}{\lambda}\right)^{k+1}\frac{s^{k+1}}{\left(s^{\alpha+1}+\frac{\nu\xi^2}{\lambda}\right)^{k+1}}\cdot\frac{1}{s^{\alpha}+\frac{1}{\lambda}}\frac{s}{s^2+\omega^2}\mathrm{d}\xi$$
$$-\frac{2U}{\pi}\frac{\mu}{\lambda}\int_0^{\infty}\cos(y\xi)\sum_{k=0}^{\infty}\left(\frac{-1}{\lambda}\right)^{k}\frac{s^{\alpha+k+1}}{\left(s^{\alpha+1}+\frac{\nu\xi^2}{\lambda}\right)^{k+1}}\cdot\frac{1}{s^{\alpha}+\frac{1}{\lambda}}\frac{s}{s^2+\omega^2}\mathrm{d}\xi, \tag{7.24}$$

or equivalently,

$$\overline{\tau}(y,s) = -\frac{2\rho A}{\pi}\int_0^{\infty}\left(\frac{1}{s}-\frac{1}{s+\nu\xi^2}\right)\cdot\frac{\cos(y\xi)}{\xi^2}\mathrm{d}\xi+\frac{2\mu A}{\pi}\int_0^{\infty}\frac{s^{\alpha-1}}{\left(s+\nu\xi^2\right)\left(s^{\alpha}+\frac{1}{\lambda}\right)}\cos(y\xi)\mathrm{d}\xi$$
$$-\frac{2A\nu}{\pi}\frac{\mu}{\lambda}\int_0^{\infty}\xi^2\cos(y\xi)\sum_{k=0}^{\infty}\sum_{j=0}^{\infty}\left(-\frac{1}{\lambda}\right)^k\left(-\frac{\nu\xi^2}{\lambda}\right)^j$$
$$\cdot\frac{(k+1)_j}{\Gamma(1+j)}\frac{1}{\left(s+\nu\xi^2\right)\left(s^{\alpha}+\frac{1}{\lambda}\right)s^{\alpha k+\alpha j+j+2}}\mathrm{d}\xi$$
$$+\frac{2U}{\pi}\frac{\mu}{\lambda}\int_0^{\infty}\cos(y\xi)\sum_{k=0}^{\infty}\sum_{j=0}^{\infty}\left(-\frac{1}{\lambda}\right)^{k+1}\left(-\frac{\nu\xi^2}{\lambda}\right)^j$$
$$\times\frac{(k+1)_j}{\Gamma(1+j)}\frac{1}{\left(s^{\alpha}+\frac{1}{\lambda}\right)s^{\alpha k+\alpha j+j+\alpha}}\frac{s}{s^2+\omega^2}\mathrm{d}\xi$$
$$-\frac{2U}{\pi}\frac{\mu}{\lambda}\int_0^{\infty}\cos(y\xi)\sum_{k=0}^{\infty}\sum_{j=0}^{\infty}\left(-\frac{1}{\lambda}\right)^{k}\left(-\frac{\nu\xi^2}{\lambda}\right)^j$$
$$\times\frac{(k+1)_j}{\Gamma(1+j)}\frac{1}{\left(s^{\alpha}+\frac{1}{\lambda}\right)s^{\alpha k+\alpha j+j}}\frac{s}{s^2+\omega^2}\mathrm{d}\xi. \tag{7.25}$$

Performing the invert transformation for this result, we obtain the shear stress as:

$$
\begin{aligned}
\tau(y,t) = & -\frac{2\rho A}{\pi}\int_0^\infty \left(1-e^{-\nu\xi^2 t}\right)\cdot\frac{\cos(y\xi)}{\xi^2}\mathrm{d}\xi \\
& +\frac{2\mu A}{\pi}\int_0^\infty\int_0^t e^{-\nu\xi^2(t-s)}\cos(y\xi)\times R_{\alpha,\alpha-1}\left(-\frac{1}{\lambda},0,s\right)\mathrm{d}s\mathrm{d}\xi \\
& -\frac{2A\nu}{\pi}\frac{\mu}{\lambda}\int_0^\infty \xi^2\cos(y\xi)\sum_{k=0}^{\infty}\sum_{j=0}^{\infty}\left(-\frac{1}{\lambda}\right)^k\left(-\frac{\nu\xi^2}{\lambda}\right)^j\cdot\frac{(k+1)_j}{\Gamma(1+j)}\int_0^t\int_0^\sigma e^{-\nu\xi^2(\sigma-s)} \\
& \times\frac{s^{\alpha k+\alpha j+j+1}}{\Gamma(\alpha k+\alpha j+j+2)}R_{\alpha,0}\left(-\frac{1}{\lambda},0,t-\sigma\right)\mathrm{d}s\mathrm{d}\sigma\mathrm{d}\xi \\
& +\frac{2U}{\pi}\frac{\mu}{\lambda}\int_0^\infty\int_0^t\int_0^\sigma\cos(y\xi)\sum_{k=0}^{\infty}\sum_{j=0}^{\infty}\left(-\frac{1}{\lambda}\right)^{k+1}\left(-\frac{\nu\xi^2}{\lambda}\right)^j \\
& \times\frac{(k+1)_j}{\Gamma(1+j)}\frac{s^{\alpha k+\alpha j+\alpha+j-1}}{\Gamma(\alpha k+\alpha j+\alpha+j)}R_{\alpha,0}\left(-\frac{1}{\lambda},0,t-\sigma\right)\cos(\omega(\sigma-s))\mathrm{d}s\mathrm{d}\sigma\mathrm{d}\xi \\
& -\frac{2U}{\pi}\frac{\mu}{\lambda}\int_0^\infty\int_0^t\int_0^\sigma\cos(y\xi)\times\sum_{k=0}^{\infty}\sum_{j=0}^{\infty}\left(-\frac{1}{\lambda}\right)^k\left(-\frac{\nu\xi^2}{\lambda}\right)^j\cdot \\
& \times\frac{(k+1)_j}{\Gamma(1+j)}\frac{s^{\alpha k+\alpha j+j-1}}{\Gamma(\alpha k+\alpha j+j)}R_{\alpha,0}\left(-\frac{1}{\lambda},0,t-\sigma\right)\cos(\omega(\sigma-s))\mathrm{d}s\mathrm{d}\sigma\mathrm{d}\xi
\end{aligned}
\tag{7.26}
$$

2. The case. $u(0,\, t) = U\sin(\omega t)$

Proceeding in a similar manner, we obtain the corresponding solution of the following form:

$$
\begin{aligned}
\tau(y,t) = & \frac{2U}{\pi}\frac{\mu}{\lambda}\int_0^\infty\int_0^t\int_0^\sigma\cos(y\xi)\sum_{k=0}^{\infty}\sum_{j=0}^{\infty}\left(-\frac{1}{\lambda}\right)^{k+1}\left(-\frac{\nu\xi^2}{\lambda}\right)^j\cdot\frac{(k+1)_j}{\Gamma(1+j)} \\
& \times\frac{s^{\alpha k+\alpha j+\alpha+j-1}}{\Gamma(\alpha k+\alpha j+\alpha+j)} \\
& \times R_{\alpha,0}\left(-\frac{1}{\lambda},0,t-\sigma\right)\sin(\omega(\sigma-s))\mathrm{d}s\mathrm{d}\sigma\mathrm{d}\xi-\frac{2U}{\pi}\frac{\mu}{\lambda}\int_0^\infty\int_0^t\int_0^\sigma\cos(y\xi) \\
& \times\sum_{k=0}^{\infty}\sum_{j=0}^{\infty}\left(-\frac{1}{\lambda}\right)^k \\
& \times\left(-\frac{\nu\xi^2}{\lambda}\right)^j\cdot\frac{(k+1)_j}{\Gamma(1+j)}\frac{s^{\alpha k+\alpha j+j-1}}{\Gamma(\alpha k+\alpha j+j)}R_{\alpha,0}\left(-\frac{1}{\lambda},0,t-\sigma\right)\sin(\omega(\sigma-s))\mathrm{d}s\mathrm{d}\sigma\mathrm{d}\xi.
\end{aligned}
\tag{7.27}
$$

7.2.5 Limiting Cases

1. **The case.** $u(0, t) = At + U\cos(\omega t)$

By letting $\alpha \to 1$ in Eqs. (7.20) and (7.26), we obtain the solutions corresponding to the classical Maxwell model:

$$\begin{aligned}
u(y,t) &= At + U\cos(\omega t) - \frac{2A}{\nu\pi}\int_0^{\infty}\left(1 - e^{-\nu\xi^2 t}\right)\cdot\frac{\sin(y\xi)}{\xi^3}\mathrm{d}\xi \\
&\quad - \frac{2A\nu}{\pi}\int_0^{\infty}\int_0^{t}\xi\sin(y\xi)\cdot e^{-\nu\xi^2(t-s)}\sum_{k=0}^{\infty}\sum_{j=0}^{\infty}\left(\frac{-1}{\lambda}\right)^k\left(-\frac{\nu\xi^2}{\lambda}\right)^j \\
&\quad \times\frac{(k+1)_j s^{k+2j+1}}{\Gamma(1+j)\Gamma(k+2j+2)}\mathrm{d}s\mathrm{d}\xi \\
&\quad + \frac{2U}{\pi}\int_0^{\infty}\int_0^{t}\frac{\sin(y\xi)}{\xi}\cdot\sum_{k=0}^{\infty}\sum_{j=0}^{\infty}\left(\frac{-1}{\lambda}\right)^{k+1}\left(-\frac{\nu\xi^2}{\lambda}\right)^j\frac{(k+1)_j}{\Gamma(1+j)} \\
&\quad \times\frac{s^{k+2j}}{\Gamma(k+2j+1)}\cos(\omega(t-s))\mathrm{d}s\mathrm{d}\xi \\
&\quad - \frac{2U}{\pi}\int_0^{\infty}\int_0^{t}\frac{\sin(y\xi)}{\xi}\sum_{k=0}^{\infty}\sum_{j=0}^{\infty}\left(\frac{-1}{\lambda}\right)^k\left(-\frac{\nu\xi^2}{\lambda}\right)^j\frac{(k+1)_j s^{k+2j-1}}{\Gamma(1+j)\Gamma(k+2j)} \\
&\quad \times\cos(\omega(t-s))\mathrm{d}s\mathrm{d}\xi
\end{aligned} \tag{7.28}$$

$$\begin{aligned}
\tau(y,t) &= -\frac{2\rho A}{\pi}\int_0^{\infty}\left(1 - e^{-\nu\xi^2 t}\right)\cdot\frac{\cos(y\xi)}{\xi^2}\mathrm{d}\xi - \frac{2\mu A}{\pi}\lambda\int_0^{\infty}\frac{\cos(y\xi)}{\nu\lambda\xi^2 - 1}\left(e^{-\nu\xi^2 t} - e^{-\frac{t}{\lambda}}\right)\mathrm{d}\xi \\
&\quad - \frac{2\nu A}{\pi}\frac{\mu}{\lambda}e^{-\frac{t}{\lambda}}\int_0^{\infty}\xi^2\cos(y\xi)\int_0^{t}\int_0^{\sigma}e^{-\nu\xi^2(\sigma-s)+\frac{\sigma}{\lambda}}\left\{\sum_{k=0}^{\infty}\sum_{j=0}^{\infty}\left(-\frac{1}{\lambda}\right)^k\left(-\frac{\nu\xi^2}{\lambda}\right)^j\right. \\
&\quad \left.\times\frac{(k+1)_j s^{k+2j+1}}{\Gamma(1+j)\Gamma(k+2j+2)}\right\}\mathrm{d}s\mathrm{d}\sigma\mathrm{d}\xi \\
&\quad + \frac{2U}{\pi}\frac{\mu}{\lambda}e^{-\frac{t}{\lambda}}\int_0^{\infty}\cos(y\xi)\int_0^{t}\int_0^{\sigma}e^{\frac{\sigma}{\lambda}}\cos(\omega(\sigma-s))\left\{\sum_{k=0}^{\infty}\sum_{j=0}^{\infty}\left(-\frac{1}{\lambda}\right)^{k+1}\left(-\frac{\nu\xi^2}{\lambda}\right)^j\right. \\
&\quad \left.\times\frac{(k+1)_j s^{k+2j}}{\Gamma(1+j)\Gamma(k+2j+1)}\right\}\mathrm{d}s\mathrm{d}\sigma\mathrm{d}\xi \\
&\quad - \frac{2U}{\pi}\frac{\mu}{\lambda}e^{-\frac{t}{\lambda}}\int_0^{\infty}\cos(y\xi)\int_0^{t}\int_0^{\sigma}e^{\frac{\sigma}{\lambda}}\cos(\omega(\sigma-s))\left\{\sum_{k=0}^{\infty}\sum_{j=0}^{\infty}\left(-\frac{1}{\lambda}\right)^k\left(-\frac{\nu\xi^2}{\lambda}\right)^j\right. \\
&\quad \left.\times\frac{(k+1)_j s^{k+2j-1}}{\Gamma(1+j)\Gamma(k+2j)}\right\}\mathrm{d}s\mathrm{d}\sigma\mathrm{d}\xi.
\end{aligned} \tag{7.29}$$

2. The case. $u(0, t) = U\sin(\omega t)$

Similarly, we can obtain the corresponding velocity field and tangential stress for classical Maxwell fluid, respectively, as:

$$u(y,t) = U\sin(\omega t) + \frac{2U}{\pi}\int_0^\infty\int_0^t \frac{\sin(y\xi)}{\xi}\cdot\sum_{k=0}^\infty\sum_{j=0}^\infty\left(\frac{-1}{\lambda}\right)^{k+1} \cdot\left(-\frac{\nu\xi^2}{\lambda}\right)^j\frac{(k+1)_j}{\Gamma(1+j)}\frac{s^{k+2j}}{\Gamma(k+2j+1)}\sin(\omega(t-s))\mathrm{d}s\mathrm{d}\xi - \frac{2U}{\pi}\int_0^\infty\int_0^t\frac{\sin(y\xi)}{\xi}\sum_{k=0}^\infty\sum_{j=0}^\infty\left(\frac{-1}{\lambda}\right)^k \cdot \left(-\frac{\nu\xi^2}{\lambda}\right)^j\frac{(k+1)_j s^{k+2j-1}}{\Gamma(1+j)\Gamma(k+2j)}\sin(\omega(t-s))\mathrm{d}s\mathrm{d}\xi, \tag{7.30}$$

$$\tau(y,t) = \frac{2U}{\pi}\frac{\mu}{\lambda}e^{-\frac{t}{\lambda}}\int_0^\infty\cos(y\xi)\int_0^t\int_0^\sigma e^{\frac{\sigma}{\lambda}}\sin(\omega(\sigma-s))\cdot \left\{\sum_{k=0}^\infty\sum_{j=0}^\infty\left(-\frac{1}{\lambda}\right)^{k+1}\left(-\frac{\nu\xi^2}{\lambda}\right)^j\frac{(k+1)_j s^{k+2j}}{\Gamma(1+j)\Gamma(k+2j+1)}\right\}\mathrm{d}s\mathrm{d}\sigma\mathrm{d}\xi - \frac{2U}{\pi}\frac{\mu}{\lambda}e^{-\frac{t}{\lambda}}\int_0^\infty\cos(y\xi)\int_0^t\int_0^\sigma e^{\frac{\sigma}{\lambda}}\sin(\omega(\sigma-s)) \cdot\left\{\sum_{k=0}^\infty\sum_{j=0}^\infty\left(-\frac{1}{\lambda}\right)^k\left(-\frac{\nu\xi^2}{\lambda}\right)^j\frac{(k+1)_j s^{k+2j-1}}{\Gamma(1+j)\Gamma(k+2j)}\right\}\mathrm{d}s\mathrm{d}\sigma\mathrm{d}\xi. \tag{7.31}$$

3. Taking the limits of Eqs. (7.20) and (7.26), as $U \to 0$, we obtain:

$$u(y,t) = At - \frac{2A}{\nu\pi}\int_0^\infty\left(1-e^{-\nu\xi^2 t}\right)\cdot\frac{\sin(y\xi)}{\xi^3}\mathrm{d}\xi - \frac{2A\nu}{\pi}\int_0^\infty\int_0^t \xi\sin(y\xi)\cdot e^{-\nu\xi^2(t-s)} \times\sum_{k=0}^\infty\sum_{j=0}^\infty\left(\frac{-1}{\lambda}\right)^k\left(-\frac{\nu\xi^2}{\lambda}\right)^j\frac{(k+1)_j s^{k\alpha+j\alpha+j+1}}{\Gamma(1+j)\Gamma(k\alpha+j\alpha+j+2)}\mathrm{d}s\mathrm{d}\xi, \tag{7.32}$$

$$
\begin{aligned}
\tau(y,t) = & -\frac{2\rho A}{\pi}\int_0^{\infty}\left(1-e^{-\nu\xi^2 t}\right)\cdot\frac{\cos(y\xi)}{\xi^2}\mathrm{d}\xi \\
& +\frac{2\mu A}{\pi}\int_0^{\infty}\int_0^{t} e^{-\nu\xi^2(t-s)}\cos(y\xi)R_{\alpha,\alpha-1}\left(-\frac{1}{\lambda},0,s\right)\mathrm{d}s\mathrm{d}\xi \\
& -\frac{2A\nu}{\pi}\frac{\mu}{\lambda}\int_0^{\infty}\xi^2\cos(y\xi)\sum_{k=0}^{\infty}\sum_{j=0}^{\infty}\left(-\frac{1}{\lambda}\right)^k\left(-\frac{\nu\xi^2}{\lambda}\right)^j\cdot\frac{(k+1)_j}{\Gamma(1+j)} \\
& \times\int_0^{t}\int_0^{\sigma} e^{-\nu\xi^2(\sigma-s)}\frac{s^{\alpha k+\alpha j+j+1}}{\Gamma(\alpha k+\alpha j+j+2)}R_{\alpha,0}\left(-\frac{1}{\lambda},0,t-\sigma\right)\mathrm{d}s\mathrm{d}\sigma\mathrm{d}\xi
\end{aligned}
\tag{7.33}
$$

In the special cases that when $\alpha \to 1$, the general solutions Eqs. (7.20) and (7.26) reduce to the classical solutions for Maxwell fluid performing the same motion.

7.2.6 Analysis and Discussion

The exact solutions for the velocity field, obtained in series forms and generalized G and R functions, are presented graphically for specified parameters. For the sake of simplicity, all the diagrams are plotted by taking $U=1$, $A=1$, $\lambda=8$.

Fig. 7.2 shows the velocity profiles of $u(y, t)$ versus y for different values of ω when other parameters are fixed. As expected, the velocity goes to zero as y increases. It can be seen that for condition of (A) $u(0, t)=At+U\cos(\omega t)$, nearing the plate, the velocity profile is a decreasing function of ω and oscillates decreasing in cosine characteristics (the frequency of oscillation of the plate). However, for condition of (B) $u(0, t)=U\sin(\omega t)$, nearing the plate, the velocity profile increases with ω in the sine oscillation characteristics. In the special cases when $\alpha \to 1$, the general solutions reduce to the classical solutions for Maxwell fluid performing the same motion. Figs. 7.3 and 7.4 show the velocity $u(y, t)$ versus y for different values of α and for different values of t when other parameters are fixed. The detailed analyses are omitted here.

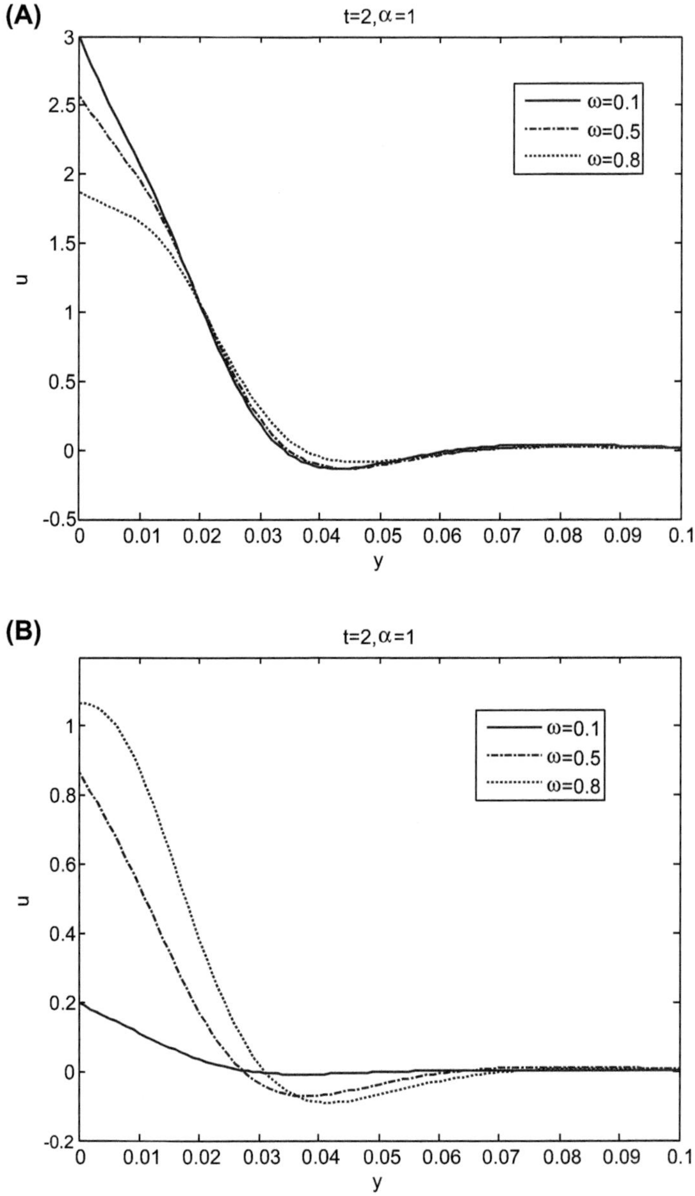

FIGURE 7.2 The velocity $u(y, t)$ versus y for values of ω when other parameters are fixed (A) $u(0, t) = At + U\cos(\omega t)$, (B) $u(0, t) = U\sin(\omega t)$.

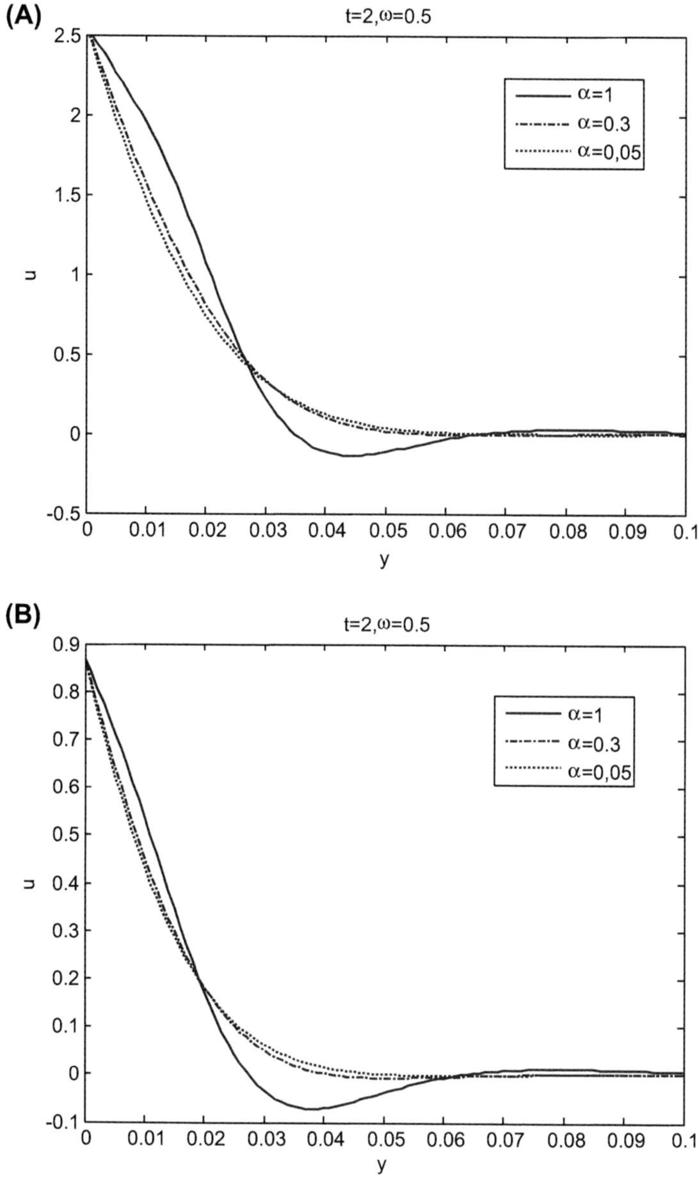

FIGURE 7.3 The velocity $u(y, t)$ versus y for values of α when other parameters are fixed (A) $u(0, t) = At + U\cos(\omega t)$, (B) $u(0, t) = U\sin(\omega t)$.

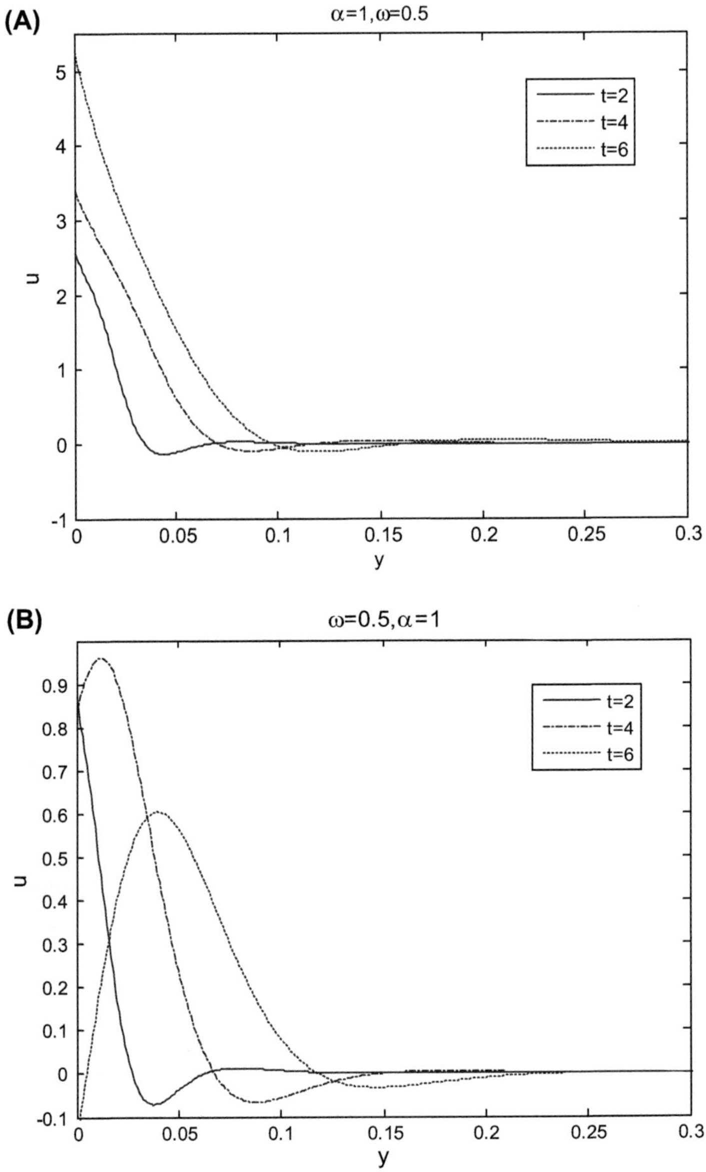

FIGURE 7.4 The velocity $u(y, t)$ versus y for values of t when other parameters are fixed (A) $u(0, t) = At + U\cos(\omega t)$, (B) $u(0, t) = U\sin(\omega t)$.

7.3 HELICAL FLOWS OF FRACTIONAL OLDROYD-B FLUID IN POROUS MEDIUM

7.3.1 Formulation of the Problem

It is well known that Darcy's law relates the pressure drop and the velocity in an unbounded porous medium. For the flows of a viscous Newtonian fluid at low speed through a porous medium, the pressure drop caused by the frictional drag is directly proportional to the velocity. However, some studies performed on the viscoelastic fluid flow in the porous medium have shown that the classical Darcy's law is not suitable for characterizing those fluids' transport in porous medium. Tan and Masuoka investigated Stokes' first problem for viscoelastic fluids in a porous half-space (Tan, 2006; Tan and Masuoka, 2005a,b, 2007). Xue and Nie (2009) discussed the exact solutions of the Rayleigh–Stokes problem for a heated generalized second grade fluid in a porous half-space.

In this section, we study the flow and heat transfer of Oldroyd-B fluid in a cylinder in porous medium. It is assumed that the fluid is initially stationary at two infinite long cylindrical walls, and the radii of the cylinder wall are R_1 and $R_2 (R_2 > R_1)$, respectively. At the moment $t = 0$, in the axial direction of θ, the outer wall of the cylinder is suddenly affected by the pressure gradient and the shock of the outer wall with time. The aim of this research is to find the analytical solution to the problem, and to analyze the flow and heat transfer characteristics of the velocity and temperature fields.

The conservation and constitutive equations of an incompressible Oldroyd-B fluid with fractional derivative is given by (Fetecau et al., 2009; Samko et al., 1993; Tan, 2006; Tan and Masuoka, 2005a,b):

$$\mathbf{T} = -P\mathbf{I} + \mathbf{S} \quad \mathbf{S} + \lambda\frac{\mathrm{D}\mathbf{S}}{\mathrm{D}t} = \mu\left[1 + \lambda_r\frac{\mathrm{D}}{\mathrm{D}t}\right]\mathbf{A}, \tag{7.34}$$

where $-P\mathbf{I}$ denotes the indeterminate spherical stress, $\mathbf{S}$ is the extra-stress tensor, $\mathbf{T}$ is the Cauchy stress tensor, $\mathbf{A} = \mathbf{L} + \mathbf{L}^T$ is the first Rivlin–Ericksen tensor with $\mathbf{L}$ the velocity gradient, μ is the dynamic viscosity of the fluid, λ and $\lambda_r(<\lambda)$ are relaxation and retardation times, $\mathrm{D}\mathbf{S}/\mathrm{D}t$ and $\mathrm{D}\mathbf{A}/\mathrm{D}t$ are defined by:

$$\frac{\mathrm{D}\mathbf{S}}{\mathrm{D}t} = \mathrm{D}_t^{\alpha}\mathbf{S} + (\mathbf{V}\cdot\nabla)\mathbf{S} - \mathbf{S}(\nabla\mathbf{V}) - (\nabla\mathbf{V})^T\mathbf{S}, \tag{7.35}$$

$$\frac{\mathrm{D}\mathbf{A}}{\mathrm{D}t} = \mathrm{D}_t^{\beta}\mathbf{S} + (\mathbf{V}\cdot\nabla)\mathbf{A} - \mathbf{A}(\nabla\mathbf{V}) - (\nabla\mathbf{V})^T\mathbf{A}, \tag{7.36}$$

where ∇ is the gradient operator, D_t^{α} and D_t^{β} are the fractional differential operators based on Riemann–Liouville's, defined as:

$$\mathrm{D}_t^p f(t) = \frac{1}{\Gamma(1-p)}\frac{\mathrm{d}}{\mathrm{d}t}\int_0^t \frac{f(\tau)}{(t-\tau)^p}\mathrm{d}\tau, \quad 0 < p < 1, \tag{7.37}$$

$\Gamma(\cdot)$ denotes the Gamma function. This model reduces to the ordinary Oldroyd-B model when $\alpha = \beta = 1$ and to the fractional Maxwell model when $\beta \neq 0$ and $\lambda_r \to 0$.

In cylindrical coordinates (r, θ, z), the helical flow velocity is given by:

$$\mathbf{V} = u(r,t)e_\theta + w(r,t)e_z, \tag{7.38}$$

where e_θ and e_z are the unit vectors in the θ- and z-directions. Since the velocity field $\mathbf{V}$ is independent of θ and z, we also assume that $\mathbf{S}$ is independent of θ and z. The fluid is assumed to be at rest at the moment $t = 0$, then

$$\mathbf{V}(r,0) = \mathbf{0}, \;\; \mathbf{S}(r,t) = \mathbf{0}. \tag{7.39}$$

We obtain:

$$\left(1 + \lambda \mathrm{D}_t^\alpha\right)\tau_{rz} = \mu\left(1 + \lambda_r \mathrm{D}_t^\beta\right)\frac{\partial w}{\partial r}, \tag{7.40}$$

$$\left(1 + \lambda \mathrm{D}_t^\alpha\right)\tau_{r\theta} = \mu\left(1 + \lambda_r \mathrm{D}_t^\beta\right)\left(\frac{\partial u}{\partial r} - \frac{u}{r}\right), \tag{7.41}$$

where $S_{rr} = S_{zz} = S_{\theta z} = S_{\theta\theta} = 0$, $\tau_{rz}(r, t) = S_{rz}(r, t)$ and $\tau_{r\theta}(r, t) = S_{r\theta}(r, t)$ are the shear stresses, which are not zero.

We consider here the modified Darcy's law of the generalized viscoelastic flow in the porous medium; the phenomenological model can be described as:

$$\left(1 + \lambda \mathrm{D}_t^\alpha\right)\nabla\mathbf{P} = -\frac{\mu\phi}{K}\left(1 + \lambda_r \mathrm{D}_t^\beta\right)\mathbf{V}, \tag{7.42}$$

where K is the permeability and ϕ is the porosity of the porous medium (Khan et al., 2007).

In the absence of body forces in the axial direction and the pressure gradient, the balance of the linear momentum leads to the relevant and meaningful equation,

$$\rho\frac{\mathrm{D}\mathbf{V}}{\mathrm{D}t} = -\nabla\mathbf{P} + \operatorname{div}\mathbf{S} + \mathbf{r}. \tag{7.43}$$

which yields

$$\left(1 + \lambda \mathrm{D}_t^\alpha\right)\mathbf{r} = -\frac{\mu\phi}{K}\left(1 + \lambda_r \mathrm{D}_t^\beta\right)\mathbf{V}, \tag{7.44}$$

and

$$\rho\left(1 + \lambda \mathrm{D}_t^\alpha\right)\frac{\mathrm{D}\mathbf{V}}{\mathrm{D}t} = -\left(1 + \lambda \mathrm{D}_t^\alpha\right)\nabla\mathbf{P} + \left(1 + \lambda \mathrm{D}_t^\alpha\right)\operatorname{div}\mathbf{S} - \frac{\mu\phi}{K}\left(1 + \lambda_r \mathrm{D}_t^\beta\right)\mathbf{V}. \tag{7.45}$$

Eliminating $\tau(r, t)$ between Eqs. (7.39), (7.40), (7.41), and (7.45), we obtain the following governing equations:

$$\left(1+\lambda D_t^{\alpha}\right)\frac{\partial w}{\partial t} = \nu\left(1+\lambda_r D_t^{\beta}\right)\left(\frac{\partial^2}{\partial r^2}+\frac{1}{r}\frac{\partial}{\partial r}\right)w - \frac{\nu\phi}{K}\left(1+\lambda_r D_t^{\beta}\right)w,$$

$$\left(1+\lambda D_t^{\alpha}\right)\frac{\partial u}{\partial t} = \nu\left(1+\lambda_r D_t^{\beta}\right)\left(\frac{\partial^2}{\partial r^2}+\frac{1}{r}\frac{\partial}{\partial r}-\frac{1}{r^2}\right)u - \frac{\nu\phi}{K}\left(1+\lambda_r D_t^{\beta}\right)u,$$

$$r\in(R_1,R_2),\ \ t>0, \tag{7.46}$$

where $\nu = \mu/\rho$ is the kinematic viscosity and ρ is the constant density of the fluid.

We consider the energy equation in the presence of radiation transfer. In terms of Fourier's law, the energy equation may be written as:

$$(\rho c)_{fs}\frac{DT}{Dt} = \lambda_e \mathrm{div}(gradT) - \mathrm{div}\,\overrightarrow{q_R} + Q_T. \tag{7.47}$$

The radiative heat flux q_r in view of the Rosseland approximation is simplified as (Raptis et al., 2004):

$$q_r = -\frac{4\sigma^*}{3k^*}\frac{\partial T^4}{\partial r}, \tag{7.48}$$

where σ^* and k^* are the Stefan–Boltzmann constant and the mean absorption coefficient, respectively. We assume that the term T^4 may be expressed as a linear function of temperature, and expand T^4 in a Taylor series about T_e, which is the environment temperature and neglecting higher-order terms, we get:

$$T^4 = 4T_e^3T - 3T_e^4, \tag{7.49}$$

$$\text{then } q_r = -\frac{16\sigma^* T_e^3}{3k^*}\frac{\partial T}{\partial r}.$$

The porous media term Q_T is chosen as $Q_T = \frac{\mu\phi^2}{K}u^2$, then the governing equation of heat transfer may be represented as:

$$\frac{\partial T}{\partial t} = \left(\frac{\lambda_e}{(\rho c)_{fs}} + \frac{16\sigma^* T_e^3}{3k^*(\rho c)_{fs}}\right)\left(\frac{\partial^2}{\partial r^2}+\frac{1}{r}\frac{\partial}{\partial r}\right)T + \frac{\mu\phi^2}{K(\rho c)_{fs}}u^2. \tag{7.50}$$

Therefore the sequential fractional heat transfer equation has the form:

$$\left(1+\lambda D_t^{\gamma}\right)\frac{\partial T}{\partial t} = \left(\frac{\lambda_e}{(\rho c)_{fs}} + \frac{16\sigma^* T_e^3}{3k^*(\rho c)_{fs}}\right)\left(\frac{\partial^2}{\partial r^2}+\frac{1}{r}\frac{\partial}{\partial r}\right)T + \frac{\mu\phi^2}{K(\rho c)_{fs}}u^2. \tag{7.51}$$

7.3.2 Helical Flow Between Coaxial Cylinders

We consider here an incompressible Oldroyd-B fluid at rest in the annular region between two infinite circular cylinders of radius R_1 and $R_2(>R_1)$. At time

$t = 0^+$, the inner cylinder suddenly begins to move by a time-dependent longitudinal shear stress and the outer cylinder suddenly begins to oscillate in the θ-direction. The appropriate initial and boundary conditions are:

$$u(r,0) = \frac{\partial u(r,0)}{\partial t} = w(r,0) = \frac{\partial w(r,0)}{\partial t} = 0, \;\; \tau_{rz}(r,0) = 0, \;\; r \in (R_1, R_2), \tag{7.52}$$

$$\left(1 + \lambda D_t^{\alpha}\right)\tau_{rz}(r,t)\Big|_{r=R_1} = \mu\left(1 + \lambda_r D_t^{\beta}\right)\frac{\partial w}{\partial r}\Big|_{r=R_1} = fe^{at}, \;\; w(R_2,t) = 0,$$

$$u(R_2,t) = U \sin \omega t, \;\; u(R_1,t) = 0, \;\; t > 0, \tag{7.53}$$

where f is constant.

Assume that the term u^2 may be expressed as a linear function of u, and in view of the best square approximation in the interval of $[0, U \cos \omega t]$, obtain

$$u^2 \approx \frac{3}{4} Uu \cos \omega t. \tag{7.54}$$

For simplicity, we only discuss the case of $\omega = 0$, so we have:

$$(1 + \lambda D_t^{\gamma})\frac{\partial T}{\partial t} = A\left(\frac{\partial^2}{\partial r^2} + \frac{1}{r}\frac{\partial}{\partial r}\right)T + \frac{3\mu\phi^2 U}{4K(\rho c)_{fs}} u, \tag{7.55}$$

and the appropriate initial and boundary conditions are:

$$T(r,0) = T_0, \;\; t = 0, \tag{7.56}$$

$$\frac{\partial T(r,t)}{\partial t}\Big|_{t=0} = 0, \tag{7.57}$$

$$T(R_1,t) = T_0 + T_2, \;\; T(R_2,t) = T_1\varphi(t) + T_0, \tag{7.58}$$

where $A = \left(\frac{\lambda_e}{(\rho c)_{fs}} + \frac{16\sigma^* T_e^3}{3k^*(\rho c)_{fs}}\right)$, $\varphi(t)$ is the linear function of t, and T_0, T_1, T_2 are all the constants.

7.3.3 Calculation of the Velocity Field

Applying the Laplace transforms to Eqs. (7.46), (7.52), and (7.53), in terms of sequential fractional derivatives Laplace transform (Zheng et al., 2011c), we find that

$$(q + \lambda q^{\alpha+1})\overline{u} = \nu(1 + \lambda_r q^{\beta})\left(\frac{\partial^2}{\partial r^2} + \frac{1}{r}\frac{\partial}{\partial r} - \frac{1}{r^2}\right)\overline{u} - \frac{\nu\phi}{K}(1 + \lambda_r q^{\beta})\overline{u}, \tag{7.59}$$

$$\left(q+\lambda q^{\alpha+1}\right)\overline{w}=\nu\left(1+\lambda_r q^{\beta}\right)\left(\frac{\partial^2}{\partial r^2}+\frac{1}{r}\frac{\partial}{\partial r}\right)\overline{w}-\frac{\nu\phi}{K}\left(1+\lambda_r q^{\beta}\right)\overline{w},\ \ r\in(R_1,R_2), \tag{7.60}$$

$$\overline{w}(R_2,q)=0,\ \ \frac{\partial\overline{w}}{\partial r}\bigg|_{r=R_1}=\frac{f}{\mu(q-a)\left(1+\lambda_r q^{\beta}\right)},\ \ \overline{u}(R_2,q)=U\frac{\omega}{q^2+\omega^2}. \tag{7.61}$$

1. θ-direction.

In the following, let us denote by (Xue and Nie, 2009)

$$\overline{u}_H=\int_{R_1}^{R_2} r\overline{u}B_1(rr_n)\mathrm{d}r,\ n=1,2,3,\ldots, \tag{7.62}$$

the finite Hankel transforms of $\overline{u}$, where r_n are the positive roots of the transcendental equation $B_1(R_2r)=0$, and

$$B_1(rr_n)=J_1(rr_n)Y_1(R_1r_n)-J_1(R_1r_n)Y_1(rr_n),$$

while $J_n(\cdot)$ and $Y_n(\cdot)$ are Bessel functions of the first and second kind of order n.

Multiplying both sides of Eq. (7.59) by $rB_1(rr_n)$, integrating with respect to r from R_1 to R_2 and taking into account the conditions of Eq. (7.61) and the identity

$$\int_{R_1}^{R_2} r\left(\frac{\partial^2}{\partial r^2}+\frac{1}{r}\frac{\partial}{\partial r}-\frac{1}{r^2}\right)\overline{u}B_1(rr_n)\mathrm{d}r=\frac{2U\omega}{\pi(q^2+\omega^2)}\frac{J_1(R_1r_n)}{J_1(R_2r_n)}-r_n^2\overline{u}_H(r_n,q), \tag{7.63}$$

we obtain

$$\begin{aligned}\overline{u}_H=&\frac{2U\omega J_1(R_1r_n)}{\pi J_1(R_2r_n)}\frac{1}{(q^2+\omega^2)}-\frac{2U\omega J_1(R_1r_n)}{\pi J_1(R_2r_n)}\frac{\phi}{Kr_n^2\left(r_n^2+\frac{\phi}{K}\right)(q^2+\omega^2)}\\&-\frac{2\nu U\omega J_1(R_1r_n)}{\pi J_1(R_2r_n)}\frac{q(1+\lambda q^{\alpha})}{\left(\nu r_n^2+\frac{\nu\phi}{K}\right)(q^2+\omega^2)\left[q(1+\lambda q^{\alpha})+\left(\nu r_n^2+\frac{\nu\phi}{K}\right)\left(1+\lambda_r q^{\beta}\right)\right]}.\end{aligned} \tag{7.64}$$

Inverting the previous results by means of the Hankel transform, using the expansion

$$\begin{aligned}&\frac{q(1+\lambda q^{\alpha})}{\left(\nu r_n^2+\frac{\nu\phi}{K}\right)\left[q(1+\lambda q^{\alpha})+\left(\nu r_n^2+\frac{\nu\phi}{K}\right)\left(1+\lambda_r q^{\beta}\right)\right]}\\&=\sum_{m=0}^{\infty}(-1)^m\left(\nu r_n^2+\frac{\nu\phi}{K}\right)^{m-1}\sum_{i=0}^{\infty}\frac{(-1)^i\lambda_r^{m-i}}{i!}\sum_{l=0}^{\infty}\frac{(-1)^l}{l!\lambda^{l+m}}\\&\times\frac{\Gamma(i-m)\Gamma(l+m)}{\Gamma(-m)\Gamma(m)}q^{(m-i)\beta-(m+l)\alpha-m},\end{aligned} \tag{7.65}$$

which yields that

$$\overline{u} = \frac{UR_2(r^2 - R_1^2)}{(R_2^2 - R_1^2)r} \frac{\omega}{q^2 + \omega^2} - \frac{U\pi\phi}{K} \sum_{n=1}^{\infty} \frac{J_1(R_2r_n)J_1(R_1r_n)B_1(rr_n)}{\left(J_1^2(R_1r_n) - J_1^2(R_2r_n)\right)\left(r_n^2 + \frac{\phi}{K}\right)} \frac{\omega}{(q^2 + \omega^2)}$$
$$- \pi\nu U \sum_{n=1}^{\infty} \frac{J_1(R_2r_n)J_1(R_1r_n)B_1(rr_n)}{J_1^2(R_1r_n) - J_1^2(R_2r_n)} \sum_{m=0}^{\infty} (-1)^m \left(\nu r_n^2 + \frac{\nu\phi}{K}\right)^{m-1}$$
$$\times \sum_{i=0}^{\infty} \frac{(-1)^i \lambda_r^{m-i}}{i!} \sum_{l=0}^{\infty} \frac{(-1)^l}{l!\lambda^{l+m}}$$
$$\times \frac{\Gamma(i-m)\Gamma(l+m)}{\Gamma(-m)\Gamma(m)} \frac{\omega}{q^2 + \omega^2} q^{(m-i)\beta - (m+l)\alpha - m}. \tag{7.66}$$

We apply the discrete inversion Laplace transform method, attaining the following expressions for the velocity field:

$$u = \frac{UR_2(r^2 - R_1^2)}{(R_2^2 - R_1^2)r} \sin(\omega t) - \frac{U\pi\phi}{K} \sum_{n=1}^{\infty} \frac{J_1(R_2r_n)J_1(R_1r_n)B_1(rr_n)}{\left(J_1^2(R_1r_n) - J_1^2(R_2r_n)\right)\left(r_n^2 + \frac{\phi}{K}\right)} \sin(\omega t)$$
$$- \pi\nu U \sum_{n=1}^{\infty} \frac{J_1(R_2r_n)J_1(R_1r_n)B_1(rr_n)}{J_1^2(R_1r_n) - J_1^2(R_2r_n)} \sum_{m=0}^{\infty} (-1)^m \left(\nu r_n^2 + \frac{\nu\phi}{K}\right)^{m-1} \sum_{i=0}^{\infty} \frac{(-1)^i \lambda_r^{m-i}}{i!} \sum_{l=0}^{\infty} \frac{(-1)^l}{l!\lambda^{l+m}}$$
$$\times \frac{\Gamma(i-m)\Gamma(l+m)}{\Gamma(-m)\Gamma(m)\Gamma[(m+l)\alpha + m - (m-i)\beta]} \int_0^t \sin(\omega(t-\tau))\tau^{(m+l)\alpha + m - (m-i)\beta - 1} d\tau. \tag{7.67}$$

2. *z*-direction.

Similarly, let us denote by (Khan et al., 2006)

$$\overline{w}_H = \int_{R_1}^{R_2} r\overline{w}B(rr_n)\mathrm{d}r, \quad n = 1, 2, 3, \ldots, \tag{7.68}$$

the finite Hankel transforms of $\overline{w}$, where r_n are the positive roots of the transcendental equation $B(R_2r) = 0$, and

$$B(rr_n) = J_0(rr_n)Y_1(R_1r_n) - J_1(R_1r_n)Y_0(rr_n).$$

Taking into account the conditions of Eq. (7.61) and the identity

$$\int_{R_1}^{R_2} r\left(\frac{\partial^2}{\partial r^2} + \frac{1}{r}\frac{\partial}{\partial r}\right)\overline{w}B(rr_n)\mathrm{d}r = \frac{2f}{\pi r_n \mu (q-a)(1 + \lambda_r q^\beta)} - r_n^2 \overline{w}_H(r_n, q), \tag{7.69}$$

we find that

$$\overline{w}_H = \frac{2f}{\pi r_n \rho (q-a)\left[q(1 + \lambda q^\alpha) + \left(\nu r_n^2 + \frac{\nu\phi}{K}\right)(1 + \lambda_r q^\beta)\right]}. \tag{7.70}$$

Considering the expression

$$\frac{1}{q(1+\lambda q^{\alpha})+\left(\nu r_n^2+\frac{\nu\phi}{K}\right)(1+\lambda_r q^{\beta})}=\sum_{m=0}^{\infty}(-1)^m\left(\nu r_n^2+\frac{\nu\phi}{K}\right)^{-m-1}\sum_{i=0}^{\infty}\frac{(-1)^k\lambda^{m-i}}{i!}\sum_{l=0}^{\infty}\frac{(-1)^l}{l!\lambda_r^{l+m+1}}\times\frac{\Gamma(i-m)\Gamma(l+m+1)}{\Gamma(-m)\Gamma(m+1)}q^{(\alpha+1)m-(m+1)\beta-i\alpha-l\beta},\tag{7.71}$$

we use the inverse Hankel transform and get the expression for $\overline{w}$ in the form

$$\overline{w}=\frac{fR_1}{\mu}\ln\left(\frac{r}{R_2}\right)\frac{1}{(q-a)(1+\lambda_r q^{\beta})}-\frac{\pi f\phi}{\mu K}\sum_{n=1}^{\infty}\frac{J_0^2(R_2r_n)B(rr_n)}{\left(J_1^2(R_1r_n)-J_0^2(R_2r_n)\right)r_n\left(r_n^2+\frac{\phi}{K}\right)}\times\frac{1}{(q-a)(1+\lambda_r q^{\beta})}+\frac{\pi f}{\rho}\sum_{n=1}^{\infty}\frac{r_nJ_0^2(R_2r_n)B(rr_n)}{J_1^2(R_1r_n)-J_0^2(R_2r_n)}\sum_{m=1}^{\infty}(-1)^m\left(\nu r_n^2+\frac{\nu\phi}{K}\right)^{-m-1}\times\sum_{i=0}^{\infty}\frac{(-1)^i\lambda^{m-i}}{i!}\sum_{l=0}^{\infty}\frac{(-1)^l}{l!\lambda_r^{l+m+1}}\frac{\Gamma(i-m)\Gamma(l+m+1)}{\Gamma(-m)\Gamma(m+1)}\frac{q^{(\alpha+1)m-(m+1)\beta-i\alpha-l\beta}}{q-a}.\tag{7.72}$$

Appling the discrete inverse Laplace transform to Eq. (7.72), we obtain the following expression for the velocity field:

$$w=\frac{fR_1}{\mu}\ln\left(\frac{r}{R_2}\right)e^{at}-\frac{fR_1}{\mu}\ln\left(\frac{r}{R_2}\right)\sum_{n=0}^{\infty}\left(-\frac{1}{\lambda_r}\right)^n\int_0^t e^{a(t-\tau)}\cdot\frac{\tau^{n\beta-1}}{\Gamma(n\beta)}\mathrm{d}\tau-\frac{\pi f\phi}{\mu K}\sum_{n=1}^{\infty}\frac{J_0^2(R_2r_n)B(rr_n)}{\left(J_1^2(R_1r_n)-J_0^2(R_2r_n)\right)r_n\left(r_n^2+\frac{\phi}{K}\right)}\left\{e^{at}-\sum_{n=0}^{\infty}\left(-\frac{1}{\lambda_r}\right)^n\int_0^t e^{a(t-\tau)}\cdot\frac{\tau^{n\beta-1}}{\Gamma(n\beta)}\mathrm{d}\tau\right\}+\frac{\pi f}{\rho}\sum_{n=1}^{\infty}\frac{r_nJ_0^2(R_2r_n)B(rr_n)}{J_1^2(R_1r_n)-J_0^2(R_2r_n)}\sum_{m=1}^{\infty}(-1)^m\left(\nu r_n^2+\frac{\nu\phi}{K}\right)^{-m-1}\sum_{i=0}^{\infty}\frac{(-1)^i\lambda^{m-i}}{i!}\sum_{l=0}^{\infty}\frac{(-1)^l}{l!\lambda_r^{l+m+1}}\times\frac{\Gamma(i-m)\Gamma(l+m+1)}{\Gamma(-m)\Gamma(m+1)\Gamma[(m+1+l)\beta+i\alpha-(\alpha+1)m]}\int_0^t e^{a(t-\tau)}\cdot\tau^{(m+1+l)\beta+i\alpha-(\alpha+1)m-1}\mathrm{d}\tau\tag{7.73}$$

7.3.4 Calculation of the Shear Stress

1. **$r\theta$-plane.**

   Applying the Laplace transform to Eq. (7.41), we find that

$$(1+\lambda q^{\alpha})\overline{\tau}_{r\theta}=\mu\left(1+\lambda_r q^{\beta}\right)\left(\frac{\partial}{\partial r}-\frac{1}{r}\right)\overline{u},\tag{7.74}$$

where

$$\overline{\tau}_{r\theta}=\frac{2\mu U R_2 R_1^2}{(R_2^2-R_1^2)r^2}\,\frac{\omega}{q^2+\omega^2}\cdot\frac{1+\lambda_r q^\beta}{1+\lambda q^\alpha}-\frac{\mu U\pi\phi}{K}\sum_{n=1}^{\infty}\frac{J_1(R_2r_n)J_1(R_1r_n)[r_nB(rr_n)-(2/r)B_1(rr_n)]}{\left(J_1^2(R_1r_n)-J_1^2(R_2r_n)\right)\left(r_n^2+\dfrac{\phi}{K}\right)}$$
$$\times\frac{\omega}{q^2+\omega^2}\,\frac{1+\lambda_r q^\beta}{1+\lambda q^\alpha}-\mu\pi\nu U\omega\sum_{n=1}^{\infty}\frac{J_1(R_2r_n)J_1(R_1r_n)[r_nB(rr_n)-(2/r)B_1(rr_n)]}{J_1^2(R_1r_n)-J_1^2(R_2r_n)}$$
$$\times\frac{q(1+\lambda q^\alpha)}{\left(\nu r_n^2+\dfrac{\nu\phi}{K}\right)(q^2+\omega^2)\left[q(1+\lambda q^\alpha)+\left(\nu r_n^2+\dfrac{\nu\phi}{K}\right)(1+\lambda_r q^\beta)\right]}\tag{7.75}$$

Applying the discrete inversion Laplace transform to the obtained result again, we find for the shear stress the expression

$$\tau_{r\theta}=\frac{2\mu U R_2 R_1^2}{(R_2^2-R_1^2)r^2}\sum_{k=0}^{\infty}\frac{(-1)^{k+1}\lambda_r^{1-k\beta}}{k(k-1)}\sum_{l=0}^{\infty}\frac{(-1)^l}{\lambda^{l+1}}\int_0^t \sin\omega(t-\tau)\frac{\tau^{(l+1)\alpha+(k-1)\beta-1}}{\Gamma[(l+1)\alpha+(k-1)\beta]}\mathrm{d}\tau$$
$$-\frac{\mu U\pi\phi}{K}\sum_{n=1}^{\infty}\frac{J_1(R_2r_n)J_1(R_1r_n)[r_nB(rr_n)-(2/r)B_1(rr_n)]}{\left(J_1^2(R_1r_n)-J_1^2(R_2r_n)\right)\left(r_n^2+\dfrac{\phi}{K}\right)}\sum_{k=0}^{\infty}\frac{(-1)^{k+1}\lambda_r^{1-k\beta}}{k(k-1)}$$
$$\times\sum_{l=0}^{\infty}\frac{(-1)^l}{\lambda^{l+1}}\int_0^t \sin\omega(t-\tau)\frac{\tau^{(l+1)\alpha+(k-1)\beta-1}}{\Gamma[(l+1)\alpha+(k-1)\beta]}\mathrm{d}\tau$$
$$-\mu\pi\nu U\sum_{n=1}^{\infty}\frac{J_1(R_2r_n)J_1(R_1r_n)[r_nB(rr_n)-(2/r)B_1(rr_n)]}{J_1^2(R_1r_n)-J_1^2(R_2r_n)}\sum_{m=0}^{\infty}(-1)^m\left(\nu r_n^2+\frac{\nu\phi}{K}\right)^{m-1}$$
$$\times\sum_{i=0}^{\infty}\frac{(-1)^i\lambda_r^{m+1-i}}{i!}\sum_{l=0}^{\infty}\frac{(-1)^l}{l!\lambda^{l+m+1}}\frac{\Gamma(i-m-1)\Gamma(l+m+1)}{\Gamma(-m-1)\Gamma(m+1)\Gamma[i\beta+l\alpha+m+(m+1)(\alpha-\beta)]}$$
$$\times\int_0^t \sin\omega(t-\tau)\cdot\tau^{i\beta+l\alpha+m+(m+1)(\alpha-\beta)-1}\mathrm{d}\tau\tag{7.76}$$

2. *rz*-plane.

In the same way, the shear stress can be described as:

$$\tau_{rz}=\frac{fR_1}{r}e^{at}-\frac{fR_1}{r}\sum_{n=0}^{\infty}\left(-\frac{1}{\lambda}\right)^n\int_0^t e^{a(t-\tau)}\cdot\frac{\tau^{n\alpha-1}}{\Gamma(n\alpha)}\mathrm{d}\tau$$
$$+\frac{\pi f\phi}{K}\sum_{n=1}^{\infty}\frac{J_0^2(R_2r_n)B_1(rr_n)}{\left(J_1^2(R_1r_n)-J_0^2(R_2r_n)\right)\left(r_n^2+\dfrac{\phi}{K}\right)}\left\{e^{at}-\sum_{n=0}^{\infty}\left(-\frac{1}{\lambda}\right)^n\int_0^t e^{a(t-\tau)}\cdot\frac{\tau^{n\alpha-1}}{\Gamma(n\alpha)}\mathrm{d}\tau\right\}$$
$$-\pi\nu f\sum_{n=1}^{\infty}\frac{r_n^2J_0^2(R_2r_n)B_1(rr_n)}{J_1^2(R_1r_n)-J_0^2(R_2r_n)}\sum_{m=1}^{\infty}(-1)^m\left(\nu r_n^2+\frac{\nu\phi}{K}\right)^{-m-1}\sum_{i=0}^{\infty}\frac{(-1)^i\lambda^{m-i-1}}{i!}\sum_{l=0}^{\infty}\frac{(-1)^l}{l!\lambda_r^{l+m}}$$
$$\times\frac{\Gamma(i-m+1)\Gamma(l+m)}{\Gamma(-m+1)\Gamma(m)\Gamma[(m+l)\beta-(m-i-1)\alpha-m]}\int_0^t e^{a(t-\tau)}\cdot\tau^{(m+l)\beta-(m-i-1)\alpha-m-1}\mathrm{d}\tau.\tag{7.77}$$

From Eq. (7.77) it is clear that

$$\left(1+\lambda D_t^{\alpha}\right)\tau_{rz}(R,t) = \frac{fR_1}{r}e^{at} - \frac{fR_1}{r}\sum_{n=0}^{\infty}\left(-\frac{1}{\lambda}\right)^n\int_0^t e^{a(t-\tau)}\frac{\tau^{n\alpha-1}}{\Gamma(n\alpha)}d\tau$$
$$+\lambda D_t^{\alpha}\left(\frac{fR_1}{r}e^{at} - \frac{fR_1}{r}\sum_{n=0}^{\infty}\left(-\frac{1}{\lambda}\right)^n\int_0^t e^{a(t-\tau)}\frac{\tau^{n\alpha-1}}{\Gamma(n\alpha)}d\tau\right)$$
$$= fe^{at}. \tag{7.78}$$

Indeed, let $a=0$, we can obtain:

$$\left(1+\lambda D_t^{\alpha}\right)\tau_{rz}(r,t)\big|_{r=R_1} = f, \tag{7.79}$$

and

$$\tau_{rz}(R_1,t) = f\left[1-e^{\frac{-t}{\lambda}}\right], \tag{7.80}$$

This is the same as the result obtained by Khan et al. (2006).

7.3.5 The Solution of Heat Transfer Equation

Let $\theta = T - T_0$, we get:

$$(1+\lambda D_t^{\gamma})\frac{\partial\theta}{\partial t} = A\left(\frac{\partial^2}{\partial r^2}+\frac{1}{r}\frac{\partial}{\partial r}\right)\theta + \frac{3\mu\phi^2 U}{4K(\rho c)_{fs}}u. \tag{7.81}$$

and

$$\theta(r,0)=0,\quad \left.\frac{\partial\theta(r,t)}{\partial t}\right|_{t=0} = 0, \tag{7.82}$$

$$\theta(R_1,t)=T_2,\quad \theta(R_2,t)=T_1\varphi(t). \tag{7.83}$$

Applying the Laplace transform of the sequential fractional derivatives to Eqs. (7.81)–(7.83), we get:

$$\left(q+\lambda q^{\gamma+1}\right)\overline{\theta} = A\left(\frac{\partial^2}{\partial r^2}+\frac{1}{r}\frac{\partial}{\partial r}\right)\overline{\theta} + \frac{3\mu\phi^2 U}{4K(\rho c)_{fs}}\overline{u}, \tag{7.84}$$

$$\overline{\theta}(R_1,q)=T_2,\quad \overline{\theta}(R_2,q)=T_1 f(q), \tag{7.85}$$

where $f(q) = L\{\varphi(t)\}$.

In terms of Hankel transform, we obtain

$$\overline{\theta}_H(r_n,q)=\frac{2AT_1f(q)}{\pi\left(q+\lambda q^{\gamma+1}+Ar_n^2\right)}\frac{J_0(R_1r_n)}{J_0(R_2r_n)}-\frac{2AT_2}{\pi\left(q+\lambda q^{\gamma+1}+Ar_n^2\right)}+\frac{3\mu\phi^2U}{4K(\rho c)_{fs}}\frac{\overline{u}_H}{q+\lambda q^{\gamma+1}+Ar_n^2}. \tag{7.86}$$

Then, we apply the inversion Laplace transforms and inversion Hankel transform, using the following formula of the Laplace transform of the function $G_{a,b,c}(d, t)$:

$$L^{-1}\left\{\frac{q^b}{(q^a-d)^c}\right\}=G_{a,b,c}(d,t). \tag{7.87}$$

Getting the approximate solution of heat transfer equation:

$$\begin{aligned}\theta=&-\pi AT_1\sum_{n=1}^{\infty}\frac{r_n^2J_0(R_2r_n)J_0(R_1r_n)B_0(rr_n)}{J_0^2(R_1r_n)-J_0^2(R_2r_n)}\sum_{i=0}^{\infty}\left(\frac{-1}{\lambda}\right)^{i+1}\\&\times\int_0^t\varphi(t-\tau)G_{\gamma+1,i,i+1}\left(-Ar_n^2\lambda^{-1},\tau\right)\mathrm{d}\tau\\&+\pi AT_2\sum_{n=1}^{\infty}\frac{r_n^2J_0^2(R_2r_n)B_0(rr_n)}{J_0^2(R_1r_n)-J_0^2(R_2r_n)}\sum_{i=0}^{\infty}\left(\frac{-1}{\lambda}\right)^{i+1}G_{\gamma+1,i,i+1}\left(-Ar_n^2\lambda^{-1},t\right)\\&-\frac{3\mu\phi^2U\pi^2}{8K(\rho c)_{fs}}\sum_{n=1}^{\infty}\frac{r_n^2J_0^2(R_2r_n)B_0(rr_n)}{J_0^2(R_1r_n)-J_0^2(R_2r_n)}\sum_{i=0}^{\infty}\left(\frac{-1}{\lambda}\right)^{i+1}\\&\times\int_0^t u_H(t-\tau)G_{\gamma+1,i,i+1}\left(-Ar_n^2\lambda^{-1},\tau\right)\mathrm{d}\tau\end{aligned} \tag{7.88}$$

7.3.6 Results and Discussion

The closed analytical solutions of the velocity, shear stress, and temperature fields distribution are obtained. In an unbound porous medium, the motion is due to the longitudinal time-dependent shear stress, which is prescribed by the boundary of the cylinder and oscillating velocity boundary conditions. The closed-form analytical solutions are obtained and presented by integral and series form in terms of the generalized functions. In the special cases, when

$K \to \infty$ or $\phi \to 0$, we obtain the corresponding solutions for the ordinary generalized Oldroyd-B fluids.

Fig. 7.5 presents the profiles of velocity components u and w for different values of time. Results display that the profiles of velocity components u and w

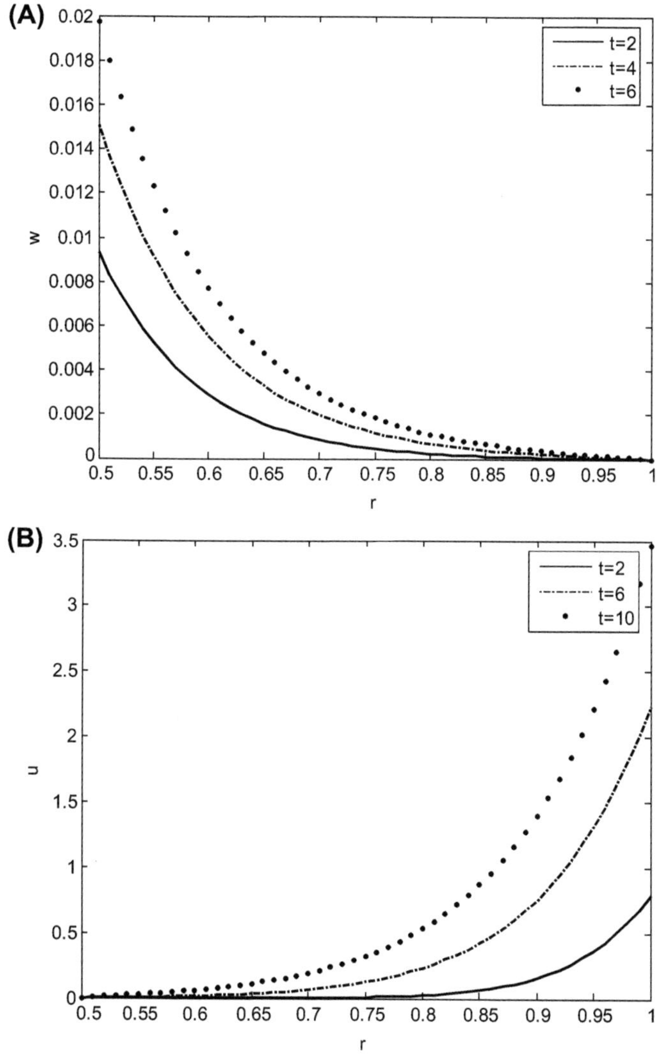

FIGURE 7.5 Velocity profiles for different values of time t ($\lambda = 10$; $\beta = 0.5$; $\alpha = 0.3$; $\nu = 0.005$; $\phi = 50$; $K = 1$; (w) $\lambda_r = 20$; (u) $\lambda_r = 7, 0 < \omega t < \pi/2$). (A) w and (B) u.

are both increasing functions of t when $f > 0$ and $0 < \omega t < \pi/2$. Moreover, the amplitude of fluctuation of the velocities increases with time. Figs. 7.6 and 7.7 show that, by increasing the fractional parameter α, the amplitude of fluctuation of velocities becomes smaller and the velocities profiles increase with the same radius. In contrast, the amplitude of fluctuation of velocities becomes bigger with the increase of the fractional parameter β, and the rate of increase of the velocities from positive to negative in the z-direction.

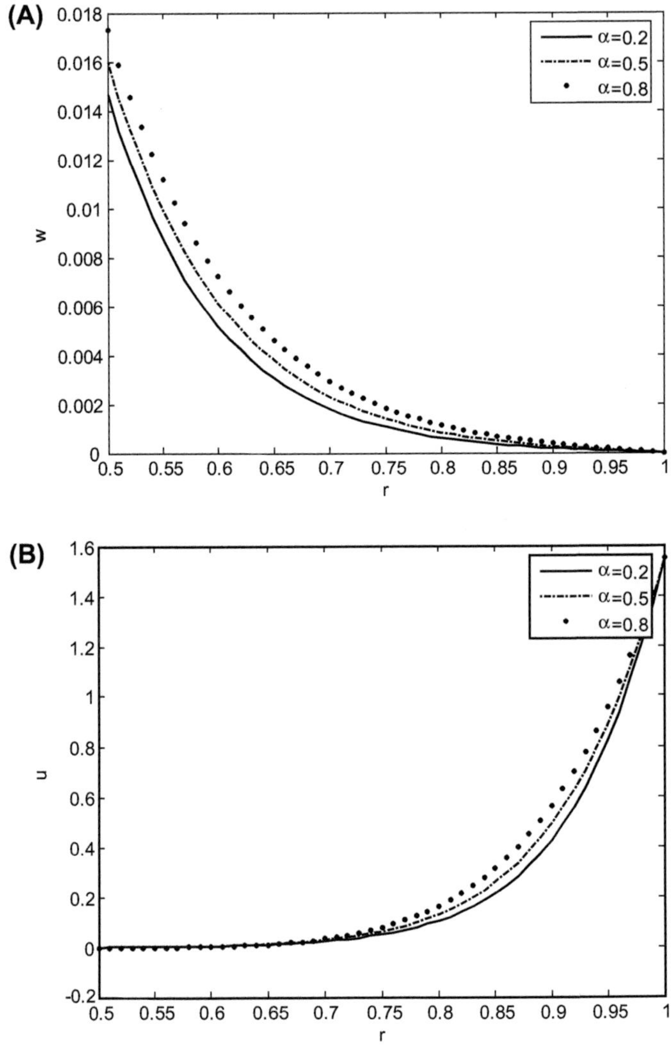

FIGURE 7.6 Velocity profiles for different values of fractional parameter α ($\lambda = 10$; $\beta = 0.5$; $t = 4$; $\nu = 0.005$; $\phi = 50$; $K = 1$; (w) $\lambda_r = 20$; (u) $\lambda_r = 7$). (A) w and (B) u.

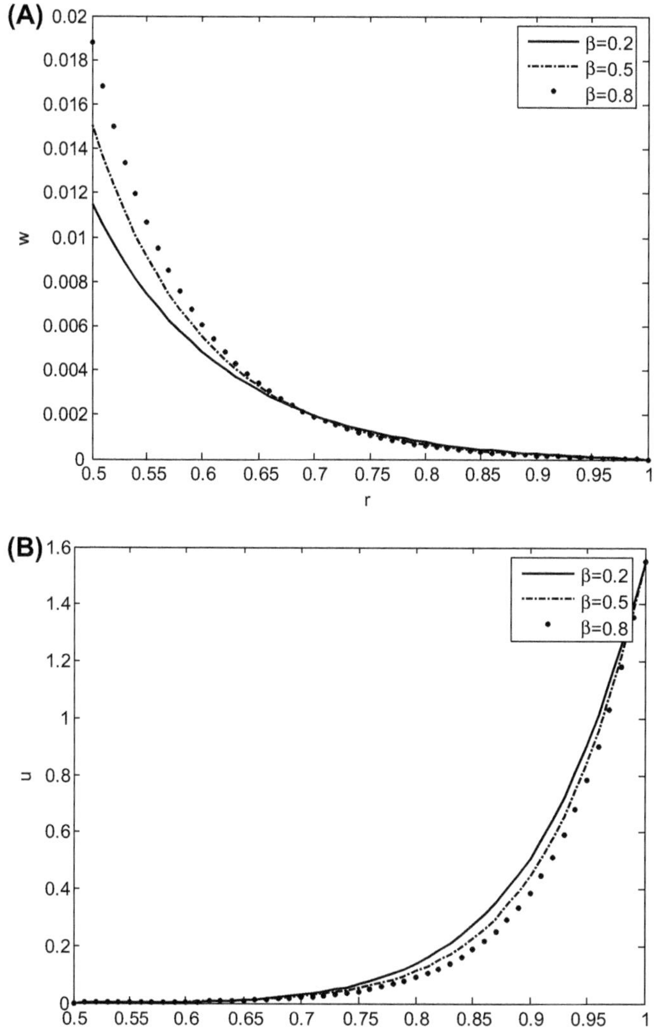

FIGURE 7.7 Velocity profiles for different values of fractional parameter β ($\lambda = 10$; $\alpha = 0.3$; $t = 4$; $\nu = 0.005$; $\phi = 50$; $K = 1$; $(w)\lambda_r = 20$; $(u)\lambda_r = 7$). (A) w and (B) u.

Fig. 7.8 shows the impacts of different values of kinematic viscosity on the profiles components u and w. One can see that, as the kinematic viscosity increases, both profiles of the velocity components u and w become bigger, but the amplitude of fluctuation of velocities decreases and the influence of rate of increase of the velocities by kinematic viscosity becomes smaller and smaller. Figs. 7.9 and 7.10 display the impacts of permeability K and porosity ϕ on

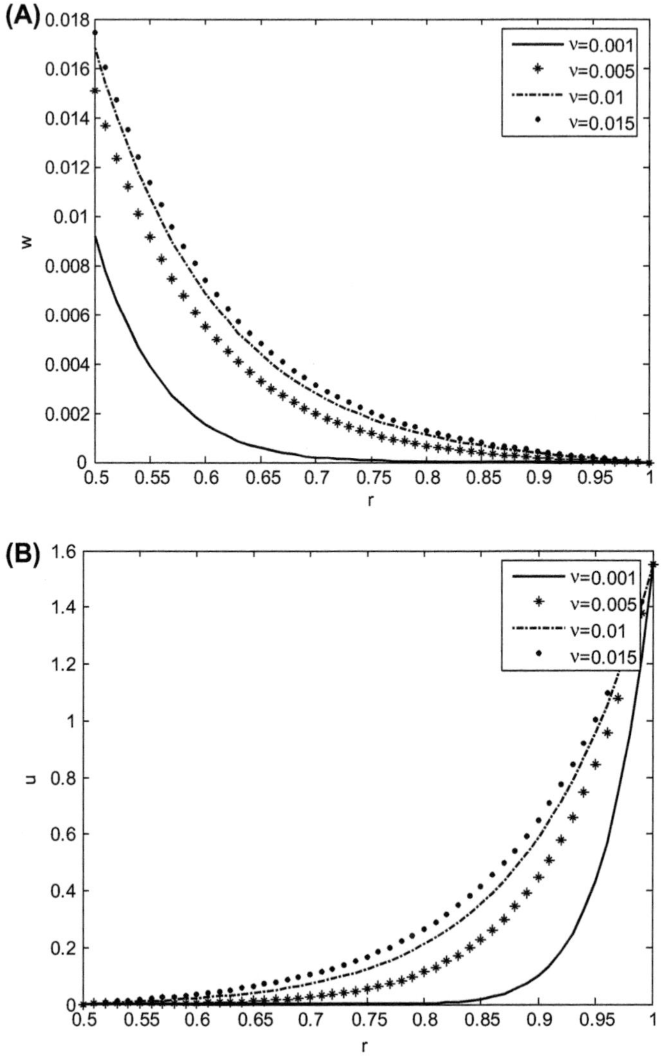

FIGURE 7.8 Velocity profiles for different values of kinematic viscosity ν ($\lambda = 10$; $\beta = 0.5$; $\alpha = 0.3$; $t = 4$; $\phi = 50$; $K = 1$; (w) $\lambda_r = 20$; (u) $\lambda_r = 7$). (A) w and (B) u.

velocity profiles of the u and w with the fluid flows in porous medium. Compared with kinematic viscosity, the permeability has a similar but more pronounced effect on velocities. Oppositely, as the porosity increases, the velocities become smaller in the same radius, but the amplitude of fluctuation of velocities increases and the influence of rate of increase of the velocities by kinematic viscosity becomes smaller and smaller.

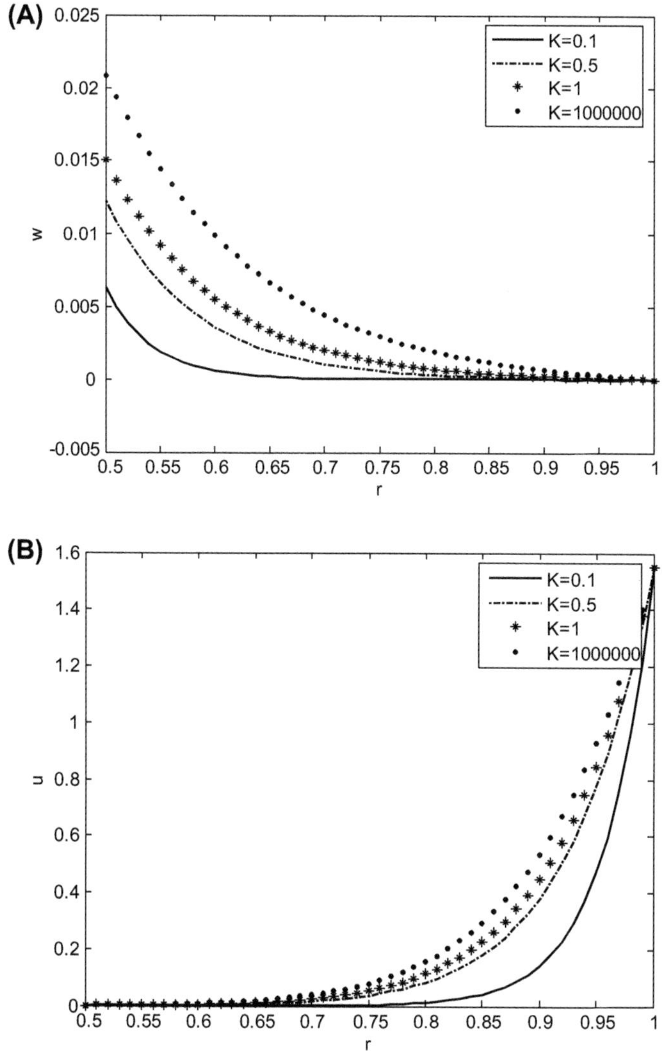

FIGURE 7.9 Velocity profiles for different values of permeability K ($\lambda = 10$; $\beta = 0.5$; $\alpha = 0.3$; $t = 4$; $\nu = 0.005$; $\phi = 50$; (w) $\lambda_r = 20$; (u) $\lambda_r = 7$). (A) w and (B) u.

Figs. 7.11 and 7.12 display the behavior of the velocity profiles for different values of material constants λ and λ_r. The figures show that by increasing magnitude of λ, the fluctuations between two circular cylinders become bigger and bigger. But with the increase of the λ_r, the fluctuations become smaller. The effects on the velocities show similar results in pertinent articles in the literature.

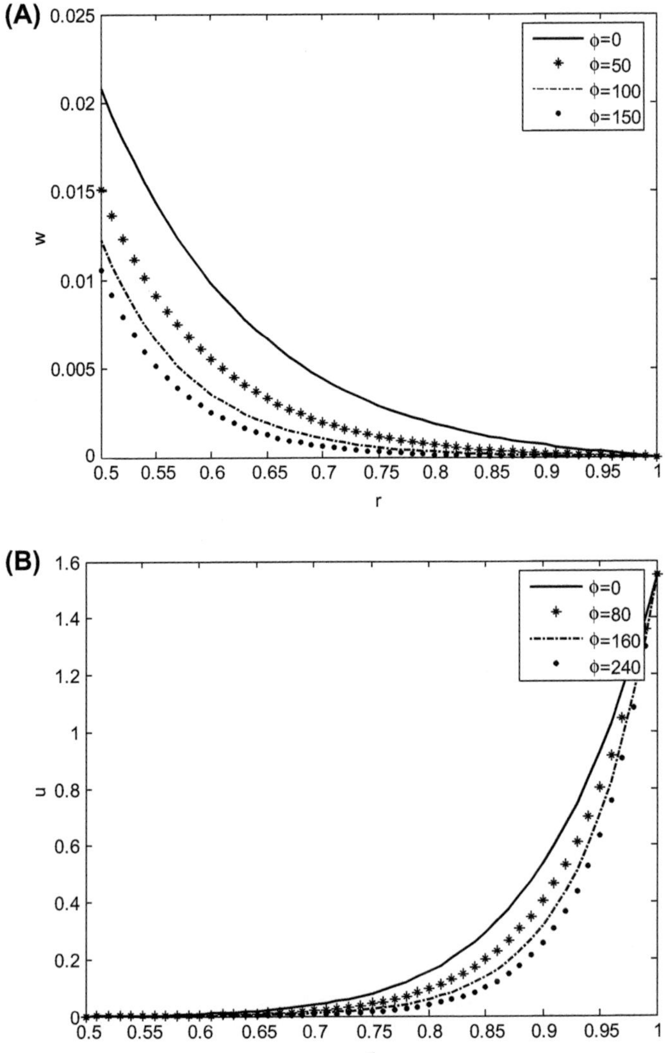

FIGURE 7.10 Velocity profiles for different values of porosity ϕ ($\lambda = 10$; $\beta = 0.5$; $\alpha = 0.3$; $t = 4$; $\nu = 0.005$; $K = 1$; (w) $\lambda_r = 20$; (u) $\lambda_r = 7$). (A) w and (B) u.

A comparison is presented for the velocity profiles, corresponding to the motions of a generalized Oldroyd-B fluid, ordinary Oldroyd-B fluid, and Newtonian fluid in concentric cylinders, and we plotted the figures as functions of t when $r = 0.7$ in Fig. 7.13. It can be observed that the non-Newtonian effects become weak for a large time. It can also be seen that the amplitude of fluctuation of generalized Oldroyd-B fluid flows smaller than ordinary

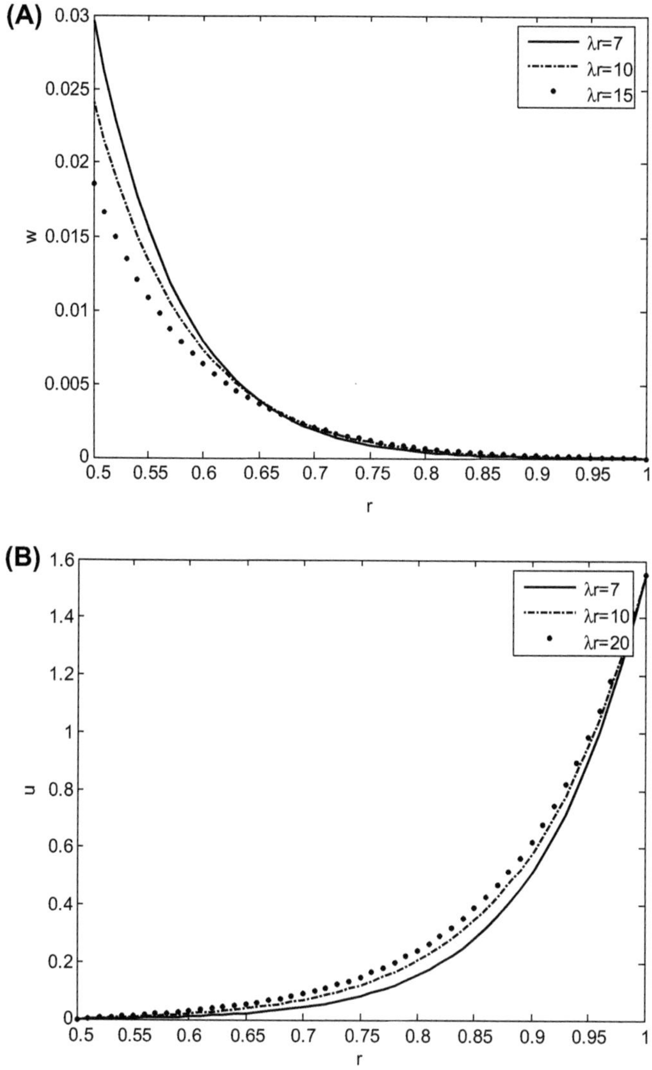

FIGURE 7.11 Velocity profiles for different values of λ_r ($\lambda = 10$; $\beta = 0.5$; $\alpha = 0.3$; $t = 4$; $\nu = 0.005$; $\phi = 50$; $K = 1$). (A) w and (B) u.

Oldroyd-B fluid and Newtonian fluid at starting time. Fig. 7.14 shows the comparisons between the exact analytical solutions and numerical results, obtained by finite difference method, for generalized Oldroyd-B fluids and Newtonian fluids. These velocity profiles can validate the correctness of the derivations.

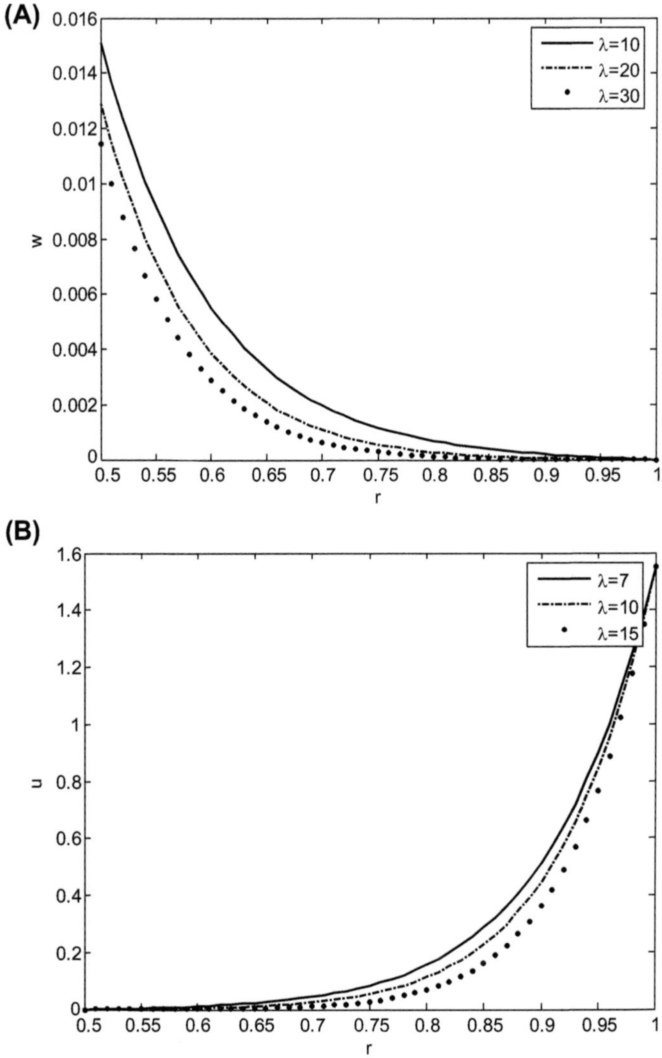

FIGURE 7.12 Velocity profiles for different values of material constant λ ($\beta = 0.5$; $\alpha = 0.3$; $t = 4$; $\nu = 0.005$; $\phi = 50$; $K = 1$; (w) $\lambda_r = 20$; (u) $\lambda_r = 7$). (A) w and (B) u.

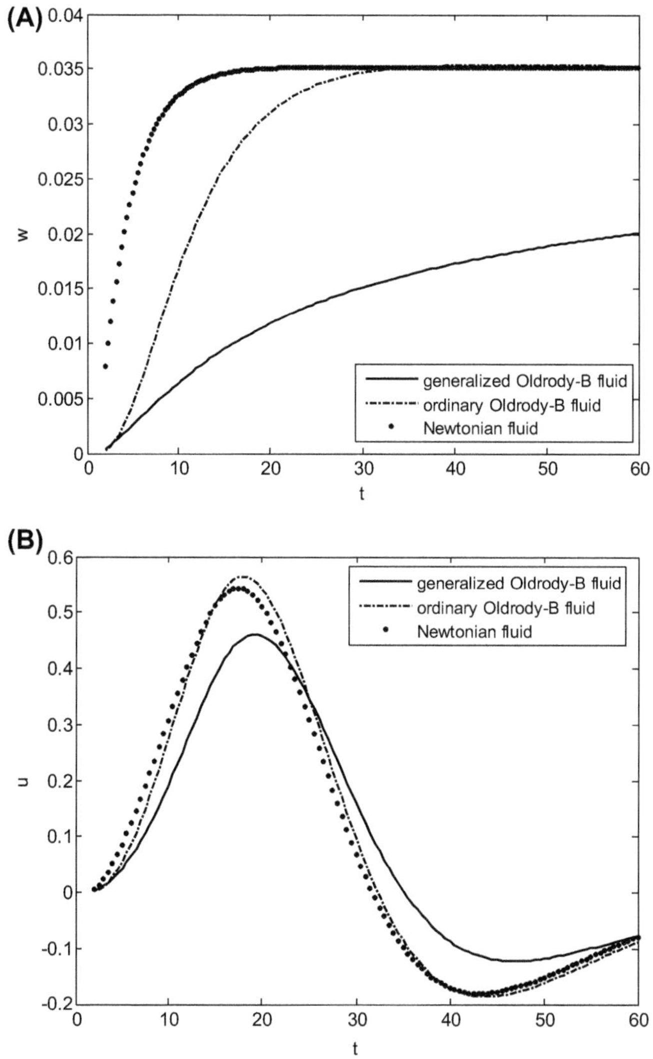

FIGURE 7.13 Velocity profiles corresponding to a generalized Oldroyd-B fluid, ordinary Oldroyd-B fluid, and Newtonian fluid at different values of time t. (A) w and (B) u.

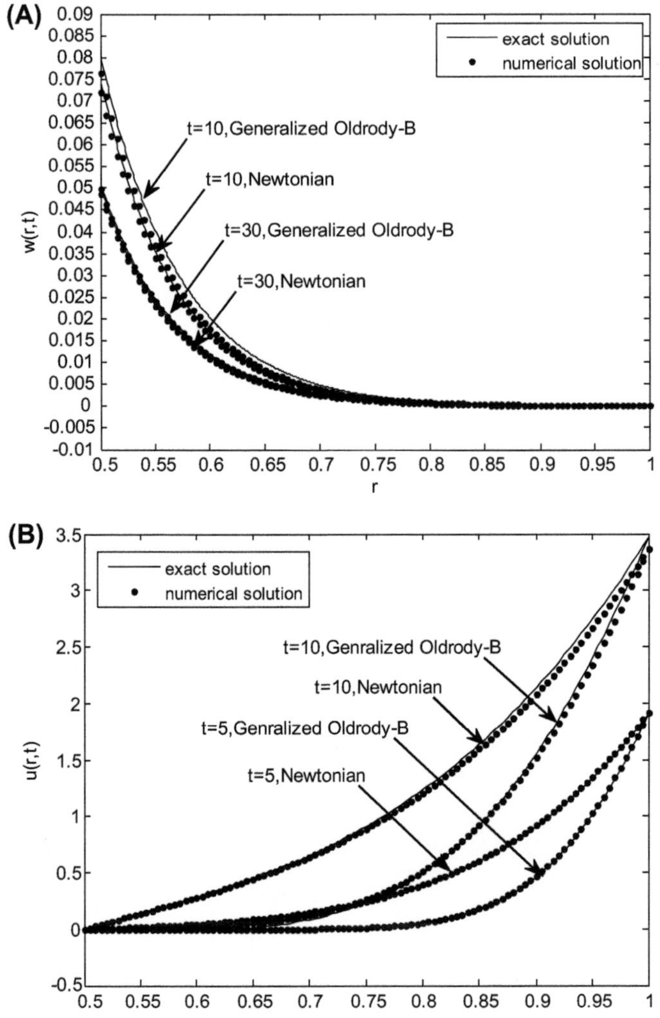

FIGURE 7.14 Velocity profiles for generalized Oldroyd-B fluids ($\alpha = 0.3$, $\beta = 0.8$, $\lambda = 3$, $\lambda_r = 0.5$) and Newtonian fluids ($\lambda = 0$, $\lambda_r = 0$). (A) w and (B) u.

7.4 MAGNETOHYDRODYNAMIC FLOW AND HEAT TRANSFER OF GENERALIZED BURGERS' FLUID

Burgers' fluid models form a subclass of the viscoelastic fluid, which has attracted the attention of many scholars in recent years. Xue and Nie (2008) and Xue et al. (2008) considered fractional generalized Burgers' fluid in a porous half-space. Hyder Ali (2010) discussed some unidirectional flows of viscoelastic fractional Burgers' fluid between two parallel plates. Khan et al. (2009a,b, 2010a,b) investigated some fractional Burgers' fluid models including oscillating flow, accelerating flow, rotating flow, etc.

Magnetohydrodynamic (MHD) fluid complex flow and radiation heat transfer, due to its wide applications in engineering and technology, has been recently investigated by many researchers (Abel et al., 2009; Ahercliff, 1965; Bandelli, 1995; Brewster, 1992; Cortell, 2008; Derek et al., 2002; EI-Aziz, 2009; Fetecau, 2001; Fetecau and Fetecau, 2002; Hayat et al., 2004; Hayat and Sajid, 2007; Lorenzo and Hartley, 2008; Mahmoud, 2007; Rahman and Sultana, 2008; Srinivas and Muthuraj, 2010). The effects of radiation on MHD fluid flow and heat transfer over an unsteady stretching surface were studied by Cortell (2008), Abel et al. (2009), and Abou Zeid (2009). However, there are a few investigations to consider the viscoelastic fluids under the effects of thermal radiation. In this section, we investigate MHD Burgers' fluid flow and radiation heat transfer due to an exponential accelerating plate in which the fluid is assumed to be a gray, absorbing–emitting radiation but nonscattering medium. Moreover, a parametric analysis is performed to illustrate the influence of these parameters.

7.4.1 Governing Equations

The constitutive equations for an incompressible fractional generalized Burgers' fluid are given by:

$$\mathbf{T} = -p\mathbf{I} + \mathbf{S}, \left(1 + \lambda_1^{\alpha}\widetilde{\mathrm{D}}_{\mathrm{t}}^{\alpha} + \lambda_2^{\alpha}\widetilde{\mathrm{D}}_{t}^{2\alpha}\right)\mathbf{S} = \mu\left(1 + \lambda_3^{\beta}\widetilde{\mathrm{D}}_{t}^{\beta}\right)\mathbf{A}, \tag{7.89}$$

where $\mathbf{T}$ is the Cauchy stress tensor, $-p\mathbf{I}$ denotes the indeterminate spherical stress, $\mathbf{S}$ is the extra-stress tensor, $\mathbf{A} = \mathbf{L} + \mathbf{L}^{\mathrm{T}}$ is the first Rivlin–Ericksen tensor, $\mathbf{L}$ is the velocity gradient, μ, λ_1, λ_3 are material constants, known as the viscosity coefficient, the relaxation and retardation times, respectively, λ_2 is a material constant, and $\widetilde{\mathrm{D}}_t^p$ the upper convected fractional derivative defined by:

$$\widetilde{\mathrm{D}}_t^p\mathbf{S} = \mathrm{D}_t^p\mathbf{S} + \mathbf{V}\cdot\nabla\mathbf{S} - \mathbf{L}\mathbf{S} - \mathbf{S}\mathbf{L}^{\mathrm{T}}, \widetilde{\mathrm{D}}_t^{2p}\mathbf{S} = \widetilde{\mathrm{D}}_t^p\left(\widetilde{\mathrm{D}}_t^p\mathbf{S}\right). \tag{7.90}$$

In the previous relations $\mathbf{V}$ is the velocity vectors, ∇ is the gradient operator, D_t^{α} and D_t^{β} are derivatives, based on Riemann–Liouville's definition (Oldham and Spanier, 1974), respectively:

$$\mathrm{D}_t^p f(t) = \frac{1}{\Gamma(1-p)}\frac{\mathrm{d}}{\mathrm{d}t}\int_0^t \frac{f(\tau)}{(t-\tau)^p}\mathrm{d}\tau, 0 \le p < 1, \tag{7.91}$$

where $\Gamma(\cdot)$ is the Gamma function.

The motion equation for an incompressible fluid is given by:

$$\rho \frac{\mathrm{d}\mathbf{V}}{\mathrm{d}t} = \nabla \cdot \mathbf{T} + \rho \mathbf{b}, \tag{7.92}$$

where ρ is the constant density of the fluid, $\mathbf{b}$ is the body force field.

Assuming the velocity field is chosen, the form:

$$\mathbf{V} = u(y, t)\mathbf{i} \tag{7.93}$$

where u is the velocity and $\mathbf{i}$ is the unit vector in the x-direction. Since u is a function of y and t, the stress field will also depend upon y and t. There we consider the conducting fluid is permeated by an imposed magnetic field $\mathbf{B} = [0, B_0, 0]$, which acts in the positive y-coordinate. In the low-magnetic Reynolds number approximation (Khan et al., 2009a,b), in which the induced magnetic field can be ignored, the magnetic body force is represented by $\sigma B_0^2 u$, where σ is the electrical conductivity of the fluid. Substituting Eq. (7.93) into Eqs. (7.89) and (7.92), and taking into account the initial conditions $\mathbf{S}(y, 0) = \mathbf{S}_t(y, 0) = \mathbf{0}$, we get the relevant equations:

$$\rho \frac{\mathrm{d}u}{\mathrm{d}t} = -\frac{\partial p}{\partial x} + \mu \frac{\partial S_{xy}}{\partial y} - \sigma B_0^2 u, \tag{7.94}$$

$$\left(1 + \lambda_1^\alpha \mathrm{D}_t^\alpha + \lambda_2^\alpha \mathrm{D}_t^{2\alpha}\right) S_{xy} = \mu \left(1 + \lambda_3^\beta \mathrm{D}_t^\beta\right) \partial_y u, \tag{7.95}$$

$$\begin{aligned} &\left(1 + \lambda_1^\alpha \mathrm{D}_t^\alpha + \lambda_2^\alpha \mathrm{D}_t^{2\alpha}\right) S_{xx} - 2S_{xy}\left[\lambda_1^\alpha + \lambda_2^\alpha \mathrm{D}_t^\alpha\right] \frac{\partial u}{\partial y} \\ &- 2\lambda_2^\alpha \frac{\partial u}{\partial y} \mathrm{D}_t^\alpha S_{xy} = -2\mu \lambda_3^\beta \left(\frac{\partial u}{\partial y}\right)^2. \end{aligned} \tag{7.96}$$

Eliminating S_{xy} between Eqs. (7.95) and (7.96), in the absence of a pressure gradient in the x-direction, we obtain:

$$\left(1 + \lambda_1^\alpha \mathrm{D}_t^\alpha + \lambda_2^\alpha \mathrm{D}_t^{2\alpha}\right) \frac{\partial u(y,t)}{\partial t} = \nu \left(1 + \lambda_3^\beta \mathrm{D}_t^\beta\right) \frac{\partial^2 u(y,t)}{\partial y^2} - M\left(1 + \lambda_1^\alpha \mathrm{D}_t^\alpha + \lambda_2^\alpha \mathrm{D}_t^{2\alpha}\right) u(y,t), \tag{7.97}$$

where $\nu = \mu/\rho$ is the kinematic viscosity and $M = \sigma B_0^2/\rho$.

The fluid is assumed to be a gray, absorbing–emitting radiation but non-scattering medium. When Fourier's law of heat conduction is considered (Bandelli, 1995; Fetecau and Fetecau, 2002; Shah, 2010), the energy equation may be written in the form:

$$\frac{\partial \theta}{\partial t} = \frac{k_T}{\rho C_p} \frac{\partial^2 \theta}{\partial y^2} + \frac{\nu}{C_p} \left[\frac{\partial u}{\partial y}\right]^2 - \frac{1}{\rho C_p} \frac{\partial q_r}{\partial y}, \tag{7.98}$$

where k_T is the thermal conductivity, C_p is the specific heat of a fluid at constant pressure, and q_r is the radiative heat flux.

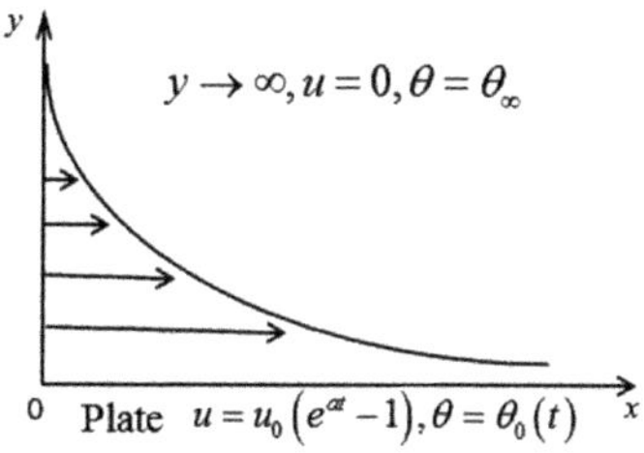

FIGURE 7.15 Sketch of the physical model and coordinates system.

7.4.2 Formulation of the Problem

Consider the generalized Burgers' fluid occupying the space above a flat plate. A schematic representation of the physical model and coordinates system is depicted in Fig. 7.15. The plate aligned with the x-axis at $y = 0$. Initially the fluid as well as the plate are at rest, and at time $t = 0^+$ the plate slides in its plane with the velocity $u_0(e^{at} - 1)$, where u_0 (>0), a are constants. Let $\theta_0(t)$ denote the temperature of the plate for$t \geq 0$, and suppose that the temperature of the fluid at the moment $t = 0$ is θ_∞. Due to the shear, the fluid speeds up gradually. Accordingly, the flow is governed by Eqs. (7.92)–(7.94), and the associated boundary and initial conditions of the motion equation are:

$$u(y,0) = \frac{\partial u(y,0)}{\partial t} = \frac{\partial^2 u(y,0)}{\partial t^2} = 0, \quad y > 0, \tag{7.99}$$

$$u(0,t) = u_0(e^{at} - 1), \quad t > 0, \tag{7.100}$$

$$u(y,t), \quad \frac{\partial u(y,t)}{\partial y} \to 0, \text{ as } y \to \infty. \tag{7.101}$$

The corresponding initial and boundary conditions of the energy equation are:

$$\theta(y,0) = \theta_\infty, \quad \text{for } y > 0, \tag{7.102}$$

$$\theta(0,t) = \theta_0(t), \quad \text{for } t \geq 0, \tag{7.103}$$

$$\theta(y,t) \to \theta_\infty, \quad \frac{\partial \theta(y,t)}{\partial y} \to 0, \quad \text{for } y \to \infty. \tag{7.104}$$

7.4.3 The Solution of Velocity Fields

Introducing the dimensionless quantities in Eqs. (7.97), (7.99)–(7.101):

$$u^* = \frac{u}{u_0},\ y^* = \frac{u_0 y}{\nu},\ t^* = \frac{u_0^2 t}{\nu},\ \lambda_1^* = \lambda_1 \frac{u_0^2}{\nu},$$
$$\lambda_2^* = \lambda_2 \left(\frac{u_0^2}{\nu}\right)^2,\ \lambda_3^* = \lambda_3 \frac{u_0^2}{\nu},\ M^* = \frac{M\nu}{u_0^2},\ a^* = \frac{a\nu}{u_0^2}. \tag{7.105}$$

Dimensionless motion equations can be given (for brevity the dimensionless mark "*" is omitted here), yields:

$$\left(1 + \lambda_1^\alpha D_t^\alpha + \lambda_2^\alpha D_t^{2\alpha}\right) \frac{\partial u(y,t)}{\partial t} = \left(1 + \lambda_3^\beta D_t^\beta\right) \frac{\partial^2 u(y,t)}{\partial y^2} - M\left(1 + \lambda_1^\alpha D_t^\alpha + \lambda_2^\alpha D_t^{2\alpha}\right) u(y,t), \tag{7.106}$$

the initial conditions become:

$$u(y,0) = \frac{\partial u(y,0)}{\partial t} = \frac{\partial^2 u(y,0)}{\partial t^2} = 0,\ y > 0, \tag{7.107}$$

the boundary conditions are:

$$u(0,t) = e^{at} - 1,\ t > 0, \tag{7.108}$$

$$u(y,t), \frac{\partial u(y,t)}{\partial y} \rightarrow 0, \quad \text{as } y \rightarrow \infty. \tag{7.109}$$

In order to solve the previous problem, we use Fourier sine transformation and Laplace transformation for fractional derivatives. Firstly, multiplying both sides of Eq. (7.106) by $\sqrt{2/\pi}\sin(\xi y)$, integrating them with respect to y from 0 to ∞ and taking account corresponding initial and boundary conditions Eqs. (7.107)–(7.109), we obtain:

$$\left(1 + \lambda_1^\alpha D_t^\alpha + \lambda_2^\alpha D_t^{2\alpha}\right) \frac{\partial u_s(\xi,t)}{\partial t} = \left(1 + \lambda_3^\beta D_t^\beta\right) \left[\sqrt{\frac{2}{\pi}}\xi(e^{at} - 1) - \xi^2 u_s(\xi,t)\right] - M\left(1 + \lambda_1^\alpha D_t^\alpha + \lambda_2^\alpha D_t^{2\alpha}\right) u_s(\xi,t), \tag{7.110}$$

where the Fourier sine transform $u_s(\xi, t)$ of $u(y, t)$ has to satisfy the conditions:

$$u_s(\xi,0) = \frac{\partial u_s(\xi,0)}{\partial t} = \frac{\partial^2 u_s(\xi,0)}{\partial t^2} = 0,\ \xi > 0. \tag{7.111}$$

Applying Laplace transform for sequential fractional derivative to Eq. (7.110) and using the initial condition Eq. (7.111), we get:

$$\overline{u}_s(\xi,s)=\sqrt{\frac{2}{\pi}}\frac{1}{\xi}\left(\frac{1}{s-a}-\frac{1}{s}\right)\frac{\xi^2\left(1+\lambda_3^\beta s^\beta\right)}{\left[(s+M)\left(1+\lambda_1^\alpha s^\alpha+\lambda_2^\alpha s^{2\alpha}\right)+\xi^2\left(1+\lambda_3^\beta s^\beta\right)\right]},\tag{7.112}$$

where $\overline{u}_s(\xi,s)$ is the Laplace transform of $u_s(\xi,t)$ with respect to t. In order to avoid the lengthy procedure of residues and contour integrals, we rewrite Eq. (7.112) into the series form:

$$\overline{u}_s(\xi,s)=\sqrt{\frac{2}{\pi}}\frac{1}{\xi}\left(\frac{1}{s-a}-\frac{1}{s}\right)-\sqrt{\frac{2}{\pi}}\frac{1}{\xi}\left(\frac{1}{s-a}-\frac{1}{s}\right)$$
$$\times\sum_{k=0}^{\infty}(-1)^k\frac{1}{\left(\xi^2\right)^{k+1}\lambda_3^{\beta(k+1)}}\sum_{m,l\geq 0}^{m+l=k+1}\frac{(k+1)!}{m!l!}M^m$$
$$\times\sum_{n,w,\eta\geq 0}^{n+w+\eta=k}\frac{k!}{n!w!\eta!}\lambda_1^{\alpha n}\lambda_2^{\alpha w}\left(1+\lambda_1^\alpha s^\alpha+\lambda_2^\alpha s^{2\alpha}\right)\frac{s^\sigma}{\left(\lambda_3^{-\beta}+s^\beta\right)^{k+1}}.\tag{7.113}$$

In which $\sigma=\alpha n+2\alpha w+l$. Taking the discrete Laplace transform method, we obtain:

$$u_s(\xi,t)=\sqrt{\frac{2}{\pi}}\frac{1}{\xi}\left(e^{at}-1\right)-\sqrt{\frac{2}{\pi}}\frac{1}{\xi}\sum_{k=0}^{\infty}(-1)^k\frac{1}{\xi^{2(k+1)}\lambda_3^{\beta(k+1)}}$$
$$\times\sum_{m,l\geq 0}^{m+l=k+1}\frac{(k+1)!}{m!l!}M^m\sum_{n,w,\eta\geq 0}^{n+w+\eta=k}\frac{k!}{n!w!\eta!}\lambda_1^{\alpha n}\lambda_2^{\alpha w}$$
$$\times\int_0^t\left(e^{a(t-\tau)}-1\right)\begin{pmatrix}G_{\beta,\sigma,k+1}\left(-\lambda_3^{-\beta},\tau\right)+\lambda_1^\alpha G_{\beta,\sigma+\alpha,k+1}\left(-\lambda_3^{-\beta},\tau\right)\\+\lambda_2^\alpha G_{\beta,\sigma+2\alpha,k+1}\left(-\lambda_3^{-\beta},\tau\right)\end{pmatrix}d\tau,\tag{7.114}$$

where

$$G_{\alpha,a,b}(d,t)=\sum_{l=0}^{\infty}\frac{(b)_l t^{\alpha(l+b)-a-1}}{\Gamma(l+1)\Gamma((l+b)\alpha-a)}(d)^l,\tag{7.115}$$

$(b)_l=b(b+1)\ldots(b+l-1)$ is the Pochhammer polynomial (Cortell, 2008). In obtaining Eq. (7.114), the following property of the G-function is used:

$$G_{a,b,c}(p,t)=L^{-1}\left\{\frac{s^b}{\left(s^a-p\right)^c}\right\};\ Re(ac-b)>0,\ Re(s)>0,\ \left|\frac{p}{s^a}\right|<1.\tag{7.116}$$

Applying the inverse Fourier sine transform to Eq. (7.114), we find the following expression for the velocity field:

$$u(y,t) = (e^{at} - 1)$$
$$-\frac{2}{\pi}\int_0^\infty \frac{\sin(\xi y)}{\xi}\sum_{k=0}^{\infty}(-1)^k \frac{1}{\xi^{2(k+1)}\lambda_3^{\beta(k+1)}}\sum_{m,l\geq 0}^{m+l=k+1}\frac{(k+1)!}{m!l!}M^m$$
$$\times \sum_{n,w,\eta\geq 0}^{n+w+\eta=k}\frac{k!}{n!w!\eta!}\lambda_1^{\alpha n}\lambda_2^{\alpha w}$$
$$\times \int_0^t \left(e^{a(t-\tau)} - 1\right)\left\{G_{\beta,\sigma,k+1}\left(-\lambda_3^{-\beta},\tau\right) + \lambda_1^\alpha G_{\beta,\sigma+\alpha,k+1}\left(-\lambda_3^{-\beta},\tau\right)\right.$$
$$\left. + \lambda_2^\alpha G_{\beta,\sigma+2\alpha,k+1}\left(-\lambda_3^{-\beta},\tau\right)\right\}d\tau d\xi. \tag{7.117}$$

7.4.4 The Solution of Temperature Fields

The radiative heat flux in the x-direction is negligible in comparison with that in the y-direction. Using the Rosseland approximation for radiation (Brewster, 1992; Mahmoud, 2007; Rahman and Sultana, 2008; Srinivas and Muthuraj, 2010), the radiative heat flux is simplified as:

$$q_r = -\frac{4\sigma^*}{3k^*}\frac{\partial \theta^4}{\partial y}, \tag{7.118}$$

where σ^* and k^* are the Stefan–Boltzmann constant and the mean absorption coefficient, respectively. We assume that the temperature differences within the flow are such that the term θ^4 may be expressed as a linear function of temperature. Hence, expanding θ^4 in a Taylor series about a free stream temperature θ_∞ and neglecting higher-order terms we get:

$$\theta^4 \cong 4\theta_\infty^3\theta - 3\theta_\infty^4. \tag{7.119}$$

It should be noted that this radiative transfer pertains to an optically thick model.

In view of Eqs. (7.118) and (7.119), Eq. (7.98) reduces to:

$$\frac{\partial\theta}{\partial t} = \frac{k_T}{\rho C_p}\frac{\partial^2\theta}{\partial y^2} + \frac{\nu}{C_p}\left[\frac{\partial u}{\partial y}\right]^2 + \frac{1}{\rho C_p}\frac{16\sigma^*\theta_\infty^3}{3k^*}\frac{\partial^2\theta}{\partial y^2}. \tag{7.120}$$

Introducing the dimensionless quantities:

$$\theta^* = \frac{\theta-\theta_\infty}{\theta_0(0)-\theta_\infty},\ u^* = \frac{u}{u_0},\ y^* = \frac{u_0 y}{\nu},\ t^* = \frac{u_0^2 t}{\nu},\ \eta^* = \frac{u_0^2}{C_p(\theta_0(0)-\theta_\infty)},\ \Pr^* = \frac{C_p\mu}{k_T}. \tag{7.121}$$

Eqs. (7.102)–(7.104), (7.120) can reduce to dimensionless equations as follows (for brevity the dimensionless mark "*" is omitted here):

$$\frac{\partial\theta(y,t)}{\partial t}=\frac{1}{\Pr}\left[\frac{3N_R+4}{3N_R}\right]\frac{\partial^2\theta(y,t)}{\partial y^2}+\eta\left[\frac{\partial u(y,t)}{\partial y}\right]^2, \tag{7.122}$$

where $N_R=\frac{k^*k_T}{4\sigma^*\theta_\infty^3}$. Letting $g(y,t)=\left[\frac{\partial u(y,t)}{\partial y}\right]^2$, $k_0=\frac{3N_R}{3N_R+4}$, Eq. (7.122) can be rewritten as:

$$\frac{\partial\theta(y,t)}{\partial t}=\frac{1}{k_0\Pr}\frac{\partial^2\theta(y,t)}{\partial y^2}+\eta g(y,t). \tag{7.123}$$

The corresponding initial and boundary conditions become:

$$\theta(y,0)=0,\quad \text{for } y>0, \tag{7.124}$$

$$\theta(0,t)=f(t),\quad \text{for } t\geq 0, \tag{7.125}$$

$$\theta(y,t)\to 0,\quad \frac{\partial\theta(y,t)}{\partial y}\to 0,\quad \text{for } y\to\infty, \tag{7.126}$$

where $f(t)=(\theta_0(t)-\theta_\infty)/(\theta_0(0)-\theta_\infty)$. In order to obtain the exact solution to Eqs. (7.123)–(7.126), using Fourier sine transform to Eq. (7.123), we obtain:

$$\frac{d\theta_s(\xi,t)}{dt}+\frac{1}{k_0\Pr}\xi^2\theta_s(\xi,t)=\sqrt{\frac{2}{\pi}}\frac{\xi}{k_0\Pr}f(t)+\eta g_s(\xi,t) \tag{7.127}$$

$$\theta_s(\xi,0)=0, \tag{7.128}$$

where $\theta_s(\xi,t)$ and $g_s(\xi,t)$ denote the Fourier sine transform of $\theta(y,t)$ and $g(y,t)$ with respect to y, respectively. The solution of the ordinary differential equation Eq. (7.127) subject to the initial condition of Eq. (7.128) is given by:

$$\theta_s(\xi,t)=e^{-\xi^2t/(k_0\Pr)}\int_0^t\left[\sqrt{\frac{2}{\pi}}\frac{\xi}{k_0\Pr}f(\tau)+\eta g_s(\xi,\tau)\right]e^{\xi^2\tau/(k_0\Pr)}d\tau. \tag{7.129}$$

Inverting Eq. (7.129) by means of Fourier sine transform, we get:

$$\theta(y,t)=\sqrt{\frac{2}{\pi}}\int_0^\infty\sin(\xi y)e^{-\xi^2t/(k_0\Pr)}\int_0^t\left[\sqrt{\frac{2}{\pi}}\frac{\xi}{k_0\Pr}f(\tau)+\eta g_s(\xi,\tau)\right]e^{\xi^2\tau/(k_0\Pr)}d\tau d\xi. \tag{7.130}$$

7.4.5 Results and Discussion

The purpose of this research is to provide exact solutions for the unsteady MHD flow of a generalized Burgers' fluid due to an exponential accelerating plate. The exact solutions for the velocity field $u(y,t)$ and temperature field $\theta(y,t)$ in terms of G-function are obtained by using Fourier sine transform and Laplace transform. Moreover, some figures are plotted to show the behavior of some different emerging parameters of interest involved in the velocity field and temperature field.

Figs. 7.16 and 7.17 show the velocity profiles with the fractional parameters and magnetic field parameters. It is clearly seen that the smaller the α, the more slowly the velocity decays. However, one can see that an increase in material parameter β has quite the opposite effect to that of α. Moreover, the magnetic body force is favorable to the velocity decays. The velocity in the case of magnetohydrodynamic fluid is less when compared with hydrodynamic fluid. This is due to the fact that applied transverse magnetic field produces a drag in the form of Lorentz force thereby decreasing the magnitude of velocity. Fig. 7.18 is the velocity profiles u vs the time t. With the increasing values of t, the velocity rapidly speeds up. Figs. 7.19 and 7.20 demonstrate the influence of t for temperature field. As expected, clearly the non-Newtonian effects are stronger at lager values of t. The greater the value of t, the higher the temperature. The temperature profiles are similar with different values of t, but not similar quantitatively. Fig. 7.21 is the graph of temperature

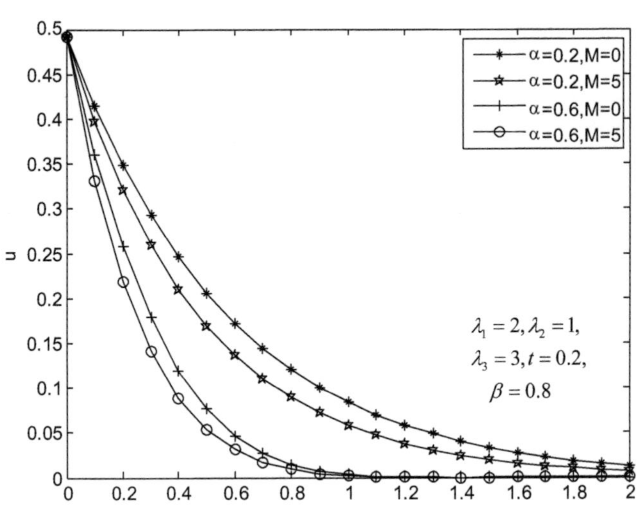

FIGURE 7.16 Velocity fields for different values of α and M when keeping other parameters fixed.

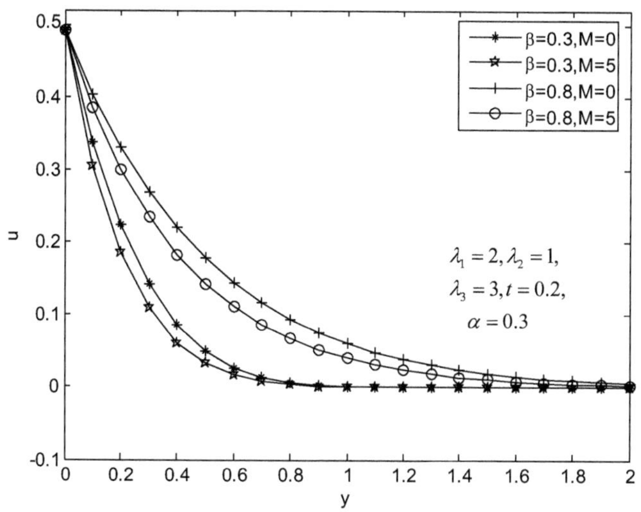

FIGURE 7.17 Velocity fields for different values of β and M when keeping other parameters fixed.

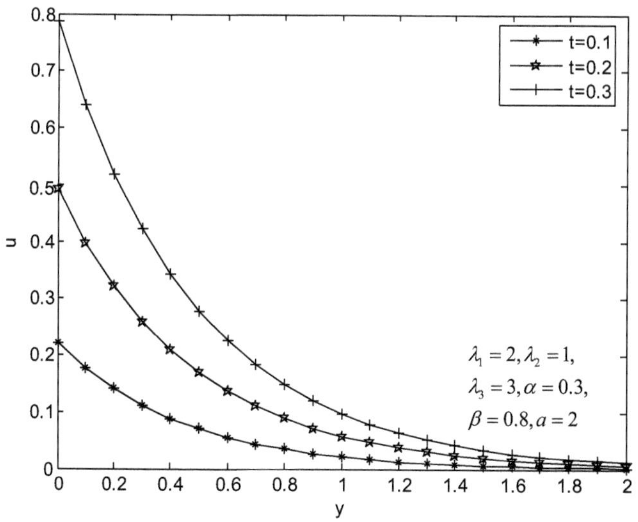

FIGURE 7.18 Velocity fields for different values of t when keeping other parameters fixed.

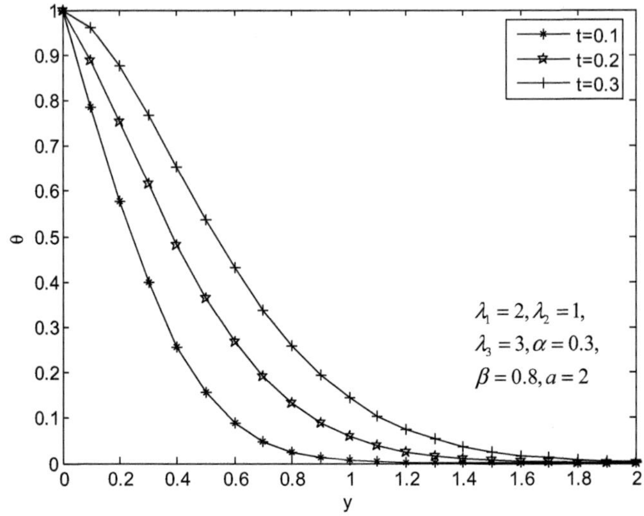

FIGURE 7.19 Temperature profiles for different values of t when $f(t) = 1$.

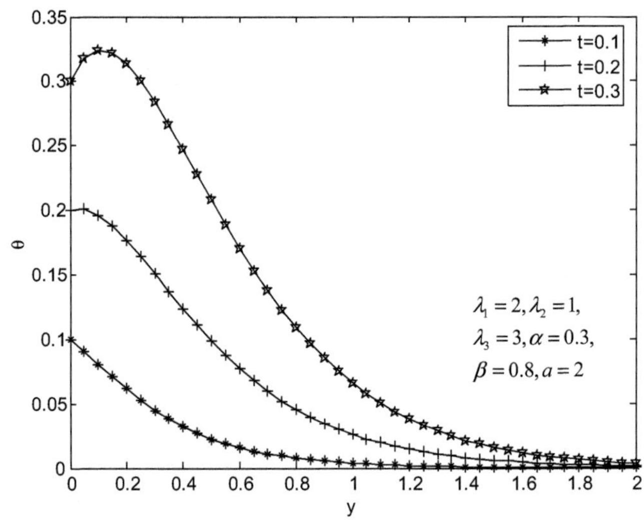

FIGURE 7.20 Temperature profiles for different values of t when $f(t) = t$.

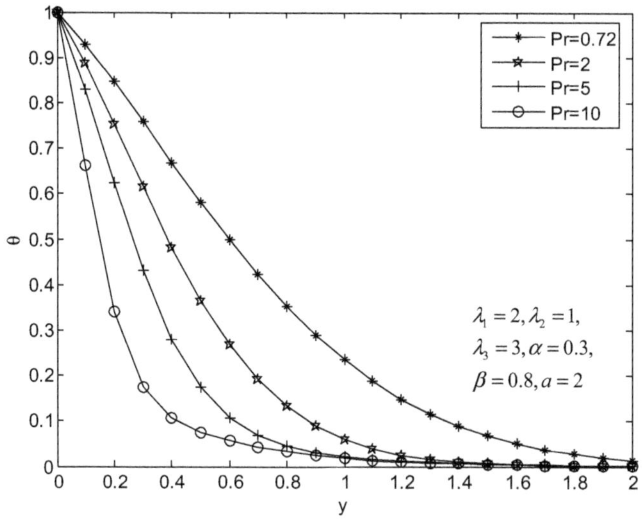

FIGURE 7.21 Temperature profiles for different values of Pr keeping other parameters fixed.

distribution θ vs the Prandtl number Pr. It is clear that there is a fall in temperature with increasing the Prandtl number. Fig. 7.22 depicts the effect of varying N_R for temperature field. The results show marked decrease in the temperature distributions with increase in N_R.

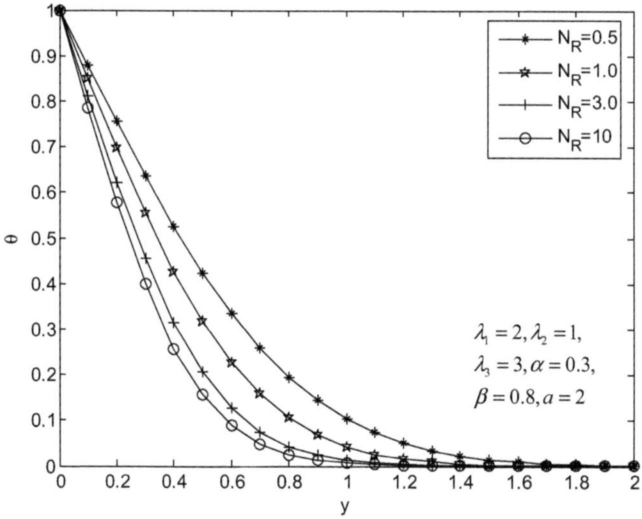

FIGURE 7.22 Temperature profiles for different values of N_R keeping other parameters fixed.

7.5 SLIP EFFECTS ON MAGNETOHYDRODYNAMIC FLOW OF FRACTIONAL OLDROYD-B FLUID

The no-slip boundary condition used in fluid mechanics is an assumption that cannot be derived from first principles. The phenomenon of wall slip has many industrial and practical applications at both macro- and microscales. Fluids exhibiting boundary slip are important in technological applications, such as the polishing of artificial heart valves. The non-Newtonian fluids, such as polymer melts, often exhibit macroscopic wall slip that in general is governed by a nonlinear and nonmonotone relation between the slip velocity and the traction (Derek et al., 2002). Also, wall slip can occur in the working fluid that contains concentrated suspensions (Soltani and Yilmazer, 1998).

Ali et al. (2008) and Ebaid (2008) studied the effects of magnetic field and slip conditions on the peristaltic transform. The slip effects on peristaltic flow of a third-order fluid and a non-Newtonian Maxwellian fluid were investigated by Hayat et al. (2008) and El-Shehawy et al. (2006). Zhang et al. (2008a,b), Mahmoud (2008), and Rao and Rajagopai (1999) discussed the two-dimensional slip flows. The effects of slip condition on flows of Oldroyd 6-constant fluid and 8-constant fluid were investigated by Hayat et al. (2007, 2008). Khaleda and Vafaib (2004) gave the exact solutions for the slip effects on Stokes and Couette flows due to oscillating wall. Zheng et al. (2012) presented an analysis for slip effects on MHD flow of a generalized Oldroyd-B fluid with fractional derivative duo to an accelerating plate. The closed-form solutions for velocity and shear stress were obtained in terms of Fox *H*-function by using the discrete Laplace transform of the sequential fractional derivatives.

7.5.1 Governing Equations

The constitutive equations for an incompressible fluid are given by:

$$\nabla\mathbf{V} = 0,\ \rho\frac{\mathrm{d}\mathbf{V}}{\mathrm{d}t} = \nabla\mathbf{T} + \rho\mathbf{b}, \tag{7.131}$$

where $\mathbf{T}$ is the Cauchy stress tensor, $\mathbf{V}$ is the velocity vector, ρ is the constant density of the fluid, and $\mathbf{b}$ is the body force field.

We consider here the fluid is permeated by an imposed magnetic field B_0, which acts in the positive y-coordinate (Zheng et al., 2012). Under the assumption of low-magnetic Reynolds number, the magnetic body force can be represented as $\sigma B_0^2 u$. Assuming the velocity and shear stress of the form:

$$\mathbf{V} = u(y,t)\mathbf{i}, \quad \mathbf{S} = \mathbf{S}(y,t), \tag{7.132}$$

where u is the velocity and $\mathbf{i}$ is the unit vector in the x-direction. Taking account of the initial condition $\mathbf{S}(y,0) = 0$. In the absence of pressure gradient in the x-direction, the governing equations of the generalized Oldroyd-B fluid are:

$$\left(1+\lambda \mathrm{D}_t^{\alpha}\right)\tau(y,t)=\mu\left(1+\lambda_r \mathrm{D}_t^{\beta}\right)\partial_y u(y,t), \tag{7.133}$$

$$\left(1+\lambda \mathrm{D}_t^{\alpha}\right)\frac{\partial u}{\partial t}=\nu\left(1+\lambda_r \mathrm{D}_t^{\beta}\right)\frac{\partial^2 u}{\partial y^2}-M\left(1+\lambda \mathrm{D}_t^{\alpha}\right)u, \tag{7.134}$$

where $\tau(y,t)=S_{xy}(y,t)$ is the shear stress that is not zero, $\nu=\mu/\rho$ is the kinematic viscosity, λ, λ_r are material constants, known as the relaxation and retardation times, and $M=\sigma B_0^2/\rho$, D_t^{α} is the fractional differential operator.

$$\mathrm{D}_t^{\alpha}f(t)=\frac{1}{\Gamma(1-\alpha)}\frac{\mathrm{d}}{\mathrm{d}t}\int_0^t\frac{f(\tau)}{t(-\tau)^{\alpha}}\mathrm{d}\tau,\ \ 0\le\alpha<1, \tag{7.135}$$

where $\Gamma(\cdot)$ is the Gamma function.

7.5.2 Formulation of the Problem

Let us consider the flow problem of an incompressible generalized Oldroyd-B fluid over an infinite plate at $y\ge 0$ with fluid that occupies the space $y>0$. Here we assume the existence of slip boundary between the velocity of the fluid at the wall $u(0,\,t)$ and the speed of the wall, the relative velocity between $u(0,\,t)$, and the wall is assumed to be proportional to the shear rate at the wall. The governing equations may be described by Eqs. (7.133)–(7.134), and the corresponding initial and boundary conditions are written as follows:

$$u(y,0)=\partial_t u(y,0)=0,\ \ y>0, \tag{7.136}$$

$$u(0,t)=At^b+\theta\frac{\partial u}{\partial y},\ \ t>0, \tag{7.137}$$

$$u(y,t),\ \ \partial_y u(y,t)\to 0\ \text{ as }\ y\to\infty,t>0, \tag{7.138}$$

where b is a constant, θ is the slip strength or slip coefficient. If $\theta=0$ then the general assumed no-slip boundary condition is obtained. If θ is finite, fluid slip occurs at the wall but its effect depends upon the length scale of the flow.

Employing the nondimensional quantities in Eqs. (7.134), (7.136)–(7.138):

$$u^*=\frac{u}{\left(A\nu^b\right)^{\frac{1}{2b+1}}},\ \ y^*=\frac{\left(A\nu^b\right)^{\frac{1}{2b+1}}y}{\nu},\ \ t^*=\frac{\left(A\nu^b\right)^{\frac{2}{2b+1}}t}{\nu},\ \ \lambda^*=\lambda\left(\frac{\left(A\nu^b\right)^{\frac{2}{2b+1}}}{\nu}\right)^{\alpha},$$
$$\lambda_r^*=\lambda_r\left(\frac{\left(A\nu^b\right)^{\frac{2}{2b+1}}}{\nu}\right)^{\beta},\ \ M^*=\frac{M\nu}{\left(A\nu^b\right)^{\frac{2}{2b+1}}},\ \ \theta^*=\frac{\left(A\nu^b\right)^{\frac{1}{2b+1}}\theta}{\nu}. \tag{7.139}$$

Dimensionless equations can be given (for brevity the dimensionless mark "*" is omitted here):

$$\left(1+\lambda D_t^{\alpha}\right)\frac{\partial u}{\partial t}=\left(1+\lambda_r D_t^{\beta}\right)\frac{\partial^2 u}{\partial y^2}-M\left(1+\lambda D_t^{\alpha}\right)u. \tag{7.140}$$

The initial and boundary conditions take the form:

$$u(y,0)=\partial_t u(y,0)=0,\quad y>0, \tag{7.141}$$

$$u(0,t)=t^b+\theta\frac{\partial u}{\partial y},\quad t>0, \tag{7.142}$$

$$u(y,t),\ \ \partial_y u(y,t)\rightarrow 0\ \text{as}\ y\rightarrow\infty, t>0. \tag{7.143}$$

7.5.3 Exact Solutions

In order to solve the previous problem, we use the Laplace transforms principle of sequential fractional derivatives, which yields:

$$\frac{\partial^2\overline{u}}{\partial y^2}-\frac{(s+M)(1+\lambda s^{\alpha})}{(1+\lambda_r s^{\beta})}\overline{u}=0, \tag{7.144}$$

subject to the boundary condition,

$$\overline{u}(0,s)=\frac{1}{s^{b+1}}+\theta\left.\frac{\partial\overline{u}}{\partial y}\right|_{y=0}, \tag{7.145}$$

$$\overline{u}(y,s),\ \ \partial_y\overline{u}(y,s)\rightarrow 0\ \text{as}\ y\rightarrow\infty, \tag{7.146}$$

where $\overline{u}(y,s)$ is the image function of $u(y,t)$ and s is a transform parameter. Solving Eqs. (7.144)–(7.146), we get:

$$\overline{u}(y,s)=\frac{1}{s^{b+1}\left(1+\theta\left[\frac{(s+M)(1+\lambda s^{\alpha})}{(1+\lambda_r s^{\beta})}\right]^{\frac{1}{2}}\right)}\exp\left[-\left[\frac{(s+M)(1+\lambda s^{\alpha})}{(1+\lambda_r s^{\beta})}\right]^{\frac{1}{2}}y\right]. \tag{7.147}$$

The shear stress can be calculated from Eq. (7.133), taking the Laplace transform of Eq. (7.133) and introducing Eq. (7.147), we obtain:

$$\overline{\tau}(y,s) = -(s+M)^{\frac{1}{2}}\left[\frac{(1+\lambda s^{\alpha})}{(1+\lambda_r s^{\beta})}\right]^{-\frac{1}{2}} \cdot \frac{1}{s^{b+1}\left(1+\theta\left[\frac{(s+M)(1+\lambda s^{\alpha})}{(1+\lambda_r s^{\beta})}\right]^{\frac{1}{2}}\right)} \cdot \exp\left(-\left[\frac{(s+M)(1+\lambda s^{\alpha})}{(1+\lambda_r s^{\beta})}\right]^{\frac{1}{2}} y\right), \tag{7.148}$$

where $\overline{\tau}(y,s)$ is the Laplace transform of $\tau(y,t) = S_{xy}/\left(\rho(A\nu^{b})^{\frac{2}{2b+1}}\right)$.

In order to avoid the burdensome calculations of residues and contour integrals, we apply discrete inverse Laplace transform to get to the velocity and the shear stress. Writing Eq. (7.147) as series forms:

$$\overline{u}(y,s) = \frac{1}{s^{b+1}} + \sum_{j=1}^{\infty}(-\theta)^{j}\sum_{l=0}^{\infty}\frac{(-M)^{l}}{l!}\left[\frac{\lambda}{\lambda_r}\right]^{\frac{j}{2}}\sum_{m=0}^{\infty}\frac{\left(-\lambda^{-1}\right)^{m}}{m!}\sum_{n=0}^{\infty}\frac{\left(-\lambda_r^{-1}\right)^{n}}{n!} \times \frac{\Gamma\left(l-\frac{j}{2}\right)}{\Gamma\left(-\frac{j}{2}\right)}\frac{\Gamma\left(m-\frac{j}{2}\right)}{\Gamma\left(-\frac{j}{2}\right)}\frac{\Gamma\left(n+\frac{j}{2}\right)}{\Gamma\left(\frac{j}{2}\right)} \times \frac{1}{s^{\frac{j}{2}(\beta-\alpha-1)+l+\alpha m+\beta n+b+1}} + \sum_{j=0}^{\infty}(-\theta)^{j}\sum_{k=1}^{\infty}\frac{(-y)^{k}}{k!}\sum_{l=0}^{\infty}\frac{(-M)^{l}}{l!}\left[\frac{\lambda}{\lambda_r}\right]^{\frac{j+k}{2}}\sum_{m=0}^{\infty}\frac{\left(-\lambda^{-1}\right)^{m}}{m!} \times \sum_{n=0}^{\infty}\frac{\left(-\lambda_r^{-1}\right)^{n}}{n!}\frac{\Gamma\left(l-\frac{j+k}{2}\right)}{\Gamma\left(-\frac{j+k}{2}\right)}\frac{\Gamma\left(m-\frac{j+k}{2}\right)}{\Gamma\left(-\frac{j+k}{2}\right)}\frac{\Gamma\left(n+\frac{j+k}{2}\right)}{\Gamma\left(\frac{j+k}{2}\right)}\frac{1}{s^{\frac{j+k}{2}(\beta-\alpha-1)+l+\alpha m+\beta n+b+1}}. \tag{7.149}$$

Applying the discrete inverse Laplace transform, we arrive at:

$$u(y,t) = t^b + \sum_{j=1}^{\infty}(-\theta)^j \sum_{l=0}^{\infty}\frac{(-M)^l}{l!}\left[\frac{\lambda}{\lambda_r}\right]^{\frac{j}{2}} \sum_{m=0}^{\infty}\frac{\left(-\lambda^{-1}\right)^m}{m!} t^{\frac{j}{2}(\beta-\alpha-1)+l+\alpha m+b}$$

$$\times \sum_{n=0}^{\infty}\frac{1}{n!}\frac{\Gamma\left(l-\frac{j}{2}\right)}{\Gamma\left(-\frac{j}{2}\right)}\frac{\Gamma\left(m-\frac{j}{2}\right)}{\Gamma\left(-\frac{j}{2}\right)}\frac{\Gamma\left(n+\frac{j}{2}\right)}{\Gamma\left(\frac{j}{2}\right)}$$

$$\times \frac{\left(-\lambda_r^{-1}t^{\beta}\right)^n}{\Gamma\left(\frac{j}{2}(\beta-\alpha-1)+l+\alpha m+\beta n+b+1\right)}$$

$$+ \sum_{j=0}^{\infty}(-\theta)^j \sum_{k=1}^{\infty}\frac{(-y)^k}{k!} \sum_{l=0}^{\infty}\frac{(-M)^l}{l!}\left[\frac{\lambda}{\lambda_r}\right]^{\frac{j+k}{2}} \sum_{m=0}^{\infty}\frac{\left(-\lambda^{-1}\right)^m}{m!} t^{\frac{j+k}{2}(\beta-\alpha-1)+l+\alpha m+b}$$

$$\times \sum_{n=0}^{\infty}\frac{1}{n!}\frac{\Gamma\left(l-\frac{j+k}{2}\right)}{\Gamma\left(-\frac{j+k}{2}\right)}\frac{\Gamma\left(m-\frac{j+k}{2}\right)}{\Gamma\left(-\frac{j+k}{2}\right)}\frac{\Gamma\left(n+\frac{j+k}{2}\right)}{\Gamma\left(\frac{j+k}{2}\right)}$$

$$\times \frac{\left(-\lambda_r^{-1}t^{\beta}\right)^n}{\Gamma\left(\frac{j+k}{2}(\beta-\alpha-1)+l+\alpha m+\beta n+b+1\right)}. \tag{7.150}$$

In terms of Fox H-function, we rewrite the previous equation as a simpler form:

$$u(y,t) = t^b + \sum_{j=1}^{\infty}(-\theta)^j \sum_{l=0}^{\infty}\frac{(-M)^l}{l!}\left[\frac{\lambda}{\lambda_r}\right]^{\frac{j}{2}} \sum_{m=0}^{\infty}\frac{\left(-\lambda^{-1}\right)^m}{m!} t^{\frac{j}{2}(\beta-\alpha-1)+l+\alpha m+b}$$

$$\times H_{3,5}^{1,3}\left[\frac{t^{\beta}}{\lambda_r}\middle|\begin{matrix}\left(1-l+\frac{j}{2},0\right),\left(1-m+\frac{j}{2},0\right),\left(1-\frac{j}{2},1\right)\\ (0,1),\left(1+\frac{j}{2},0\right),\left(1+\frac{j}{2},0\right),\left(1-\frac{j}{2},0\right),\left(\frac{j}{2}(\alpha-\beta+1)-l-\alpha m-b,\beta\right)\end{matrix}\right]$$

$$+ \sum_{j=0}^{\infty}(-\theta)^j \sum_{k=1}^{\infty}\frac{(-y)^k}{k!} \sum_{l=0}^{\infty}\frac{(-M)^l}{l!}\left[\frac{\lambda}{\lambda_r}\right]^{\frac{j+k}{2}} \sum_{m=0}^{\infty}\frac{\left(-\lambda^{-1}\right)^m}{m!} t^{\frac{j+k}{2}(\beta-\alpha-1)+l+\alpha m+b}$$

$$\times H_{3,5}^{1,3}\left[\frac{t^{\beta}}{\lambda_r}\middle|\begin{matrix}\left(1-l+\frac{j+k}{2},0\right),\left(1-m+\frac{j+k}{2},0\right),\left(1-\frac{j+k}{2},1\right)\\ (0,1),\left(1+\frac{j+k}{2},0\right),\left(1+\frac{j+k}{2},0\right),\left(1-\frac{j+k}{2},0\right),\left(\frac{j+k}{2}(\alpha-\beta+1)-l-\alpha m-b,\beta\right)\end{matrix}\right], \tag{7.151}$$

where the property of the Fox H-function is:

$$\sum_{n=0}^{\infty}\frac{(-z)^n\Pi_{j=1}^{p}\Gamma(a_j+A_jn)}{n!\Pi_{j=1}^{q}\Gamma(b_j+B_jn)}=H_{p,q+1}^{1,p}\left[z\Big|_{(0,1),(1-b_1,B_1),\ldots,(1-b_q,B_q)}^{(1-a_1,A_1),\ldots,(1-a_p,A_p)}\right]. \tag{7.152}$$

From Eq. (7.152), we easily obtain:

$$u(0,t)=t^b+\sum_{j=1}^{\infty}(-\theta)^j\sum_{l=0}^{\infty}\frac{(-M)^l}{l!}\left[\frac{\lambda}{\lambda_r}\right]^{\frac{j}{2}}\sum_{m=0}^{\infty}\frac{\left(-\lambda^{-1}\right)^m}{m!}t^{\frac{j}{2}(\beta-\alpha-1)+l+\alpha m+b}$$
$$\times H_{3,5}^{1,3}\left[\frac{t^\beta}{\lambda_r}\Big|_{(0,1),\left(1+\frac{j}{2},0\right),\left(1+\frac{j}{2},0\right),\left(1-\frac{j}{2},0\right),\left(\frac{j}{2}(\alpha-\beta+1)-l-\alpha m-b,\beta\right)}^{\left(1-l+\frac{j}{2},0\right),\left(1-m+\frac{j}{2},0\right),\left(1-\frac{j}{2},1\right)}\right], \tag{7.153}$$

and

$$\left.\frac{\partial u(y,t)}{\partial y}\right|_{y=0}=-\sum_{j=0}^{\infty}(-\theta)^j\sum_{l=0}^{\infty}\frac{(-M)^l}{l!}\left[\frac{\lambda}{\lambda_r}\right]^{\frac{j+1}{2}}\sum_{m=0}^{\infty}\frac{\left(-\lambda^{-1}\right)^m}{m!}t^{\frac{j+1}{2}(\beta-\alpha-1)+l+\alpha m+b}$$
$$\times H_{3,5}^{1,3}\left[\frac{t^\beta}{\lambda_r}\Big|_{(0,1),\left(1+\frac{j+1}{2},0\right),\left(1+\frac{j+1}{2},0\right),\left(1-\frac{j+1}{2},0\right),\left(\frac{j+1}{2}(\alpha-\beta+1)-l-\alpha m-b,\beta\right)}^{\left(1-l+\frac{j+1}{2},0\right),\left(1-m+\frac{j+1}{2},0\right),\left(1-\frac{j+1}{2},1\right)}\right], \tag{7.154}$$

$$\theta\left.\frac{\partial u(y,t)}{\partial y}\right|_{y=0}=-\theta\sum_{j=0}^{\infty}(-\theta)^j\sum_{l=0}^{\infty}\frac{(-M)^l}{l!}\left[\frac{\lambda}{\lambda_r}\right]^{\frac{j+1}{2}}\sum_{m=0}^{\infty}\frac{\left(-\lambda^{-1}\right)^m}{m!}t^{\frac{j+1}{2}(\beta-\alpha-1)+l+\alpha m+b}$$
$$\times H_{3,5}^{1,3}\left[\frac{t^\beta}{\lambda_r}\Big|_{(0,1),\left(1+\frac{j+1}{2},0\right),\left(1+\frac{j+1}{2},0\right),\left(1-\frac{j+1}{2},0\right),\left(\frac{j+1}{2}(\alpha-\beta+1)-l-\alpha m-b,\beta\right)}^{\left(1-l+\frac{j+1}{2},0\right),\left(1-m+\frac{j+1}{2},0\right),\left(1-\frac{j+1}{2},1\right)}\right],$$
$$=\sum_{j=1}^{\infty}(-\theta)^j\sum_{l=0}^{\infty}\frac{(-M)^l}{l!}\left[\frac{\lambda}{\lambda_r}\right]^{\frac{j}{2}}\sum_{m=0}^{\infty}\frac{\left(-\lambda^{-1}\right)^m}{m!}t^{\frac{j}{2}(\beta-\alpha-1)+l+\alpha m+b}$$
$$\times H_{3,5}^{1,3}\left[\frac{t^\beta}{\lambda_r}\Big|_{(0,1),\left(1+\frac{j}{2},0\right),\left(1+\frac{j}{2},0\right),\left(1-\frac{j}{2},0\right),\left(\frac{j}{2}(\alpha-\beta+1)-l-\alpha m-b,\beta\right)}^{\left(1-l+\frac{j}{2},0\right),\left(1-m+\frac{j}{2},0\right),\left(1-\frac{j}{2},1\right)}\right]. \tag{7.155}$$

We can see that the exact solution satisfies the boundary condition Eq. (7.142).

Adopting the similar procedure in Eq. (7.148), we obtain the shear stress:

$$\tau(y,t) = -\sum_{j=0}^{\infty}(-\theta)^j\sum_{k=0}^{\infty}\frac{(-y)^k}{k!}\sum_{l=0}^{\infty}\frac{(-M)^l}{l!}\left[\frac{\lambda}{\lambda_r}\right]^{\frac{j+k-1}{2}}\sum_{m=0}^{\infty}\frac{\left(-\lambda^{-1}\right)^m}{m!}$$
$$\times\sum_{n=0}^{\infty}\frac{1}{n!}t^{\frac{j+k+1}{2}(\beta-\alpha-1)+l+\alpha m+b-1}$$
$$\times H_{3,5}^{1,3}\left[\frac{t^\beta}{\lambda_r}\middle|\begin{matrix}\left(1-l+\frac{j+k+1}{2},0\right),\left(1-m+\frac{j+k-1}{2},0\right),\left(1-\frac{j+k-1}{2},1\right)\\(0,1),\left(1+\frac{j+k+1}{2},0\right),\left(1+\frac{j+k-1}{2},0\right),\left(1-\frac{j+k-1}{2},0\right),\left(\frac{j+k-1}{2}(\alpha-\beta+1)-l-\alpha m-b+1,\beta\right)\end{matrix}\right]. \tag{7.156}$$

7.5.4 Special Cases

1. In the special case when $\theta=0$, Eqs. (7.151) and (7.156) are simplified into:

$$u(y,t) = t^b + \sum_{k=1}^{\infty}\frac{(-y)^k}{k!}\sum_{l=0}^{\infty}\frac{(-M)^l}{l!}\left[\frac{\lambda}{\lambda_r}\right]^{\frac{k}{2}}\sum_{m=0}^{\infty}\frac{\left(-\lambda^{-1}\right)^m}{m!}t^{\frac{k}{2}(\beta-\alpha-1)+l+\alpha m+b}$$
$$\times H_{3,5}^{1,3}\left[\frac{t^\beta}{\lambda_r}\middle|\begin{matrix}\left(1-l+\frac{k}{2},0\right),\left(1-m+\frac{k}{2},0\right),\left(1-\frac{k}{2},1\right)\\(0,1),\left(1+\frac{k}{2},0\right),\left(1+\frac{k}{2},0\right),\left(1-\frac{k}{2},0\right),\left(\frac{k}{2}(\alpha-\beta+1)-l-\alpha m-b,\beta\right)\end{matrix}\right], \tag{7.157}$$

$$\tau(y,t) = -\sum_{k=0}^{\infty}\frac{(-y)^k}{k!}\sum_{l=0}^{\infty}\frac{(-M)^l}{l!}\left[\frac{\lambda}{\lambda_r}\right]^{\frac{k-1}{2}}\sum_{m=0}^{\infty}\frac{\left(-\lambda^{-1}\right)^m}{m!}\sum_{n=0}^{\infty}\frac{1}{n!}t^{\frac{k-1}{2}(\beta-\alpha-1)+l+\alpha m+b-1}$$
$$\times H_{3,5}^{1,3}\left[\frac{t^\beta}{\lambda_r}\middle|\begin{matrix}\left(1-l+\frac{k+1}{2},0\right),\left(1-m+\frac{k-1}{2},0\right),\left(1-\frac{k-1}{2},1\right)\\(0,1),\left(1+\frac{k+1}{2},0\right),\left(1+\frac{k-1}{2},0\right),\left(1-\frac{k-1}{2},0\right),\left(\frac{k-1}{2}(\alpha-\beta+1)-l-\alpha m-b+1,\beta\right)\end{matrix}\right], \tag{7.158}$$

which are the solutions for the MHD flow of a generalized Oldroyd-B fluid without slip effect.

2. In the special case when $M = 0$, Eqs. (7.151) and (7.156) reduce to:

$$\begin{aligned} u(y,t) &= t^b + \sum_{j=1}^{\infty}(-\theta)^j\left[\frac{\lambda}{\lambda_r}\right]^{\frac{j}{2}}\sum_{m=0}^{\infty}\frac{\left(-\lambda^{-1}\right)^m}{m!}t^{\frac{j}{2}(\beta-\alpha-1)+\alpha m+b} \\ &\times H_{2,4}^{1,2}\left[\frac{t^\beta}{\lambda_r}\middle|\begin{matrix}\left(1-m+\frac{j}{2},0\right),\left(1-\frac{j}{2},1\right)\\ (0,1),\left(1+\frac{j}{2},0\right),\left(1-\frac{j}{2},0\right),\left(\frac{j}{2}(\alpha-\beta+1)-\alpha m-b,\beta\right)\end{matrix}\right] \\ &+\sum_{j=0}^{\infty}(-\theta)^j\sum_{k=1}^{\infty}\frac{(-y)^k}{k!}\left[\frac{\lambda}{\lambda_r}\right]^{\frac{j+k}{2}}\sum_{m=0}^{\infty}\frac{\left(-\lambda^{-1}\right)^m}{m!}t^{\frac{j+k}{2}(\beta-\alpha-1)+\alpha m+b} \\ &\times H_{2,4}^{1,2}\left[\frac{t^\beta}{\lambda_r}\middle|\begin{matrix}\left(1-m+\frac{j+k}{2},0\right),\left(1-\frac{j+k}{2},1\right)\\ (0,1),\left(1+\frac{j+k}{2},0\right),\left(1-\frac{j+k}{2},0\right),\left(\frac{j+k}{2}(\alpha-\beta+1)-\alpha m-b,\beta\right)\end{matrix}\right], \end{aligned} \tag{7.159}$$

$$\begin{aligned} \tau(y,t) &= -\sum_{j=0}^{\infty}(-\theta)^j\sum_{k=0}^{\infty}\frac{(-y)^k}{k!}\left[\frac{\lambda}{\lambda_r}\right]^{\frac{j+k-1}{2}}\sum_{m=0}^{\infty}\frac{\left(-\lambda^{-1}\right)^m}{m!}\sum_{n=0}^{\infty}\frac{1}{n!}t^{\frac{j+k-1}{2}(\beta-\alpha)+\alpha m+b} \\ &\times H_{2,4}^{1,2}\left[\frac{t^\beta}{\lambda_r}\middle|\begin{matrix}\left(1-m+\frac{j+k-1}{2},0\right),\left(1-\frac{j+k-1}{2},1\right)\\ (0,1),\left(1+\frac{j+k-1}{2},0\right),\left(1-\frac{j+k-1}{2},0\right),\left(\frac{j+k-1}{2}(\alpha-\beta)-\alpha m-b,\beta\right)\end{matrix}\right], \end{aligned} \tag{7.160}$$

which correspond to the flow of a generalized Oldroyd-B fluid without magnetic field effect.

3. Letting $\lambda \to 0$ into Eqs. (7.159)–(7.160), as expected, the similar solutions for the generalized second grade fluid are recovered. When $\alpha = \beta = 1$, Eqs. (7.159)–(7.160) reduce to the solutions of an ordinary Oldroyd-B model.

7.5.5 Results and Discussion

The purpose of this research is to provide closed-form expressions for the velocity $u(y, t)$ and the shear stress $\tau(y, t)$ corresponding to the unsteady flow of an incompressible generalized Oldroyd-B fluid with slip effect in uniform magnetic field. The exact solutions are obtained in terms of Fox H-function by means of Laplace transform. In order to study the effects of physical parameters on velocity field and shear stress, some figures are plotted to show the dynamic characteristics of the generalized Oldroyd-B fluid.

From those figures, we can see that when increasing the slip at the wall the velocity decreases at the wall. Figs. 7.23 and 7.24 are depicted to show the

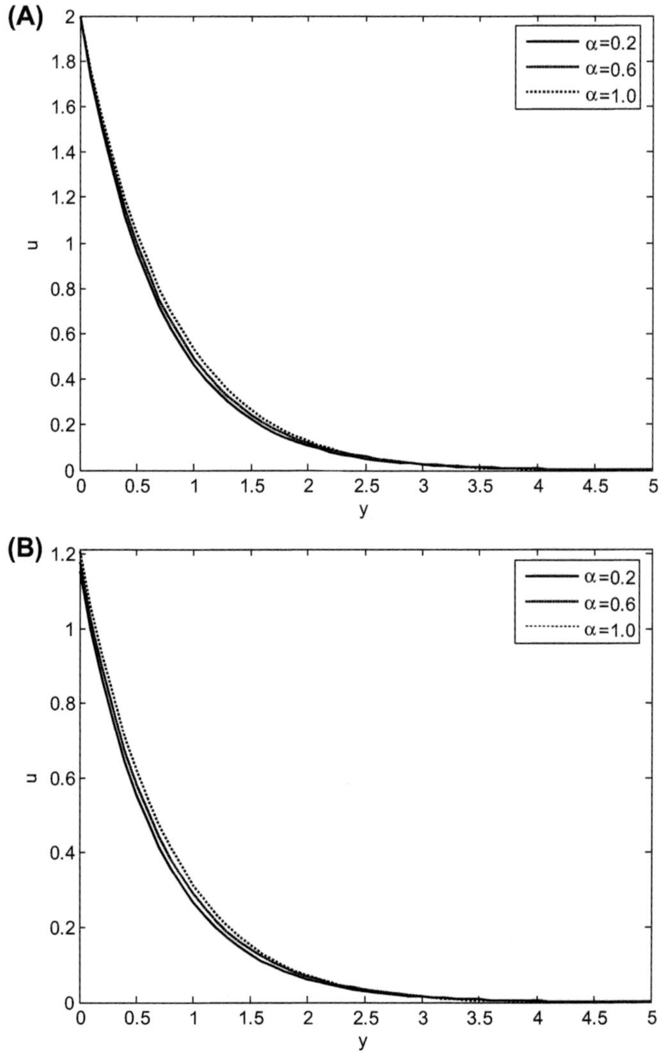

FIGURE 7.23 Profiles of the dimensionless velocity $u(y,t)$ with various values of the fractional parameter α. (A) $\theta = 0$; (B) $\theta = 0.5$.

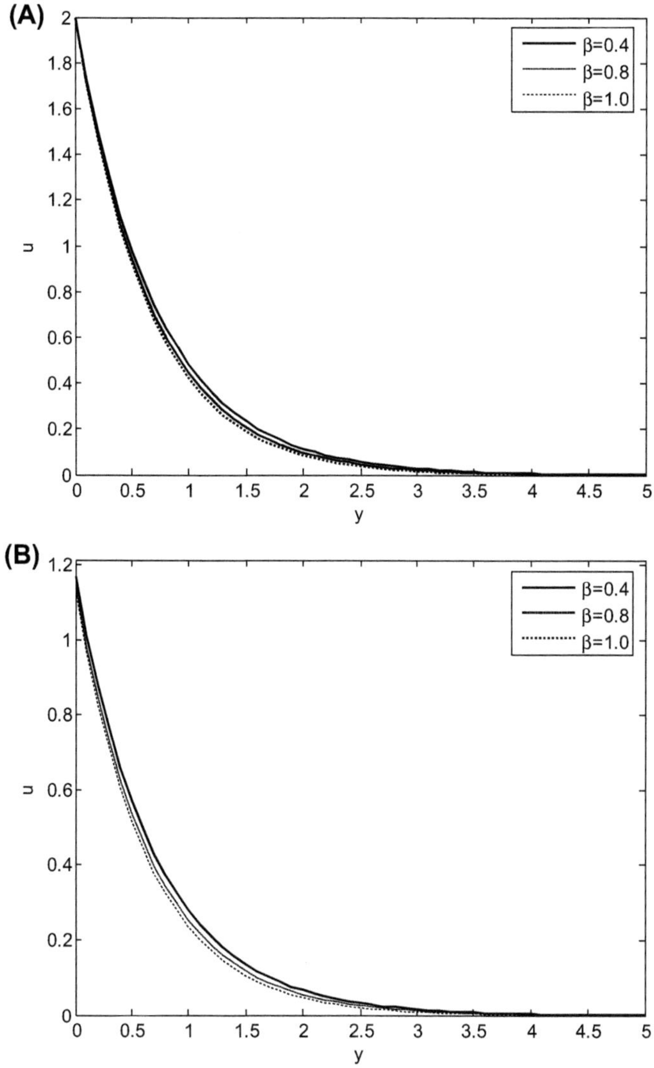

FIGURE 7.24 Profiles of the dimensionless velocity $u(y,t)$ with various values of the fractional parameter β. (A) $\theta = 0$; (B) $\theta = 0.5$.

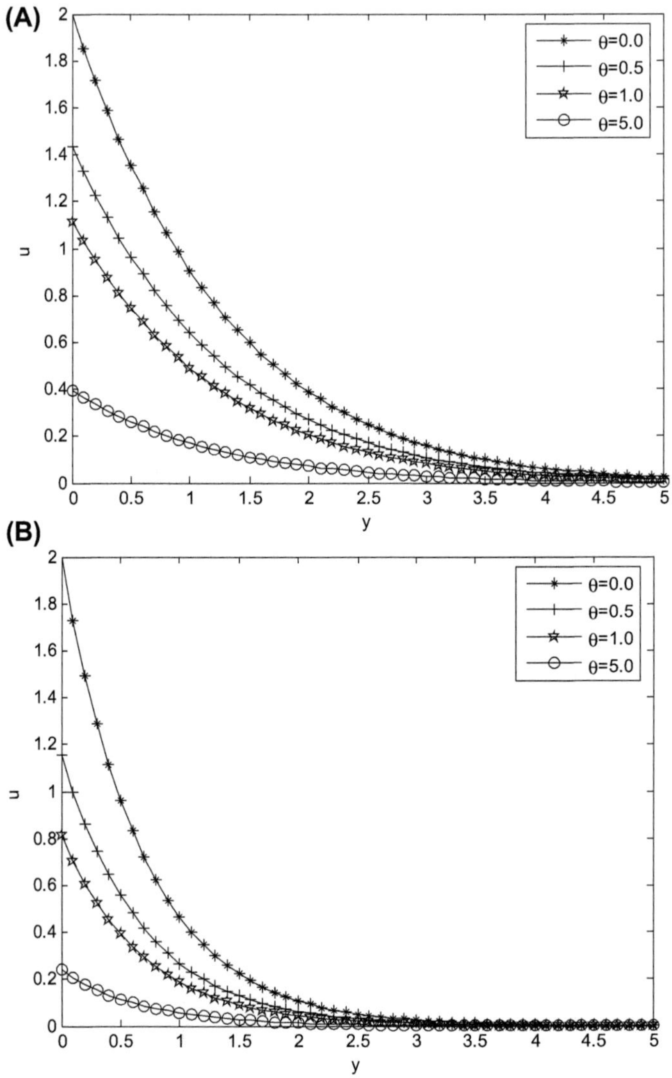

FIGURE 7.25 Profiles of the dimensionless velocity $u(y,t)$ with various values of the slip coefficient θ. (A) $M = 0$; (B) $M = 2$.

changes of the velocity with the fractional parameters α and β. It is clearly seen that the smaller the values of α, the more rapidly the velocity decays. The effect of β on the velocity is contrary to that of α. Fig. 7.25 demonstrates the influence of slip coefficient on the velocity. Panel (A) shows the results of magnetic parameter $M = 0$. Panel (B) is $M = 2$. Fig. 7.26 shows the influence of magnetic parameter on the velocity. Panel (A) shows the results of slip

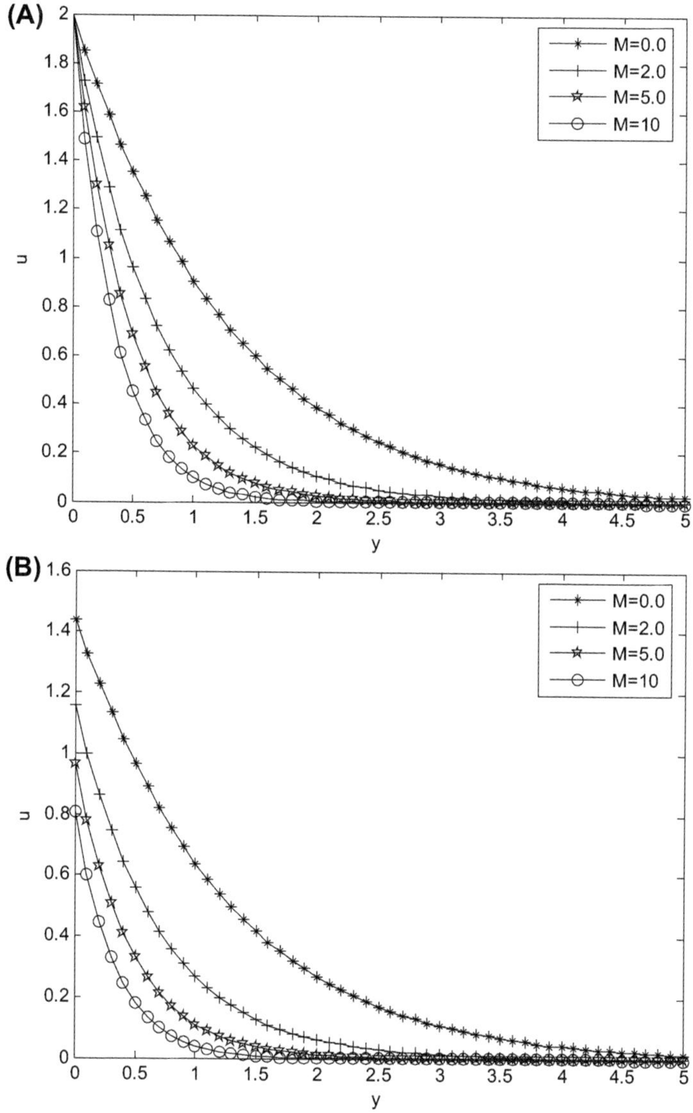

FIGURE 7.26 Profiles of the dimensionless velocity $u(y,t)$ with various values of the magnetic parameter M. (A) $\theta = 0$; (B) $\theta = 0.5$.

coefficient $\theta = 0$. Panel (B) is $\theta = 0.5$. Results indicate that the velocity profiles decrease with increasing slip coefficient θ.

Figs. 7.27 and 7.28 provide the graphic illustrations for the effects of magnetic parameter on velocity fields with power law index b. The velocity is increasing with the increase of the power law index. It also can be noticed that

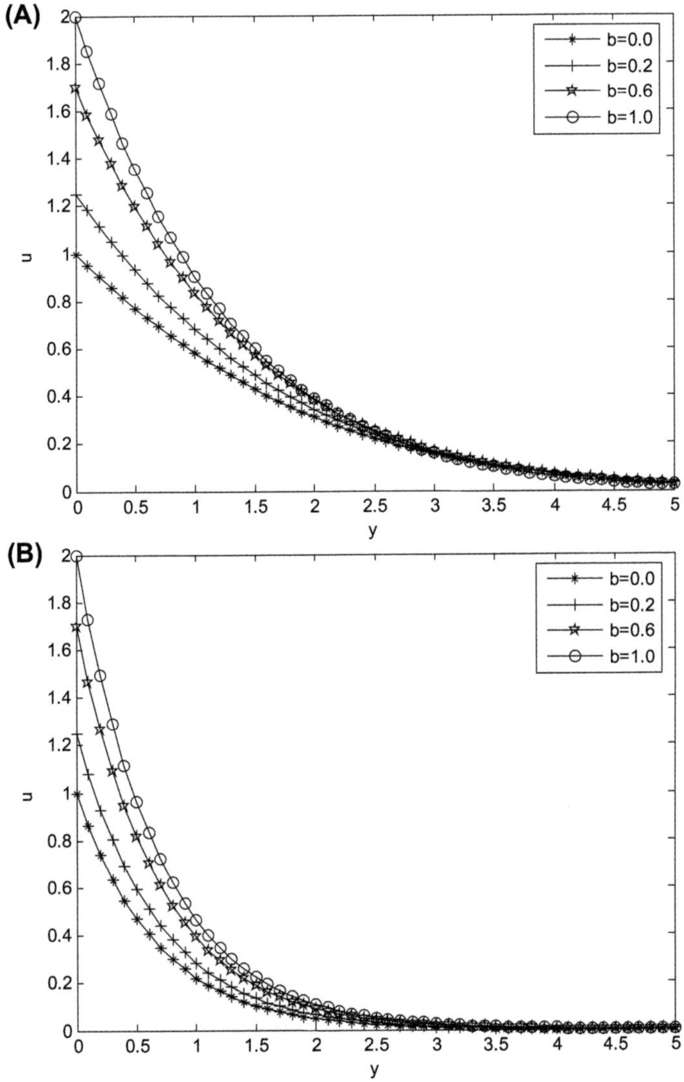

FIGURE 7.27 Profiles of the dimensionless velocity $u(y,t)$ with various values of the power index b for $\theta = 0$. (A) $M = 0$; (B) $M = 2$.

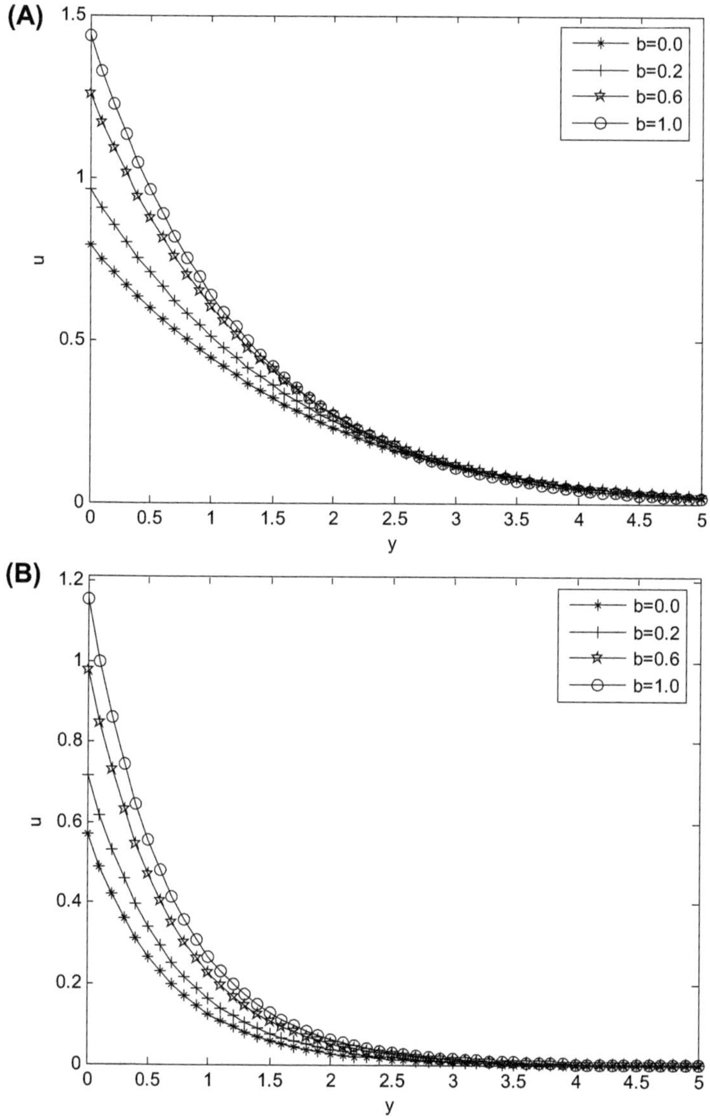

FIGURE 7.28 Profiles of the dimensionless velocity $u(y,t)$ with various values of the power index b for $\theta = 0.5$. (A) $M = 0$; (B) $M = 2$.

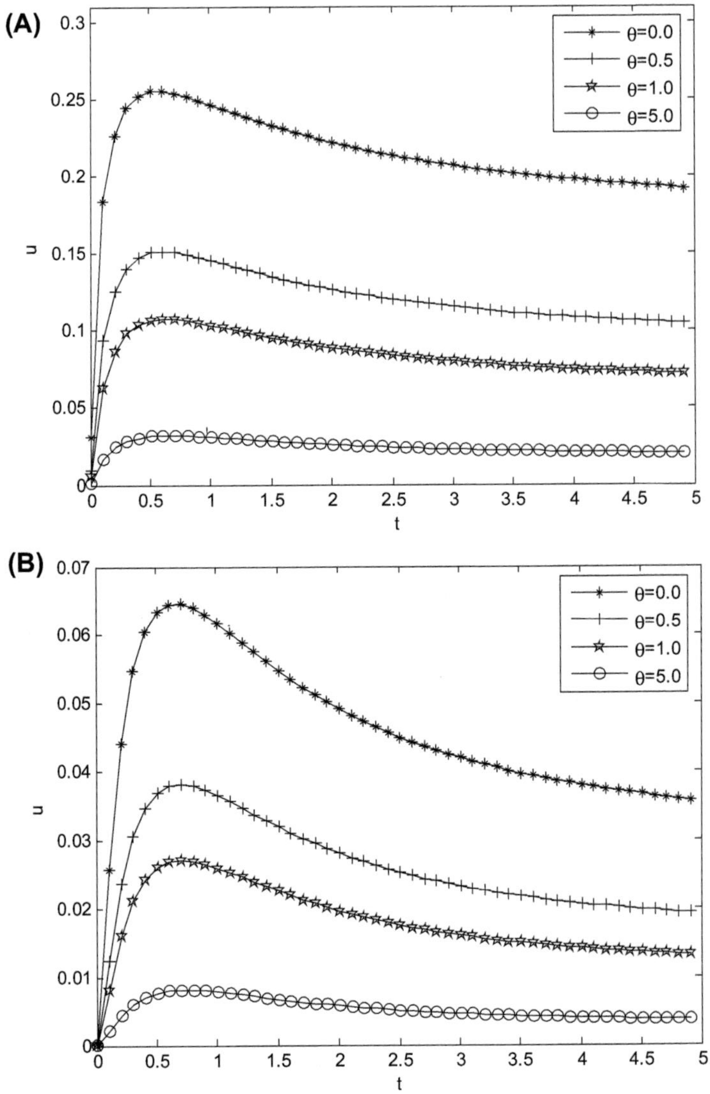

FIGURE 7.29 Profiles of the dimensionless velocity $u(y, t)$ with various values of slip coefficient θ for $b = 0$. (A) $y = 1$; (B) $y = 2$.

the effect of magnetic parameter is to decrease the velocity. This implies that the magnetic field creates a resistive force similar to the drag force, which acts in the opposite direction of the fluid motion; this force causes a decrease in fluid velocity.

Figs. 7.29 and 7.30 demonstrate the velocity changes with time at given points ($y = 1$ and $y = 2$) for power index $b = 1$ and 2, respectively. It can be

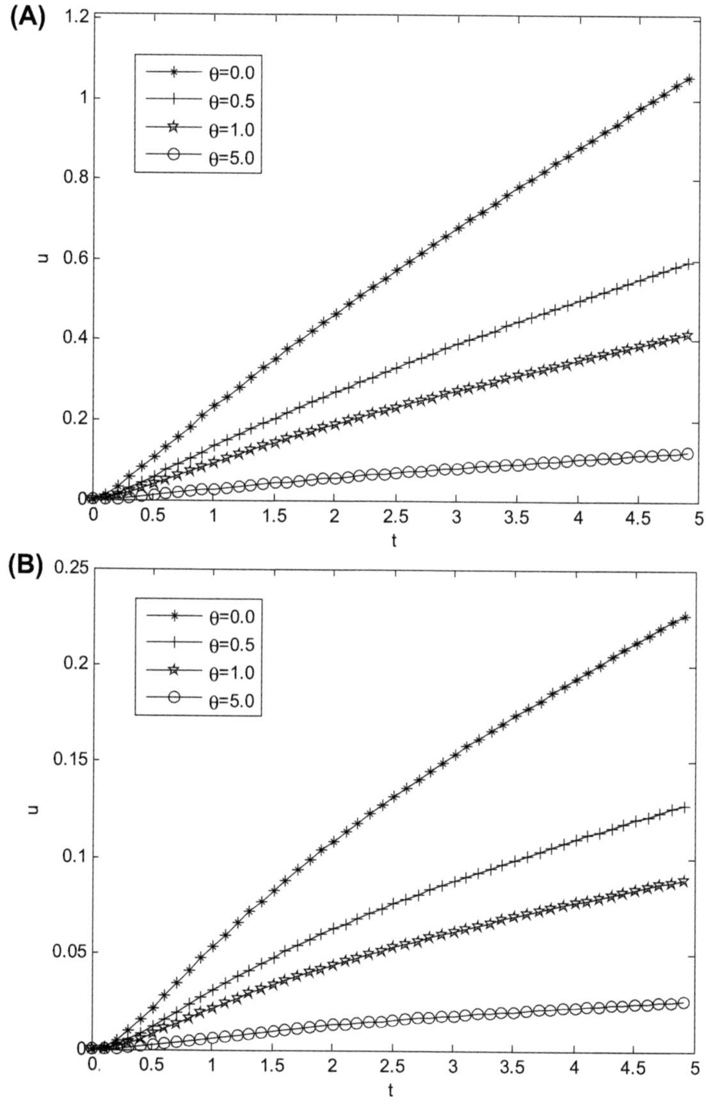

FIGURE 7.30 Profiles of the dimensionless velocity $u(y, t)$ with various values of slip coefficient θ for $b = 1$. (A) $y = 1$; (B) $y = 2$.

seen that with increasing the distance from the plate, the flow velocity decrease rapidly. Fig. 7.31 shows the changes of shear stress with different values of slip coefficient and magnetic parameter; the results indicated that the strongest shear stress occurs near the plate and the shear stress decreases rapidly with the increase of distance from the plate.

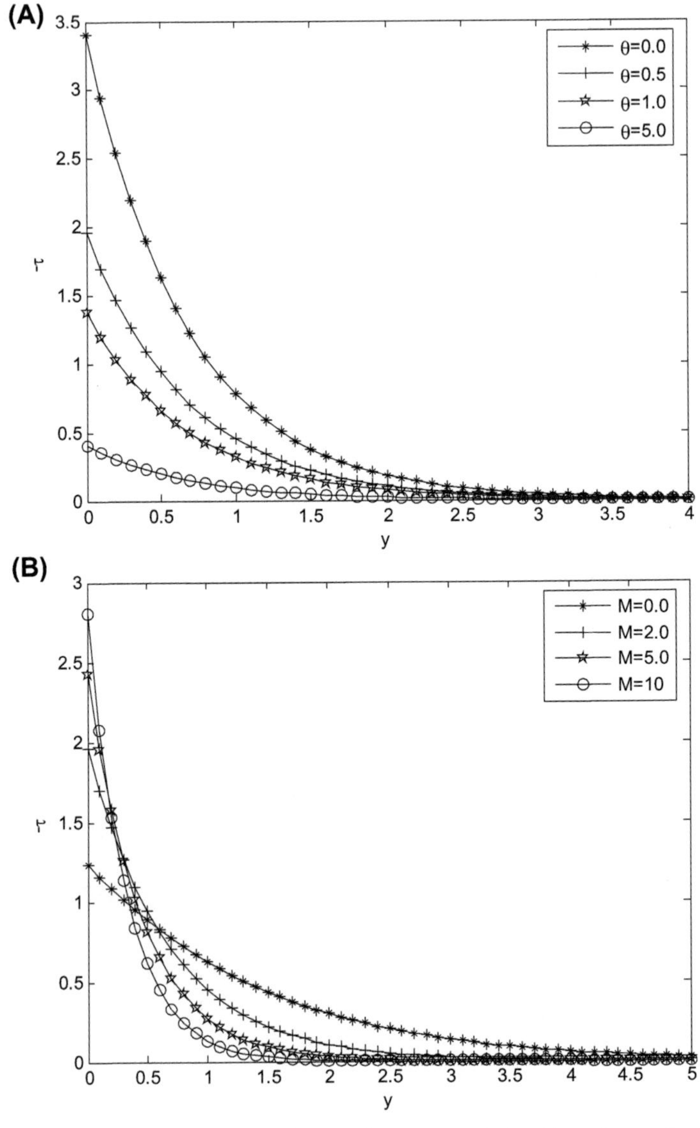

FIGURE 7.31 Profiles of the dimensionless shear stress $\tau(y,t)$ with various values of slip coefficient and magnetic parameter. (A) $\theta = 0.5$; (B) $M = 2$.

7.6 THE 3D FLOW OF GENERALIZED OLDROYD-B FLUID

This section deals with the 3D flow of a generalized Oldroyd-B fluid due to a constant pressure gradient between two side walls perpendicular to the plate. The fractional calculus approach is used to establish the constitutive relationship of non-Newtonian fluid model.

7.6.1 Governing Equation

The constitutive equations of an incompressible generalized Oldroyd-B are:

$$\mathbf{T} = -p\mathbf{I} + \mathbf{S}, \tag{7.161}$$

$$\left(1 + \lambda \frac{\mathrm{D}^{\alpha}}{\mathrm{D}t^{\alpha}}\right)\mathbf{S} = \mu\left(1 + \theta \frac{\mathrm{D}^{\beta}}{\mathrm{D}t^{\beta}}\right)\mathbf{A}_1, \tag{7.162}$$

where $\mathbf{T}$ denotes the Cauchy stress tensor, $-p\mathbf{I}$ is the indeterminate spherical stress, $\mathbf{S}$ is the extra-stress tensor, $\mathbf{A}_1 = \mathbf{L} + \mathbf{L}^T$ is the first Rivlin–Ericksen tensor with the velocity gradient $\mathbf{L} = grad\ \mathbf{V}$, μ the dynamic viscosity of the fluid, λ and θ are relaxation and retardation times, α and β the fractional calculus parameters such that $0 \le \alpha \le \beta \le 1$, and

$$\frac{\mathrm{D}^{\alpha}\mathbf{S}}{\mathrm{D}t^{\alpha}} = \mathrm{D}_t^{\alpha}\mathbf{S} + (\mathbf{V}\cdot\nabla)\mathbf{S} - \mathbf{L}\mathbf{S} - \mathbf{S}\mathbf{L}^T, \tag{7.163}$$

$$\frac{\mathrm{D}^{\beta}\mathbf{A}_1}{\mathrm{D}t^{\beta}} = \mathrm{D}_t^{\beta}\mathbf{A}_1 + (\mathbf{V}\cdot\nabla)\mathbf{A}_1 - \mathbf{L}\mathbf{A}_1 - \mathbf{A}_1\mathbf{L}^T, \tag{7.164}$$

in which D_t^{α} and D_t^{β} are the fractional differentiation operators of order α and β based on Riemann–Liouville's definition is defined as (Miller and Ross, 1993; Samko et al., 1993):

$$\mathrm{D}_t^{p}[f(t)] = \frac{1}{\Gamma(1-p)}\frac{\mathrm{d}}{\mathrm{d}t}\int_0^t \frac{f(\tau)}{(t-\tau)^p}\mathrm{d}\tau, \quad 0 \le p \le 1. \tag{7.165}$$

$\Gamma(\cdot)$ denotes the Gamma function. This model reduces to the ordinary Oldroyd-B model when $\alpha = \beta = 1$.

In the following we shall determine a velocity field and an extra stress of the form:

$$\mathbf{V} = (u(y,z,t), 0, 0), \quad \mathbf{S} = \mathbf{S}(y,z,t), \tag{7.166}$$

where $u(y,z,t)$ is the velocity in the x-coordinate direction.

Substituting Eq. (7.166) into (7.162) and taking into account the initial condition $\mathbf{S}(y,z,0) = \mathbf{0}$ (the fluid being at rest up to the moment $t = 0$), we obtain $\mathbf{S}_{yy} = \mathbf{S}_{yz} = \mathbf{S}_{zz} = 0$ (Havlin and Ben-Avraham, 2002; Tan and Masuoka, 2005a) and the relevant partial differential equations:

$$\left(1 + \lambda D_t^{\alpha}\right)\tau_1 = \mu\left(1 + \theta D_t^{\beta}\right)\partial_y u(y, z, t), \tag{7.167}$$

$$\left(1 + \lambda D_t^{\alpha}\right)\tau_2 = \mu\left(1 + \theta D_t^{\beta}\right)\partial_z u(y, z, t), \tag{7.168}$$

in which $\tau_1 = \mathbf{S}_{xy}$ and $\tau_2 = \mathbf{S}_{xz}$ are tangential stresses.

The balance of linear momentum, in the absence of body forces, reduces to (Havlin and Ben-Avraham, 2002):

$$\partial_y \tau_1 + \partial_z \tau_2 - \partial_x p = \rho \partial_t u, \tag{7.169}$$

$$\partial_y p = \partial_z p = 0, \tag{7.170}$$

where ρ is the constant density of the fluid and $\partial p/\partial x$ is the pressure gradient along x-axis. Eliminating τ_1 and τ_2 between Eqs. (7.167) and (7.168) and assuming that there is a constant pressure gradient in the flow direction, we find the governing equation under form:

$$\left(1 + \lambda D_t^{\alpha}\right)\frac{\partial u(y, z, t)}{\partial t} = \nu\left(1 + \theta D_t^{\beta}\right)\left(\frac{\partial^2}{\partial y^2} + \frac{\partial^2}{\partial z^2}\right)u(y, z, t) - \frac{1}{\rho}\left(1 + \lambda D_t^{\alpha}\right)\frac{\partial p}{\partial x}, \tag{7.171}$$

in which $\nu = \mu/\rho$ is the kinematic viscosity of the fluid.

7.6.2 Formulation of the Problem

Let us consider an incompressible generalized Oldroyd-B fluid occupying the space above a flat plate perpendicular to the y-axis and between two side walls perpendicular to the plate. Initially, the fluid is at rest and at time $t = 0^+$ it is suddenly set into motion by a constant pressure gradient in the x-direction. Its velocity and stress field are of the form Eq. (7.166) and the associate initial and boundary conditions are:

$$u(y, z, 0) = \frac{\partial u(y, z, 0)}{\partial t} = 0, \quad y \geq 0 \text{ and } 0 \leq z \leq h, \tag{7.172}$$

$$u(0, z, t) = 0, \quad t \geq 0 \text{ and } 0 \leq z \leq h, \tag{7.173}$$

$$u(y, z, t) = \frac{\partial u(y, z, t)}{\partial y} = 0, \quad y, t \geq 0 \text{ and } z = 0, h, \tag{7.174}$$

where h is the distance between the two side walls. Moreover, the natural condition is:

$$\frac{\partial u(y,z,t)}{\partial y} \to 0, \quad \text{as } y \to \infty \text{ and } t \geq 0, \quad 0 < z < h, \tag{7.175}$$

in which is the consequence of the fact that the velocity of fluid becomes steady along y-axis as $y \to \infty$ and there is no shear along y-axis (Tan and Masuoka, 2005b).

7.6.3 Calculation of the Velocity Field

Multiplying both sides of Eq. (7.171) by $\sin\left(\frac{n\pi z}{h}\right)$, integrating the result with respect to z from 0 to h, respectively, we attain to the differential equations:

$$\begin{aligned}\left(1+\lambda D_t^{\alpha}\right)\frac{\partial u_n(y,n,t)}{\partial t} &= \nu\left(1+\theta D_t^{\beta}\right)\frac{\partial^2}{\partial y^2}u_n(y,n,t) \\ &\quad + A\frac{h}{n\pi}\left(1+\lambda\frac{t^{-\alpha}}{\Gamma(1-\alpha)}\right)(1-(-1)^n) \\ &\quad - \nu\left(\frac{n\pi}{h}\right)^2\left(1+\theta D_t^{\beta}\right)u_n(y,n,t).\end{aligned} \tag{7.176}$$

Applying the Laplace transform to Eq. (7.176) and using the Laplace transform formula for sequential fractional derivatives,

$$\overline{u}_s(y,z,s) = \int_0^{\infty} u(y,t)e^{-st}\mathrm{d}t, \quad s > 0, \tag{7.177}$$

we find that the image functions of $\overline{u}_{sn}(y,n,s)$ are given by:

$$\begin{aligned}&\frac{\partial^2}{\partial y^2}\overline{u}_{sn}(y,n,s) - \left[\xi^2 + \frac{s(1+\lambda s^{\alpha})}{\nu(1+\theta s^{\beta})}\right]\overline{u}_{sn}(y,n,s) \\ &\quad + \frac{A}{\nu\xi}(1-(-1)^n)\frac{(1+\lambda s^{\alpha})}{s(1+\theta s^{\beta})} = 0,\end{aligned} \tag{7.178}$$

where $\xi = \frac{n\pi}{h}$.

The solution of Eq. (7.178) satisfying the boundary conditions Eqs. (7.173) and (7.175) is of the following form:

$$\overline{u}_{sn} = \frac{-A(1-(-1)^n)}{\xi s^2\left[\nu\xi^2\frac{(1+\theta s^{\beta})}{s(1+\lambda s^{\alpha})}+1\right]}\left[\exp\left(-y\sqrt{\xi^2+\frac{s(1+\lambda s^{\alpha})}{\nu(1+\theta s^{\beta})}}\right)-1\right]. \tag{7.179}$$

In order to obtain an analytical solution for this problem and to avoid lengthy calculations of residues and contour integrals, we will apply the discrete inverse Laplace transform method (Khan et al., 2007) to get the velocity fields. Now rewriting Eq. (7.179) in series form as

$$\overline{u}_{sn} = -A(1-(-1)^n)\sum_{i=0}^{\infty}\sum_{j=0}^{\infty}\sum_{k=0}^{\infty}\sum_{l=0}^{\infty}\sum_{m=0}^{\infty}\frac{(-1)^{i+j+k+l+m}y^j}{j!k!l!m!}\nu^{i-k}\xi^{2i+j-2k-1}$$

$$\times\frac{\Gamma\left(-\frac{j}{2}+k\right)\Gamma(i-k+l)\Gamma(-i+k+m)}{\Gamma\left(-\frac{j}{2}\right)\Gamma(i-k)\Gamma(-i+k)}\frac{\lambda^{-i+k-l}\theta^{i-k-m}}{s^{i-k+(i-k+l)\alpha-(i-k-m)\beta+2}}$$

$$+A(1-(-1)^n)\sum_{i=0}^{\infty}\sum_{p=0}^{\infty}\sum_{n=0}^{\infty}\frac{(-1)^{i+p+q}}{p!q!}\nu^i\xi^{2i-1}\frac{\Gamma(i+p)\Gamma(-i+q)}{\Gamma(i)\Gamma(-i)}\frac{\lambda^{-i-p}\theta^{i-q}}{s^{i+(i+p)\alpha-(i-q)\beta+2}}. \tag{7.180}$$

Applying the discrete inverse Laplace transform to Eq. (7.180), we can get;

$$u_n = -A(1-(-1)^n)\sum_{i=0}^{\infty}\sum_{j=0}^{\infty}\sum_{k=0}^{\infty}\sum_{l=0}^{\infty}\sum_{m=0}^{\infty}\frac{(-1)^{i+j+k+l+m}y^j}{j!k!l!m!}\nu^{i-k}\xi^{2i+j-2k-1}\lambda^{-i+k-l}\theta^{i-k-m}$$

$$\times\frac{\Gamma\left(-\frac{j}{2}+k\right)\Gamma(i-k+l)\Gamma(-i+k+m)s^{i-k+(i-k+l)\alpha-(i-k-m)\beta+1}}{\Gamma\left(-\frac{j}{2}\right)\Gamma(-i+k)\Gamma(i-k)\Gamma(i-k+(i-k+l)\alpha-(i-k-m)\beta+2)}$$

$$+A(1-(-1)^n)\sum_{i=0}^{\infty}\sum_{p=0}^{\infty}\sum_{q=0}^{\infty}\frac{(-1)^{i+p+q}}{p!q!}\nu^i\xi^{2i-1}\lambda^{-i-p}\theta^{i-q}$$

$$\times\frac{\Gamma(i+p)\Gamma(-i+q)s^{i+(i+p)\alpha-(i-q)\beta+1}}{\Gamma(i)\Gamma(-i)\Gamma(i+(i+p)\alpha-(i-q)\beta+2)}. \tag{7.181}$$

Finally, the inverse finite Fourier sine transform gives the analytical solution of velocity distribution:

$$u = \frac{2}{h}\sum_{n=1}^{\infty}\sin\left(\frac{n\pi z}{h}\right)u_n$$

$$= -\frac{2A}{h}\sum_{n=1}^{\infty}\sin\left(\frac{n\pi z}{h}\right)(1-(-1)^n)\sum_{i=0}^{\infty}\sum_{j=0}^{\infty}\sum_{k=0}^{\infty}\sum_{l=0}^{\infty}\sum_{m=0}^{\infty}$$

$$\times\frac{(-1)^{i+j+k+l+m}y^j}{j!k!l!m!}\nu^{i-k}\xi^{2i+j-2k-1}\lambda^{-i+k-l}\theta^{i-k-m}$$

$$\times\frac{\Gamma\left(-\frac{j}{2}+k\right)\Gamma(i-k+l)\Gamma(-i+k+m)s^{i-k+(i-k+l)\alpha-(i-k-m)\beta+1}}{\Gamma\left(-\frac{j}{2}\right)\Gamma(-i+k)\Gamma(i-k)\Gamma(i-k+(i-k+l)\alpha-(i-k-m)\beta+2)}$$

$$+\frac{2A}{h}\sum_{n=1}^{\infty}\sin\left(\frac{n\pi z}{h}\right)(1-(-1)^n)\sum_{i=0}^{\infty}\sum_{p=0}^{\infty}\sum_{q=0}^{\infty}\frac{(-1)^{i+p+q}}{p!q!}\nu^i\xi^{2i-1}\lambda^{-i-p}\theta^{i-q}$$

$$\times\frac{\Gamma(i+p)\Gamma(-i+q)s^{i+(i+p)\alpha-(i-q)\beta+1}}{\Gamma(i)\Gamma(-i)\Gamma(i+(i+p)\alpha-(i-q)\beta+2)} \tag{7.182}$$

In terms of the Fox H-function, Eq. (7.182) takes the simpler forms of:

$$u = -\frac{2}{h}\sum_{n=1}^{\infty}\sin\left(\frac{n\pi z}{h}\right)A(1-(-1)^n)\sum_{i=0}^{\infty}\sum_{j=0}^{\infty}\sum_{k=0}^{\infty}\sum_{l=0}^{\infty}\frac{(-1)^{i+j+k+l}y^j}{j!k!l!}$$

$$\times\ \nu^{i-k}\xi^{2i+j-2k-1}\lambda^{-i+k-m}\theta^{i-k}s^{(i-k)(1+\alpha-\beta)+m\alpha+1}$$

$$\times H_{3,5}^{1,3}\left[\frac{s^\beta}{\theta}\left|\begin{array}{c}(1+i-k,1),\left(1+\frac{j}{2}-k,0\right),(1-i+k-l,0)\\ (1,0),((i-k)(\beta-\alpha-1)-1-l\alpha,\beta),\left(1+\frac{j}{2},0\right),\\ (1+i-k,0),(1-i+k,0)\end{array}\right.\right]$$

$$+\frac{2}{h}\sum_{n=1}^{\infty}\sin\left(\frac{n\pi z}{h}\right)A(1-(-1)^n)\sum_{i=0}^{\infty}\sum_{p=0}^{\infty}\frac{(-1)^{i+p}}{p!}\nu^i\xi^{2i-1}\lambda^{-i-p}\theta^{i-q}s^{i(1+\alpha-\beta)+p\alpha+1}$$

$$\times H_{2,4}^{1,2}\left[\frac{s^\beta}{\theta}\left|\begin{array}{c}(1+i,1),(1-i-p,0)\\ (1,0),(i(\beta-\alpha-1)-1-p\alpha,\beta),(1+i,0),(1-i,0)\end{array}\right.\right]$$

(7.183)

To obtain Eq. (7.183), the following property of the Fox H-function (Fetecau et al., 2009) is used:

$$\mathrm{H}_{p,q+1}^{1,p}\left[-\chi\left|\begin{array}{c}(1-a_1,A_1),\ldots,(1-a_p,A_p)\\ (1,0),(1-b_1,B_1),\ldots,(1-b_q,B_q)\end{array}\right.\right]=\sum_{k=0}^{\infty}\frac{\Gamma(a_1+A_1k),\ldots,\Gamma(a_p+A_pk)}{k!\Gamma(b_1+B_1k),\ldots,\Gamma(b_q+B_qk)}\chi^k$$

(7.184)

7.6.4 Calculation of the Shear Stress

Applying the Laplace transform to Eqs. (7.167) and (7.168), we find that

$$\overline{\tau}_1=\frac{\mu\left(1+\theta s^\beta\right)}{1+\lambda s^\alpha}\frac{\partial\overline{u}(y,z,s)}{\partial y}, \tag{7.185}$$

$$\overline{\tau}_2=\frac{\mu\left(1+\theta s^\beta\right)}{1+\lambda s^\alpha}\frac{\partial\overline{u}(y,z,s)}{\partial z}. \tag{7.186}$$

The image function $\overline{u}(y,z,s)$ of $u(y,z,t)$ can be easy obtained from Eq. (7.183). Consequently, evaluating $\frac{\partial \overline{u}(y,z,s)}{\partial y}$ from the mentioned equations and introducing it into Eq. (7.185), the result is that

$$\overline{\tau}_1 = \frac{2\rho\nu}{h}\sum_{n=1}^{\infty}\sin\left(\frac{n\pi z}{h}\right)\frac{A(1-(-1)^n)}{\nu^2\xi^3\left[1+\frac{s(1+\lambda s^{\alpha})}{\nu\xi^2(1+\theta s^{\beta})}\right]^{\frac{1}{2}}}\exp\left(-y\sqrt{\xi^2+\frac{s(1+\lambda s^{\alpha})}{\nu(1+\theta s^{\beta})}}\right). \tag{7.187}$$

Finally, the stress field in terms of the Fox H-function is:

$$\tau_1 = \frac{A\rho}{h}\sum_{n=1}^{\infty}\sin\left(\frac{n\pi z}{h}\right)(1-(-1)^n)\sum_{i=0}^{\infty}\sum_{j=0}^{\infty}\sum_{k=0}^{\infty}\sum_{l=0}^{\infty}\sum_{m=0}^{\infty}\frac{(-1)^{i+j+k+l}y^j\xi^{-2i-2k+j-3}}{2^{2i-1}\nu^{i+k+1}i!j!k!l!}$$
$$\times\ \lambda^{i+k-l}\theta^{-i-k}s^{(i+k)(-\alpha+\beta-1)+l\alpha-1}$$
$$\times\, \mathrm{H}^{1,4}_{4,6}\left[\frac{s^{\beta}}{\theta}\left|\begin{matrix}\left(1+\frac{j}{2}-k,0\right),(1-2i,0),(1+i+k-l,0),(1-i-k,m)\\(1,0),\left(1+\frac{j}{2},0\right),(1+i+k,0),(1-i-k,0),(1-i,0),[1+(i+k)(\alpha-\beta+1)-l\alpha,\beta]\end{matrix}\right.\right] \tag{7.188}$$

As regards $\tau_2(y,z,t)$, giving the shear stresses on the side walls, it can be immediately obtained from Eqs. (7.183) and (7.186).

7.6.5 Special Cases

Making $\lambda \to 0$ and $\alpha \neq 0$ into Eqs. (7.183) and (7.188), we attain similar solutions for a generalized second grade fluid, performing the same motion. Thus the velocity and the stress fields reduce to:

$$u = -\frac{2}{h}\sum_{n=1}^{\infty}\sin\left(\frac{n\pi z}{h}\right)A(1-(-1)^n)\sum_{i=0}^{\infty}\sum_{j=0}^{\infty}\sum_{k=0}^{\infty}\frac{(-1)^{i+j+k}y^j}{j!k!}\nu^{i-k}\xi^{2i+j-2k-1}\theta^{i-k}s^{(i-k)(1-\beta)+1}$$
$$\times\, H^{1,2}_{2,4}\left[\frac{s^{\beta}}{\theta}\left|\begin{matrix}(1+i-k,1),\left(1+\frac{j}{2}-k,0\right)\\(1,0),[-1-i+k+(i-k)\beta,\beta],\left(1+\frac{j}{2},0\right),(1+i-k,0)\end{matrix}\right.\right]$$
$$+\frac{2}{h}\sum_{n=1}^{\infty}\sin\left(\frac{n\pi z}{h}\right)A(1-(-1)^n)\sum_{i=0}^{\infty}(-1)^i\nu^i\xi^{2i-1}\theta^{i-q}s^{i-i\beta+1}$$
$$\times\, H^{1,1}_{1,3}\left[\frac{s^{\beta}}{\theta}\left|\begin{matrix}(1+i,1)\\(1,0),(-1-i+i\beta,\beta),(1+i,0)\end{matrix}\right.\right], \tag{7.189}$$

$$\tau_1 = \frac{A\rho}{h}\sum_{n=1}^{\infty}\sin\left(\frac{n\pi z}{h}\right)(1-(-1)^n)\sum_{i=0}^{\infty}\sum_{j=0}^{\infty}\sum_{k=0}^{\infty}\sum_{m=0}^{\infty}\frac{(-1)^{i+j+k}y^j\xi^{-2i-2k+j-3}}{2^{2i-1}\nu^{i+k+1}i!j!k!}$$

$$\times\,\theta^{-i-k}s^{(i+k)(\beta-1)-1}$$

$$\times\,\mathrm{H}^{1,3}_{3,5}\left[\frac{s^\beta}{\theta}\left|\frac{\left(1+\frac{j}{2}-k,0\right),(1-2i,0),(1-i-k,m)}{(1,0),\left(1+\frac{j}{2},0\right),(1-i-k,0),(1-i,0),[1+(i+k)(\beta+1),\beta]}\right.\right]. \tag{7.190}$$

If one sets $\theta \to 0$ and $\beta \neq 0$, Eqs. (7.183) and (7.188) can be simplified as:

$$u = -\frac{2}{h}\sum_{n=1}^{\infty}\sin\left(\frac{n\pi z}{h}\right)A(1-(-1)^n)\sum_{i=0}^{\infty}\sum_{j=0}^{\infty}\sum_{k=0}^{\infty}\sum_{l=0}^{\infty}\frac{(-1)^{i+j+k+l}y^j}{j!k!}$$

$$\times\,\nu^{i-k}\xi^{2i+j-2k-1}s^{i-k+(i-k)\alpha+1}\lambda^{-i+k}$$

$$\times\,H^{1,2}_{2,4}\left[\frac{s^\alpha}{\lambda}\left|\begin{matrix}\left(1+\frac{j}{2}-k,0\right),(1+i-k,1)\\(1,0),\left(1+\frac{j}{2},0\right),(1-i+k,0),(-1-i+k-(i-k)\alpha,\alpha)\end{matrix}\right.\right]$$

$$+\frac{2}{h}\sum_{n=1}^{\infty}\sin\left(\frac{n\pi z}{h}\right)A(1-(-1)^n)\sum_{i=0}^{\infty}\sum_{p=0}^{\infty}(-1)^i\nu^i\xi^{2i-1}\lambda^{-i}s^{i+i\alpha+1}$$

$$\times\,H^{1,2}_{2,3}\left[\frac{s^\alpha}{\lambda}\left|\begin{matrix}(1-i,1)\\(1,0),(1-i,0),(-1-i-i\alpha,\alpha)\end{matrix}\right.\right], \tag{7.191}$$

$$\tau_1 = \frac{A\rho}{h}\sum_{n=1}^{\infty}\sin\left(\frac{n\pi z}{h}\right)(1-(-1)^n)\sum_{i=0}^{\infty}\sum_{j=0}^{\infty}\sum_{k=0}^{\infty}\sum_{l=0}^{\infty}\frac{(-1)^{i+j+k}y^j\xi^{-2i-2k+j-3}}{2^{2i-1}\nu^{i+k+1}i!j!k!}$$

$$\times\,\lambda^{i+k}s^{-(i+k)(\alpha+1)-1}$$

$$\times\,H^{1,4}_{4,5}\left[\frac{s^\alpha}{\lambda}\left|\frac{\left(1+\frac{j}{2}-k,0\right),(1-2i,0),(1+i+k,1)}{(1,0),\left(1+\frac{j}{2},0\right),(1+i+k,0),(1-i,0),[1+(i+k)(\alpha+1),\alpha]}\right.\right], \tag{7.192}$$

which represent the velocity field and stress field corresponding to the flow of a generalized Maxwell fluid performing the same motion.

Letting now α, $\beta \to 1$ in Eqs. (7.183) and (7.188), we obtain the exact solutions corresponding to an ordinary Oldroyd-B fluid performing the same motion:

$$u = \frac{2}{h}\sum_{n=1}^{\infty}\sin\left(\frac{n\pi z}{h}\right)\tilde{u} = -\frac{2}{h}\sum_{n=1}^{\infty}\sin\left(\frac{n\pi z}{h}\right)A(1-(-1)^n)\sum_{i=0}^{\infty}\sum_{j=0}^{\infty}\sum_{k=0}^{\infty}\sum_{l=0}^{\infty}\frac{(-1)^{i+j+k+l}y^j}{j!k!l!}$$
$$\times\, \nu^{i-k}\xi^{2i+j-2k-1}\lambda^{-i+k-m}\theta^{i-k}s^{i-k+m+1}$$
$$\times\, H_{3,5}^{1,3}\left[\frac{s}{\theta}\left|\begin{matrix}(1+i-k,1),\left(1+\frac{j}{2}-k,0\right),(1-i+k-l,0)\\(1,0),(k-i-l-1,1),\left(1+\frac{j}{2},0\right),(1+i-k,0),(1-i+k,0)\end{matrix}\right.\right],$$
$$+\frac{2}{h}\sum_{n=1}^{\infty}\sin\left(\frac{n\pi z}{h}\right)A(1-(-1)^n)\sum_{i=0}^{\infty}\sum_{p=0}^{\infty}\frac{(-1)^{i+p}}{p!}\nu^i\xi^{2i-1}\lambda^{-i-p}\theta^{i-q}s^{i+p+1}$$
$$\times\, H_{2,4}^{1,2}\left[\frac{s}{\theta}\left|\begin{matrix}(1+i,1),(1-i-p,0)\\(1,0),(-i-1-p,1),(1+i,0),(1-i,0)\end{matrix}\right.\right] \tag{7.193}$$

$$\tau_1 = \frac{A\rho}{h}\sum_{n=1}^{\infty}\sin\left(\frac{n\pi z}{h}\right)(1-(-1)^n)\sum_{i=0}^{\infty}\sum_{j=0}^{\infty}\sum_{k=0}^{\infty}\sum_{l=0}^{\infty}\sum_{m=0}^{\infty}\frac{(-1)^{i+j+k+l}y^j\xi^{-2i-2k+j-3}}{2^{2i-1}\nu^{i+k+1}i!j!k!l!}$$
$$\times\, \lambda^{i+k-l}\theta^{-i-k}s^{-i-k+l-1}$$
$$\times\, H_{4,6}^{1,4}\left[\frac{s}{\theta}\left|\begin{matrix}\left(1+\frac{i}{2}-k,0\right),(1-2i,0),(1+i+k-l,0),(1-i-k,m)\\(1,0),\left(1+\frac{i}{2},0\right),(1+i+k,0),(1-i-k,0),(1-i,0),[1+i+k-l,1]\end{matrix}\right.\right]. \tag{7.194}$$

7.6.6 Results and Discussion

We studied 3D flow of a generalized Oldroyd-B fluid due to a constant pressure gradient between two side walls perpendicular to the plate. By means of the finite Fourier sine transform and the Laplace transform, the exact analytical solutions are obtained in the form of series and special functions. In the limiting cases when $\alpha = \beta = 1$ or $\alpha \neq 0$ and $\lambda \to 0$ or $\beta \neq 0$ and $\theta \to 0$, our solutions reduce to those corresponding to an ordinary Oldroyd-B fluid or a generalized second grade fluid and a generalized Maxwell one, respectively. Hence, this flow model is more useful compared with the ordinary Oldroyd-B model. Some 3D surfaces are also presented to show the characteristics of solutions for variation of emerging parameters. Furthermore, comparisons are performed between generalized Oldroyd-B fluid, ordinary Oldroyd-B fluid, generalized second grade fluid, and generalized Maxwell fluid. For the sake of simplicity all the 3D surfaces are plotted by assuming $\rho = \mu = 1$.

Figs. 7.32 and 7.33 show the variations of material parameter λ and θ. The effect of increasing λ (or decreasing θ) results in the increase of the velocity surfaces for generalized Oldroyd-B fluids.

In Figs. 7.34 and 7.35, the variations of noninteger fractional parameters α and β are shown. The effect of increasing α is to delay the increase of velocity

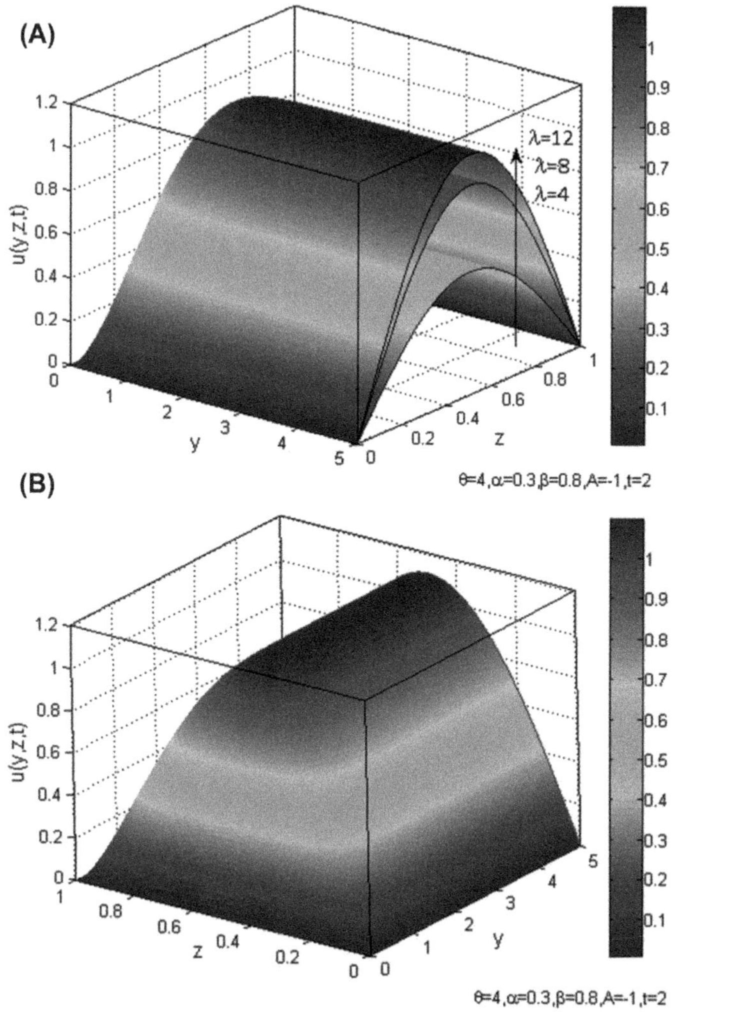

FIGURE 7.32 The 3D surface of velocity $u(y,z,t)$ for different values of λ keeping other parameters fixed (in two views). (A) $y-z$ and (B) $z-y$.

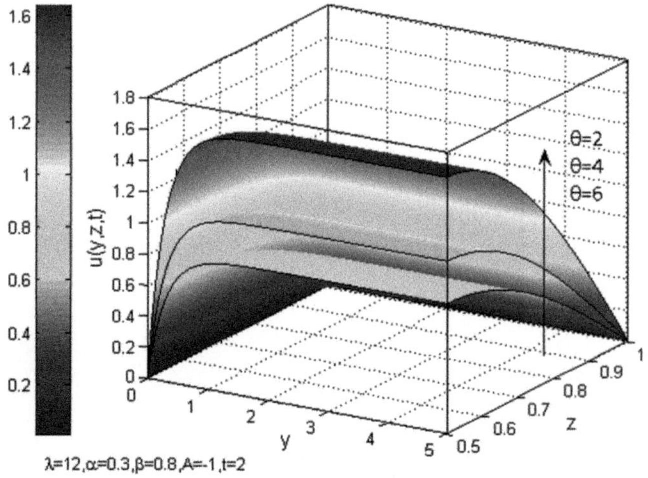

FIGURE 7.33 The 3D surface of velocity $u(y,z,t)$ for different values of θ keeping other parameters fixed.

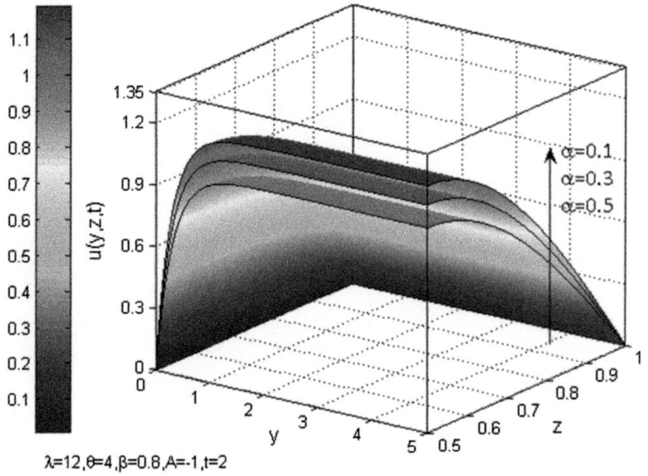

FIGURE 7.34 The 3D surface of velocity $u(y,z,t)$ for different values of α keeping other parameters fixed.

surfaces for generalized Oldroyd-B fluids, however, an increase in β shows an opposite trend for the fluids. Fig. 7.36 shows the variations of the velocity surfaces for different values of time. It is obvious to note that the flow velocity increases with increasing time.

In Fig. 7.37, the effect of the constant pressure gradient is shown for the generalized Oldroyd-B fluids. It is shown that, with other parameters keeping

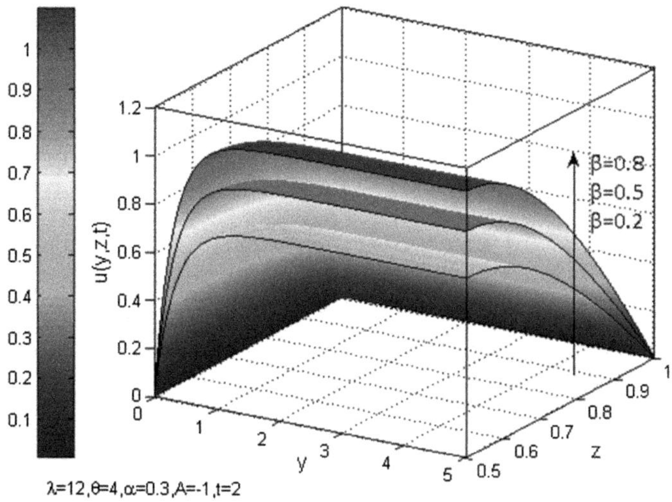

FIGURE 7.35 The 3D surface of velocity $u(y,z,t)$ for different values of β keeping other parameters fixed.

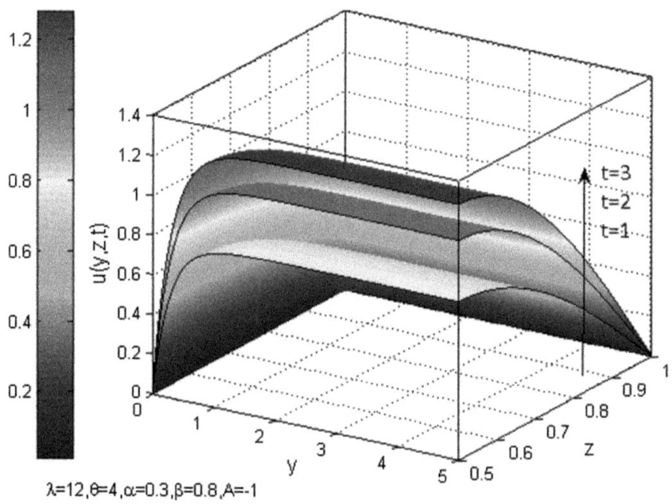

FIGURE 7.36 The 3D surface of velocity profiles $u(y,z,t)$ for different values of t keeping other parameters fixed.

fixed, the velocity of fluid is increased by increasing pressure gradient. Furthermore, the velocity becomes steady with the increasing y, which denotes that the boundary condition Eq. (7.171) is satisfied.

Fig. 7.38 represents for comparison the variations of the velocity fields corresponding to an ordinary Oldroyd-B, generalized Oldroyd-B, generalized

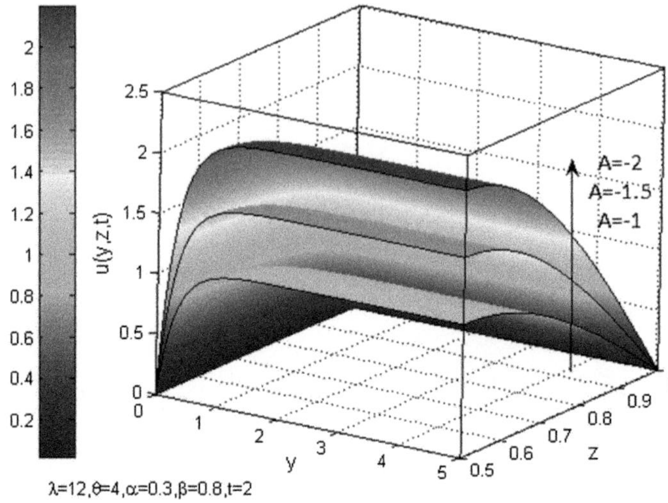

FIGURE 7.37 The 3D surface of velocity $u(y,z,t)$ for different values of A keeping other parameters fixed.

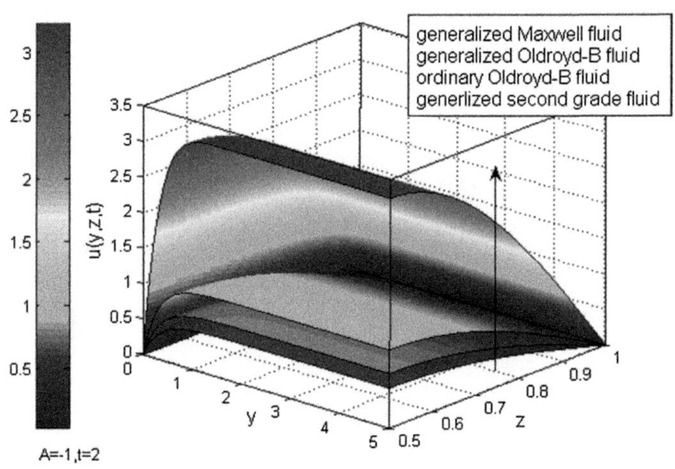

FIGURE 7.38 The 3D surface of velocity $u(y,z,t)$ for different kinds of fluids.

Maxwell, and generalized second grade fluid in the two side walls perpendicular to the palate. It should be noted that the velocity of generalized Maxwell fluid is the greatest in magnitude than those of any others fluids, which is then followed by the velocity in magnitude of generalized Oldroyd-B fluid and ordinary Oldroyd-B fluid, and generalized second grade fluid has the smallest velocity in magnitude.

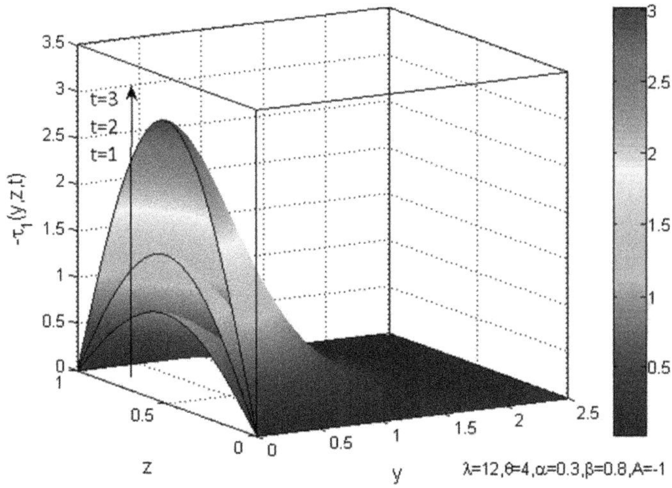

FIGURE 7.39 The 3D surface of shear stress $\tau_1(y,z,t)$ for different values of t keeping other parameters.

Moreover, some strong effects on the velocity field, resulting from the relaxation and retardation times and the orders of fractional parameters, can be easily detected through these figures.

Fig. 7.39 shows the variations of shear stress for different values of t for the generalized Oldroyd-B fluids. As it was expected the strongest shear stress occurs near the plate, however, with the increase of the distance from the plate, the shear stress of the generalized Oldroyd-B fluid decays rapidly, which also denotes that the boundary condition Eq. (7.171) is satisfied according to Eq. (7.167).

7.7 SUMMARY

Fractional calculus approaches are introduced in characterizing the constitutive relationship of viscoelastic non-Newtonian fluid. Fractional Maxwell fluid flow due to accelerating plate with the helical flows of fractional Oldroyd-B fluid in porous medium, the Oldroyd-B fluid between two infinite circular cylinders, the MHD flow and heat transfer of a generalized Burgers' fluid with radiative heat flux, the velocity slip effects on MHD flow of fractional Oldroyd-B fluid. The 3D flow of a generalized Oldroyd-B fluid, etc. are studied. Exact solutions are obtained by using the integral transformation techniques. The effects of pertinent parameters on the velocity and temperature fields and the shear stress distributions are presented graphically and analyzed in detail.

REFERENCES

Abel, M.S., Mahesha, N., Tawade, J., 2009. Heat transfer in a liquid film over an unsteady stretching surface with viscous dissipation in presence of external magnetic field. Applied Mathematical Modelling 33, 3430–3441.

Abou Zeid, M.Y., 2009. Numerical treatment of heat and mass transfer of MHD flow of Carreau fluid with diffusion and chemical reaction through a non Darcy porous medium. Open Mathematics Journal 2, 22–35.

Ahercliff, J.A., 1965. A Textbook of Magnetohydrodynamics. Oxford Pergamon Press.

Ali, N., Hussain, Q., Hayat, T., Asghar, S., 2008. Slip effects on the peristaltic transport of MHD fluid with variable viscosity. Physics Letters A 372, 1477–1489.

Bandelli, R., 1995. Unsteady unidirectional flows of second grade fluids in domains with heated boundaries. International Journal of Non-Linear Mechanics 30, 263–269.

Bellout, H., Bloom, F., 2014. Incompressible Bipolar and Non-Newtonian Viscous Fluid Flow. Springer, Cham–Heidelberg–New York–Dordrecht–London.

Brewster, M.Q., 1992. Thermal Radiative Transfer Properties. Wiley, Canada.

Cortell, R., 2008. Effects of viscous dissipation and radiation on the thermal boundary layer over a nonlinearly stretching sheet. Physics Letters A 372, 631–636.

Derek, C., Tretheway, D.C., Meinhart, C.D., 2002. Apparent fluid slip at hydrophibic microchannel walls. Physics of Fluids 14, 9–12.

Deville, M.O., Gatski, T.B., 2012. Mathematical Modeling for Complex Fluids and Flows. Springer, Heidelberg–Dordrecht–London–New York.

Ebaid, A., 2008. Effects of magnetic field and wall slip conditions on the peristaltic transport of a Newtonian fluid in an asymmetric channel. Physics Letters A 372, 4493–4499.

EI-Aziz, M.A., 2009. Radiation effect on the flow and heat transfer over an unsteady stretching sheet. International Communications in Heat and Mass Transfer 36, 521–524.

El-Shehawy, E.F., El-Dabe, N.T., El-Desoki, I.M., 2006. Slip effects on the peristaltic flow of a non-Newtonian Maxwellian fluid. Acta Mechanica 186, 141–159.

Fetecau, C., Fetecau, C., 2002. The Rayleigh-Stokes problem for heated second grade fluids. International Journal of Non-Linear Mechanics 37, 1011–1015.

Fetecau, C., Fetecau, C., Kamran, M., Vieru, D., 2009. Exact solution for the flow of a generalized Oldroyd-B fluid induced by a constantly accelerating plate between side walls perpendicular to the plate. Journal of Non-Newtonian Fluid Mechanics 156, 189–201.

Fetecau, C., Mahmood, A., Jamil, M., 2010. Exact solutions for the flow of a viscoelastic fluid induced by a circular cylinder subject to a time dependent shear stress. Communications in Nonlinear Science and Numerical Simulation 15, 3931–3938.

Fetecau, C., 2001. Unsteady Couette flows of second grade fluids in heated cylindrical domains. Acta Mechanica 150, 127–134.

Havlin, S., Ben-Avraham, D., 2002. Diffusion in disordered media. Advances in Physics 51, 187–292.

Hayat, T., Sajid, M., 2007. Homotopy analysis of MHD boundary layer flow of an upper-convected Maxwell fluid. International Journal of Engineering Science 45, 393–401.

Hayat, T., Khan, M., Ayub, M., 2004. Exact solutions of flow problems of an Oldroyd-B fluid. Applied Mathematics and Computation 151, 105–119.

Hayat, T., Khan, M., Ayub, M., 2007. The effect of the slip condition on flows of an Oldroyd 6-constant fluid. Journal of Computational and Applied Mathematics 202, 402–413.

Hayat, T., Qureshi, M.U., Ali, N., 2008. The influence of slip on the peristaltic motion of a third order fluid in an asymmetric channel. Physics Letters A 372, 2653–2664.

Hyder Ali, S., 2010. Unsteady flows of a viscoelastic fluid with the fractional Burgers' model. Nonlinear Analysis: Real World Applications 11, 1714–1721.

Khaleda, A.R.A., Vafaib, K., 2004. The effect of the slip condition on Stokes and Couette flows due to an oscillating wall: exact solutions. International Journal of Non-Linear Mechanics 39, 795–809.

Khan, M., Hayat, T., Asghar, S., 2006. Exact solution for MHD flow of a generalized Oldroyd-B fluid with modified Darcy's law. International Journal of Engineering Science 44, 333–339.

Khan, M., Saleem, M., Fetecau, C., Hayat, T., 2007. Transient oscillatory and constantly accelerated non-Newtonian flow in a porous medium. International Journal of Non-Linear Mechanics 42, 1224–1239.

Khan, M., Hyder Ali, S., Qi, H.T., 2009a. Exact solutions for some oscillating flows of a second grade fluid with a fractional derivative model. Mathematical and Computer Modelling 49, 1519–1530.

Khan, M., Hyder Ali, S., Qi, H.T., 2009b. Some accelerated flows for a generalized Oldroyd-B fluid. Nonlinear Analysis: Real World Applications 10, 980–991.

Khan, M., Hyder Ali, S., Qi, H.T., 2009c. On accelerated flows of a viscoelastic fluid with the fractional Burgers' model. Nonlinear Analysis: Real World Applications 10, 2286–2296.

Khan, M., Hyder Ali, S., Qi, H.T., 2009d. Exact solutions of starting flows for a fractional Burgers' fluid between coaxial cylinders. Nonlinear Analysis: Real World Applications 10, 1775–1783.

Khan, M., Anjum, A., Fetecau, C., Qi, H.T., 2010a. Exact solutions for some oscillating motions of a fractional Burgers' fluid. Mathematical and Computer Modelling 51, 682–692.

Khan, M., Anjum, A., Qi, H.T., Fetecau, C., 2010b. On exact solutions for some oscillating motions of a generalized Oldroyd-B fluid. Zeitschrift Für Angewandte Mathematik Und Physik 61, 133–145.

Khan, M., Arshad, M., Aujum, A., 2012. On exact solutions of Stokes second problem for MHD Oldroyd-B fluid. Nuclear Engineering and Design 243, 20–32.

Li, C.R., Zheng, L.C., Zhang, X.X., Chen, G., 2016. Flow and heat transfer of a generalized Maxwell fluid with modified fractional Fourier's law and Darcy's law. Computers and Fluids 125, 25–38.

Li, C.R., 2014. Study on the Flow and Heat Transfer of Viscoelastic Fluid Based on the Theory of Fractional Calculus (Ph.D. thesis). University of Science and Technology Beijing, China.

Liu, Y.Q., 2011. Abnormal Flow, Heat and Mass Transfer Behavior Research Based on Fractal Medium Dynamics (Ph.D. thesis). University of Science and Technology Beijing, China.

Lorenzo, C.F., Hartley, T.T., 2008. Generalized Functions for the Fractional Calculus (NASA/TP-1999–209424).

Mahmood, A., Parveen, S., Ara, A., Khan, N.A., 2009. Exact analytic solutions for the unsteady flow of a non-Newtonian fluid between two cylinders with fractional derivative model. Communications in Nonlinear Science and Numerical Simulation 14, 3309–3319.

Mahmoud, M.A.A., 2007. Thermal radiation effects on MHD flow of a micropolar fluid over a stretching surface with variable thermal conductivity. Journal of Physics A 375, 401–410.

Mahmoud, E.H., 2008. Slip effects on flow and heat transfer of a non-Newtonian fluid on a stretching surface with thermal radiation. International Journal of Chemical Reactor Engineering 6, 92–113.

Miller, K.S., Ross, B., 1993. An Introduction to Fractional Calculus and Fractional Differential Equations. John Wiley, New York.

Oldham, K.B., Spanier, J., 1974. The Fractional Calculus. Academic Press, New York.

Podlubny, I., 1999. Fractional Differential Equations. Academic Press, San Diego.

Qi, H.T., Xu, M.Y., 2007. Stokes' first problem for a viscoelastic fluid with the generalized Oldroyd-B model. Acta Mechanica Sinica 23, 463–469.

Rahman, M.M., Sultana, T., 2008. Radiative heat transfer flow of micropolar fluid with variable heat flux in a porous medium. Nonlinear Analysis-Modelling and Control 13, 71–87.

Rao, I.J., Rajagopai, K.R., 1999. The effect of the slip boundary condition on the flow of fluids in a channel. Acta Mechanica 135, 113–126.

Raptis, A., Perdikis, C., Takhar, H.S., 2004. Effect of thermal radiation on MHD flow. Applied Mathematics and Computation 153, 645–649.

Samko, S.G., Kilbas, A.A., Marichev, O.I., 1993. Fractional Integrals and Derivatives: Theory and Applications. Gordon and Breach, USA.

Shah, S.H.A.M., 2010. Unsteady flows of a viscoelastic fluid with the fractional Burgers' model. Nonlinear Analysis: Real World Applications 11, 1714–1721.

Soltani, F., Yilmazer, U., 1998. Slip velocity and slip layer thickness in flow of concentrated suspensions. Journal of Applied Polymer Science 70, 515–522.

Srinivas, S., Muthuraj, R., 2010. Effect of thermal radiation and space porosity on MHD mixed convection flow in a vertical channel using homotopy analysis method. Communications in Nonlinear Science and Numerical Simulation 15, 2098–2108.

Tan, W.C., Masuoka, T., 2005a. Stokes' first problem for a second grade fluid in a porous half-space with heated boundary. International Journal of Non-Linear Mechanics 40, 515–522.

Tan, W., Masuoka, T., 2005b. Stokes' first problem for an Oldroyd-B fluid in a porous half-space. Physics of Fluids 17, 256–277.

Tan, W.C., Masuoka, T., 2007. Stability analysis of a Maxwell fluid in a porous medium heated from below. Physics Letters A 360, 454–460.

Tan, W.C., 2006. Velocity overshoot of start-up flow for a Maxwell fluid in a porous half-space. Chinese Physics 15, 2644–2650.

Vieru, D., Fetecau, C., Fetecau, C., 2008. Flow of a viscoelastic fluid with the fractional Maxwell model between two side walls perpendicular to a plate. Applied Mathematics and Computation 200, 459–464.

Weigand, B., 2015. Analytical Methods for Heat Transfer and Fluid Flow Problems. Springer, Heidelberg–Dordrecht–London–New York.

Xue, C.F., Nie, J.X., 2008. Exact solutions of Stokes' first problem for heated generalized Burgers' fluid in a porous half-space. Nonlinear Analysis: Real World Applications 9, 1628–1637.

Xue, C.F., Nie, J.X., 2009. Exact solutions of the Rayleigh-Stokes problem for a heated generalized second grade fluid in a porous half-space. Applied Mathematical Modelling 33, 524–531.

Xue, C.F., Nie, J.X., Tan, W.C., 2008. An exact solution of start-up flow for the fractional generalized Burgers' fluid in a porous half-space. Nonlinear Analysis: Real World Applications 69, 2086–2094.

Zhang, T.T., Jia, L., Wang, Z.C., 2008a. Validation of Navier–Stokes equations for slip flow analysis within transition region. International Journal of Heat and Mass Transfer 51, 6323–6327.

Zhang, T.T., Jia, L., Wang, Z.C., Li, X., 2008b. The application of homotopy analysis method for 2-dimensional steady slip flow in microchannels. Physics Letters A 372, 3223–3227.

Zheng, L.C., Zhang, X.X., 2011. Modern Analysis Methods in Nonlinear Partial Differential Equations. Science Press, China.

Zheng, L.C., Li, C.R., Zhang, X.X., Gao, Y.T., 2011a. Exact solutions for the unsteady rotating flows of a generalized Maxwell fluid with oscillating pressure gradient between coaxial cylinders. Computers and Mathematics with Applications 62, 1105–1115.

Zheng, L.C., Liu, Y.Q., Zhang, X.X., 2011b. Exact solutions for MHD flow of generalized Oldroyd-B fluid due to an infinite accelerating plate. Mathematical and Computer Modelling 54, 780–788.

Zheng, L.C., Wang, K.N., Gao, Y.T., 2011c. Unsteady flow and heat transfer of a generalized Maxwell fluid due to a hyperbolic sine accelerating plate. Computers and Mathematics with Applications 61, 2209–2212.

Zheng, L.C., Liu, Y.Q., Zhang, X.X., 2012. Slip effects on MHD flow of a generalized Oldroyd-B fluid with fractional derivative. Nonlinear Analysis: Real World Applications 13, 513–523.

Chapter 8

Numerical Methods

ABSTRACT

In classical investigations on non-Newtonian fluid problems, the non-Newtonian mechanism is only taken into account in momentum equation while energy equation taken same form as in Newtonian fluid. In this chapter, we present some numerical studies for transport characteristics and mechanism of non-Newtonian fluids. Several modified Fourier's heat conduction laws and modified Darcy's diffusion laws are proposed for power law non-Newtonian fluids and fractional Maxwell viscoelastic fluid subject to various physical regimes and the fractional convection diffusion in a comb-like structure with Cattaneo–Christov flux. The governing equations are formulated and numerical solutions are obtained, the influences of pertinent parameters on the velocity, temperature, and concentration fields transport characteristics are analyzed and discussed in detail.

8.1 REVIEW OF NUMERICAL METHODS

Numerical methods have become important means for solving nonlinear differential equations of fluid problems. Many complex problems that could be solved in the past by analysis methods now can be resolved quickly by numerical simulations. In this chapter, we present some works for solving numerically the heat transfer and fluid flow problems arising in power law non-Newtonian fluids or fractional viscoelastic fluids subject to various nonclassical conditions. In the following, we first review some numerical methods used in this book (Ames, 1977; Ciarlet, 1978; Fletcher, 1988; Samarskii, 2001; Smith, 1985).

8.1.1 Numerical Methods for Linear System of Equations

Linear systems of equations are associated with many problems in engineering and science, as well as with applications of mathematics to the social sciences

Mathematics in Science and Engineering. http://dx.doi.org/10.1016/B978-0-12-811753-8.00008-6

and the quantitative study of business and economic problems. Consider to solve the linear system $AX = B$:

$$A = \begin{bmatrix} a_{11} & a_{12} & \dots & a_{1n} \\ a_{21} & a_{22} & \dots & a_{2n} \\ a_{31} & a_{32} & \dots & a_{3n} \\ \dots & \dots & \dots & \dots \\ a_{n1} & a_{n2} & \dots & a_{nn} \end{bmatrix}, \quad X = \begin{bmatrix} x_1 \\ x_2 \\ x_3 \\ \vdots \\ x_n \end{bmatrix}, \quad B = \begin{bmatrix} b_1 \\ b_2 \\ b_3 \\ \vdots \\ b_n \end{bmatrix}$$

where A is an $n \times n$ matrix, X and B are both $n \times 1$ column vector, respectively. The determinant of A denoted by det$A$ or $|A|$, which provides existence and uniqueness results for linear systems when $|A| \neq 0$.

There are many numerical methods for solving linear systems of equations, such as Gaussian elimination, pivoting strategies, matrix inversion, matrix factorization, iterative techniques, etc. As an example, we present matrix factorization used in this book to illustrate applications, and one can find other methods in any textbook of numerical analysis.

Gaussian elimination is the principal tool in the direct solution of linear systems of equations. From study on the Gaussian elimination element method for $Ax = b$, we know that the essence of the eliminating process is to perform $\frac{n}{2}(n-1)$ times sequential of the elementary row transformation on coefficient matrix A to transform the matrix into an upper triangular matrix. If Gaussian elimination can be performed on the linear system $AX = B$ without row interchanges, then the matrix A can be factored into the product of a lower-triangular matrix L and an upper-triangular matrix U. The factorization is particularly useful when it has the form $A = LU$, where L is lower triangular and U is upper triangular, defined as follows:

$$L = \begin{bmatrix} 1 & & & & \\ l_{21} & 1 & & & \\ l_{31} & l_{32} & 1 & & \\ \dots & \dots & \dots & & \\ l_{n1} & l_{n2} & \dots & \dots \; l_{nn-1} & 1 \end{bmatrix} \text{ and } U = \begin{bmatrix} u_{11} & u_{12} & u_{13} & \dots & u_{1n} \\ & u_{22} & u_{23} & \dots & u_{2n} \\ & & u_{33} & \dots & u_{3n} \\ & & & \dots & \dots \\ & & & & u_{nn} \end{bmatrix},$$

then

$$a_{ij} = \sum_{k=1}^{n} l_{ik}u_{kj} = \begin{cases} \sum\limits_{k=1}^{i-1} l_{ik}u_{kj} + u_{ij} & j \geq i \\ \sum\limits_{k=1}^{j-1} l_{ik}u_{kj} + l_{ij}u_{jj} & j < i \end{cases} \qquad i,j = 1,2,\dots,n \tag{8.1}$$

For $i = 1$, $a_{1j} = u_{1j}$, $j = 1, 2, \dots, n$; and for $j = 1$, $a_{i1} = l_{i1}u_{11}$, $i = 2, 3, \dots, n$

We obtain

$$u_{1j} = a_{1j}, \quad j = 1, 2, \ldots, n, \tag{8.2}$$

$$l_{i1} = a_{i1}/u_{11}, \quad i = 2, 3, \ldots, n, \tag{8.3}$$

$$u_{ij} = a_{ij} - \sum_{k=1}^{i-1} l_{ik}u_{kj}, \quad i \le j, \tag{8.4}$$

$$l_{ij} = \left(a_{ij} - \sum_{k=1}^{j-1} l_{ik}u_{kj}\right) \bigg/ u_{jj}, \quad i > j, \tag{8.5}$$

The factorization of the matrix can be divided into two kinds: the present lower triangular matrix is the unit of the triangular matrix, known as the Doolittle decomposition; and when the unit is on the upper triangular matrix it is called Crout decomposition.

8.1.2 Numerical Methods for Ordinary/Partial Differential Equations

Numerical methods for differential equations are used to find numerical approximations to the solutions of ordinary/partial differential equation (ODE/PDE), they are also known as "numerical integration." There are many numerical methods for solving ordinary/partial differential equations. Here, we only present three methods used in this book to illustrate the applications.

8.1.2.1 Runge—Kutta Method

Runge—Kutta method is an effective and widely used method for solving the initial-value problems of differential equations. Runge—Kutta method can be used to construct high order accurate numerical method by functions' self without needing the high order derivatives of functions.

Consider first-order initial-value problem:

$$\begin{cases} y' = f(x,y), & a \le x \le b \\ y(a) = y_0 \end{cases} \tag{8.6}$$

To derive the Runge—Kutta method, we divide the interval $[a,b]$ into N subintervals as $[x_n,x_{n+1}]$ ($n = 0, 1,\ldots N-1$), integrating $y' = f(x,y)$ over $[x_n,x_{n+1}]$ and utilizing the mean value theorem for integrals, obtain

$$y(x_{n+1}) - y(x_n) = \int_{x_n}^{x_{n+1}} f(x, y(x))dx = hf(\xi, y(\xi)) \tag{8.7}$$

where $h = x_{n+1} - x_n$, $\xi \in [x_n, x_{n+1}]$, i.e.,

$$y(x_{n+1}) = y(x_n) + hf(\xi, y(\xi)) \tag{8.8}$$

If we approximate $f(\xi, y(\xi))$ by the linear combination values $f(\xi_1, y(\xi_1))$, $f(\xi_2, y(\xi_2))$, ..., $f(\xi_m, y(\xi_m))$ of $f(x, y(x))$ on the interval $[x_n, x_{n+1}]$, then arrive at the general form of Runge– Kutta method:

$$y_{n+1} = y_n + h\sum_{i=1}^{m} c_i f(\xi_i, y(\xi_i)) \tag{8.9}$$

Choose different values of parameters m, c_i and ξ_i, we can obtain different form computation formula of Runge–Kutta; one can obtain higher order Runge–Kutta computation formula by choosing suitable values of parameters. The most widely used Runge–Kutta formula is

$$\begin{cases} y_{n+1} = y_n + \dfrac{1}{6}(K_1 + 2K_2 + 2K_3 + K_4) \\ K_1 = hf(x_n, y_n) \\ K_2 = hf\left(x_n + \dfrac{1}{2}h, y_n + \dfrac{1}{2}K_1\right) \\ K_3 = hf\left(x_n + \dfrac{1}{2}h, y_n + \dfrac{1}{2}K_2\right) \\ K_4 = hf(x_n + h, y_n + K_3) \end{cases} \tag{8.10}$$

It needs four values of function in each step iteration, also called four-order Runge–Kutta method.

In the following, we consider the higher order differential equations, for example,

$$\begin{cases} y'' = f(x, y, y') \\ y(x_0) = y_0, y'(x_0) = u_0 \end{cases}. \tag{8.11}$$

Let $y' = u$, Eq. (8.11) is transformed into the system of first-order differential equations:

$$\begin{cases} y' = u \\ u' = f(x, y, u) \\ y(x_0) = y_0, u(x_0) = u_0 \end{cases}. \tag{8.12}$$

Then, we can solve the initial values for the system of first-order differential equations, and the computation formula is written as following:

$$\begin{cases} y_{n+1} = y_n + \dfrac{1}{6}(m_1 + 2m_2 + 2m_3 + m_4) \\ u_{n+1} = u_n + \dfrac{1}{6}(k_1 + 2k_2 + 2k_3 + k_4) \end{cases}. \tag{8.13}$$

with

$$\begin{cases} m_1 = hu_n, \\ m_2 = h\left(u_n + \frac{1}{2}k_1\right), \\ m_3 = h\left(u_n + \frac{1}{2}k_2\right), \\ m_4 = h(u_n + k_3). \end{cases} \begin{cases} k_1 = hf(x_n, y_n, u_n), \\ k_2 = hf\left(x_n + \frac{1}{2}h, y_n + \frac{1}{2}m_1, u_n + \frac{1}{2}k_1\right), \\ k_3 = hf\left(x_n + \frac{1}{2}h, y_n + \frac{1}{2}m_2, u_n + \frac{1}{2}k_2\right), \\ k_4 = hf(x_n + h, y_n + m_3, u_n + k_3). \end{cases} \tag{8.14}$$

8.1.2.2 Shooting Method

Shooting method is a commonly used method for solving nonlinear boundary-value problem of differential equation, which is a method by transforming the boundary-value problem into an initial-value problem. The basic idea of shooting method is to transform the boundary-value problem of differential equations into initial-value problem with initial parameter to be established. This can be performed as follows: firstly, to give an initial value of parameter, one can solve the initial-value problem in terms of the Runge–Kutta method; secondly, to adjust the given initial value of parameter, such that the solution of initial value will satisfy the boundary conditions. For example, we consider the shooting technique for the nonlinear second-order boundary-value problem:

$$\begin{cases} y'' = f(x, y, y'), \quad a < x < b \\ y(a) = \alpha, y(b) = \beta \end{cases}. \tag{8.15}$$

Firstly, to give an initial value, by supposing $y'(a) = S$, we obtain an initial-value problem of differential equation with initial parameter of S:

$$\begin{cases} y'' = f(x, y, y'), \quad a \le x \le b \\ y(a) = \alpha, y'(a) = S \end{cases}. \tag{8.16}$$

Eq. (8.16) can be solved by using Runge–Kutta method, denoting the solution by $y(x; S)$, if the solution satisfies the boundary conditions, i.e., $y(b; S) = \beta$, then $y(x; S)$ is the solution of boundary-value problem (Eq. 8.15). If this is not true, then we approximate the solution to the boundary-value problem by using the solutions to a sequence of initial-value problems involving a parameter S, we do this by choosing the parameters $S = S_k$ and then solve the sequence of initial-value problems as

$$\begin{cases} y'' = f(x, y, y') \quad a \le x \le b \\ y(a) = \alpha, y'(a) = S_k, \quad k = 0, 1, 2, \cdots \end{cases}, \tag{8.17}$$

in a manner to ensure that

$$\lim_{k \to \infty} y(b; S_k) = \beta, \tag{8.18}$$

where $y(x;S_k)$ is the solution of initial-value problem Eq. (8.17).

This technique is called a "shooting" method, by analogy to the procedure of firing objects at a stationary target. We start with a parameter S_0 that determines the initial elevation at which the object is fired from the point (a, α) and along the curve described by the solution to the initial-value problem:

$$\begin{cases} y'' = f(x, y, y') & a \le x \le b \\ y(a) = \alpha, y'(a) = S_0 \end{cases}. \tag{8.19}$$

Denote the solution of Eq. (8.19) by $y(x;S_0)$. If $y(x; S_0)$ satisfies

$$y(b; S_0) = \beta \quad \text{or} \quad |y(b; S_0) - \beta| < \varepsilon \ . \tag{8.20}$$

Then we can take $y(x; S_0)$ as the solution of boundary-value problem Eq. (8.15). Otherwise, if $y(b; S_0)$ is not sufficiently close to β, then we correct our approximation by choosing elevations y_1,y_2 and so on, until $y(b; S_k)$ is sufficiently close to "hitting" β, such that

$$|y(b; S_k) - \beta| < \varepsilon.$$

where ε is a small tolerance constant, usually $\varepsilon = 10^{-6}$.

To determine the parameters S_k, we need to determine the parameter S, such that

$$\varphi(S) = y(b; S) - \beta = 0. \tag{8.21}$$

Eq. (8.21) can be solved by Newtonian iteration method, to generate the sequence $\{S_k\}$, here only one initial approximation S_0 is needed, i.e.,

$$S_{k+1} = S_k - \frac{\varphi(S_k)}{\varphi'(S_k)} = S_k - \frac{y(b; S_k) - \beta}{\partial y(b; S_k)/\partial S} \quad k = 0, 1, 2... \tag{8.22}$$

Until $\varphi(S_k) = 0$. We can stop the iteration processes by $|\varphi(S_k)| \le \varepsilon$ with a given tolerance of parameter ε.

For higher order differential equations, or equations solving boundary-value problems, multiple shooting method is needed. Multiple shooting method is much more complex than a single-parameter shooting method. The method divides the interval over which a solution is sought into several smaller intervals, solves an initial value problem in each of the smaller intervals, and imposes additional matching conditions to form a solution on the whole interval. The method constitutes a significant improvement in distribution of nonlinearity and numerical stability over single shooting methods. Multiple shooting method with "smoothness condition" not only can be used to solve boundary-value problem, but also can be used to calculate multidimensional solutions of nonlinear elliptic boundary-value problems.

8.1.2.3 Control Volume Method

The control volume method is also known as the finite volume method. The basic ideas and methods are: the calculation area is divided into a series of unoverlapping control volume, and each grid point is contained in a control volume, the differential equation to be solved for each control volume integral, then a set of discrete equations is obtained. The unknown is the dependent variable on the grid points. In order to obtain the integral of the control volume, it is necessary to assume the variation law of the value between the grid points. From the choice of the integral region, the control volume method is a subregion method in the weighted residual method. In short, the subregion method belongs to the basic method of finite volume class. The basic idea of the control volume method is easy to understand, and can get a direct physical interpretation. The physical meaning of the discrete equation is the conservation principle of the finite size of the control volume, which is the same as that of the differential equation. Control volume method is derived from the discrete equations, the integral conservation of the variables and the arbitrary set of control volume are met, so the entire computational area, naturally also be met.

There are some discrete methods, such as finite difference method, where only when the grid is extremely fine, the discrete equations can satisfy the integral conservation. The control volume method, even in the case of rough grid, also shows the exact integral conservation. In the case of the discrete method, the control volume method can be regarded as the intermediate between the finite element method and the finite difference method. The finite element method must assume the variation of the value between the grid points (both interpolation function), and the finite element method is used as the approximate solution. The finite difference method can only consider the numerical value on the grid point without considering the variation of the value between the grid points. The control volume method is only for the numerical value of the nodes, which is similar to the finite difference method. But the control volume method must assume the distribution of the points in the grid, which is similar to the finite element method. In the control volume method, the interpolation function is only used to calculate the integral of the control volume, and the interpolation function is not considered.

The control volume method is the most widely used numerical discretization method in computational fluid dynamics and computational heat transfer. It usually consists of five parts as: (1) grid generation; (2) the discretization of the convection term; (3) discretization of boundary conditions; (4) the coupling of pressure and velocity; and (5) the solution of the discrete equation.

8.1.3 Numerical Methods for Fractional Differential Equations

In recent years, fractional differential equations and its application have gotten extensive attention. The main reason is due to the rapid development of the theory of fractional calculus itself and is widely used in mathematics, physics, chemistry, biology, medicine, mechanics, control theory, signal and image processing, environmental science, financial, and other various disciplines.

Fractional order differential equations are generalized and noninteger order differential equations, which can be obtained in time and space with a power law memory kernel of the nonlocal relationships; they provide a powerful tool to describing the memory of different substances and the nature of the inheritance. All of these studies have a clear physical background, which open up a new field of scientific research, including a new theoretical analysis and numerical methods for fractional order dynamical systems.

The numerical calculation of the fractional differential equations has been successfully applied in many research fields. A list of the key references is given in Refs. (Chen et al., 2010; Liu et al., 2009, 2013a, 2015).

8.2 HEAT TRANSFER OF POWER LAW FLUID IN A TUBE WITH DIFFERENT FLUX MODELS

In this section, we present numerical research on forced convection heat transfer of power law non-Newtonian fluids in a circle duct with different thermal conductivity models. A hydrodynamic entrance length is neglected, which allows establishing a fully developed flow. Four thermal conductivity models are considered: constant thermal conductivity model, linear thermal conductivity varying with temperature, thermal conductivity varying as a function of velocity gradient, and thermal conductivity varying as a function of temperature gradient. The energy equation is solved by using an LU decomposition coupled with control volume technique–based finite difference method. Results are compared and the physical characteristics are also analyzed.

8.2.1 Background of the Problem

For half a century, considerable attention has been paid to predict the transport behavior of non-Newtonian fluid flows. Many constitutive models have been proposed. Among these models, the Ostwaald–de Waele fluid model, i.e., the so-called power law model (Acrivos et al., 1960; Schowalter, 1960), gained much acceptance (Abel et al., 2009; Arunachalam and Rajappa, 1978; Chaim, 1998; Crosan and Pop, 2001; Gorla et al., 1992; Hady, 1995; Hossain et al., 2000a,b; Hossain et al., 2001; Howell et al., 1997; Kays, 1966; Kumari et al., 1997; Rao et al., 1999; Salem, 2007; Wang, 1995a,b). For an incompressible power law non-Newtonian fluid in cylindrical coordinates, its shear stress is characterized as $\tau = K\left(\left|\frac{\partial u}{\partial r}\right|^{n-1}\frac{\partial u}{\partial r}\right)$ and the kinematic viscosity is $\nu = \gamma\left|\frac{\partial u}{\partial r}\right|^{n-1}$ (K and $\gamma = K/\rho$ are positive constants). The case $n = 1$ corresponds to a Newtonian fluid, $0 < n < 1$ is pseudoplastic non-Newtonian fluids while $n > 1$ describes dilatant fluids.

In classical research works for power law non-Newtonian fluids, the influence of power law kinematic viscosity is only considered in momentum transport of velocity field, which has been ignored in energy transfer of temperature field, i.e., the thermal conductivity is treated the same as in Newtonian fluid. Obviously, this is inconsistent with the fact that the changing viscosity should affect both the velocity field and the temperature as well. Some researchers have paid attention to this inconsistency.

Pop et al. (1991) and Pop (1993) had suggested that the thermal conductivity of power law non-Newtonian fluids is a power law dependence on the velocity gradient. Ece and Buyuk (2002) performed a boundary-layer analysis for the free convection flow over a vertical flat plate embedded in a porous medium saturated by a power law non-Newtonian fluid and gave the similarity solution to the classical boundary-layer equations. A modified Fourier's law of heat conduction for power law fluids is proposed. Recently, Zheng et al. (2006', 2008), based on the assumption that the thermal diffusion in temperature field is similar to the viscosity diffusion in velocity field for power law viscosity, proposed a modified Fourier's heat conduction law for power law non-Newtonian fluid in which a constitutive model for thermal conductivity is written as

$$k(T) = \begin{cases} k_0\left|\dfrac{\partial T}{\partial y}\right|^{n-1}, & \dfrac{\partial T}{\partial y} \neq 0 \\ k_0, & \dfrac{\partial T}{\partial y} = 0 \end{cases},$$

where k_0 is a positive constant. Li et al. (2010a,b,c) presented an investigation on steady conduction of heat and diffusion of power law fluids in a circular duct. In this chapter, we focus our attention on studying the following four types of thermal conductivity constitutive models:

1. Thermal conductivity is a constant.
2. Thermal conductivity is a linear function of temperature.
3. Thermal conductivity is a power law function of velocity gradient.
4. Thermal conductivity is a power law function of temperature gradient (proposed by Zheng).

8.2.2 Formulation of the Problems and Numerical Algorithms

Cases (1) and (2) The thermal conductivity is assumed a constant or a linear function of temperature.

We consider first the Cases (1) and (2), where the thermal conductivity is assumed a constant or a linear function of temperature and the duct wall is maintained at a constant temperature lower than the uniform temperature of the fluid at the entrance, and temperature is symmetrical about the axis. The fluid axial conduction, viscous dissipation, thermal energy sources, and

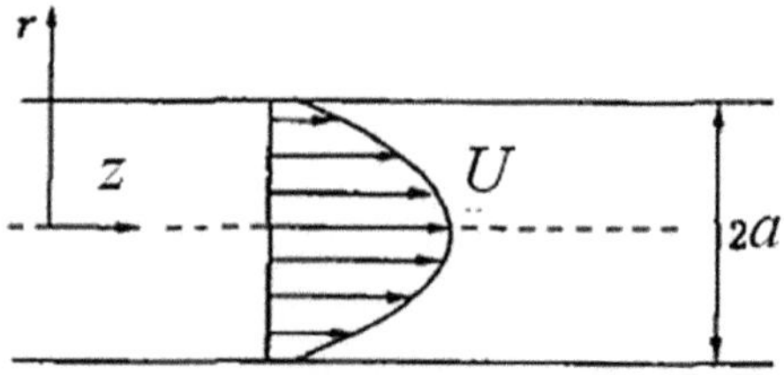

FIGURE 8.1 Physical description of the problem.

conduction in the direction of flow are negligible and laminar. Except for a few kinds of fluids, the hydrodynamic entrance region is much shorter than the thermal entrance one. The last assumption implies that a hydrodynamic entrance length is neglected, which allows establishing a fully developed flow. Take z and r to be coordinate axes parallel and perpendicular to the channel walls, and the axial component of the velocity field is given by (Barletta et al., 1997)

$$U(r) = \frac{3n+1}{n+1} U_m \left[1 - \left(\frac{r}{D_r} \right)^{(n+1)/n} \right] \tag{8.23}$$

U_m and D_r are the mean velocity and the radius of the duct and n is the power law index, the case $n = 1$ corresponds to a Newtonian fluid problem known as the Graetz problem (see Fig. 8.1).

The energy equation becomes:

$$\rho c_p U \frac{\partial T}{\partial z} = \frac{1}{r} \frac{\partial}{\partial r} \left(r k(T) \frac{\partial T}{\partial r} \right) \tag{8.24}$$

ρ and c_p are the density and fluid specific heat, respectively. $k(T)$ is the thermal conductivity and it is of the form $k(T) = k_0 \left[1 + \varepsilon \left(\frac{T - T_w}{T_{in} - T_w} \right) \right]$ with ε as a small parameter and k_0 is positive constant, and T_w and T_{in} are the wall and the inlet temperature, respectively. This research introduces the following dimensionless quantities:

Dimensionless temperature:

$$\Theta = \frac{T - T_w}{T_{in} - T_w}. \tag{8.25}$$

We assume $T_{in} > T_w$ to keep Θ positive.

Dimensionless radial coordinate:

$$R = \frac{r}{D_r} \tag{8.26}$$

Dimensionless velocity:

$$U^* = \frac{U}{4U_m}. \tag{8.27}$$

Dimensionless axial coordinate:

$$Z^* = \frac{k_0 z}{\rho c_p D_r^2 U_m}. \tag{8.28}$$

Applying the dimensionless quantities to the energy equation, the problem under consideration is transformed in dimensionless form as:

$$U^* \frac{\partial \Theta}{\partial Z^*} = \frac{1}{R} \frac{\partial}{\partial R} \left[R(1 + \varepsilon \Theta) \frac{\partial \Theta}{\partial R} \right]. \tag{8.29}$$

The dimensionless velocity profile becomes:

$$U^*(R) = \frac{3n+1}{n+1} \left[1 - R^{(n+1)/n} \right]. \tag{8.30}$$

The corresponding boundary conditions are:

$$\Theta|_{Z^*=0} = 1, \tag{8.31}$$

$$\Theta|_{R=1} = 0, \tag{8.32}$$

$$\left. \frac{\partial \Theta}{\partial R} \right|_{R=0} = 0. \tag{8.33}$$

In order to obtain the numerical solution, we transfer the problem Eqs. (8.29)–(8.33) to a system of algebraic equations. The calculation domain is divided into a few nonoverlapping control volumes each of which surrounds a grid point (shown in Fig. 8.2).

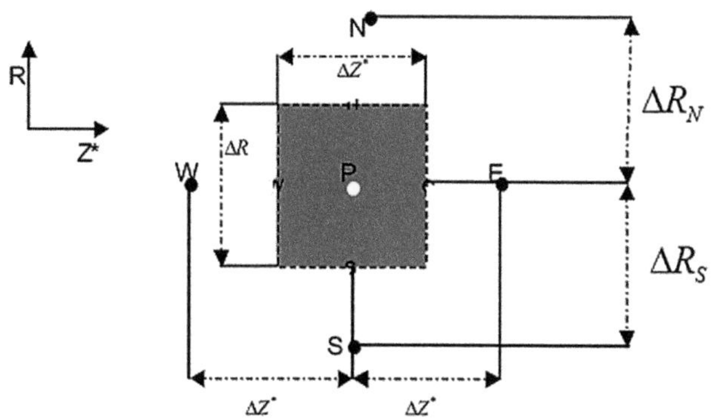

FIGURE 8.2 The sketch of control volume.

The energy Eq. (8.29) is integrated over each control volume. $\frac{\partial \Theta}{\partial R} < 0$, so we have as follows:

$$U^*(\Theta_P - \Theta_W)\frac{R_N^2 - R_S^2}{2} = \left[R_N(1+\varepsilon\Theta_P)\frac{\Theta_N - \Theta_P}{\Delta R_N} - R_S(1+\varepsilon\Theta_P)\frac{\Theta_P - \Theta_S}{\Delta R_S}\right]\Delta Z^* \tag{8.34}$$

Suppose that Θ_P is the P-th nonoverlapping control volume in R direction.

$$\varepsilon\Theta_P = \varepsilon[(\Theta_1 - \Theta_0) + (\Theta_2 - \Theta_1) + \cdots + (\Theta_P - \Theta_{P-1})] = \varepsilon\sum_1^P \Delta\Theta_Q \tag{8.35}$$

Assuming the calculation domain is divided into many control volumes that are small enough, we have

$$\lim_{\Delta\Theta_Q \to 0} \varepsilon\sum_1^P \Delta\Theta_Q = \varepsilon P \Delta\Theta = \varepsilon^* P \tag{8.36}$$

Eq. (8.34) can be written as:

$$U^*(\Theta_P - \Theta_W)\frac{R_N^2 - R_S^2}{2} = \left[R_N(1+\varepsilon^* P)\frac{\Theta_N - \Theta_P}{\Delta R_N} - R_S(1+\varepsilon^* P)\frac{\Theta_P - \Theta_S}{\Delta R_S}\right]\Delta Z^* \tag{8.37}$$

Rearranging the equation in the form as:

$$a_P\Theta_P = a_W\Theta_W + a_N\Theta_N + a_S\Theta_S \tag{8.38}$$

$$a_W = \frac{U^*\left(R_N^2 - R_S^2\right)}{2\Delta Z^*} \tag{8.39}$$

$$a_N = \frac{R_N(1+\varepsilon^* P)}{\Delta R_N} \tag{8.40}$$

$$a_S = \frac{R_S(1+\varepsilon^* P)}{\Delta R_S} \tag{8.41}$$

$$a_P = a_W + a_N + a_S \tag{8.42}$$

Eqs. (8.38)–(8.42) are a linear system of equations that can be solved by using LU decomposition method.

Case (3) The thermal conductivity is a power law function of velocity gradient.

With the same assumptions, we consider laminar power law flow with a neglecting entrance region in which the velocity profile develops. The energy equation is:

$$\rho c_p U\frac{\partial T}{\partial z} = \frac{1}{r}\frac{\partial}{\partial r}\left(rk(T)\frac{\partial T}{\partial r}\right) \tag{8.43}$$

The thermal conductivity is proposed as a form of $k(T) = k_0\left|\frac{\partial U}{\partial r}\right|^{n-1}$ with k_0 as positive constant.

Dimensionless temperature Θ, dimensionless radial coordinate R, and dimensionless velocity U^* are defined as Eqs. (8.25)–(8.27). The dimensionless axial coordinate is introduced as:

$$Z^* = \frac{k_0 U_m^{n-2} z}{\rho c_p D_r^{n+1}} \tag{8.44}$$

Applying the dimensionless quantities to the energy equation, the problem under consideration is given in dimensionless form as:

$$U^* \frac{\partial \Theta}{\partial Z^*} = \frac{1}{R}\frac{\partial}{\partial R}\left(R\left|\frac{\partial U^*}{\partial R}\right|^{n-1}\frac{\partial \Theta}{\partial R}\right) \tag{8.45}$$

The energy Eq. (8.45) is integrated over each control volume, and we have as follows:

$$\begin{aligned} U^*(\Theta_P - \Theta_W)\frac{R_N^2 - R_S^2}{2} = \Bigg[& R_N\left(\frac{\Theta_N - \Theta_P}{\Delta R_N}\right)\left|\frac{U_N^* - U_P^*}{\Delta R_N}\right|^{n-1} \\ & - R_S\left(\frac{\Theta_P - \Theta_S}{\Delta R_S}\right)\left|\frac{U_P^* - U_S^*}{\Delta R_S}\right|^{n-1}\Bigg]\Delta Z^* \end{aligned} \tag{8.46}$$

Rearranging the equation in the form as:

$$a_P\Theta_P = a_W\Theta_W + a_N\Theta_N + a_S\Theta_S \tag{8.47}$$

$$a_W = \frac{U^*\left(R_N^2 - R_S^2\right)}{2\Delta Z^*} \tag{8.48}$$

$$a_N = \frac{R_N}{\Delta R_N}\left|\frac{U_N^* - U_P^*}{\Delta R_N}\right|^{n-1} \tag{8.49}$$

$$a_S = \frac{R_S}{\Delta R_S}\left|\frac{U_P^* - U_S^*}{\Delta R_S}\right|^{n-1} \tag{8.50}$$

$$a_P = a_W + a_N + a_S \tag{8.51}$$

The linear system of Eqs. (8.47)–(8.51) can be solved by using LU decomposition method.

Case (4) The thermal conductivity is proposed to be a power law function of temperature gradient. The energy equation is:

$$\rho c_p U\frac{\partial T}{\partial z} = \frac{1}{r}\frac{\partial}{\partial r}\left(rk(T)\frac{\partial T}{\partial r}\right) \tag{8.52}$$

We assume that the heat transfer is affected by velocity fields with modified Fourier' law, the thermal conductivity is dependent on gradient of temperature and has the form of $k(T) = k_0\left|\frac{\partial T}{\partial r}\right|^{n-1}$ with k_0 as positive constant. The dimensionless temperature Θ, dimensionless radial coordinate R, and dimensionless velocity U^* are as Eqs. (8.25)–(8.27) while the dimensionless axial coordinate is:

$$Z^* = \frac{k_0 z}{\rho c_p D_r^2 U_m}\left(\frac{T_{in} - T_w}{D_r}\right)^{n-1} \tag{8.53}$$

Applying the dimensionless quantities to the energy equation, the problem under consideration is given in dimensionless form as:

$$U^* \frac{\partial \Theta}{\partial Z^*} = \frac{1}{R}\frac{\partial}{\partial R}\left(R\frac{\partial \Theta}{\partial R}\left|\frac{\partial \Theta}{\partial R}\right|^{n-1}\right) \tag{8.54}$$

The boundary conditions are stated as before.

The energy Eq. (8.54) is integrated over each control volume, so we have as follows:

$$U^*(\Theta_P - \Theta_W)\frac{R_N^2 - R_S^2}{2} = \left[R_S\left(\frac{\Theta_S - \Theta_P}{\Delta R_S}\right)^n - R_N\left(\frac{\Theta_P - \Theta_N}{\Delta R_N}\right)^n\right]\Delta Z^* \tag{8.55}$$

Suppose the radius of the pipe, which is one in dimensionless form, is divided into M-nodes. We have $\Delta R_S = \Delta R_N = 1/M$. As the fluid flows in the pipe, it gradually becomes cooler. In view of Eqs. (8.31) and (8.32), we get the conclusion that $0 < \frac{\Theta_S - \Theta_P}{\Delta R_S} < \frac{1/M}{1/M} = 1$ and $0 < \frac{\Theta_P - \Theta_N}{\Delta R_N} < \frac{1/M}{1/M} = 1$. Suppose that a linear function $y = \alpha x$ on [0,1] would minimize the error,

$$\int_0^1 \left[\left(\frac{\Theta_S - \Theta_P}{\Delta R_S}\right)^n - \alpha\frac{\Theta_S - \Theta_P}{\Delta R_S}\right]^2 d\left(\frac{\Theta_S - \Theta_P}{\Delta R_S}\right),$$

and

$$\int_0^1 \left[\left(\frac{\Theta_P - \Theta_N}{\Delta R_N}\right)^n - \alpha\frac{\Theta_P - \Theta_N}{\Delta R_N}\right]^2 d\left(\frac{\Theta_P - \Theta_N}{\Delta R_N}\right).$$

$\alpha = \frac{3}{n+2}$ is satisfied. Then, to put $\frac{\alpha\Theta_S - \alpha\Theta_P}{\Delta R_S}$ and $\frac{\alpha\Theta_P - \alpha\Theta_N}{\Delta R_N}$ in the place of $\left(\frac{\Theta_S - \Theta_P}{\Delta R_S}\right)^n$ and $\left(\frac{\Theta_P - \Theta_N}{\Delta R_N}\right)^n$, the problem now is:

$$U^*(\Theta_P - \Theta_W)\frac{R_N^2 - R_S^2}{2} = \left[R_S\frac{\alpha\Theta_S - \alpha\Theta_P}{\Delta R_S} - R_N\frac{\alpha\Theta_P - \alpha\Theta_N}{\Delta R_N}\right]\Delta Z^* \tag{8.56}$$

Rearranging the equation in the form as:

$$a_P\Theta_P = a_W\Theta_W + a_N\Theta_N + a_S\Theta_S \tag{8.57}$$

$$a_W = \frac{U^*\left(R_N^2 - R_S^2\right)}{2\Delta Z^*} \tag{8.58}$$

$$a_N = \frac{R_N \alpha}{\Delta R_N} \tag{8.59}$$

$$a_S = \frac{R_S \alpha}{\Delta R_S} \tag{8.60}$$

$$a_P = a_W + a_N + a_S \tag{8.61}$$

Eqs. (8.57)–(8.61) are solved by using LU decomposition method.

8.2.3 Results and Discussion

Commercial software Compaq Visual FORTRAN Professional Edition 6.6.0 has been used to solve the problem mentioned before. Table 8.1 briefly introduced parameters used in this research. It is worth mentioning that the Reynolds number is not obviously shown in this research as we use the dimensionless form of equations. However, to keep the fluid flow steady in the duct, the Reynolds number should be no more than 2300.

Figs. 8.3 and 8.4 depict the temperature profiles of different dimensionless axial coordinates with linear temperature-dependent thermal conductivity and effects of dimensionless axial coordinate Z^* and power law index n on the temperature profiles Θ with a linear temperature-dependent thermal conductivity model.

Fig. 8.3A displays the dimensionless temperature profiles of different dimensionless axial coordinates. The curves labeled as $n = 0.6$ correspond to shear-thinning non-Newtonian fluid flow. Note in Fig. 8.36A the increase in the dimensionless temperature as the dimensionless axial coordinate decreases. Fig. 8.3B shows the dimensionless temperature profiles versus dimensionless axial coordinate for $n = 1.3$ corresponding to shear-thickening fluid flow. Results indicate that the temperature increases toward the center of the pipe. The thermal wave of the inlet temperature has less penetration near the center with the increasing axial coordinate. It is really interesting to find that Fig. 8.3A seems to be identical to Fig. 8.3B. But we should notice that $\varepsilon^* = -0.001$ in Fig. 8.3A while $\varepsilon^* = 0.001$ in Fig. 8.36B. It is a known fact

TABLE 8.1 The Parameters Covered in This Research

Parameters	Part 2	Part 3	Part 4
Re	<2300	<2300	<2300
n	0.1~2.5	0.6~2.5	0.1~2.5
ε^*	−0.01~0.01	–	–
Z^*	0~0.1	0~0.1	0~0.1

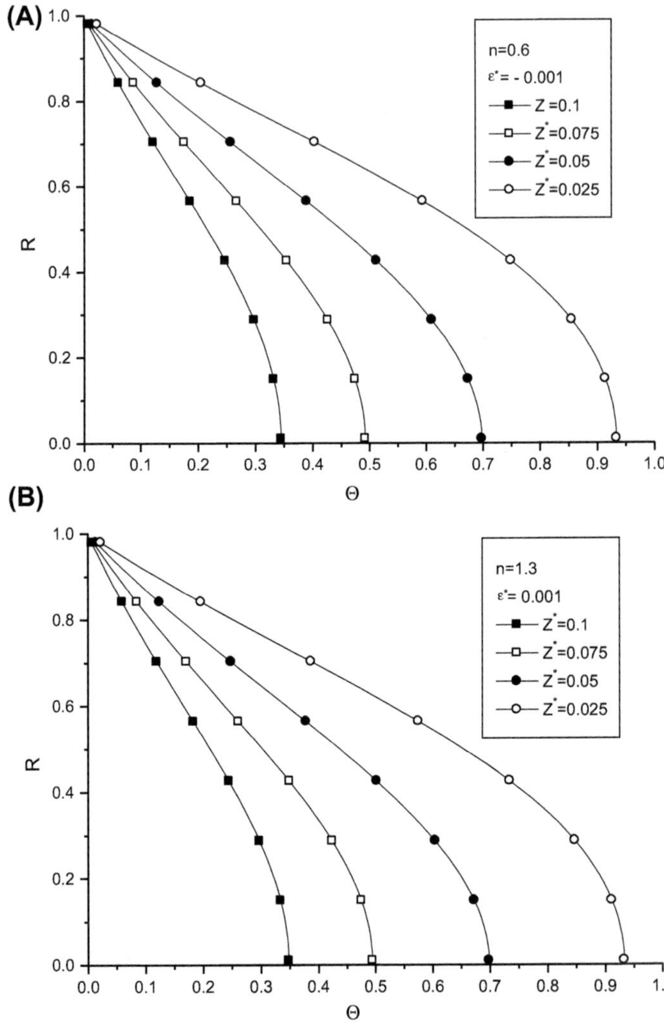

FIGURE 8.3 Temperature profiles of different dimensionless axial coordinates with linear temperature-dependent thermal conductivity. (A) $n = 0.6$ and (B) $n = 1.3$.

that different values of n imply tremendous change in temperature field distribution. The same is reiterated by Fig. 8.4.

Figs. 8.5 and 8.6 show the dimensionless temperature profiles with a velocity-dependent thermal conductivity model. It is found that the behaviors of solutions are remarkably affected by the power law index. Fig. 8.5 indicates the same phenomenon with Fig. 8.3. Fig. 8.6 shows the effects of different power law index on dimensionless temperature profiles; results indicate that, at the center of the pipe, the bigger the power law index, the higher the

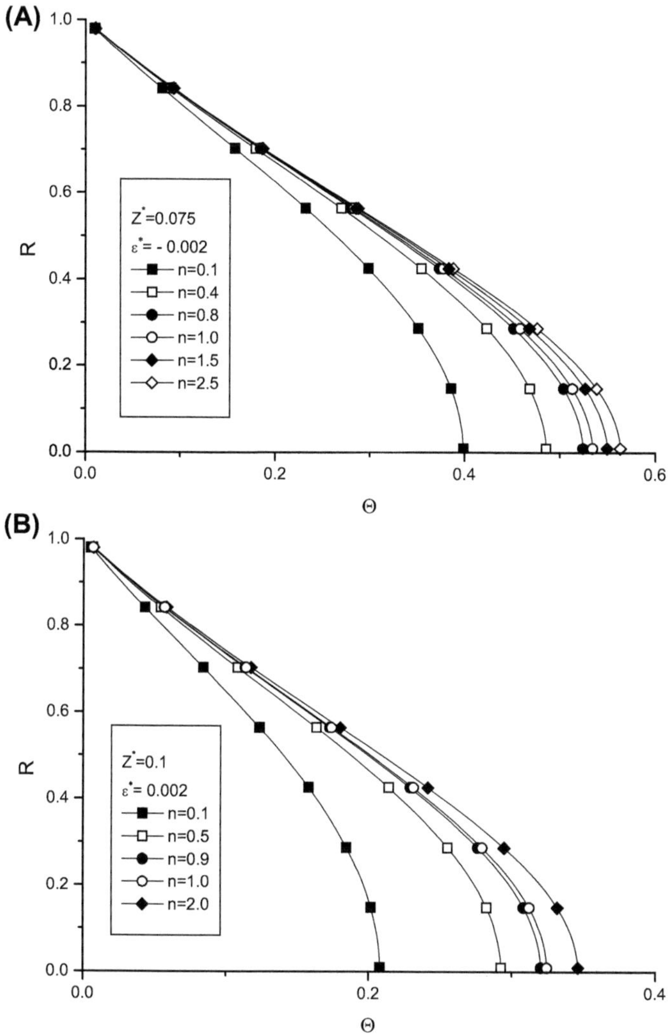

FIGURE 8.4 Temperature profiles of different power law index with linear temperature-dependent thermal conductivity. (A) $Z^* = 0.075$ and (B) $Z^* = 0.1$.

temperature is. '+' presents the Graetz solution in vivid contrast with $n = 1$, which is obtained by using the method mentioned in this research. The results agree well with those obtained by Graetz and the transfer characteristics for values of different parameters correspond to facts, which demonstrates the effectiveness of the model and the robustness of the algorithm.

Figs. 8.7 and 8.8 describe the dimensionless temperature profiles of different axial coordinates for power law index $n = 0.5$ and $n = 1.5$. It is seen that the behaviors of solutions with a power law temperature

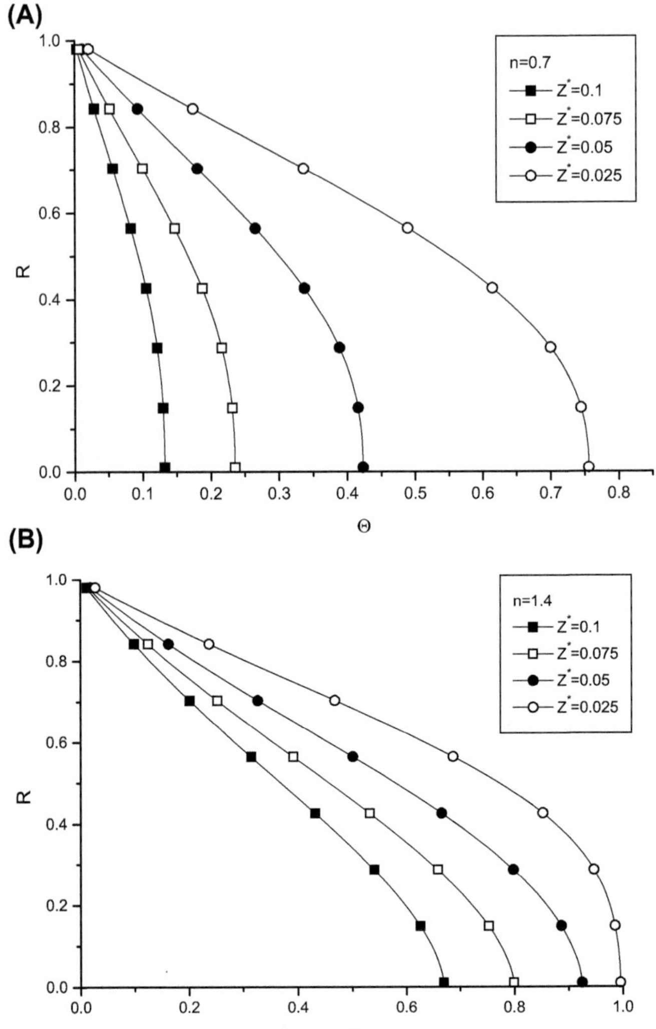

FIGURE 8.5 Temperature profiles of different dimensionless axial coordinate with velocity-dependent thermal conductivity. (A) $n = 0.7$ and (B) $n = 1.4$.

gradient−dependent Fourier's heat conductivity model are very similar with the ones with velocity temperature gradient−dependent Fourier's heat conductivity model.

Fig. 8.7 shows that the flows contribute to reduce heating effects with the increasing axial coordinate. Fig. 8.8 shows the effects of different power law index on dimensionless temperature profiles, and results indicate that, at the center of the pipe, the bigger the power law index is, the higher the

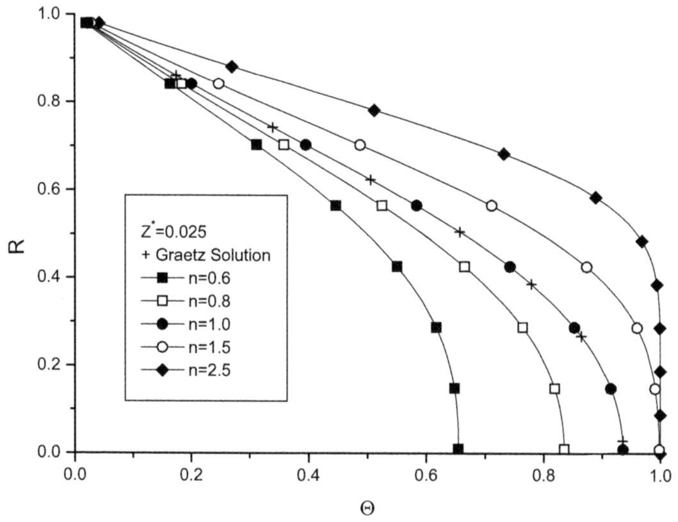

FIGURE 8.6 Temperature profiles of different power law index with velocity-dependent thermal conductivity.

temperature is, which is similar to Fig. 8.6. The temperature profile is flatter as the power law index increases. '+' presents the Graetz solution in vivid contrast with the fluid $n = 1$ which is obtained in the research. Those results also agree well with each other.

Fig. 8.9 shows the effects of linear temperature-dependent thermal conductivity model parameter ε^* on temperature profiles for both shear-thinning ($n = 0.5$) and shear-thickening ($n = 1.5$) fluids. It is observed from these pictures that the increasing values of ε^* result in decreasing the temperature. Fig. 8.10 presents a comparison of different solutions obtained by using the four different Fourier heat conductivity models. The curves with square marks are classical results with a constant thermal conductivity; the profiles with circle marks, with a thermal conductivity assumed to be a function of velocity gradient; the curves with diamond marks; with a power law temperature-dependent thermal conductivity. With the traditional model, the differences of temperature between varying power law index are small; with a temperature-dependent model, the differences are big; with a velocity-dependent model, the differences are biggest.

We now study the effects of pertinent parameters on Nusselt number. The local Nusselt number is defined by

$$Nu = -2\frac{\left.\frac{\partial\Theta}{\partial R}\right|_{R=1}}{\left.\Theta\right|_{R=1} - \Theta_b} \tag{8.62}$$

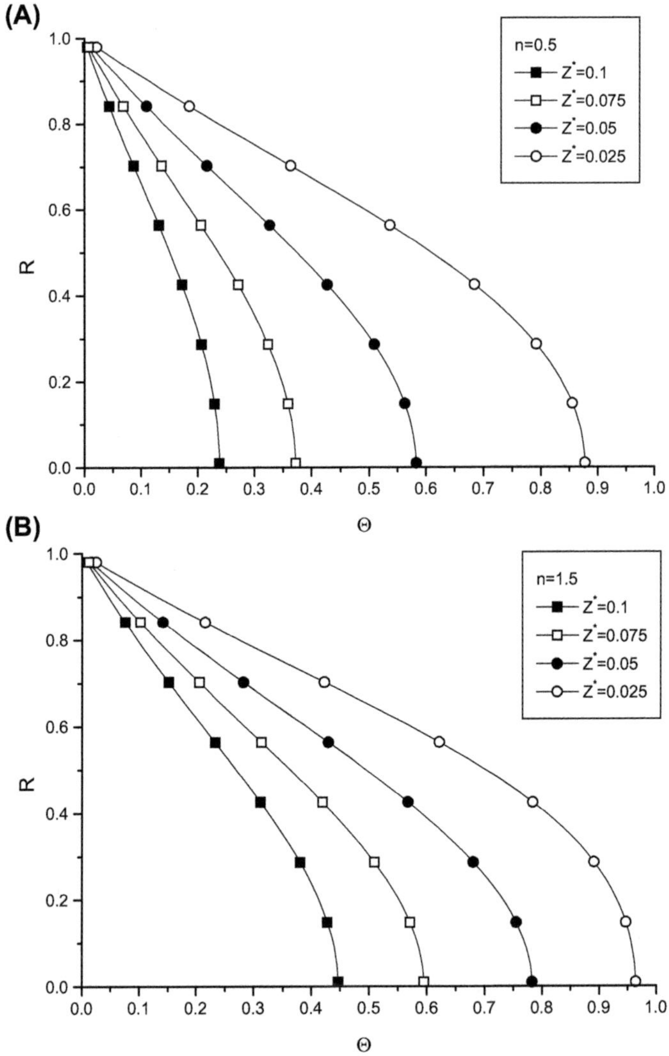

FIGURE 8.7 Temperature profiles of different dimensionless axial coordinate with power law temperature-dependent thermal conductivity. (A) $n = 0.5$ and (B) $n = 1.5$.

where Θ_b is the dimensionless bulk mean temperature, which is defined as

$$\Theta_b = \frac{\int_0^1 U^* \Theta R dR}{\int_0^1 U^* R dR} \tag{8.63}$$

In Table 8.2, local Nusselt numbers in the fully developed region from Irvine et al. (1988) and Kim et al. (1995) are in good agreement with

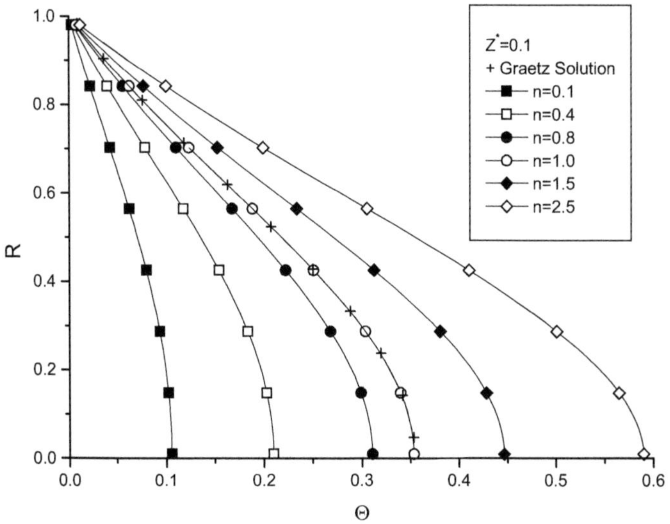

FIGURE 8.8 Temperature profiles of different power law index with power law temperature-dependent thermal conductivity.

calculated values obtained in this research. This comparison with constant model could prove the reliability of the results.

The local Nusselt number increases with the increase in thermal conductivity parameter ε^* as displayed in Fig. 8.11. The local Nusselt number from the entrance to the fully developed region is getting lower, and the Nusselt number of dilatant fluid is smaller than the one of shear-thinning non-Newtonian flow. Fig. 8.12 reveals the difference of local Nusselt number between thermal conductivity models. It shows there is obvious difference between the constant model and the velocity-dependent model. However, Nusselt number varies little from constant model to temperature-dependent model.

From the previous discussion, we can obtain the following conclusions:

1. With traditional constant model, the differences of temperature between varying power law index are small; with a temperature-dependent model, the differences are obviously big; with a velocity gradient-dependent model, the differences are biggest.
2. There is obvious difference in the local Nusselt number between the constant model and the velocity gradient-dependent model. However, Nusselt number varies little from constant model to power law temperature gradient–dependent model.
3. The results show that the heat transfer behaviors strongly depend on the value of the power law index in all the models.
4. The increasing values of thermal conductivity parameter ε^* result in decreasing the temperature and increasing the local Nusselt number when the thermal conductivity is a linear one.

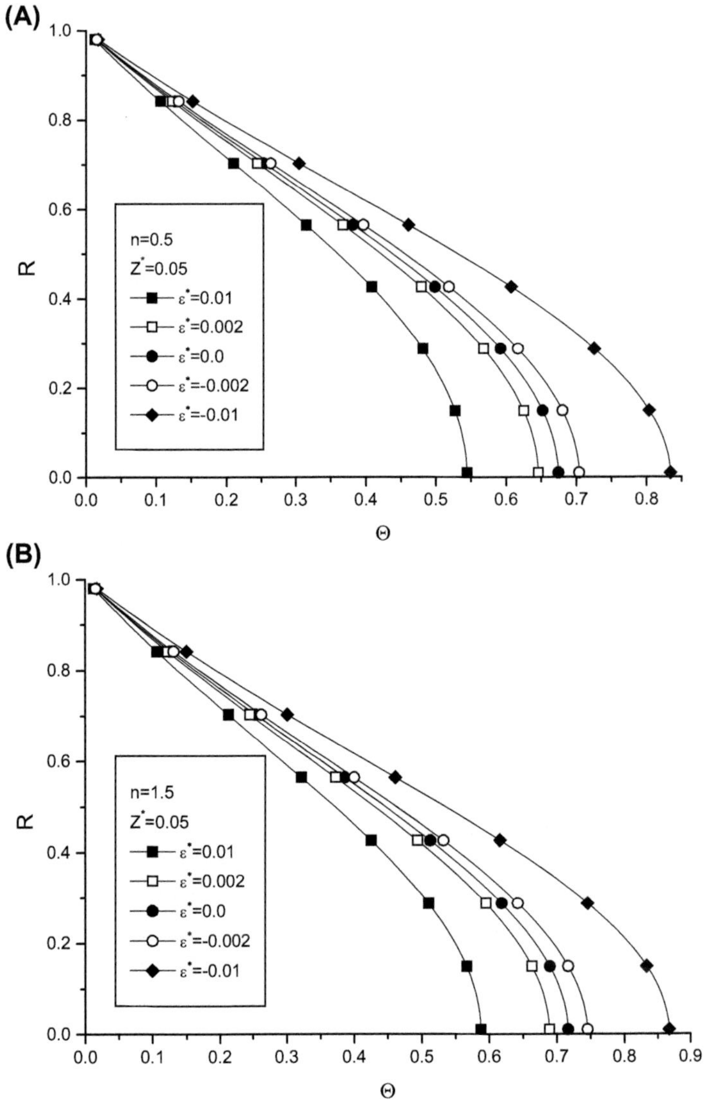

FIGURE 8.9 Effects of thermal conductivity parameter ε^* on temperature profiles in a linear temperature-dependent thermal conductivity model. (A) $n = 0.5$ and (B) $n = 1.5$.

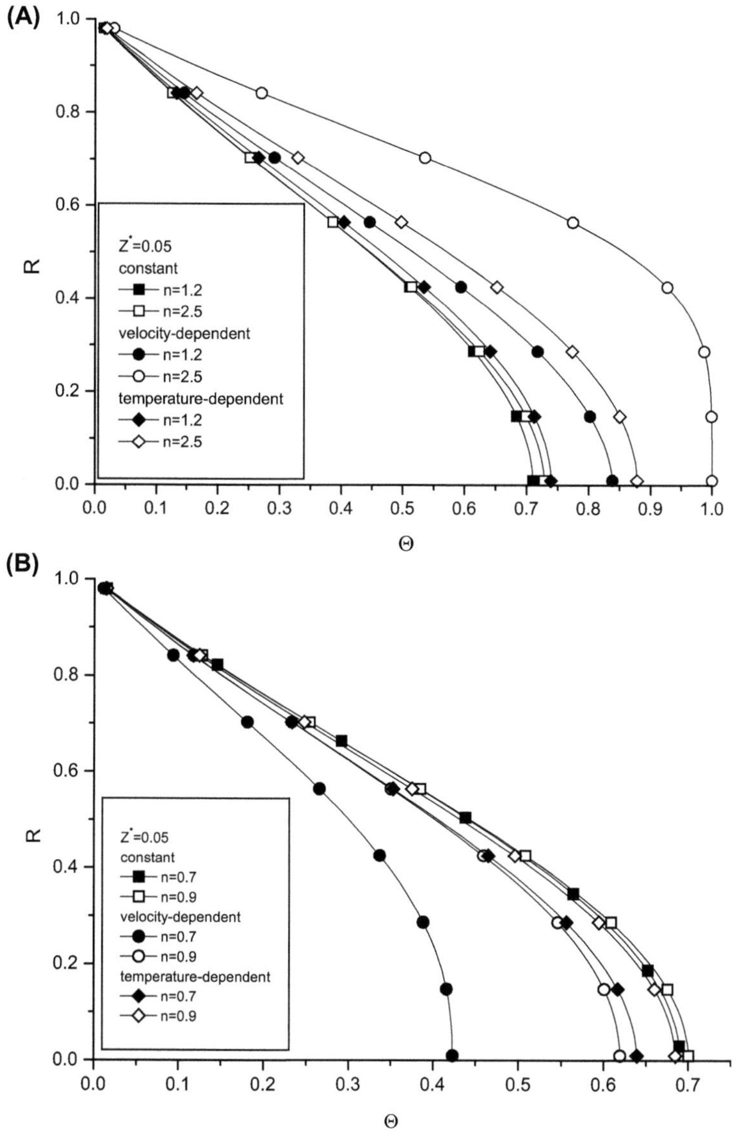

FIGURE 8.10 Comparison of solutions obtained by four heat conductivity models. (A) $n = 0.5$ and (B) $n = 1.5$.

TABLE 8.2 Comparison of Calculated Fully Developed Nusselt Numbers With Constant Model

n	Previous Works	Present Results	Differences
1.4	3.5564 (Kim, 1995)	3.558571	0.061%
1.2	3.5993 (Kim, 1995)	3.600869	0.044%
1.0	3.6568 (Irvine et al., 1988)	3.657589	0.022%
0.9	3.6934 (Irvine et al., 1988)	3.693878	0.013%
0.8	3.7377 (Irvine et al., 1988)	3.731622	0.002%
0.7	3.7921 (Irvine et al., 1988)	3.791383	0.019%
0.6	3.8605 (Irvine et al., 1988)	3.859031	0.038%
0.5	3.9494 (Irvine et al., 1988)	3.946730	0.068%

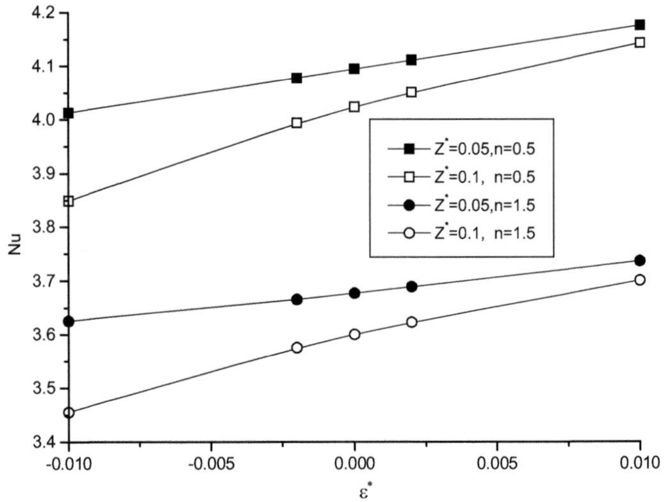

FIGURE 8.11 Effects of thermal conductivity parameter ε^* on local Nusselt number in a linear temperature-dependent thermal conductivity model.

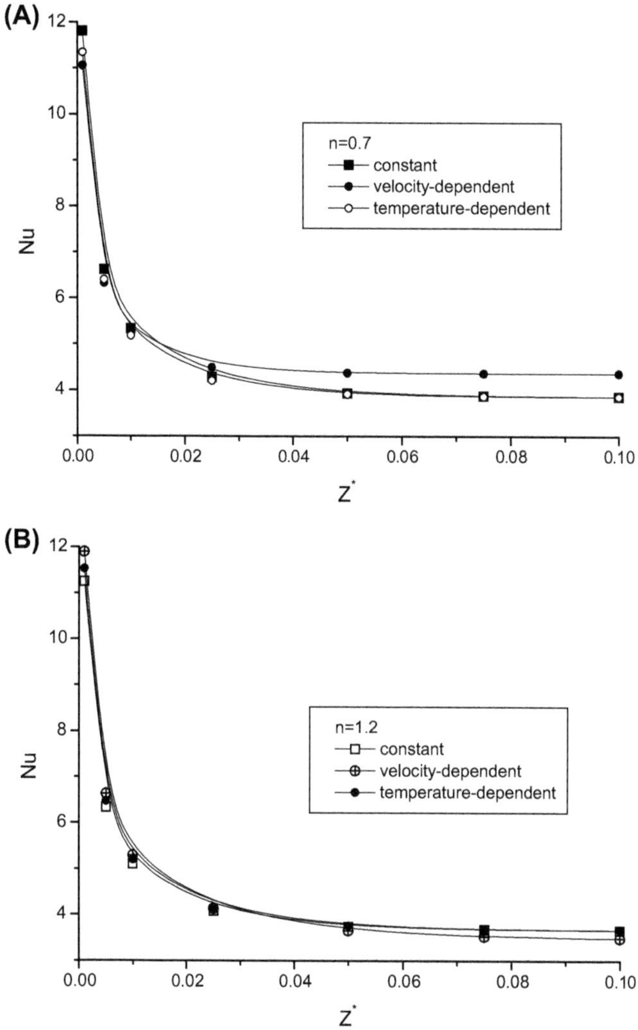

FIGURE 8.12 Comparison of local Nusselt number between models. (A) $n = 0.7$ and (B) $n = 1.2$.

8.3 HEAT TRANSFER OF THE POWER LAW FLUID OVER A ROTATING DISK

8.3.1 Background of the Problem

The fluid flow, heat, and mass transfer over rotating bodies have been studied in many industrial, geothermal, geophysical, technological, and engineering. For the complex of real machines, this type of flow and heat transfer can be modeled by the rotating disk system. The steady flow of Newtonian fluid over rotating disk was first discussed by von Kármán in 1921. Von Kármán introduced an elegant transformation of reducing the Navier–Stokes equations to a set of ordinary differential equations (ODEs), the approximate solution was obtained by using momentum integral method. In 1934, Cochran and Goldstein (1934) calculated more accurate values by numerical integration of the ODEs. Rogers and Lance (1960) and Benton (1966) obtained improved solutions. The problem of heat transfer over a rotating disk maintained at a constant temperature was firstly considered by Millsaps and Pohlhausen (1952) with a variety of Prandtl numbers in the range of $0.5 < (c_v/c_p)\mathrm{Pr} < 10$. Sparrow and Gregg (1959) obtained the results for any values of Prandtl numbers. Since the pioneering work of Von Kármán, the flow and heat transfer problems due to a rotating disk have attracted increasing attentions. A comprehensive review of the problem was reported in 1987 by Zandbergen and Dijkstra (1987).

Recently, the study of flow and heat transfer of non-Newtonian fluid over rotating disk has attracted a great deal of attention due to its many applications. In 1964, Mitschka generalized the Von Kármán's similarity transformation to power law fluid. In 2001, Andersson et al. analyzed the flow in the boundary layer of the power law fluid systematically for power law index in range of $1.5 < n < 2.0$. Attia (2008) studied rotating disk flow and heat transfer through a porous medium of a non-Newtonian fluid with suction and injection. Sahoo (2009) investigated effects of partial slip, viscous dissipation and Joule heating on Von Kármán flow, and heat transfer of an electrically conducting non-Newtonian fluid. Osalusi et al. (2007) considered viscous dissipation and Joule heating on steady MHD flow and heat transfer of a Bingham fluid over a porous rotating disk in the presence of Hall and ion-slip currents. Rashaida (2005) studied the flow of a non-Newtonian Bingham plastic fluid over a rotating disk.

We discuss in this section the steady flow and heat transfer of power law fluid over a free rotating disk. A modified Fourier's heat conduction law for power law fluids is introduced by assuming that the heat diffusion of temperature is affected by velocity field power law viscosity of non-Newtonian fluid. Suitable similarity transformation is introduced to reduce the coupled governing equations into ODEs, which is solved numerically by using multishooting method technique.

8.3.2 Formulation of the Problem and Governing Equations

Let us consider the laminar flow driven solely by an infinite disk rotating steadily with angular velocity Ω about the z-axis. The fluid occupies the infinite region on one side of the disk ($z > 0$) and the motion is rotationally symmetric. The cylindrical polar coordinate system is (r,φ,z), where r is the radial axis, φ is the tangential axis, and z is the vertical axis. The fluid near the disk surface is thrown outward radically, and this in turn induces an axial flow. u,v and w are velocity components in the directions of increasing r,φ and z, respectively. The surface of the rotating disk is maintained at a uniform temperature T_w, and the free stream is kept at a constant temperature T_∞.

The physical model and geometrical coordinates are shown in Fig. 8.13. Unless we give special interpretation, throughout this chapter we make the following assumptions:

1. The density of the fluid ρ is constant.
2. The flow is steady and axial-symmetric, that is, $\frac{\partial}{\partial \varphi} = 0, \frac{\partial}{\partial t} = 0$.
3. The motion and heat transfer of the fluid are all boundary problems (at sufficiently high Reynolds, not very slow flow).
4. The viscosity μ obeys the power law $\mu = \mu_0 \left\{ \left(\frac{\partial u}{\partial z}\right)^2 + \left(\frac{\partial v}{\partial z}\right)^2 \right\}^{(n-1)/2}$,

where μ_0 is the consistency coefficient of the fluid, n is the power law index. Particularly $n = 1$ is for Newtonian fluid and $\mu = \mu_0$.

From the assumptions and the conservation theory, we have $\frac{\partial p}{\partial z} = 0, \frac{\partial p}{\partial r} = 0, \frac{\partial p}{\partial \varphi} = 0,$ which imply that the pressure p is a constant. So the conservation equations for mass and momentum and energy are written as follows:

$$\frac{\partial u}{\partial r} + \frac{u}{r} + \frac{\partial w}{\partial z} = 0, \tag{8.64}$$

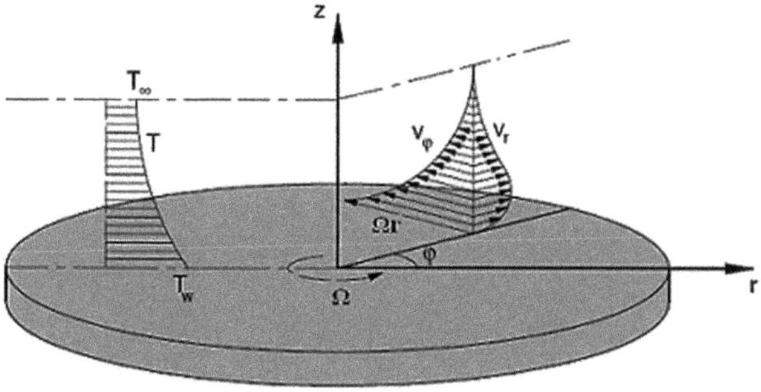

FIGURE 8.13 The sketch model of rotating disk.

$$\rho\left(u\frac{\partial u}{\partial r}-\frac{v^2}{r}+w\frac{\partial u}{\partial z}\right)=\frac{\partial}{\partial z}\left(\mu\frac{\partial u}{\partial z}\right), \tag{8.65}$$

$$\rho\left(u\frac{\partial v}{\partial r}+\frac{uv}{r}+w\frac{\partial u}{\partial z}\right)=\frac{\partial}{\partial z}\left(\mu\frac{\partial v}{\partial z}\right), \tag{8.66}$$

$$\rho c_p\left(u\frac{\partial T}{\partial r}+w\frac{\partial T}{\partial z}\right)=\frac{\partial}{\partial z}\left(\kappa\frac{\partial T}{\partial z}\right), \tag{8.67}$$

where T is the temperature of fluid over the disk, c_p is the specific heat at constant pressure, μ is the viscosity, κ is the thermal conductivity of the power law non-Newtonian fluid with modified Fourier heat conduction law chosen as $\kappa=\kappa_0\left\{\left(\frac{\partial u}{\partial z}\right)^2+\left(\frac{\partial v}{\partial z}\right)^2\right\}^{(n-1)/2}$ (Ming et al., 2009; Zheng et al., 2008). $\Pr=\mu c_p/\kappa=\mu_0 c_p/\kappa_0$ is the Prandtl number.

The corresponding boundary conditions are:

$$z=0:\quad u=0,\quad v=\Omega r,\quad w=0,\quad T=T_w; \tag{8.68a}$$

$$z\to\infty:\quad u=0,\quad v=0,\quad T=T_\infty, \tag{8.68b}$$

where $u=0$, $v=\Omega r$ present the no-slip boundary condition, and $w=0$ presents no suction or induction boundary condition at $z=0$.

8.3.3 Generalized Karman Transformation

Mitschka focused on the boundary layer equations of power law fluid and proposed a generalized dimensionless Karman similarity variable defined as

$$\xi=z\left(\frac{\Omega^{2-n}}{\mu_0/\rho}\right)^{\frac{1}{n+1}}r^{\frac{1-n}{n+1}}. \tag{8.69}$$

The motion of the fluid in radial and azimuthal direction are aroused by the rotation of the disk, so let

$$u=\Omega r\cdot F(\xi),\quad v=\Omega r\cdot G(\xi),\quad w=\left(\frac{\Omega^{1-2n}}{\mu_0/\rho}\right)^{-\frac{1}{n+1}}\cdot r^{\frac{n-1}{n+1}}\cdot H(\xi), \tag{8.70}$$

and here F, G and H denote the nondimensional velocity components in the direction of radial, azimuthal and axial, respectively.

In addition, a dimensionless temperature is defined as

$$\Theta(\xi) = (T - T_\infty)/(T_w - T_\infty) \tag{8.71}$$

where T_w and T_∞, respectively, represent the surface temperature (a constant) and the temperature of the fluid far from the disk.

Since,

$$\frac{\partial \xi}{\partial r} = \frac{1-n}{n+1} z \left(\frac{\Omega^{2-n}}{\mu_0/\rho}\right)^{\frac{1}{n+1}} r^{\frac{1-n}{n+1}-1} = \frac{1-n}{n+1}\frac{\xi}{r}, \tag{8.72}$$

$$\frac{\partial \xi}{\partial z} = \left(\frac{\Omega^{2-n}}{\mu_0/\rho}\right)^{\frac{1}{n+1}} r^{\frac{1-n}{n+1}} = \frac{\xi}{z}, \tag{8.73}$$

$$\frac{\partial u}{\partial r} = \frac{\partial(\Omega r F(\xi))}{\partial r} = \Omega F(\xi) + \Omega r F'(\xi)\frac{\partial \xi}{\partial r} = \Omega\left(F(\xi) + \frac{1-n}{n+1}\xi F'(\xi)\right), \tag{8.74}$$

$$\frac{\partial u}{\partial z} = \frac{\partial(\Omega r F(\xi))}{\partial z} = \Omega r F'(\xi)\frac{\partial \xi}{\partial z} = \Omega r \frac{\xi}{z} F'(\xi), \tag{8.75}$$

$$\frac{\partial v}{\partial r} = \frac{\partial(\Omega r G(\xi))}{\partial r} = \Omega G(\xi) + \Omega r G'(\xi)\frac{\partial \xi}{\partial r} = \Omega\left(G(\xi) + \frac{1-n}{n+1}\xi G'(\xi)\right), \tag{8.76}$$

$$\frac{\partial v}{\partial z} = \frac{\partial(\Omega r G(\xi))}{\partial z} = \Omega r G'(\xi)\frac{\partial \xi}{\partial z} = \Omega r \frac{\xi}{z} G'(\xi), \tag{8.77}$$

$$\frac{\partial w}{\partial z} = \left(\frac{\Omega^{1-2n}}{\mu_0/\rho}\right)^{-\frac{1}{n+1}} \cdot r^{\frac{n-1}{n+1}} \cdot H'(\xi)\frac{\partial \xi}{\partial z} = \Omega H'(\xi), \tag{8.78}$$

$$\frac{\partial T}{\partial r} = \frac{\partial \Theta}{\partial \xi}\cdot\frac{\partial \xi}{\partial r} = (T_w - T_\infty)\frac{1-n}{1+n}\left(\frac{\xi}{r}\right)\Theta'(\xi), \tag{8.79}$$

$$\frac{\partial T}{\partial z} = \frac{\partial \Theta}{\partial \xi}\cdot\frac{\partial \xi}{\partial z} = (T_w - T_\infty)\frac{\xi}{z}\Theta'(\xi). \tag{8.80}$$

Then

$$\mu = \mu_0 \Omega^{n-1} r^{n-1}\left(\frac{\xi}{z}\right)^{n-1}\left[(F'(\xi))^2 + (G'(\xi))^2\right]^{(n-1)/2}, \tag{8.81}$$

$$\kappa = \kappa_0 \Omega^{n-1} r^{n-1}\left(\frac{\xi}{z}\right)^{n-1}\left[(F'(\xi))^2 + (G'(\xi))^2\right]^{(n-1)/2}, \tag{8.82}$$

and

$$\frac{\partial}{\partial z}\left[\mu\frac{\partial u}{\partial z}\right]=\rho\Omega^2 r\left\{F'(\xi)\left[(F'(\xi))^2+(G'(\xi))^2\right]^{(n-1)/2}\right\}'_{\xi},\tag{8.83}$$

$$\frac{\partial}{\partial z}\left[\mu\frac{\partial v}{\partial z}\right]=\rho\Omega^2 r\left\{G'(\xi)\left[(F'(\xi))^2+(G'(\xi))^2\right]^{(n-1)/2}\right\}'_{\xi},\tag{8.84}$$

$$\frac{\partial}{\partial z}\left[\kappa\frac{\partial T}{\partial z}\right]=\rho\frac{\kappa_0}{\mu_0}\Omega(T_w-T_\infty)\left\{\Theta'(\xi)\left[(F'(\xi))^2+(G'(\xi))^2\right]^{(n-1)/2}\right\}'_{\xi}.\tag{8.85}$$

Therefore, the partial differential Eqs. (8.64)–(8.67) become the following ordinary differential equations:

$$H'=-2F-\frac{1-n}{1+n}\xi F',\tag{8.86a}$$

$$F^2-G^2+\left(H+\frac{1-n}{1+n}\xi F\right)F'=\left\{\left[(F')^2+(G')^2\right]^{(n-1)/2}F'\right\}',\tag{8.86b}$$

$$2FG+\left(H+\frac{1-n}{1+n}\xi F\right)G'=\left\{\left[(F')^2+(G')^2\right]^{(n-1)/2}G'\right\}',\tag{8.86c}$$

$$\left(\frac{1-n}{1+n}\xi F+H\right)\Theta'=\frac{1}{\Pr}\left\{\left[(F')^2+(G')^2\right]^{(n-1)/2}\Theta'\right\}'.\tag{8.86d}$$

The prime symbol denotes differentiation with respect to ξ and Pr is the Prandtl number of the fluid defined as $\Pr=\frac{\mu_0 c_p}{\kappa_0}$. The transformed boundary conditions are given by:

$$\xi=0:\quad F(0)=0, G(0)=1, H(0)=0,\quad \Theta(0)=0;\tag{8.87a}$$

$$\xi\to\infty: F=0,\ G=0,\ \Theta=1.\tag{8.87b}$$

8.3.4 Multiple Shooting Method

In order to solve the two-point boundary-value problem, finite difference and shooting method are applied by many researchers. In this research, the multishooting method is adopted to solve the two-point boundary-value problem, Eqs. (8.86a–d) and (8.87a and b). Firstly, we transform the equations to first-order system of ordinary differential equations. Let

$$y_1=F, y_2=F', y_3=G, y_4=G', y_5=H, y_6=\Theta, y_7=\Theta',$$

then

$$
\begin{aligned}
y_1' &= y_2,\\
y_2' &= \frac{1}{n}\cdot\left(y_2^2+y_4^2\right)^{\frac{1-n}{2}}\cdot\left\{\left[1+(n-1)\left(y_2^2+y_4^2\right)^{-1}y_4^2\right]\left[y_1^2-y_3^2+\left(y_5+\frac{1-n}{1+n}\xi y_1\right)y_2\right]\right.\\
&\quad\left.-(n-1)\left(y_2^2+y_4^2\right)^{-1}y_2y_4\left[2y_1y_3+\left(y_5+\frac{1-n}{1+n}\xi y_1\right)y_4\right]\right\},\\
y_3' &= y_4,\\
y_4' &= \frac{1}{n}\cdot\left(y_2^2+y_4^2\right)^{\frac{1-n}{2}}\cdot\left\{\left[1+(n-1)\left(y_2^2+y_4^2\right)^{-1}y_2^2\right]\left[2y_1y_3+\left(y_5+\frac{1-n}{1+n}\xi y_1\right)y_4\right]\right.\\
&\quad\left.-(n-1)\left(y_2^2+y_4^2\right)^{-1}y_2y_4\left[y_1^2-y_3^2+\left(y_5+\frac{1-n}{1+n}\xi y_1\right)y_2\right]\right\},\\
y_5' &= -2y_1-\frac{1-n}{1+n}\xi y_2,\\
y_6' &= y_7,\\
y_7' &= \left(y_2^2+y_4^2\right)^{\frac{1-n}{2}}y_7\left\{\left(\Pr-\frac{n-1}{n}\right)\left(y_5+\frac{1-n}{1+n}\xi y_1\right)-\frac{n-1}{n}\left(y_2^2+y_4^2\right)^{-1}\right.\\
&\quad\left.\times\left(y_1^2y_2-y_2y_3^2+2y_1y_3y_4\right)\right\}.
\end{aligned}
\tag{8.88}
$$

Boundary conditions are:

$$
y_1(0)=0,\ \ y_3(0)=1,\ \ y_5(0)=0,\ \ y_6(0)=1;\ \ y_1(\infty)=0,\ \ y_3(\infty)=0,\ \ y_6(\infty)=0,
\tag{8.89}
$$

where the value of ξ_∞ is affected by the parameters n and Pr. Set $y_2(0)=a$, $y_4(0)=b$, and $y_7(0)=c$, where a, b, c are adjusted by the shooting technique to satisfy the right boundary conditions $y_1(\infty)=0$, $y_3(\infty)=0$, $y_6(\infty)=0$. Thus the boundary-value problem turns into initial-value problem and can be numerically integrated by using Runge–Kutta method.

8.3.5 Results and Discussion

Numerical results are as follows:

Fig. 8.14 shows the distribution of the velocity and temperature with the variable ξ at $n=0.2$ and $\Pr=1.0$ satisfying the right boundary conditions $F(\infty)=0.01$, $G(\infty)=0.01$, $H(\infty)=0.01$, $\Theta(\infty)=0.01$. F represents the component of radial velocity, G represents the component of azimuthal velocity, and H represents the axial velocity. It is indicated that the magnitude of F and H are very small for shear-thinning fluid.

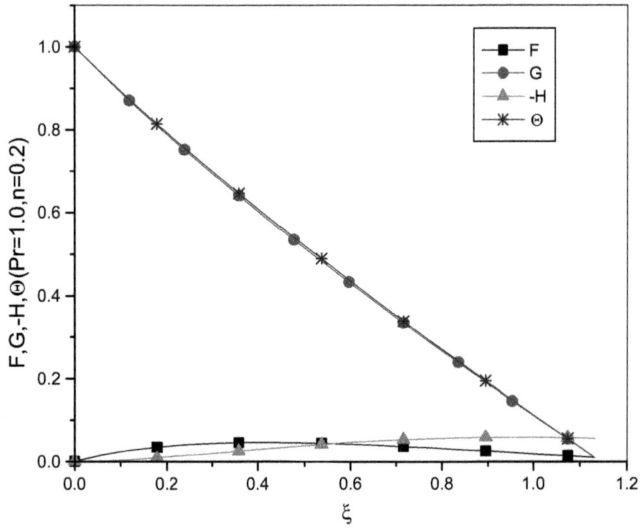

FIGURE 8.14 Velocity components and temperature for a shear-thinning fluid with $n = 0.2$ and $\mathrm{Pr} = 1.0$. Radial (F), azimuthal (G), axial (H), and temperature (Θ).

Fig. 8.15 shows the distribution of the velocity and temperature with the variable ξ at $n = 1.0$ and $\mathrm{Pr} = 1.0$ satisfying the right boundary conditions $F(\infty) = 0.01$, $G(\infty) = 0.01$, $H(\infty) = 0.01$, $\Theta(\infty) = 0.01$, where the fluid is Newtonian fluid. Fig. 8.16 shows the distribution of the velocity and temperature with the variable ξ at $n = 10$ and $\mathrm{Pr} = 1.0$ satisfying the right

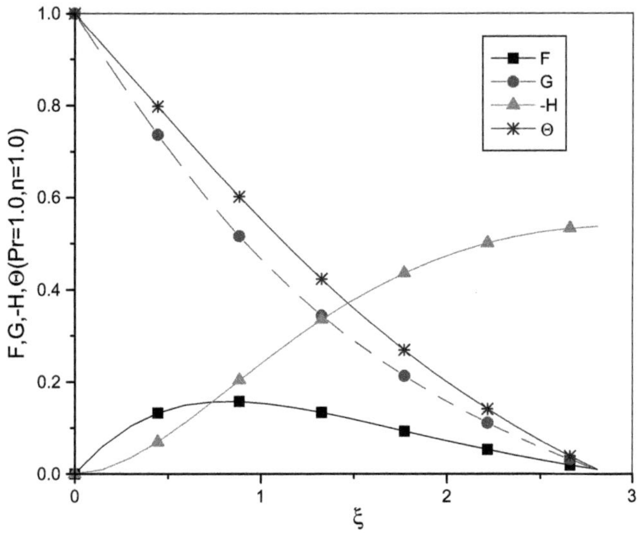

FIGURE 8.15 Velocity components and temperature for Newtonian fluid with $n = 1$ and $\mathrm{Pr} = 1.0$. Radial (F), azimuthal (G), axial (H), and temperature (Θ).

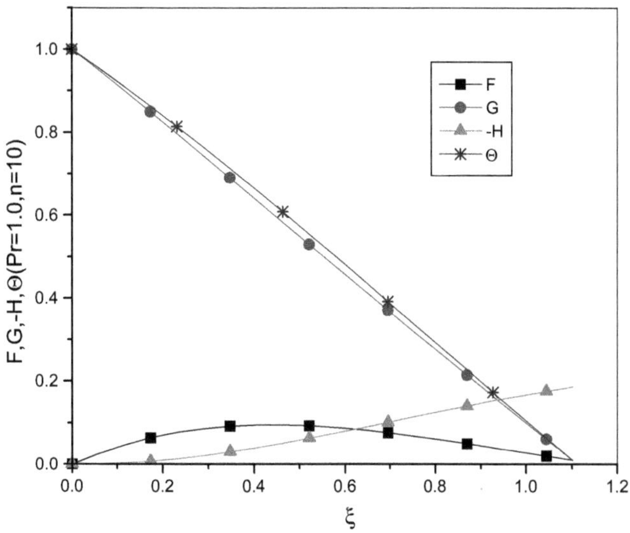

FIGURE 8.16 Velocity components and temperature for a shear-thickening fluid with $n = 10$ and $n = 10$, $\mathrm{Pr} = 1.0$. Radial (F), azimuthal (G), axial (H), and temperature (Θ).

boundary conditions $F(\infty) = 0.01$, $G(\infty) = 0.01$, $H(\infty) = 0.01$, $\Theta(\infty) = 0.01$. The fluid is shear-thickening fluid.

Fig. 8.17A shows the component F and the thickness of the boundary layer increase with n for $0.2 < n < 1.0$, $\mathrm{Pr} = 1.0$. Fig. 8.17B shows the maximum of F increases with n, but the thickness of boundary layer decreases for $1.0 < n < 2.0$, $\mathrm{Pr} = 1.0$, and Fig. 8.17C shows that the component F and the thickness of the boundary layer decreases with n for $2.0 < n < 5.0$, $\mathrm{Pr} = 1.0$. Fig. 8.18A shows that the component G and the thickness of the boundary layer increase with n when $0.2 < n < 1.0$, $\mathrm{Pr} = 1.0$. Fig. 8.18B and C show G decreases with n for $n > 1.0$, $\mathrm{Pr} = 1.0$.

Fig. 8.19A shows that the magnitude of axial velocity increases with n for $0.2 < n < 1.0$. Fig. 8.19B shows that the magnitude of axial velocity vibrates with n for $1.0 < n < 2.0$ and Fig. 8.19C shows that the magnitude of axial velocity decreases with n for $2.0 < n < 5.0$. Fig. 8.20A shows that the temperature increases with n for $0.2 < n < 1.0$. Fig. 8.20B and C show that the temperature decreases with n for $2.0 < n < 5.0$ (Table 8.3). It indicates that the ability of thermal conduction is much stronger for the thickening fluid (Figs. 8.21–8.23).

In this section, the effects of power law index n and Pr on velocity and temperature fields are studied. The boundary layer thickness obtains the maximum value for $n = 1.0$ and decreases with the increase of Pr. The purpose of the present research is to investigate the temperature and velocity distributions and gain a better understanding of the behavior of power law fluid in the laminar boundary layer on a rotating disk.

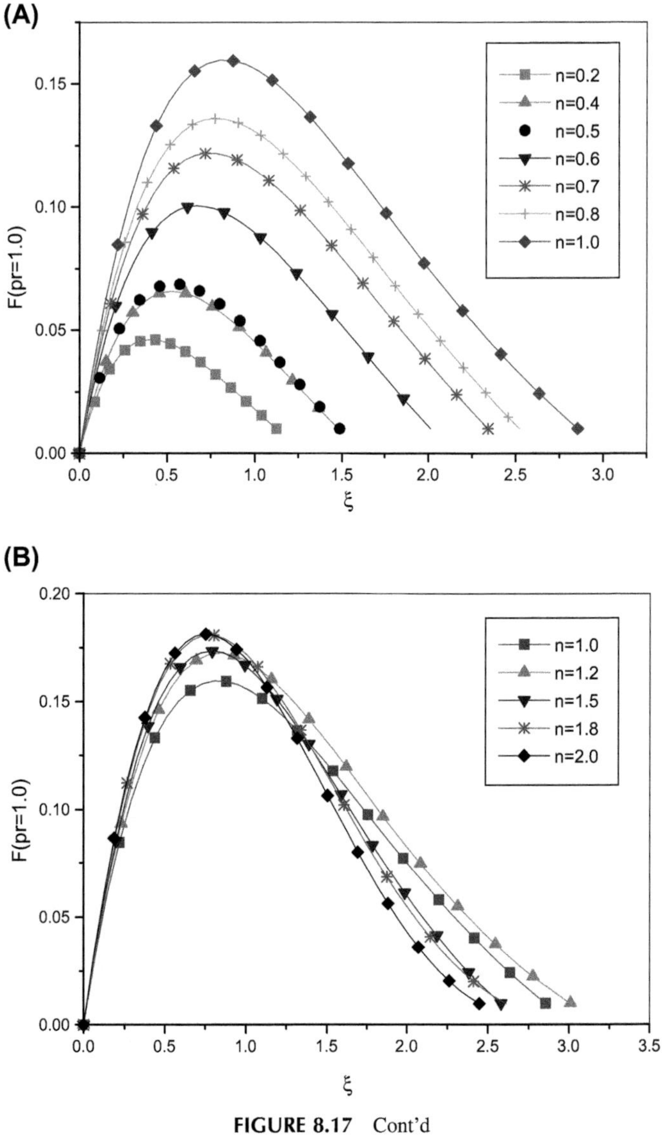

FIGURE 8.17 Cont'd

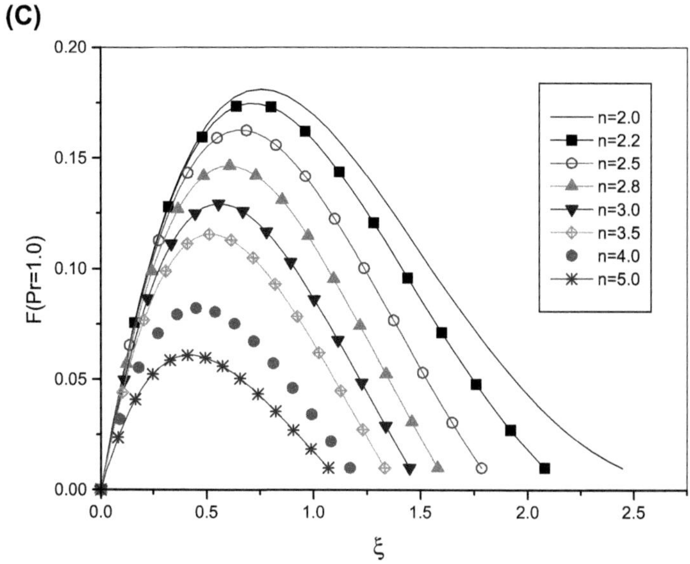

FIGURE 8.17 The distribution of velocity component F for $\mathrm{Pr} = 1.0$ (A) $0.2 < n < 1.0$, (B) $1.0 < n < 2.0$, (C) $2.0 < n < 5.0$.

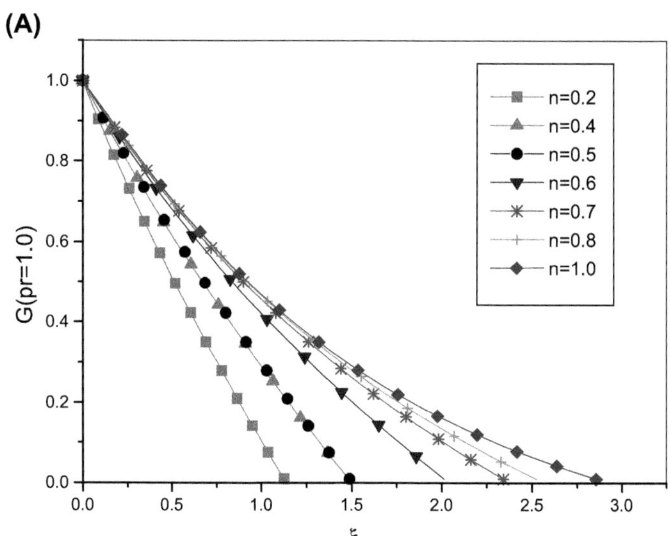

FIGURE 8.18 Cont'd

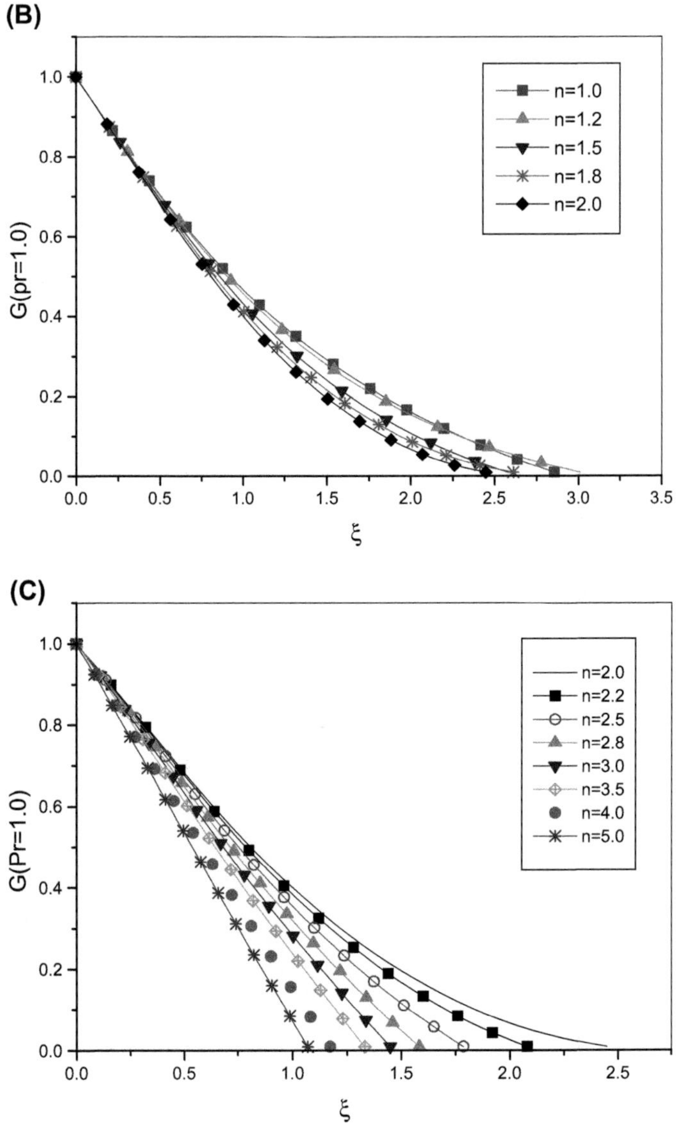

FIGURE 8.18 The distribution of velocity components G with values of power law exponent n for Pr = 1.0. (A) $0.2 < n < 1.0$, (B) $1.0 < n < 2.0$, (C) $2.0 < n < 5.0$.

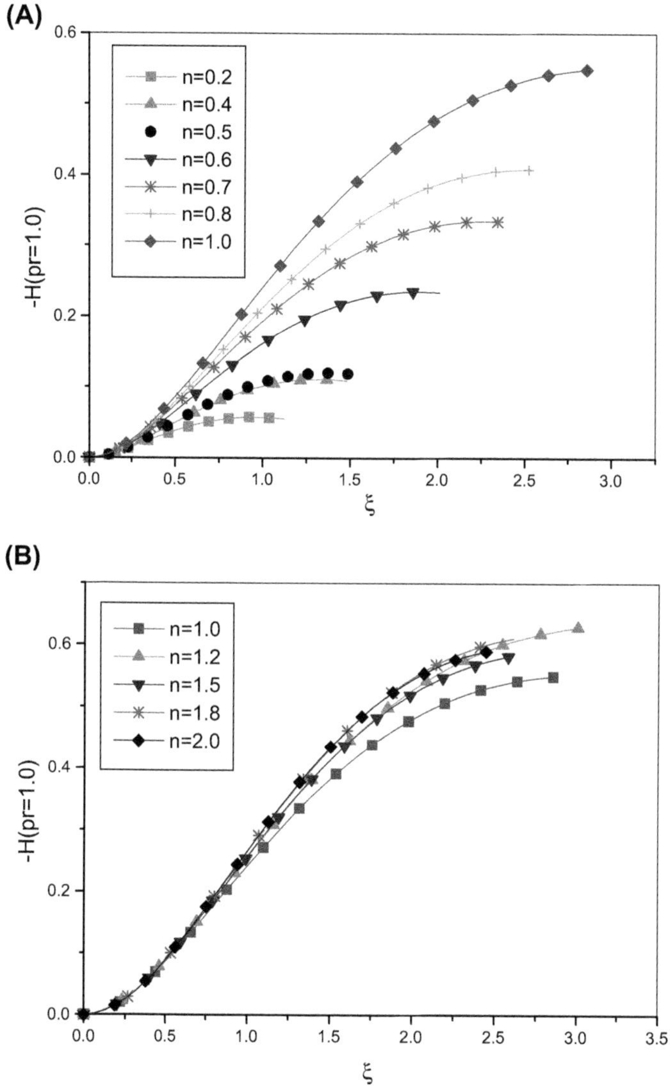

FIGURE 8.19 Cont'd

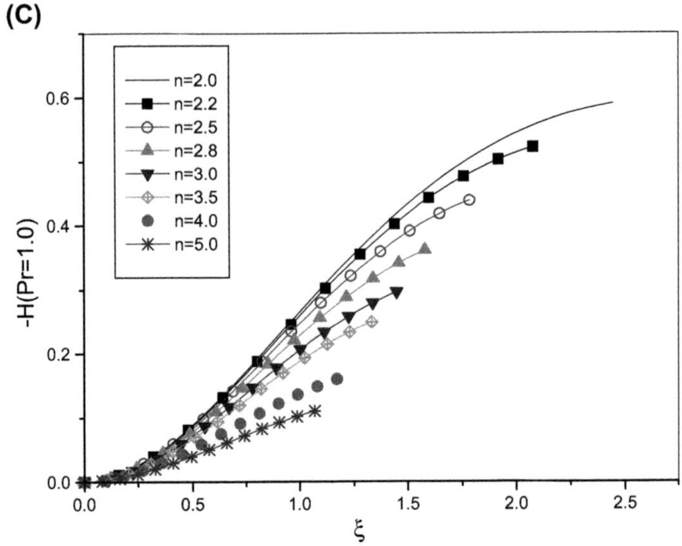

FIGURE 8.19 The distribution of axial velocity H with values of power law exponent n for $Pr = 1.0$. (A) $0.2 < n < 1.0$, (B) $1.0 < n < 2.0$, (C) $2.0 < n < 5.0$.

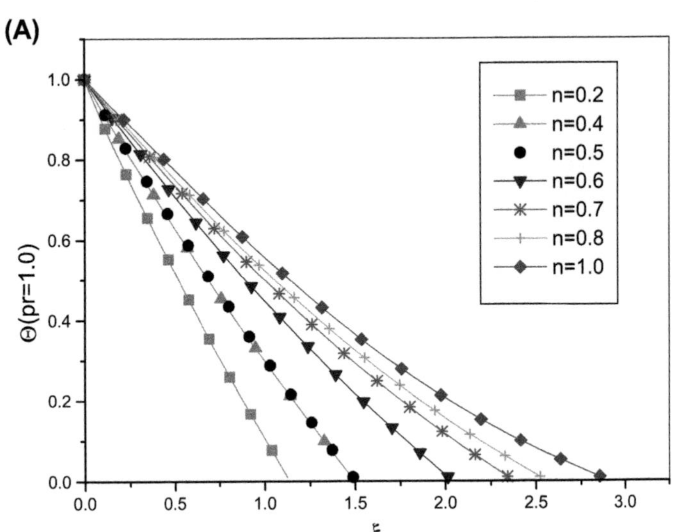

FIGURE 8.20 Cont'd

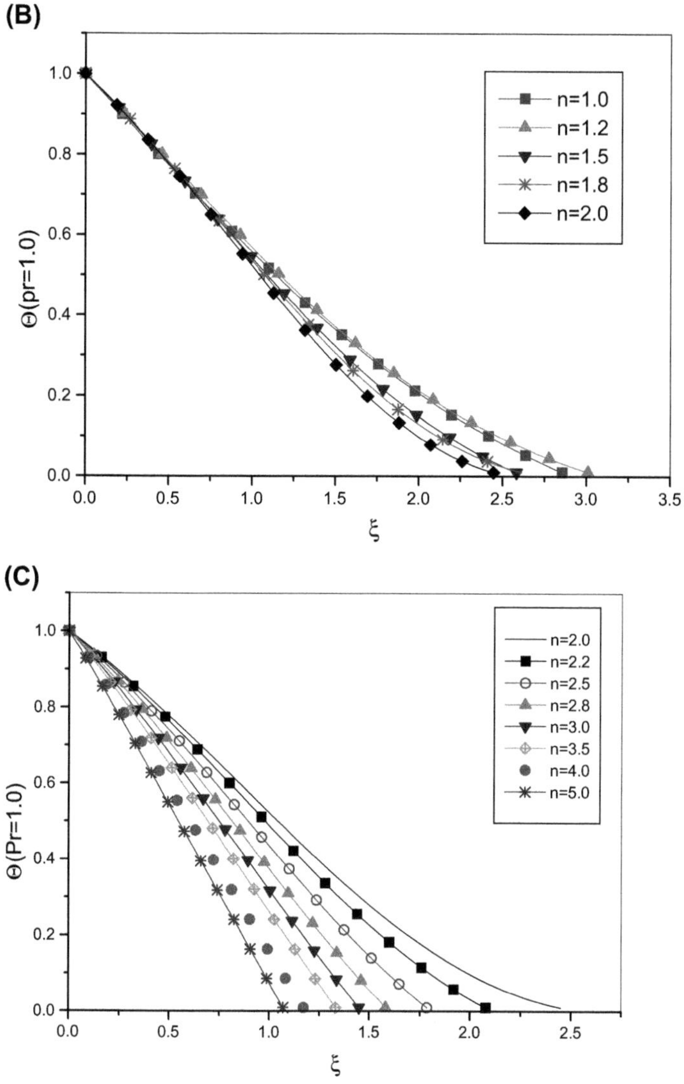

FIGURE 8.20 The distribution of temperature with power law exponent n for Pr = 1.0. (A) $0.2 < n < 1.0$, (B) $1.0 < n < 2.0$, (C) $2.0 < n < 5.0$.

TABLE 8.3 The Numerical Solutions for $n = 1.0$, $\mathrm{Pr} = 1.0$

ξ	F	F'	G	G'	$-H$	Θ	Θ'
0.000	0.000	0.482	1.000	−0.616	0.000	1.000	−0.457
0.146	0.061	0.349	0.910	−0.607	0.009	0.933	−0.456
0.293	0.103	0.239	0.823	−0.584	0.034	0.866	−0.455
0.439	0.132	0.150	0.739	−0.553	0.068	0.800	−0.452
0.586	0.148	0.080	0.661	−0.517	0.110	0.734	−0.446
0.732	0.156	0.025	0.588	−0.480	0.154	0.669	−0.437
0.879	0.156	−0.016	0.520	−0.442	0.200	0.606	−0.426
1.025	0.151	−0.047	0.458	−0.407	0.245	0.545	−0.412
1.172	0.143	−0.068	0.401	−0.373	0.289	0.485	−0.397
1.318	0.132	−0.083	0.349	−0.341	0.329	0.429	−0.379
1.465	0.119	−0.091	0.301	−0.313	0.366	0.374	−0.360
1.758	0.091	−0.097	0.217	−0.263	0.427	0.275	−0.321
1.904	0.077	−0.095	0.180	−0.242	0.452	0.229	−0.301
2.051	0.063	−0.092	0.146	−0.223	0.472	0.187	−0.281
2.197	0.050	−0.088	0.114	−0.206	0.489	0.147	−0.262
2.343	0.038	−0.083	0.085	−0.190	0.502	0.110	−0.243
2.490	0.026	−0.077	0.059	−0.176	0.511	0.075	−0.226
2.636	0.015	−0.072	0.034	−0.163	0.517	0.044	−0.210
2.783	0.005	−0.067	0.011	−0.151	0.520	0.014	−0.194
2.856	0.000	−0.064	0.000	−0.145	0.520	0.000	−0.187

TABLE 8.4 Comparison of Present Solutions With the Results of Andersson et al. (2001).

	$F'(0)$		$-G'(0)$		$-H(\infty)$		$\Theta'(0)$
Power Law Index n	Present	Andersson	Present	Andersson	Present	Andersson	Present
2.5	0.54318	–	0.62184	–	0.44030	–	−0.44752
2.2	0.54563	–	0.60927	–	0.51105	–	−0.41677
2.0	0.54571	0.547	0.60360	0.603	0.57695	0.586	−0.39699
1.7	0.53301	0.537	0.60165	0.600	0.59417	0.633	−0.40002
1.5	0.51832	0.529	0.60319	0.601	0.56188	0.676	−0.41661
1.3	0.50494	0.522	0.60577	0.603	0.54938	0.735	−0.42921
1.1	0.49592	0.514	0.61001	0.610	0.57219	0.822	−0.43502
1.0	0.48234	0.510	0.61619	0.616	0.52038	0.883	−0.45668
0.8	0.44282	0.504	0.64014	0.636	0.38118	1.089	−0.52028

From Table 8.4 we can see that the difference of component of $G'(0)$ is less than 0.002 compared with the result of Andersson's. And the $F'(0)$ and $H(\infty)$ is less than 0.01 near $n=2.0$).

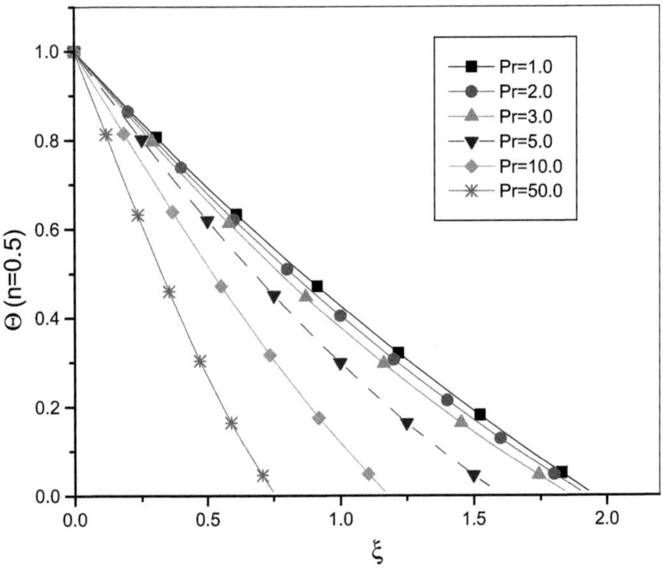

FIGURE 8.21 The variation of the temperature Θ with Pr: 1.0~50 for shear-thinning fluid $n = 0.5$.

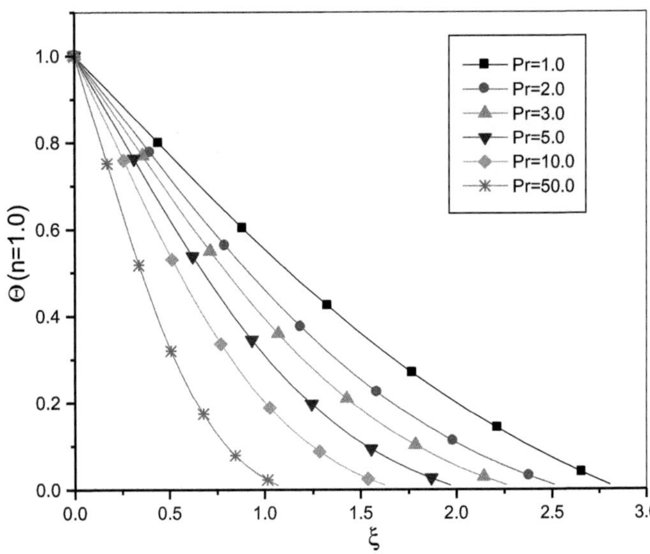

FIGURE 8.22 The variation of the temperature Θ with Pr: 1.0~50 for Newtonian fluid $n = 1.0$.

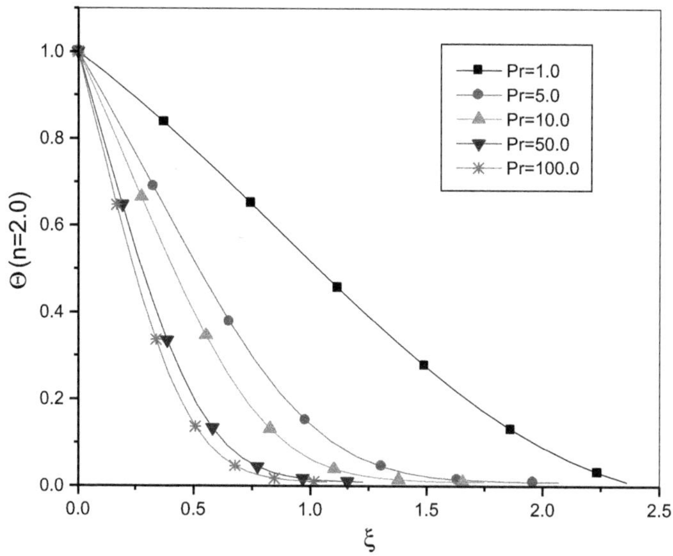

FIGURE 8.23 The variation of the temperature Θ with Pr: 1.0~50 for shear-thickening fluid $n = 2.0$.

8.4 MAXWELL FLUID WITH MODIFIED FRACTIONAL FOURIER'S LAW AND DARCY'S LAW

8.4.1 Background of the Problem

Recently, considerable attention has been devoted to study the transport behavior of fractional viscoelastic fluids. Governing equations are derived from the classical equations by replacing the derivatives of integer-order with noninteger-order integrals or derivatives. Exact analytical solutions of fractional equations are usually obtained by using Laplace transforms, Fourier transforms, and Hankel transforms and presented in terms of Green functions, Fox functions, or H-functions. In doing so, for all of these investigations, the physical property parameters of non-Newtonian fluids are assumed constants (Fetecau et al., 2009; Iomin, 2006; Jamil et al., 2012; Khan et al., 2009, 2012; Nazar et al., 2010; Vieru et al., 2008; Zheng et al., 2011, 2012).

However, experiments had indicated that the assumptions that the physical property parameters of non-Newtonian fluids are constants can only hold if temperature does not change rapidly or impulsively in any particular way (Makinde and Onyejekwe, 2011). Along this line, Seddeek (2002), Jayanthi and Kumari (2007), Cheng (2009), Malekzadeh et al. (2011), and Srinivasa and Eswara (2013) considered the variable viscosity problems for Newtonian fluids, they assumed that the viscosity varied as a linear function of temperature. Later, Malik et al. (2013) and Rundora and Makinde (2013) investigated that the viscosity is an exponential function that decreases with the temperature.

For the flow and heat transfer of non-Newtonian fluids, the generalized Fourier's law with fractional derivative was taken into consideration by many researchers. Ezzat (2010) derived a new constitutive relation for heat conduction with time fractional derivative of order α for generalized thermoelectric MHD fluid, where the boundary layer theory was applied to Stocks' first problem of an electroconducting viscoelastic flow over an infinite plate with heat source and in the presence of a transverse magnetic field. Li el al. (2012) presented an investigation for helical flows of a heated generalized Oldroyd-B fluid subject to a time-dependent shear stress in porous medium. It is well known that in the flows of viscous Newtonian fluid at low speed through a porous medium, the pressure drop caused by the frictional drag is proportional to the velocity.

Based on a modified Darcy's law for viscoelastic fluid, Tan and Masuoka (2005a,b, 2007) extended Stokes' first problem to that of a second grade fluid, Oldroyd-B fluid and Maxwell fluid in a porous half-space with a heated boundary. Xue and Nie (2009) investigated the Rayleigh–Stokes problem for a heated generalized second grade fluid in a porous half-space with a heated flat plate, and the exact solutions were obtained by using the Fourier sine transform and the fractional Laplace transform.

The purpose of the current work is to present an investigation for coupled flow and heat transfer of a generalized fractional Maxwell fluid in a porous medium between two infinite parallel plates. Unlike most classical works, the temperature-dependent fluid properties (variable fluid viscosity and thermal conductivity) are taken into account by modified fractional Fourier's law and Darcy's law to describe the constitutive relations in highly coupled velocity and temperature fields in porous medium, where the fluid viscosity is assumed to decrease exponentially with temperature. The fractional governing equations are solved numerically using implicit finite difference method based on nonshifted Grünwald formula. The effects of pertinent physical parameters on the velocity and temperature fields are presented graphically and analyzed.

8.4.2 Mathematical Formulation and Governing Equations

The conservation, constitutive, and energy equations of an incompressible fluid are given by (Zheng et al., 2012):

$$\mathbf{T1} = -P\mathbf{I} + \mathbf{S}, \quad \mathbf{S} + \lambda\frac{\mathrm{D}\mathbf{S}}{\mathrm{D}t} = \mu\mathbf{A}, \tag{8.90}$$

$$\rho c\frac{D\mathbf{T}}{Dt} = -div\mathbf{q} - \mathbf{p}div\overrightarrow{V} + Q_T, \tag{8.91}$$

where **T1** is the Cauchy stress tensor, **V** is the velocity vector, **S** is the extra-stress tensor, ρ is the constant density of the fluid, **T** is the temperature, **q** is the heat flux, **p** is the surface tensor, Q_T is the heat source, μ is the dynamic viscosity of the fluid, and λ is relaxation times. Assuming the velocity and shear stress of the form

$$\mathbf{V} = u(y,t)e_x, \quad \mathbf{S} = \mathbf{S}(y,t), \tag{8.92}$$

where e_x is the unit vector in the x-axis.

Consider the initial condition $\mathbf{S}(y,0) = \mathbf{0}$, the coefficient of variable viscosity and the Fourier's law with fractional derivative, we obtain the constitutive relations with temperature-dependent variable fluid properties for the shear stress equation of generalized Maxwell fluid and the generalized Fourier's law for heat transport:

$$\left(1 + \lambda \mathrm{D}_t^{\alpha}\right)\tau_{xy} = \mu(T)\frac{\partial u}{\partial y}, \tag{8.93}$$

$$\left(1 + \lambda_r D_t^{\beta}\right)\mathbf{q} = -\lambda_0 \frac{\partial T}{\partial y}, \tag{8.94}$$

where, $\tau(y,t) = S_{xy}(y,t)$ is the shear stress, $\mu(T)$ is the dynamic viscosity of the fluid, the fractional differential operators D_t^{α} and D_t^{β} based on Riemann–Liouville's are defined as (Podlubny, 1999):

$$\mathrm{D}_t^{*} f(t) = \frac{1}{\Gamma(1-*)}\frac{d}{dt}\int_0^t \frac{f(\tau)}{(t-\tau)^{*}}d\tau, \quad 0 < * < 1, \tag{8.95}$$

and $\Gamma(\cdot)$ denotes the Gamma function.

In this research, we introduce the modified Darcy's law of the generalized viscoelastic flow in the porous medium. The phenomenological model can be described as:

$$\left(1 + \lambda D_t^{\alpha}\right)\nabla \mathbf{P} = -\frac{\mu\phi}{K}\left(1 + \lambda_r D_t^{\beta}\right)\mathbf{V}, \tag{8.96}$$

where K is the permeability and ϕ is the porosity of the porous medium (Xue and Nie, 2009).

In the absence of body forces in the axial direction and the pressure gradient, the balance of the linear momentum leads to the relevant and meaningful equation,

$$\rho\frac{D\mathbf{V}}{Dt} = -\nabla\mathbf{P} + div\mathbf{S} + \mathbf{r}. \tag{8.97}$$

Therefore

$$\left(1 + \lambda D_t^{\alpha}\right)\mathbf{r} = -\frac{\mu\phi}{K}\mathbf{V}, \tag{8.98}$$

and

$$\rho\left(1+\lambda D_t^{\alpha}\right)\frac{D\mathbf{V}}{Dt}=-\left(1+\lambda D_t^{\alpha}\right)\nabla\mathbf{P}+\left(1+\lambda D_t^{\alpha}\right)div\mathbf{S}-\frac{\mu\phi}{K}\mathbf{V}. \tag{8.99}$$

Then, we get the following governing equation of generalized Maxwell fluid:

$$\left(1+\lambda D_t^{\alpha}\right)\frac{\partial u}{\partial t}=\frac{1}{\rho}\frac{\partial}{\partial y}\left(\mu(T)\frac{\partial u}{\partial y}\right)-\frac{\mu(T)\phi}{\rho K}u, \tag{8.100}$$

$$\left(1+\lambda_T \mathrm{D}_t^{\beta}\right)\frac{\partial \mathrm{T}}{\partial t}=\frac{\lambda_0}{\rho c}\frac{\partial^2 \mathrm{T}}{\partial y^2}+\frac{\mu(T)\phi^2}{\rho c K}\left(1+\lambda_T \mathrm{D}_t^{\beta}\right)u^2. \tag{8.101}$$

In this section, we consider dynamic viscosity of the fluid to be the form

$$\mu(T)=\mu_0\exp\left(\frac{-b}{T_0-T_\infty}T+\frac{bT_\infty}{T_0-T_\infty}\right). \tag{8.102}$$

Assume that the fluid is at rest between two infinite parallel plates. At time $t=0^+$, the one plate is suddenly moved by a time-dependent velocity in x axial direction. In this plate, we consider continuous heating with cosine function. The associated initial and boundary conditions are

$$u(y,0)=\frac{\partial u(y,0)}{\partial t}=0, \tag{8.103}$$

$$u(0,t)=U,\ \ u(L,t)=0, \tag{8.104}$$

$$T(y,0)=T_\infty,\frac{\partial T(y,0)}{\partial t}=0, \tag{8.105}$$

$$T(0,t)=T_\infty+A\cos(\omega t),\ \ T(L,t)=T_\infty, \tag{8.106}$$

where U, A, and ω are constants, T_∞ is the temperature of environment.

We assume that the term u^2 can be expressed as a linear function of u, in view of the best square approximation in the interval of $[0,U]$, we obtain

$$u^2\approx\frac{3}{4}Uu. \tag{8.107}$$

Employing the nondimensional quantities,

$$y^*=\frac{\rho U}{\mu_0}y,\ t^*=\frac{U^2\rho}{\mu_0}t,\ u^*=\frac{u}{U},\ \lambda^*=\left(\frac{U^3\rho}{\mu_0}\right)^{\alpha}\lambda,\ \theta=\frac{T-T_\infty}{T_0-T_\infty},\ \lambda_T^*=\left(\frac{U^2\rho}{\mu_0}\right)^{\beta}\lambda_T,$$

$$\Pr=\frac{\mu_0 c}{\lambda_0},\ \omega^*=\frac{\mu_0}{\rho U^2}\omega,\ Da=\left(\frac{\rho U}{\mu_0}\right)^2\cdot K,\ \theta_0=\frac{A}{T_0-T_\infty},\ L^*=\frac{\rho U}{\mu_0}L,$$

where T_0 is fluid initial temperature. Omit the dimensionless mark "*" and we obtain the dimensionless motion equation as follows:

$$\left(1+\lambda D_t^{\alpha}\right)\frac{\partial u}{\partial t}=\frac{\partial}{\partial y}\left(e^{-b\theta}\frac{\partial u}{\partial y}\right)-\frac{\phi}{Da}e^{-b\theta}u, \tag{8.108}$$

$$\Pr\left(1+\lambda_r \mathrm{D}_t^{\beta}\right)\frac{\partial \theta}{\partial t}=\frac{\partial^2\theta}{\partial y^2}+\frac{M\phi^2}{\Pr Da}e^{-b\theta}\left(1+\lambda_r \mathrm{D}_t^{\beta}\right)u. \tag{8.109}$$

Initial and boundary conditions become:

$$u(y,0)=\frac{\partial u(y,0)}{\partial t}=0, \tag{8.110}$$

$$u(0,t)=1, \quad u(L,t)=0, \tag{8.111}$$

$$\theta(y,0)=\frac{\partial \theta(y,0)}{\partial t}=0, \tag{8.112}$$

$$\theta(0,t)=\theta_0\cos(\omega t), \quad \theta(L,t)=0. \tag{8.113}$$

8.4.3 Numerical Algorithms

In this section, the Grünwald–Letnikov approximation, which is based on a numerical approximation of noninteger derivative operator, is used to describe the solutions. Divide the time interval subintervals of equal size h (also called the step size). Approximate $\mathrm{D}_t^{\alpha}f(t)$ at node k as

$$\mathrm{D}_t^{\alpha}f(t)\approx h^{-\alpha}\sum_{k=0}^{\left[\frac{t}{h}\right]}w_k^{(\alpha)}f(t-kh), \tag{8.114}$$

where the normalized Grünwald weights coefficients are $w_k^{(\alpha)}=(-1)^j\binom{\alpha}{k}$, $k=0,1,2,\cdots$ and are defined by

$$w_0^{(\alpha)}=1, \quad w_k^{(\alpha)}=\left(1-\frac{\alpha+1}{k}\right)w_{k-1}^{(\alpha)}, \quad k=1,2,3,\cdots. \tag{8.115}$$

Define $\Delta t=h$ to be a grid size in time dimension where $h=t/M$, $t_i=ih$ for $i=0, \ldots, M$, $y_j=j\Delta y$ where $\Delta y=L/N$ for $j=0, \ldots, N$. Let $\theta_{i,j}$ and $u_{i,j}$ be the numerical approximation to $\theta(t_i,y_j)$ and $u(t_i,y_j)$, respectively.

A universal numerical discretization method, the explicit finite difference method, is used for Eqs. (8.108) and (8.109), yielding:

$$\frac{u_{i+1,j}-u_{i,j}}{h}+\lambda h^{-(\alpha+1)}\sum_{k=0}^{i}\omega_k^{(\alpha+1)}u_{i-k+1,j}=e^{-b\theta_{i,j}}\frac{u_{i+1,j+1}-2u_{i+1,j}+u_{i+1,j-1}}{\Delta y^2}$$
$$-be^{-b\theta_{i,j}}\frac{\theta_{i,j+1}-\theta_{i,j}}{\Delta y}\frac{u_{i+1,j+1}-u_{i+1,j}}{\Delta y}-\frac{\phi}{Da}e^{-b\theta_{i,j}}u_{i+1,j} \tag{8.116}$$

$$\Pr\frac{\theta_{i+1,j}-\theta_{i,j}}{h}+\Pr\lambda_T h^{-(\beta+1)}\sum_{k=0}^{i}\omega_k^{(\beta+1)}\theta_{i-k+1,j}$$
$$=\frac{\theta_{i+1,j+1}-2\theta_{i+1,j}+\theta_{i+1,j-1}}{\Delta y^2}+Me^{-b\theta_{i,j}}\left(u_{i+1,j}+\lambda_T h^{-\beta}\sum_{k=0}^{i}\omega_k^{(\beta)}\theta_{i-k+1,j}\right) \tag{8.117}$$

In view of Eqs. (8.116) and (8.117), we obtain:

$$\frac{e^{-b\theta_{i,j}}(b\theta_{i,j+1}-b\theta_{i,j}-1)}{\Delta y^2}u_{i+1,j+1}$$
$$+\left(\frac{1}{h}+\lambda h^{-(\alpha+1)}+\frac{e^{-b\theta_{i,j}}\left(2-b\theta_{i,j+1}+b\theta_{i,j}+\frac{\phi}{Da}\Delta y^2\right)}{\Delta y^2}\right)u_{i+1,j} \tag{8.118}$$
$$-\frac{e^{-b\theta_{i,j}}}{\Delta y^2}u_{i+1,j-1}=\frac{1}{h}u_{i,j}-\lambda h^{-(\alpha+1)}\sum_{k=1}^{i}\omega_k^{(\alpha+1)}u_{i-k+1,j}$$

$$-\frac{1}{\Delta y^2}\theta_{i+1,j+1}+\left(\frac{\Pr}{h}+\Pr\lambda_T h^{-(\beta+1)}\omega_0^{(\beta+1)}+\frac{2}{\Delta y^2}\right)\theta_{i+1,j}-\frac{1}{\Delta y^2}\theta_{i+1,j-1}$$
$$=\frac{\Pr}{h}\theta_{i,j}-\Pr\lambda_T h^{-(\beta+1)}\sum_{k=1}^{i}\omega_k^{(\beta+1)}\theta_{i-k+1,j}+Me^{-b\theta_{i,j}}\left(u_{i+1,j}+\lambda_T h^{-\beta}\sum_{k=0}^{i}\omega_k^{(\beta)}\theta_{i-k+1,j}\right), \tag{8.119}$$

where $\alpha=0.5$, $\beta=0.7$, $\omega=3$, $\Pr=15$, $\lambda=5$, $\lambda_T=3$, $\theta_0=5$, $b=0.1$, $M=450$, $\phi=0.1$, $Da=5$ subject to the initial and boundary conditions:

$$u_{0,j}=0,\quad \theta_{0,j}=0\quad \text{for}\quad j=0,\ldots,N, \tag{8.120}$$

$$u_{i,0}=1,\quad u_{i,N}=0,\quad \text{for}\quad i=0,\ldots,M, \tag{8.121}$$

$$\theta_{i,0}=\theta_0\cos(\omega\cdot hi),\quad \theta_{i,N}=0,\quad \text{for } i=0,\ldots,M. \tag{8.122}$$

8.4.4 Results and Discussion

The coupled governing nonlinear partial differential equations of the flow and heat transfer of a generalized fractional Maxwell fluid in porous medium are investigated. The motion is due to the longitudinal speed and the energy transmission is due to oscillating boundary condition. The modified Fourier's law, modified Darcy's law, and temperature-dependent variable fluid viscosity are taken into account in coupled velocity and temperature fields. Numerical computations are performed by using an efficient implicit finite difference method based on nonshifted Grünwald formula. In order to study the characteristics of velocity and temperature fields, some numerical results are plotted and presented in Figs. 8.24–8.33.

Fig. 8.24 shows the effects of variable viscosity parameter b on velocity and temperature fields distributions. The results indicate that, for all other parameters fixed, increasing b reduces fluid viscosity, which in turn results in an increasing fluid flow (Fig. 8.24A). In the same periods of fluctuation interval, the fluid viscosity is low if the initial temperature is positive, and therefore the drag effect from lower layer to the upper fluid is low. The difference of speed is very large in the internal close to the left end point, such as in the range of about 0–0.02. On the contrary, the viscosity is relatively large when the initial temperature is negative, the drag effects from lower layer to the upper fluid are high, and therefore the difference of speed is not too large, the flow rate is relatively stable near the position of the left end point, such as for internal of 0.3–0.4. As b increases, the performance of this phenomenon will be more obvious. Fig. 8.24B shows the heat transfer slows down with increasing viscosity effects.

Fig. 8.25 indicates temperature and velocity fields characteristics with increasing time. In this research, we consider the cycle is $2\pi/3$ in the boundary of temperature equation. The initial values of temperature are positive when $t = 2, 6, 10$, which belongs to $[3\pi/2 + 2k\pi, 2\pi + 2k\pi]$. It can be seen from Fig. 8.25A, the velocity increases as time increases, which satisfies the boundary conditions. In Fig. 8.25B, the heat transfer is oscillatory attenuated with the increase of time.

Fig. 8.26 shows the effects of different values of fractional parameter α on the velocity field and temperature field distributions. It is seen that the velocity profile decreases rapidly at the position close to $y = 0$ where the influence of temperature is strong. But, the fluid velocity decreases slowly as α increases when the fluid is away from $y = 0$. Fig. 8.27 displays the effects of fractional parameter β on temperature and velocity profiles. It indicates that the amplitude of fluctuation of temperature profile increases with increasing β in Fig. 8.27B. Meanwhile, the velocity shows a jump decreasing trend due to the increase of temperature oscillation amplitude. When the temperature is negative, the fluid viscosity is strong and the velocity tends to decrease slightly for the upper fluid can move with the

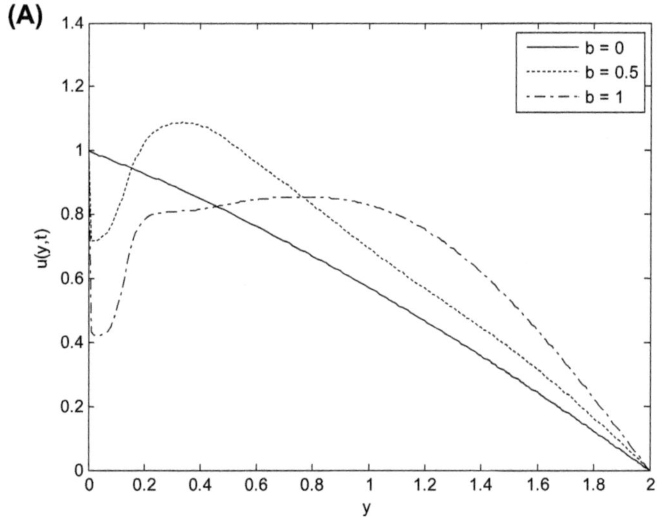

Velocity field

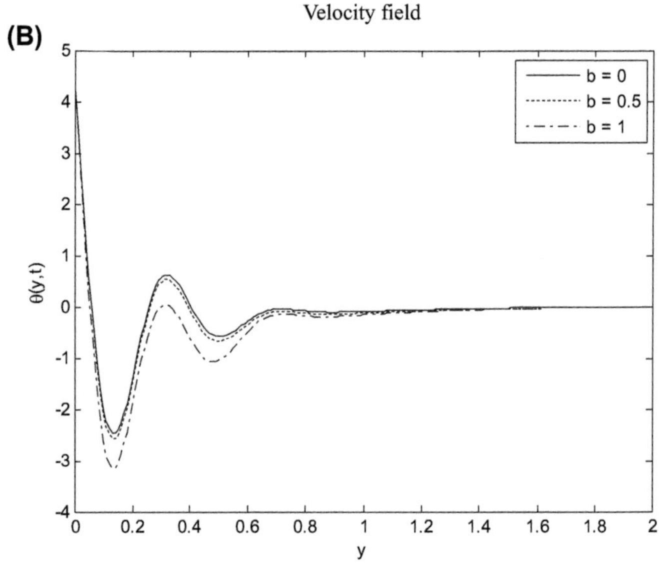

Temperature field

FIGURE 8.24 (A and B) Velocity and temperature profiles for different values of parameter *b*.

lower fluid. Whereas, it is obvious that when the temperature is positive, the fluid viscosity decreases and the velocity tends to increase with the temperature.

The effects of material constant λ and λ_T on the coupled velocity and temperature fields are presented in Figs. 8.28 and 8.29. It is seen from the

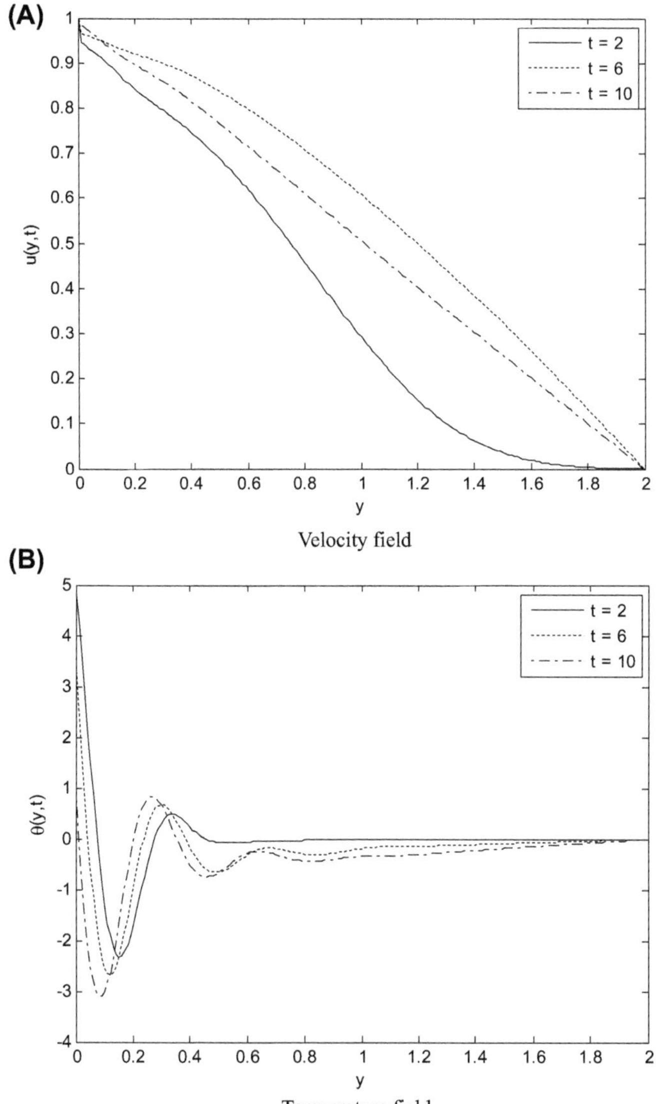

FIGURE 8.25 (A and B) Velocity and temperature profiles for values of time t.

figures that the transfer processes of velocity and heat transfer are accelerative with two parameters increasing, particularly near the boundary.

Fig. 8.30 displays the impact of porosity ϕ on velocity and temperature profiles. It can be seen that the velocity and temperature become bigger in the same position as the porosity decreases. On the contrary, Fig. 8.31 indicates

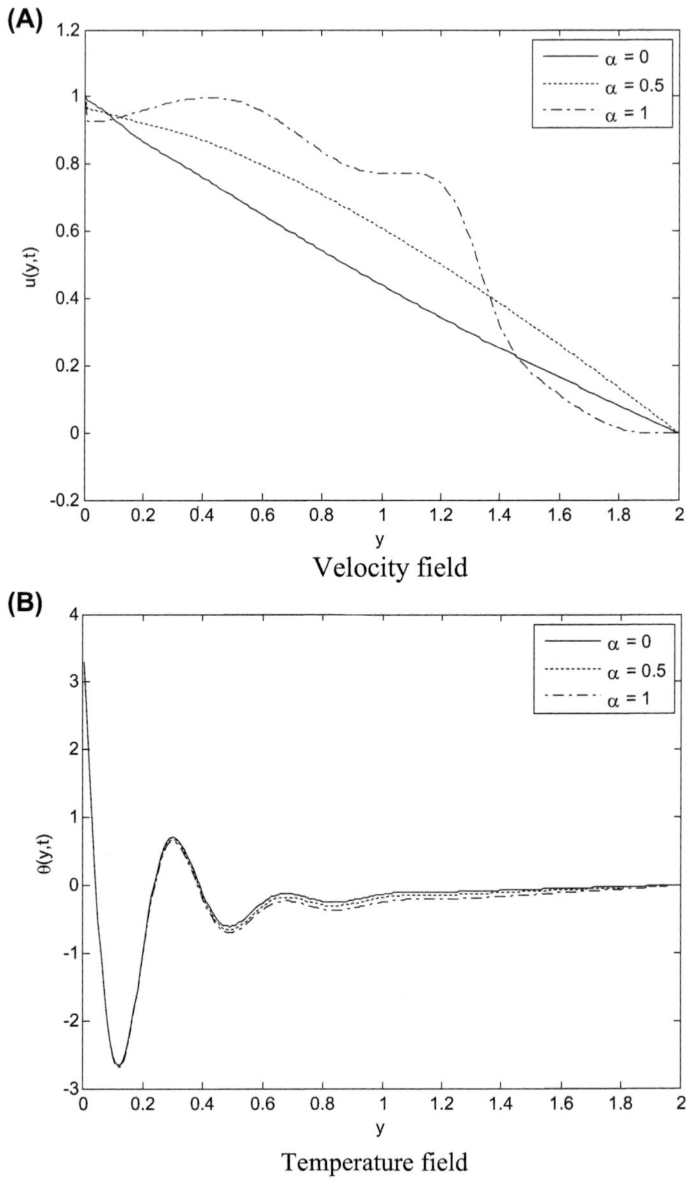

FIGURE 8.26 (A and B) Velocity and temperature profiles for values of fractional parameter α.

that the Darcy's parameter *Da* has an opposite effects on velocity and temperature profiles comparing with Fig. 8.30. It is seen from Fig. 8.32, comparing to the permeability, the Prandtl number has a similar but more pronounced effect on velocity and temperature distributions. Fig. 8.33 shows

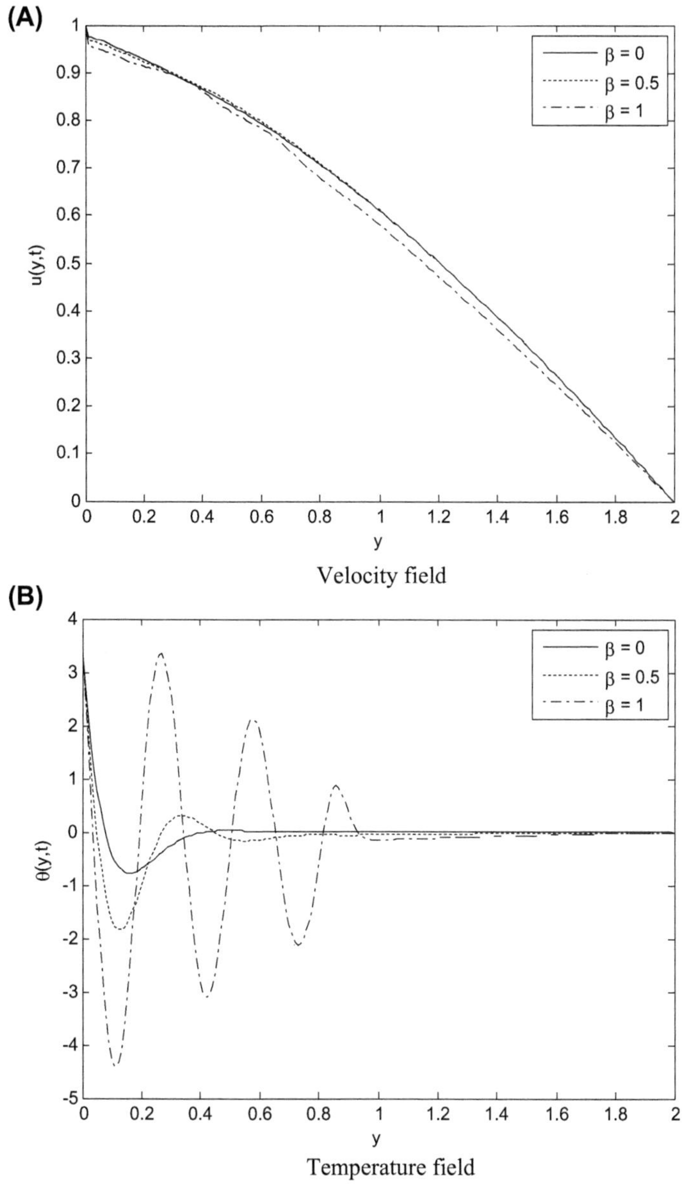

FIGURE 8.27 (A and B) Velocity and temperature profiles for values of fractional parameter β.

the effects of boundary temperature oscillations on velocity and temperature fields. With the frequency increases, the temperature fluctuations show an increasing fluctuation and the flow of fluid shows a more obvious jumping phenomenon.

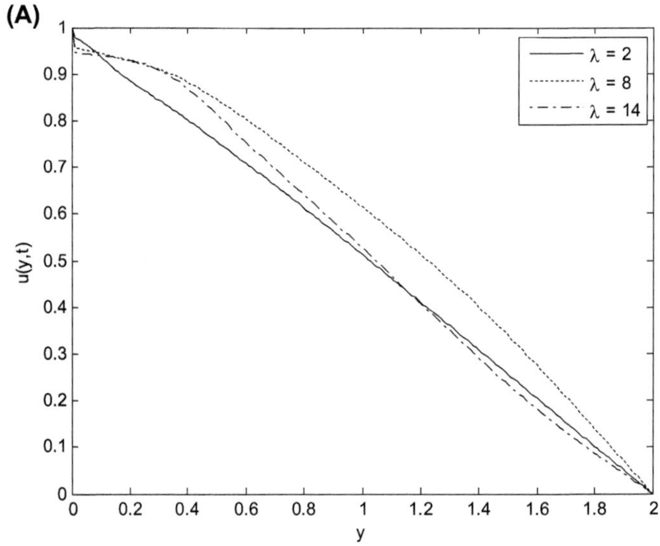

Velocity field

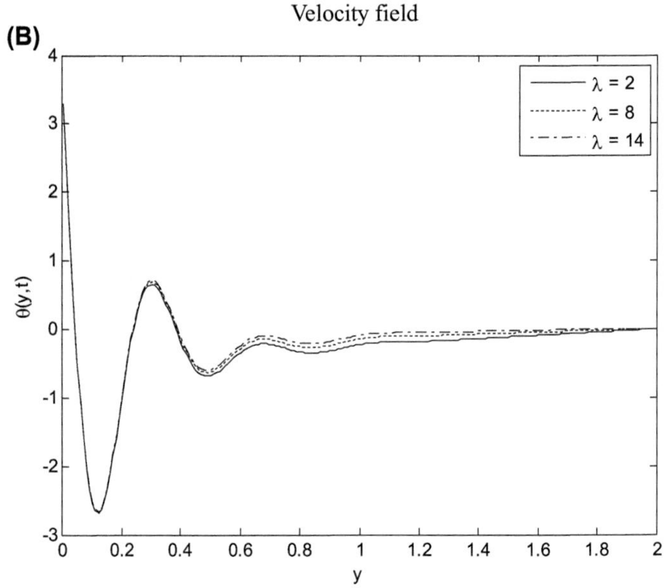

Temperature field

FIGURE 8.28 (A and B) Velocity and temperature fields for values parameter λ.

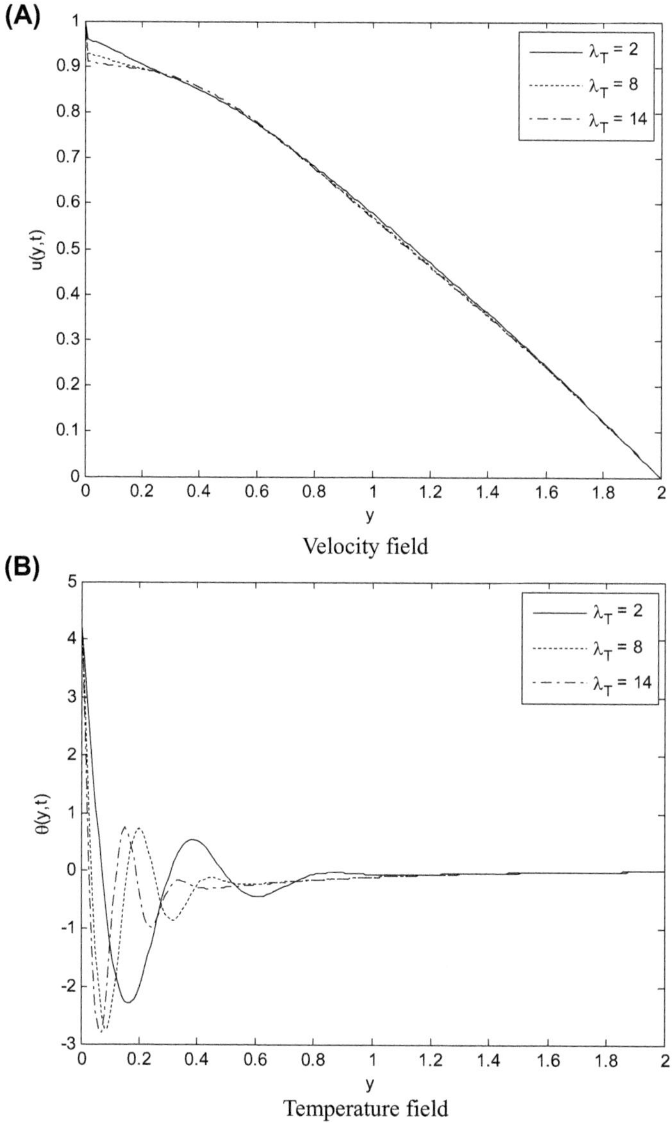

FIGURE 8.29 (A and B) Velocity and temperature profiles for values of parameter λ_T.

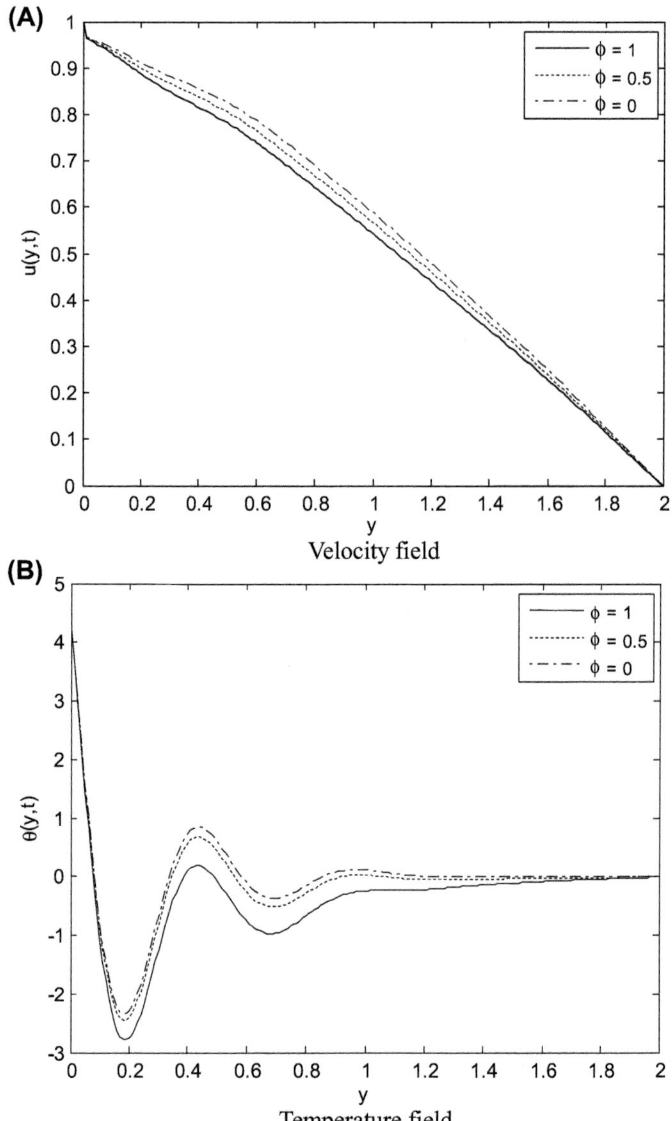

FIGURE 8.30 (A and B) Velocity and temperature profiles for different values of parameter ϕ.

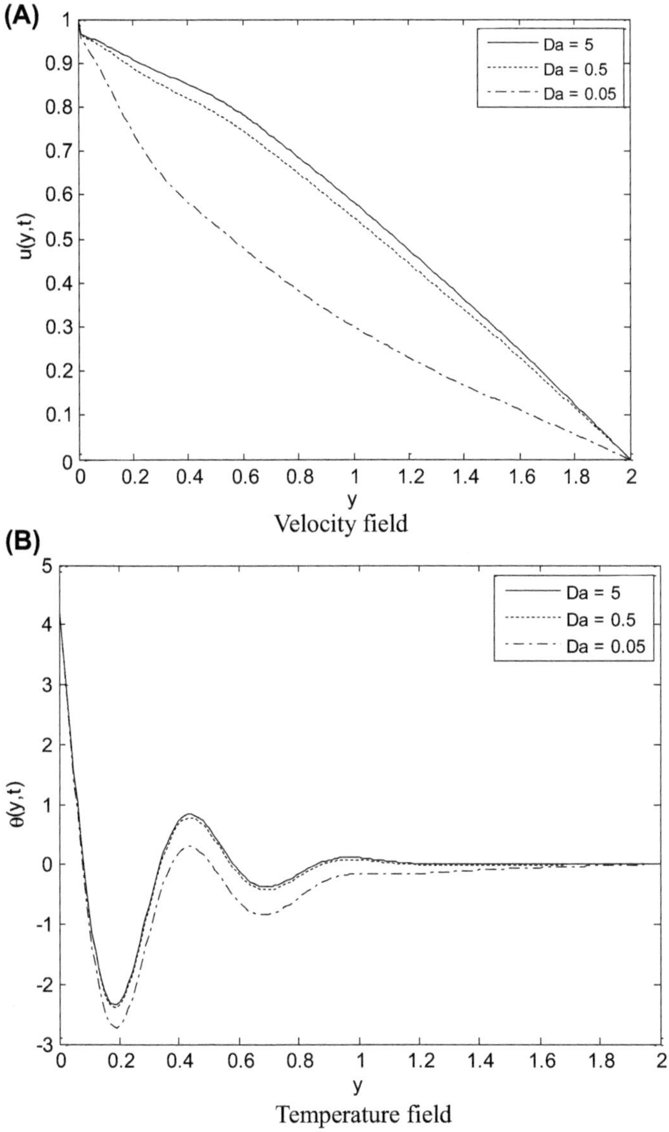

FIGURE 8.31 (A and B) Velocity and temperature profiles for different values of parameter *Da*.

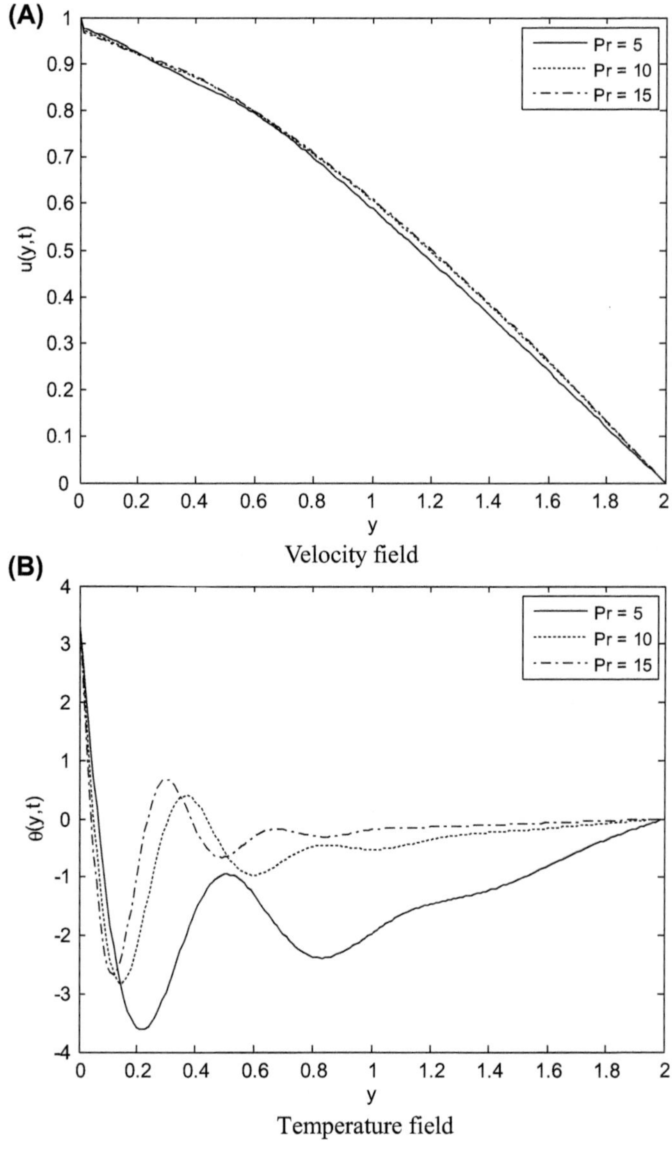

FIGURE 8.32 (A and B) Velocity and temperature profiles for different values of parameter Pr.

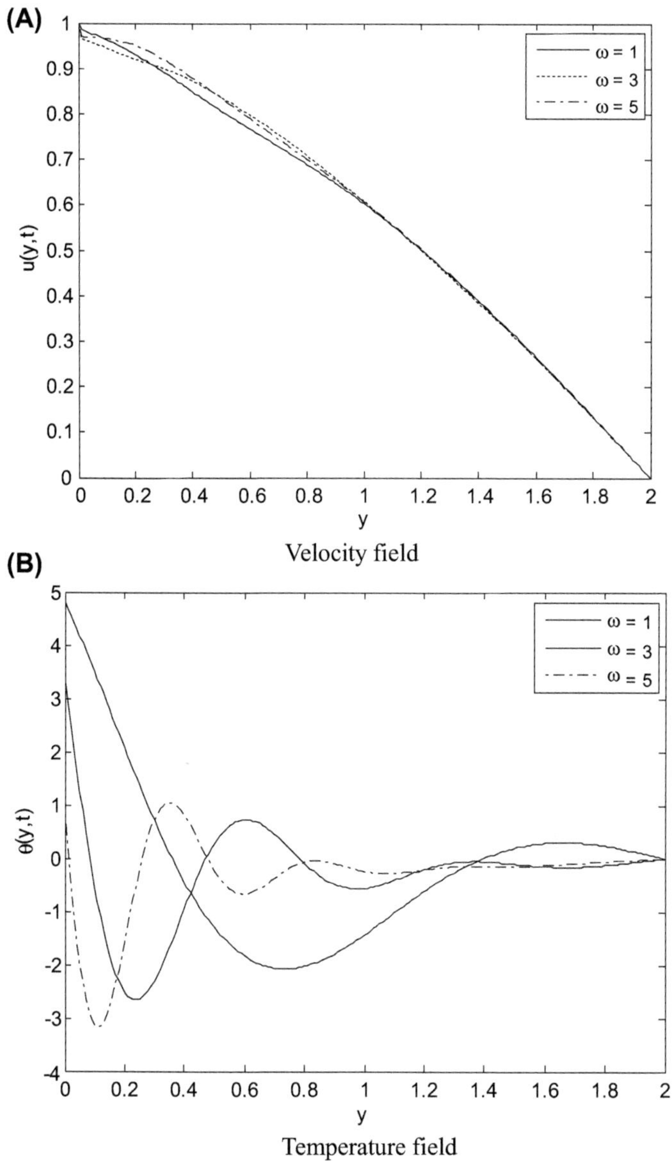

FIGURE 8.33 (A and B) Velocity and temperature profiles for different values of parameter ω.

8.5 UNSTEADY NATURAL CONVECTION HEAT TRANSFER OF FRACTIONAL MAXWELL FLUID

8.5.1 Background of the Problem

Natural convection is a mechanism of heat transportation in which the fluid motion is not generated by an external source. Instead the fluid motion is caused by buoyancy, the difference in fluid density occurring due to temperature gradients. Recently, natural convection has attracted much attentions from researchers because its wide applications both in nature and engineering technology.

In classical study on viscoelastic fluids, researchers commonly ignored the effects of nonlinear convection and dealt only with the cases when the governing equations are linear. In this research, we present an investigation for unsteady boundary layer natural convection heat transfer of Maxwell viscoelastic fluid over a vertical plate. The fractional boundary layer governing equations are first formulated and derived. From such derivation, the model constitutes nonlinear coupled equations with mixed time-space derivatives in the convection terms, which are solved by a newly developed finite difference method combined with an L1-algorithm. The effects of fractional derivative orders on velocity and temperature fields are graphically presented and analyzed in detail. Results show that both the velocity and temperature boundary layer manifest a short-term memory and basic relaxation time characteristics.

8.5.2 Mathematical Formulation

The constitutive equation of viscoelastic fluid with fractional Maxwell model is given by Podlubny (1999), Friedrich (1991), and Friedrich and Braun (1992):

$$\sigma + \lambda^\alpha \frac{d^\alpha \sigma}{dt^\alpha} = \mu \frac{d^\gamma \varepsilon}{dt^\gamma}, \quad 0 \le \alpha \le \gamma \le 1, \tag{8.123}$$

where σ, ε, λ, and μ are the shear stress, shear strain, relaxation time, and viscosity, respectively, and α and γ are fractional parameters. d^α/dt^α and d^γ/dt^γ are the Caputo fractional derivative operators and the fractional derivative of order α is defined as (Podlubny, 1999):

$$\frac{d^\alpha}{dt^\alpha} f(t) = \frac{1}{\Gamma(1-\alpha)} \int_0^t \frac{\partial f(\eta)}{\partial \eta} \frac{d\eta}{(t-\eta)^\alpha}, \tag{8.124}$$

where $\Gamma(\cdot)$ is the Gamma function. Friedrich and Braun (1992) proved that this kind of rheological constitutive equation shows fluid-like behavior only in the case of $\gamma = 1$ and α is variable, i.e.:

$$\sigma + \lambda^\alpha \frac{d^\alpha \sigma}{dt^\alpha} = \mu \frac{d\varepsilon}{dt}, \quad 0 \le \alpha \le 1. \tag{8.125}$$

Eq. (8.125) is known as the constitutive relation of fractional Maxwell fluid. It should be noted that while $\alpha = 0$ and $\alpha = 1$, it is simplified as the classical Newtonian fluid and ordinary Maxwell model, respectively.

We consider here the two-dimensional unsteady laminar boundary layer natural convection flow and heat transfer over a vertical plate, with x being the coordinate along the plate and y perpendicular to the plate. The viscous dissipation is assumed to be neglected. The constitutive relation becomes (Zhao et al., 2016):

$$\sigma_{xy} + \lambda^{\alpha}\frac{\partial^{\alpha}\sigma_{xy}}{\partial t^{\alpha}} = \mu\frac{\partial u}{\partial y}. \tag{8.126}$$

The momentum equation is:

$$\rho\frac{\partial u}{\partial t} + \rho u\frac{\partial u}{\partial x} + \rho v\frac{\partial u}{\partial y} = \frac{\partial \sigma_{xy}}{\partial y} + \rho g\beta_f(T - T_{\infty}). \tag{8.127}$$

Combining Eq. (8.126) and (8.127), we obtained the basic fractional boundary layer governing equation:

$$\begin{aligned}&\frac{\partial u}{\partial t} + u\frac{\partial u}{\partial x} + v\frac{\partial u}{\partial y} + \lambda_1^{\alpha}\frac{\partial^{\alpha+1}u}{\partial t^{\alpha+1}} + \lambda_1^{\alpha}\frac{\partial^{\alpha}}{\partial t^{\alpha}}\left(u\frac{\partial u}{\partial x}\right) + \lambda_1^{\alpha}\frac{\partial^{\alpha}}{\partial t^{\alpha}}\left(v\frac{\partial u}{\partial y}\right)\\ &= \nu\frac{\partial^2 u}{\partial y^2} + \lambda_1^{\alpha}\frac{\partial^{\alpha}}{\partial t^{\alpha}}(\beta_f g(T - T_{\infty})) + \beta_f g(T - T_{\infty}).\end{aligned} \tag{8.128}$$

Similarly, the energy equation can be described with fractional derivatives as:

$$\frac{\partial T}{\partial t} + u\frac{\partial T}{\partial x} + v\frac{\partial T}{\partial y} + \lambda_2^{\beta}\frac{\partial^{\beta+1}T}{\partial t^{\beta+1}} + \lambda_2^{\beta}\frac{\partial^{\beta}}{\partial t^{\beta}}\left(u\frac{\partial T}{\partial x}\right) + \lambda_2^{\beta}\frac{\partial^{\beta}}{\partial t^{\beta}}\left(v\frac{\partial T}{\partial y}\right) = \alpha_f\frac{\partial^2 T}{\partial y^2}, \tag{8.129}$$

where variables u, v, and T are velocity components and temperature, respectively; β_f, ν, and α_f are the volumetric thermal expansion coefficient, dynamic viscosity, and thermal diffusion coefficient of the viscoelastic fluid, respectively; λ_2 is the relaxation time of heat conduction; and α and β are the velocity and temperature fractional derivative parameter, respectively.

The initial and boundary conditions are:

$$t \le 0:\ u = 0,\ v = 0,\ T = T_{\infty};\ t > 0:\ u = 0,\ v = 0,\ T = T_w \text{ at } x = 0;$$

$$u = 0,\ T = T_{\infty} \text{ at } y = 0;\ u \to 0,\ T \to T_{\infty} \text{ as } y \to \infty.$$

Introducing the following dimensionless variables:

$$x^* = \frac{x}{L},\ y^* = \frac{y}{L}Gr^{1/4},\ u^* = \frac{uL}{\nu}Gr^{-1/2},\ v^* = \frac{vL}{\nu}Gr^{-1/4},\ t^* = \frac{\nu t}{L^2}Gr^{1/2},\ \theta = \frac{T - T_\infty}{T_w - T_\infty},$$

$$Gr = \frac{gL^3\beta_f(T_w - T_\infty)}{\nu^2},\quad \Pr = \frac{\nu}{\alpha_f},\quad \lambda_1^* = \frac{\nu\lambda_1}{L^2}Gr^{1/2},\quad \lambda_2^* = \frac{\nu\lambda_2}{L^2}Gr^{1/2},$$

where L is the length of the vertical plate.

We obtained the dimensionless governing equations (for brevity the dimensionless mart "*" is omitted hereafter):

$$\frac{\partial u}{\partial x} + \frac{\partial v}{\partial y} = 0, \tag{8.130}$$

$$\frac{\partial u}{\partial t} + u\frac{\partial u}{\partial x} + v\frac{\partial u}{\partial y} + \lambda_1^\alpha \frac{\partial^{\alpha+1} u}{\partial t^{\alpha+1}} + \lambda_1^\alpha \frac{\partial^\alpha}{\partial t^\alpha}\left(u\frac{\partial u}{\partial x}\right) + \lambda_1^\alpha \frac{\partial^\alpha}{\partial t^\alpha}\left(v\frac{\partial u}{\partial y}\right) = \frac{\partial^2 u}{\partial y^2} + \lambda_1^\alpha \frac{\partial^\alpha \theta}{\partial t^\alpha} + \theta. \tag{8.131}$$

$$\frac{\partial \theta}{\partial t} + u\frac{\partial \theta}{\partial x} + v\frac{\partial \theta}{\partial y} + \lambda_2^\beta \frac{\partial^{\beta+1} \theta}{\partial t^{\beta+1}} + \lambda_2^\beta \frac{\partial^\beta}{\partial t^\beta}\left(u\frac{\partial \theta}{\partial x}\right) + \lambda_2^\beta \frac{\partial^\beta}{\partial t^\beta}\left(v\frac{\partial \theta}{\partial y}\right) = \frac{1}{\Pr}\frac{\partial^2 \theta}{\partial y^2}. \tag{8.132}$$

The dimensionless initial and boundary conditions are:

$$t \le 0:\ u = 0,\ v = 0,\ \theta = 0;\ t > 0:\ u = 0,\ v = 0,\ \theta = 0 \text{ at } x = 0;$$

$$u = 0,\ v = 0,\ \theta = 1 \text{ at } y = 0;\ u \to 0,\ \theta = 0 \text{ as } y \to \infty.$$

It is seen that, when the fractional Maxwell model is introduced, the formulated boundary layer governing equations have nonlinear fractional mixed derivative in the convection terms, which are difficult to solve. We need to develop new numerical method and technique.

8.5.3 Numerical Algorithms

In this section we present the method to solve the nonlinear coupled equations with fractional mixed derivative and compare the particular cases derived from our model with classical results of Ostrach (1953).

8.5.3.1 Discretization Method

Define $x_i = i\Delta x,\ i = 0,\ 1,\ 2,\cdots,\ M,\ y_j = j\Delta y,\ j = 0,\ 1,\ 2,\cdots,\ N;\ t_k = k\Delta t,$ $k = 0, 1, 2,\cdots, R$, where $\Delta x = L/M$ and $\Delta y = Y_{max}/N$ are space steps, Δt is the time step. $u_{i,j}^k$ is the numerical solution of Eqs. (8.130)−(8.132) at the mesh point (x_i,y_j,t_k).

First, L1-algorithm is introduced to discretize the time fractional derivative ($0 < \alpha < 1$) as (Liu et al., 2007):

$$\begin{aligned}\frac{\partial^\alpha f(t_k)}{\partial t^\alpha} &= \frac{\Delta t^{-\alpha}}{\Gamma(2-\alpha)}\sum_{s=0}^{k-1}\alpha_s[f(t_{k-s})-f(t_{k-s-1})]+O\left(\Delta t^{2-\alpha}\right)\\ &= \frac{\Delta t^{-\alpha}}{\Gamma(2-\alpha)}\left[f(t_k)-\alpha_{k-1}f(t_0)-\sum_{s=1}^{k-1}(\alpha_{s-1}-\alpha_s)f(t_{k-s})\right]\\ &\quad+O\left(\Delta t^{2-\alpha}\right),\end{aligned} \tag{8.133}$$

where $\alpha_s=(s+1)^{1-\alpha}-s^{1-\alpha}$, $s=0, 1, 2,\ldots, R$.

Referring to properties of Caputo time fractional derivatives, the following relationship is established:

$$\partial^{\alpha+1}f(t_k)/\partial t^{\alpha+1}=\partial^\alpha f'(t_k)/\partial t^\alpha, \tag{8.134}$$

where $f'(t_k)=[f(t_k)-f(t_{k-1})]/\Delta t$ is the backward difference of time.

Second, the integer-order terms in the governing equations are discretized as (Ganesan and Palani, 2004):

$$\left.\frac{\partial u}{\partial t}\right|_{t=t_k}=\frac{u\left(x_i,y_j,t_k\right)-u\left(x_i,y_j,t_{k-1}\right)}{\Delta t}+O(\Delta t), \tag{8.135}$$

$$\left.u\frac{\partial u}{\partial x}\right|_{t=t_k}=u\left(x_i,y_j,t_{k-1}\right)\frac{u\left(x_i,y_j,t_k\right)-u\left(x_{i-1},y_j,t_k\right)}{\Delta x}+O(\Delta x), \tag{8.136}$$

$$\left.v\frac{\partial u}{\partial y}\right|_{t=t_k}=v\left(x_i,y_j,t_{k-1}\right)\frac{u\left(x_i,y_j,t_k\right)-u\left(x_i,y_{j-1},t_k\right)}{\Delta y}+O(\Delta y), \tag{8.137}$$

$$\left.\frac{\partial^2 u}{\partial y^2}\right|_{t=t_k}=\frac{u_{i,j+1}^k-2u_{i,j}^k+u_{i,j-1}^k+u_{i,j+1}^{k-1}-2u_{i,j}^{k-1}+u_{i,j-1}^{k-1}}{2\Delta y^2}+O\left(\Delta y^2\right), \tag{8.138}$$

Third, Eqs. (8.135)–(8.137) are substituted into the fractional derivatives (Lynch et al., 2003):

$$\frac{\partial^{\alpha+1}u\left(x_i,y_j,t_k\right)}{\partial t^{\alpha+1}}=\frac{\Delta t^{-1-\alpha}}{\Gamma(2-\alpha)}\left[u_{i,j}^k-u_{i,j}^{k-1}-\sum_{s=1}^{k-1}(\alpha_{s-1}-\alpha_s)\left(u_{i,j}^{k-s}-u_{i,j}^{k-s-1}\right)\right], \tag{8.139}$$

$$\frac{\partial^\alpha}{\partial t^\alpha}\left(u\frac{\partial u}{\partial x}\right)=\frac{\Delta t^{-\alpha}}{\Delta x\Gamma(2-\alpha)}\left[u_{i,j}^{k-1}\left(u_{i,j}^{k}-u_{i-1,j}^{k}\right)-\sum_{s=1}^{k-1}(\alpha_{s-1}-\alpha_s)u_{i,j}^{k-s-1}\left(u_{i,j}^{k-s}-u_{i-1,j}^{k-s}\right)\right], \tag{8.140}$$

where the truncation error is $O(\Delta t^{2-\alpha}+\Delta x)$.

$$\frac{\partial^\alpha}{\partial t^\alpha}\left(v\frac{\partial u}{\partial y}\right)=\frac{\Delta t^{-\alpha}}{\Delta y\Gamma(2-\alpha)}\left[v_{i,j}^{k-1}\left(u_{i,j}^{k}-u_{i,j-1}^{k}\right)-\sum_{s=1}^{k-1}(\alpha_{s-1}-\alpha_s)v_{i,j}^{k-s-1}\left(u_{i,j}^{k-s}-u_{i,j-1}^{k-s}\right)\right], \tag{8.141}$$

where the truncation error is $O(\Delta t^{2-\alpha}+\Delta y)$.

Similar difference process is dealt with dimensionless temperature terms. Note,

$$r_1=\frac{\lambda_1^\alpha\Delta t^{-\alpha}}{\Gamma(2-\alpha)},\quad r_2=\frac{\Delta t}{2\Delta y^2},\quad r_3=\frac{\lambda_2^\beta\Delta t^{-\beta}}{\Gamma(2-a)},\quad r_4=\frac{\Delta t}{2\mathrm{Pr}\Delta y^2},$$

$$A_1=\sum_{s=1}^{k-1}(\alpha_{s-1}-\alpha_s)\left(u_{i,j}^{k-s}-u_{i,j}^{k-s-1}\right),\quad A_2=\sum_{s=1}^{k-1}(\alpha_{s-1}-\alpha_s)u_{i,j}^{k-s-1}\left(u_{i,j}^{k-s}-u_{i-1,j}^{k-s}\right),$$

$$A_3=\sum_{s=1}^{k-1}(\alpha_{s-1}-\alpha_s)v_{i,j}^{k-s-1}\left(u_{i,j}^{k-s}-u_{i,j-1}^{k-s}\right),\quad A_4=\theta_{i,j}^{k}-\sum_{s=1}^{k-1}(\alpha_{s-1}-\alpha_s)\theta_{i,j}^{k-s},$$

$$C_1=\sum_{q=1}^{k-1}(b_{q-1}-b_q)\left(\theta_{i,j}^{k-q}-\theta_{i,j}^{k-q-1}\right),\quad C_2=\sum_{q=1}^{k-1}(b_{q-1}-b_q)u_{i,j}^{k-q-1}\left(\theta_{i,j}^{k-q}-\theta_{i-1,j}^{k-q}\right),$$

$$C_3=\sum_{q=1}^{k-1}(b_{q-1}-b_q)v_{i,j}^{k-q-1}\left(\theta_{i,j}^{k-q}-\theta_{i,j-1}^{k-q}\right),$$

where $b_q=(q+1)^{1-\beta}-q^{1-\beta}$, $q=0, 1, 2, \ldots, R$.

Thus, the iteration equations are achieved at last in the following form:

$$\begin{aligned}&-(1+r_3)\frac{\Delta t}{\Delta x}u_{i,j}^{k-1}\theta_{i-1,j}^{k}-\left[(1+r_3)\frac{\Delta t}{\Delta y}v_{i,j}^{k-1}+r_4\right]\theta_{i,j-1}^{k}\\&+\left[(1+r_3)\left(1+\frac{\Delta t}{\Delta x}u_{i,j}^{k-1}\right)+(1+r_3)\frac{\Delta t}{\Delta y}v_{i,j}^{k-1}+2r_4\right]\theta_{i,j}^{k}-r_4\theta_{i,j+1}^{k}\\&=r_4\theta_{i,j-1}^{k-1}+(1+r_3-2r_4)\theta_{i,j}^{k-1}+r_4\theta_{i,j+1}^{k-1}+r_3C_1+\frac{\Delta t}{\Delta x}r_3C_2+\frac{\Delta t}{\Delta y}r_3C_3.\end{aligned} \tag{8.142}$$

$$
\begin{aligned}
&-(1+r_1)\frac{\Delta t}{\Delta x}u_{i,j}^{k-1}u_{i-1,j}^{k}-\left[(1+r_1)\frac{\Delta t}{\Delta y}v_{i,j}^{k-1}+r_2\right]u_{i,j-1}^{k}\\
&+\left[(1+r_1)\left(1+\frac{\Delta t}{\Delta x}u_{i,j}^{k-1}\right)-(1+r_1)\frac{\Delta t}{\Delta y}v_{i,j}^{k-1}+2r_2\right]u_{i,j}^{k}-r_2u_{i,j+1}^{k}\\
&=r_2u_{i,j-1}^{k-1}+(1+r_1-2r_2)u_{i,j}^{k-1}+r_2u_{i,j+1}^{k-1}+r_1A_1+\frac{\Delta t}{\Delta x}r_1A_2\\
&+\frac{\Delta t}{\Delta y}r_1A_3+\Delta t r_1A_4+\frac{\Delta t\left(\theta_{i,j}^{k-1}+\theta_{i,j}^{k}\right)}{2},
\end{aligned}
\tag{8.143}
$$

$$
v_{i,j}^{k}=v_{i,j-1}^{k}+v_{i,j-1}^{k-1}-v_{i,j}^{k-1}+\frac{\Delta y}{2\Delta x}\begin{pmatrix}u_{i-1,j-1}^{k-1}-u_{i,j-1}^{k-1}+u_{i-1,j}^{k-1}-u_{i,j}^{k-1}\\ +u_{i-1,j-1}^{k}-u_{i,j-1}^{k}+u_{i-1,j}^{k}-u_{i,j}^{k}\end{pmatrix}.
\tag{8.144}
$$

In the ordinary integer-order systems, the local skin friction coefficient and the Nusselt number, which can be used to measure the surface shear stress and heat transfer effect, respectively, are defined as (Pal and Mandal, 2015):

$$
C_f=\frac{\mu}{\rho u_m^2}\left(\frac{\partial u}{\partial y}\right)_{y=0},\quad Nu_x=-\frac{x}{T_w-T_\infty}\left(\frac{\partial T}{\partial y}\right)_{y=0}.
\tag{8.145}
$$

In the fractional Maxwell model, the local skin friction coefficient and the Nusselt number are derived from the fractional constitutive Eq. (8.125):

$$
C_f+\lambda_1^{\alpha}\frac{\partial^{\alpha}}{\partial t^{\alpha}}C_f=Gr^{-1/4}\left(\frac{\partial u}{\partial y}\right)_{y=0},\quad Nu+\lambda_2^{\beta}\frac{\partial^{\beta}}{\partial t^{\beta}}Nu=-xGr^{1/4}\left(\frac{\partial \theta}{\partial y}\right)_{y=0}.
\tag{8.146}
$$

By solving (8.146) at each time step, the local skin friction coefficient and the Nusselt number in steady state are obtained as t_k reaches the convergence time:

$$
C_f=\left[r_1\sum_{s=1}^{k-1}(\alpha_{s-1}-\alpha_s)C_f(t_{k-s})+Gr^{-1/4}\left(\frac{\partial u}{\partial y}\right)_{y=0}\right]\Big/(1+r_1).
\tag{8.147}
$$

$$
Nu=\left[r_3\sum_{q=1}^{k-1}(\beta_{q-1}-\beta_q)Nu(t_{k-q})-xGr^{1/4}\left(\frac{\partial \theta}{\partial y}\right)_{y=0}\right]\Big/(1+r_3).
\tag{8.148}
$$

Similarly, the average skin friction coefficient and Nusselt number satisfy:

$$
\overline{C_f}+\lambda_1^{\alpha}\frac{\partial^{\alpha}}{\partial t^{\alpha}}\overline{C_f}=Gr^{-1/4}\int_0^1\left(\frac{\partial u}{\partial y}\right)_{y=0}dx,\quad \overline{Nu}+\lambda_2^{\beta}\frac{\partial^{\beta}}{\partial t^{\beta}}\overline{Nu}=-Gr^{1/4}\int_0^1\left(\frac{\partial \theta}{\partial y}\right)_{y=0}dx,
\tag{8.149}
$$

$$\overline{C_f} = \left[r_1 \sum_{s=1}^{k-1} (\alpha_{s-1} - \alpha_s) \overline{C_f}(t_{k-s}) + Gr^{-1/4} \int_0^1 \left(\frac{\partial u}{\partial y} \right)_{y=0} dx \right] \Big/ (1 + r_1). \tag{8.150}$$

$$\overline{Nu} = \left[r_3 \sum_{q=1}^{k-1} (\beta_{q-1} - \beta_q) \overline{Nu}(t_{k-q}) - Gr^{1/4} \int_0^1 \left(\frac{\partial \theta}{\partial y} \right)_{y=0} dx \right] \Big/ (1 + r_3). \tag{8.151}$$

8.5.3.2 Iteration Algorithm

The values of u, v, and θ in the designated domain at time $t = 0$ are obtained from the initial conditions. During any one-time step, the variables of previous time levels, which appear as coefficients or in summation, are treated as constants. The calculations of u, v, and θ at each time level k are conducted in the following order:

Eq. (8.142) at each internal nodal point on a particular i-level, which constitutes a tri-diagonal system of equations, is solved by Thomas algorithm (Carnahan et al., 1969). The (i-1)-level makes a contribution to the right side of the linear equations, where the starting point is from the plate. The values of θ are obtained at each internal nodal point on a particular i at the kth time level. Utilizing the values of θ at the kth time level, together with θ and u in the previous time levels tracing back to the initial time, the values of u at the kth time level are acquired. Then the values of v are solved from Eq. (8.144) at each internal nodal point on a particular i-level. This process is repeated until the designated boundary is reached. Thus, the values of u, v, and θ are obtained at all grid points in the computational domain. The values of C_f, Nu, $\overline{C_f}$, and $\overline{Nu}$ are evaluated in a similar manner, where the integrals are calculated using compound trapezoidal rule. The time process is also repeated until the steady state is reached when the absolute differences between values of velocity u and temperature θ at two consecutive time steps are less than 10^{-4} at all grid points.

8.5.3.3 Reliability Comparison

The computational region is treated as a rectangle with sides $X_{max} = 1$ and $Y_{max} = 12$ (Y_{max} corresponds to $y \to \infty$, which lies very well outside the momentum and thermal boundary layers). The solution matrix is $M \times N$ during one time step, but the whole storage matrix is $M \times N \times R$. As time iteration goes on, the running time increases by geometric ratio for the existence of fractional derivative summations. In comprehensive consideration of solution accuracy and calculation time, the mesh sizes are fixed as

$\Delta x = 0.05$, $\Delta y = 0.1$, and $\Delta t = 0.1$. It has been tested that the finite difference method in this research is conditionally stable, but solutions in the selected mesh sizes are factually stable and convergent.

Fig. 8.34 shows three-dimensional distributions of temperature and velocity by one set of parameters, which demonstrate good stability and convergence in space and time. The special case as $\lambda_1 = \lambda_2 = 0$ is compared with steady integer-order natural convection in Ostrach (1953). The solutions, for three values of Prandtl numbers, are showed in Fig. 8.35. At $X_{max} = 1$, the dimensionless variable u is double the similarity transformation variables f' while the ratio of y/η is $\sqrt[4]{4}$, which is actually assigned to 1.5 in the graphic. The results are in good agreement, which proves that the chosen mesh sizes are relatively reliable and can be utilized in the following computations.

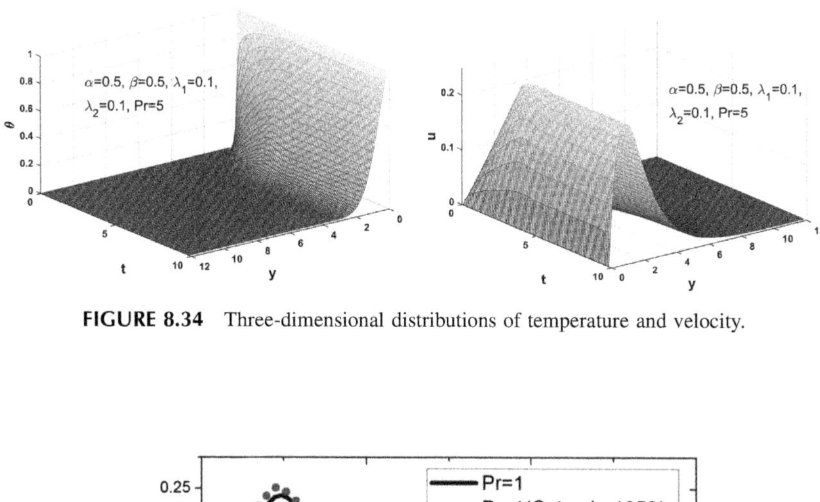

FIGURE 8.34 Three-dimensional distributions of temperature and velocity.

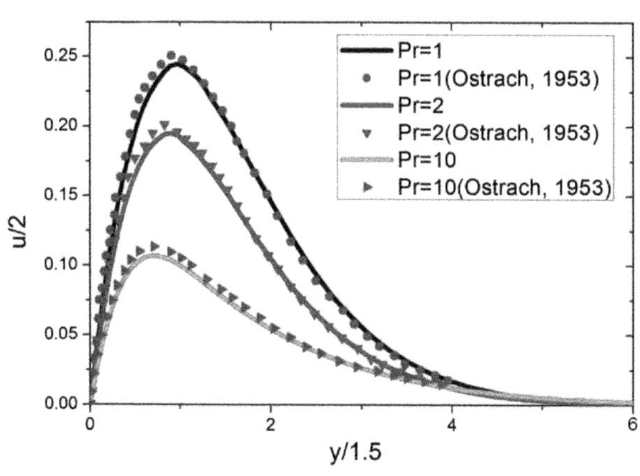

FIGURE 8.35 Comparison with classical solutions of steady natural convection.

8.5.4 Results and Discussion

Without loss of generality, the involved parameters are given as follows: the fractional order α and β are 0, 0.1, 0.2, 0.3, 0.5, and 1, respectively; the relaxation times λ_1 and λ_2 are 0.1, the Prandtl number is fixed to 5, and the Grashof number doesn't take part in the computation and is ignored. The results are compared with classical Newtonian fluid and ordinary Maxwell model when the fractional orders equal 0 and 1, respectively.

8.5.4.1 Effects on Temperature Field

Dimensionless temperature is firstly taken into consideration because the natural convection is driven by temperature gradient. Fig. 8.36 illustrates the temperature profiles at $X_{max} = 1$ in steady state with different values of fractional derivative parameter α. The monotonic decreasing smooth profile can be used to establish the thickness of the thermal boundary layer. It indicates that with the increase of α, the temperature rises and the difference between two adjacent profiles decreases remarkably, while the thermal boundary layer becomes thicker.

The effect of fractional derivative parameter β on temperature field is similar to α, which is shown in Fig. 8.37. It is found that the classical Newtonian fluid has the thinnest thermal boundary layer, while ordinary Maxwell fluid behaves in its weakest capacity in heat conduction. These results demonstrate that fractional order weakens the effect of heat conduction because viscoelasticity strengthens the heat transfer resistance.

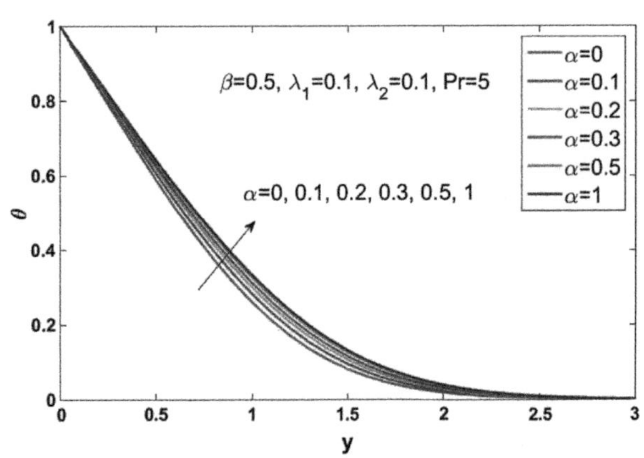

FIGURE 8.36 Temperature distributions for different α.

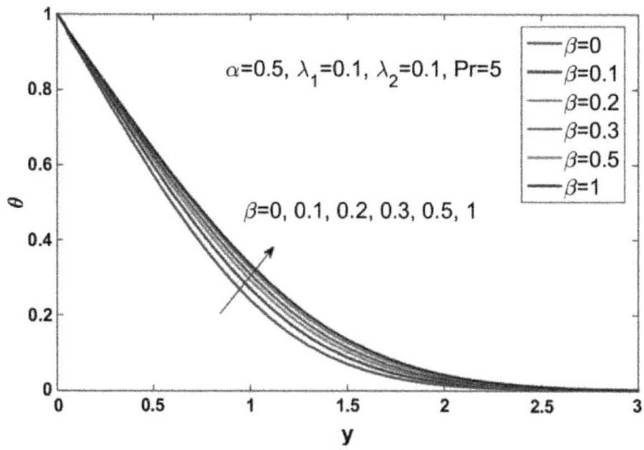

FIGURE 8.37 Temperature distributions for different β.

8.5.4.2 Effects on Velocity Field

Fig. 8.38 shows the effects of fractional derivative parameter α on velocity profiles. It indicates that the velocity profile is not a monotonic function of fractional derivative parameter. For each fractional derivative parameter, the velocity profile has a maximum, which is strongly depended on the parameter, i.e., the smaller parameter the fractional derivative parameter is, the bigger peak value of velocity profile is, and the left position to the vertical plate goes. We can also find that the velocity profiles intersect each other for different values of fractional derivative parameter, and the intersection point moves from a far-field region to the vertical plate, which implies that the thickness of the velocity boundary layer rises as the velocity fractional derivative parameter α increases.

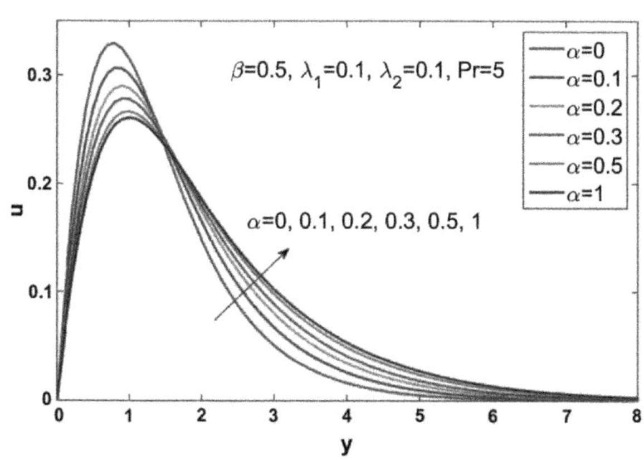

FIGURE 8.38 Velocity distributions for different α.

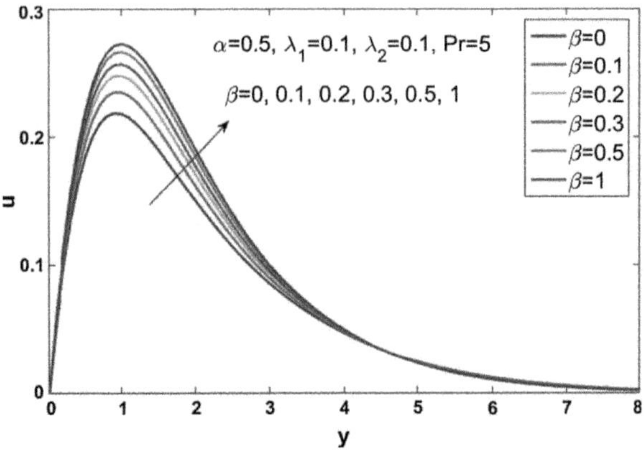

FIGURE 8.39 Velocity distributions for different β.

Fig. 8.39 shows the velocity profiles for different values of temperature fractional derivative parameter β. Similar to Fig. 8.38, it is found that for each β, the velocity profile has a maximum and strongly depends on β. However, the behavior is opposite to cases of the velocity fractional derivative parameter α. It indicates that the smaller parameter β is, the lower the velocity profile is, and the left position to vertical plate goes, which implies the loss of thickness of the velocity boundary layer as the parameter β decreases. Moreover, no intersection points exist for different temperature fractional derivative parameter β.

It is found that the ordinary Maxwell fluid velocity boundary layer has largest thickness, while the classical Newtonian fluid performs best in the natural convection. These results demonstrate that the fractional order weakens the effect of natural convection because viscoelasticity enhances the dynamic viscosity. The effects of α also imply that the fractional equation with relaxation time has a short memory of obvious states and responds slowly to external body force.

8.5.4.3 Effects on Skin Friction Coefficient

Figs. 8.40 and 8.41 illustrate the effects of α and β on average skin friction coefficient, respectively. Results indicate that, for each fractional derivative α, the average skin friction coefficient profile is monotonic increasing. The local skin friction coefficient at $X_{max} = 1$ for different α and β is presented in Tables 8.5 and 8.6. It is found that local and average skin friction coefficient rise with the increase of fractional order α and β. The effect of fractional order is irrelevant with the velocity gradient near the plate in steady state because Figs. 8.38 and 8.39 present contradictory tendency. Actually, from Eqs. (8.147) and (8.150), we know that the skin friction is a comprehensive function of velocity gradient.

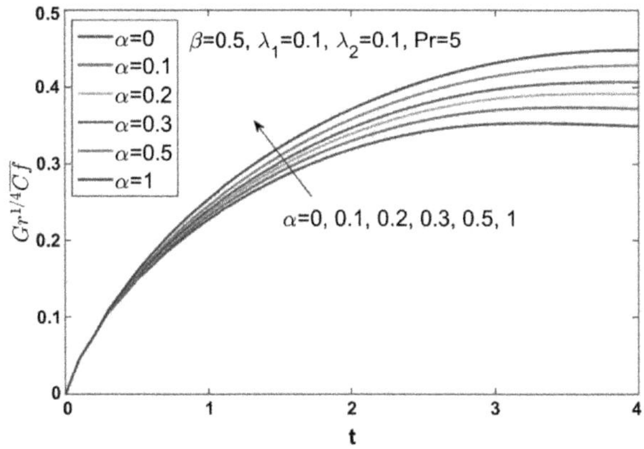

FIGURE 8.40 Average skin friction distributions for different α.

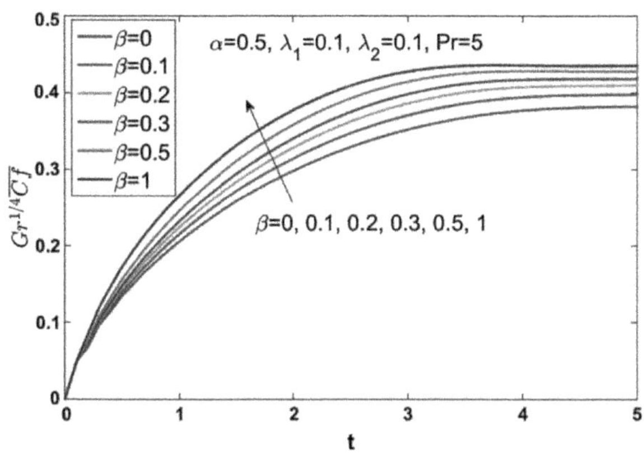

FIGURE 8.41 Average skin friction distributions for different β.

TABLE 8.5 Local Skin Friction Coefficient and Nusselt Number for Different α at $X_{max} = 1$

α	0	0.1	0.2	0.3	0.5	1
$Gr^{1/4}Cf$	0.4648	0.5019	0.5316	0.5531	0.5777	0.5903
$Nu/Gr^{1/4}$	0.7834	0.7556	0.7318	0.7144	0.6951	0.6833

TABLE 8.6 Local Skin Friction Coefficient and Nusselt Number for Different β at $X_{max} = 1$

β	0	0.1	0.2	0.3	0.5	1
$Gr^{1/4}Cf$	0.5126	0.5360	0.5533	0.5652	0.5777	0.5852
$Nu/Gr^{1/4}$	0.4344	0.5144	0.5821	0.6342	0.6951	0.7230

8.5.4.4 Effects on Nusselt Number

Figs. 8.42 and 8.43 illustrate the effects of α and β on average Nusselt number, respectively. Similarly, it is seen that, for each fractional derivative β, the average Nusselt number is monotonic decreasing, while for each fractional derivative α is not monotone, declining in early time and increasing until the steady state. The local Nusselt number at $X_{max} = 1$ for different α and β is presented in Tables 8.5 and 8.6. Results demonstrate that the local and average Nusselt number reduce with the increase of fractional derivative parameter α, but rises with the increase of β. This contrary effect of α and β showed that average Nusselt number is no more relevant with the temperature gradient near the plate in Figs. 8.36 and 8.37, but a comprehensive function of temperature gradient tracing back to initial conditions. Moreover, Figs. 8.42 and 8.43 demonstrate that fractional order α has a slight memory of heat conduction and attempts to recover in obvious states, but β degenerates this effect.

Here we study unsteady boundary layer natural convection flow and heat transfer of Maxwell viscoelastic fluid over a vertical plate. The fractional derivative is introduced in Maxwell constitutive model. The fractional

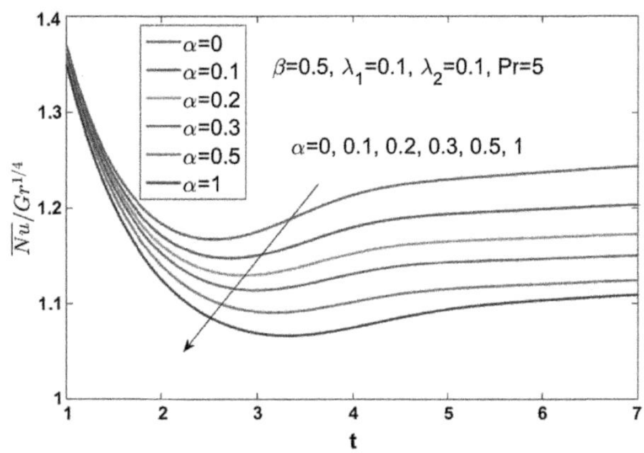

FIGURE 8.42 Average Nusselt number for different α.

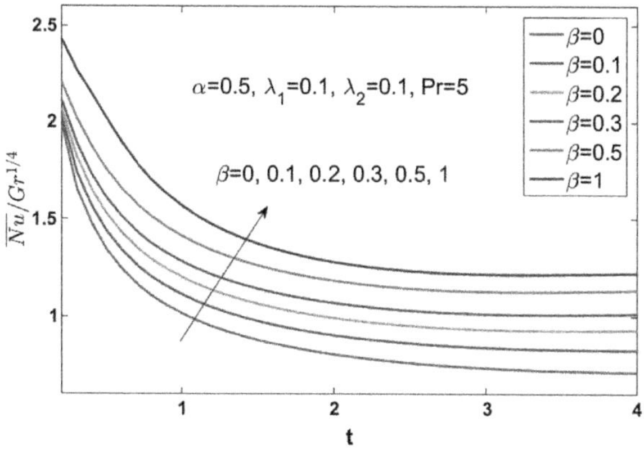

FIGURE 8.43 Average Nusselt number for different β.

boundary layer governing equations are firstly formulated and solved by a newly developed finite difference method combined with an L1-algorithm. The effects of fractional orders α and β are discussed in detail on temperature, velocity, skin friction coefficient, and Nusselt number. Results indicate that the fractional order increases the thickness of velocity and thermal boundary layer and weakens the effect of natural convection and heat conduction. Besides, skin friction coefficient and Nusselt number rise with the augment of α. As β increases, skin friction coefficient rises, but Nusselt number declines.

8.6 FRACTIONAL CONVECTION DIFFUSION WITH CATTANEO–CHRISTOV FLUX

This section presents an investigation for fractional anomalous transport of particles in a comb-like structure. The effects of Cattaneo–Christov flux are taken into account. Formulated fractional governing equation displays a parabolic character for α in (0, 0.5) and the coexisting characteristics of the parabolic and hyperbolic for $\alpha \to 1$ with relaxation parameter effect. When the relaxing parameter equals to zero, the equation reduces a parabolic equation, which is derived from the classical Fick's first law of diffusion. Solutions are obtained numerically by using L1-and L2-approximations for fractional derivative. The effects of the involved parameters on particles distribution behavior are shown graphically and analyzed. Results indicate that the anomalous transport of particles possesses both diffusion and wave characteristics with the existence of relaxation time. Meanwhile, the formation of cusps has been discussed in detail.

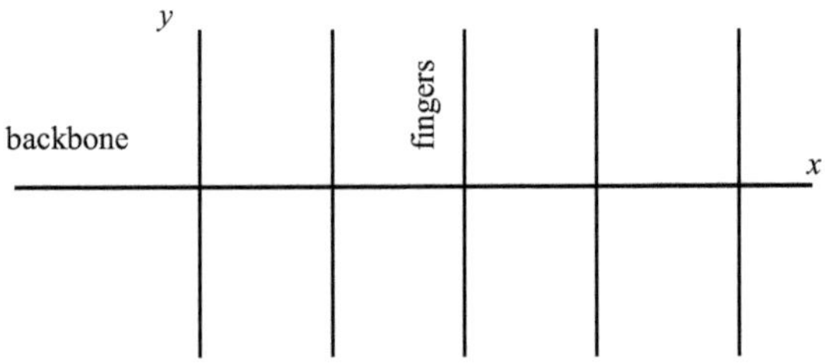

FIGURE 8.44 The sketch framework of comb model.

8.6.1 Fractional Anomalous Diffusion

The research of anomalous diffusion of particles with new constitutive relations has attracted much attention in recent years. A large number of mathematical models have been proposed, among them, the comb structure model (Arkhincheev, 2007; Iomin, 2013) shown in Fig. 8.44 is one of the most significant models that can be used to simulate the transport process of particles well. The model displays two special characteristics: one is that the migration (particles transport) and proliferation (the increase or decrease of particles number) are independent of each other, namely the migration-proliferation dichotomy (Fedotov and Iomin, 2007) is considered; the other is that the diffusion and convection velocity in the x-direction only happen along the x-axis and the diffusion in the y-direction is perpendicular to the x-axis. The comb structure is a widely used model to simulate various pertinent situations, such as cell transport (Iomin, 2005), spiny dendrites (Iomin and Méndez, 2013), percolation clusters (Arkhincheev and Baskin, 1991), and so on.

The anomalous transport of particles in comb structure can be seen as a special case of continuous time random walk and the 1-D diffusion in comb model is described by the time fractional Fokker–Planck equation (Iomin, 2006) with the time fractional derivative of order α—the classical one corresponds to the time fractional derivative of order 1/2. Iomin (2006) discussed the fractional transport of cancer cells due to self-entrapping. His work indicated that the distribution function of the fractional transport depends on the scaled proliferation rate and the order of the fractional derivative. Baskin and Iomin (2004) studied specific properties of particles transported by comb model and the diffusive transport of particles leads to subdiffusion, which corresponds $0 < \beta < 1$ in the relationship between the mean square displacement and parameter t: $\langle x^2(t)\rangle \sim t^{\beta}$. Moreover, the log-normal distribution with exponentially fast spreading was obtained for the cases that transport exponent μ in $< x^2(t) > \sim t^{\mu}$ approaches infinity. Lenzi et al. (2013) investigated a diffusive process in the comb model by considering the effects of drift terms, which represents an external force acting on the system; their results showed that an anomalous spreading may present different diffusive regimes connected to anomalous diffusion and stationary states. For more papers about the comb model, see Refs. (Arkhincheev, 2010; Arkhincheev et al., 2011; Iomin and Baskin, 2005).

The 1-D constitutive relationship (Méndeza and Iomin, 2013) to describe the comb model is derived by the Fick's first law of diffusion:

$$j = -D\frac{\partial P(x,t)}{\partial x}, \tag{8.152}$$

the corresponding time fractional continuity equation is:

$$\frac{\partial^\alpha P(x,t)}{\partial t^\alpha} = -div\, j, \tag{8.153}$$

where j refers to the diffusion flux, D denotes the diffusion coefficient, and $\alpha = 1/2$ corresponds to the classical comb model. $P(x,t)$ is the distribution function defined by $P(x,t) = \int_{-\infty}^{\infty} P_1(x,y,t)dy$ where $P_1(x,y,t)$ refers to the particles distribution at the special location (x,y) and time t. The symbol $\frac{\partial^\alpha}{\partial t^\alpha}$ stands for the time fractional derivative based on the Caputo's definition (Podlubny, 1999), given by:

$$\frac{\partial^\alpha j}{\partial t^\alpha} = \frac{1}{\Gamma(1-\alpha)}\int_0^t \frac{1}{(t-\tau)^\alpha}\frac{\partial j(x,\tau)}{\partial \tau}d\tau, \tag{8.154}$$

The classical Fick's first law of diffusion has been the most successful law for studying diffusion problem in various pertinent situations. However, an unphysical property is that it issues an infinite velocity of propagation due to the fact that there is a finite amount of particles at larger distances from the origin even for very small times.

Cattaneo model (Cattaneo, 1948; Gómez et al., 2007) overcomes the shortcoming of the Fick's first law of diffusion well by introducing the relaxation time term. The modified Fick's first law of diffusion is as follows:

$$j + \xi\frac{\partial j}{\partial t} = -D\frac{\partial P(x,t)}{\partial x}, \tag{8.155}$$

where the propagation velocity (Compte and Metzler, 1997) $v = (D/\xi)^{1/2}$, ξ is a nonnegative constant and refers particularly to the relaxation time of diffusion. The limit $\xi \to 0$ corresponds to the classical Fick's first law of diffusion with an infinite velocity of propagation.

Compte and Metzler (1997) proposed three possible generalized Cattaneo equations, each one supported by a different scheme, and the properties of these generalizations are studied in both the long-time and the short-time regimes. One generalized Cattaneo constitutive relationship is as follows:

$$j + \xi\frac{\partial^\alpha j}{\partial t^\alpha} = -D\frac{\partial P(x,t)}{\partial x}. \tag{8.156}$$

Qi and Jiang (2011) presented exact solution for the space fractional Cattaneo diffusion equation:

$$j + \xi \frac{\partial^\alpha j}{\partial t^\alpha} = -D \frac{\partial^{\mu-1} P(x,t)}{\partial |x|^{\mu-1}}, \quad 1 < \mu \leq 2. \tag{8.157}$$

Here, the symbol $\frac{\partial^{\mu-1}}{\partial |x|^{\mu-1}}$ stands for the space Riesz fractional operator and the corresponding definition is given in Ref. (Qi and Jiang, 2011). For more recent studies about the Cattaneo models, see Refs. (Atanackovic et al., 2007; Liu et al., 2013b; Xu et al., 2013; Qi and Guo, 2014).

However, the model only involves the partial time derivative and is considered only as a "place holder" for a more complete formulation. The frame-indifferent generalization of Cattaneo's law was developed by Christov (2009) by the implementation of the Oldroyd upper-convective derivative. Christov (2007) proposed a generalized Fourier's law model by incorporating spatial memory into the constitutive relation; the integral and differential versions of the memory terms in the constitutive relation are discussed. The model also considers the material invariance and has been applied well in analyzing the flow and heat transfer of various fluids (Han et al., 2014; Mustafa, 2015).

Motivated by the previously mentioned works, we study fractional anomalous diffusion in comb structure; the higher spatial gradient is introduced in the constitutive relation between the flux and particles distribution. In addition, the effects of Cattaneo–Christov flux diffusion and convection are also taken into consideration. The new constitutive relation is described by:

$$j + \xi \left[\frac{\partial j}{\partial t} + u \cdot \nabla j - j \cdot \nabla u + (\nabla \cdot u) j \right] = -D \frac{\partial P(x,t)}{\partial x} \tag{8.158}$$

The numerical discretization method with the L1-and L2-approxiamations for fractional derivative is used to obtain the numerical solution. The effects of the involved parameters on the particles distribution are analyzed and discussed in detail.

8.6.2 Mathematical Formulation

We consider convection velocity as a constant for simplicity and introduce the time fractional operator into the Cattaneo–Christov constitutive relation, Eq. (8.158), and the modified model is given by Liu et al. (2016):

$$j + \xi \left(\tau_1^{\alpha-1} \frac{\partial^\alpha j}{\partial t^\alpha} + u \frac{\partial j}{\partial x} \right) = -D \frac{\partial P(x,t)}{\partial x}, \tag{8.159}$$

where $0 < \alpha \leq 1$ and τ_1 is a newly introduced quantity whose dimension is s.

The corresponding fractional mass conservation equation is given by:

$$\tau_1^{\alpha-1}\frac{\partial^\alpha P(x,t)}{\partial t^\alpha}+u\frac{\partial P(x,t)}{\partial x}=-div\ j. \tag{8.160}$$

Combining Eqs. (8.159) and (8.160), we obtain the time fractional Cattaneo–Christov convection-diffusion equation:

$$\xi\tau_1^{2\alpha-2}\frac{\partial^{2\alpha}P}{\partial t^{2\alpha}}+\tau_1^{\alpha-1}\frac{\partial^\alpha P}{\partial t^\alpha}+2u\xi\tau_1^{\alpha-1}\frac{\partial^{\alpha+1}P}{\partial x\partial t^\alpha}+u\frac{\partial P}{\partial x}+\left(u^2\xi-D\right)\frac{\partial^2 P}{\partial x^2}=0, \tag{8.161}$$

subject to the initial and boundary conditions:

$$P(x,0)=\delta(x),\quad \left.\frac{\partial P(x,t)}{\partial t}\right|_{t=0}=0, \tag{8.162}$$

and

$$P(\pm\infty,t)=0, \tag{8.163}$$

where $\delta(x)$ refers to the Dirac distribution.

Changing the involved coefficients of Eq. (8.161), we can get various kinds of equations. By setting $\xi\neq 0$ and $u=0$, Eq. (8.161) reduces to the time fractional Cattaneo diffusion equation (Langlands and Henry, 2005). By setting $\xi=0$ and $u\neq 0$, Eq. (8.161) reduces to the time fractional convection-diffusion equation. By setting $\xi=0$ and $u=0$, Eq. (8.161) reduces to the time fractional diffusion equation, which has been analyzed and discussed by Iomin (2006).

8.6.3 Numerical Algorithms

Firstly, we define h as the space step and $x_i=-\frac{L}{2}+ih$ $(i=0, 1, 2, \ldots, M)$, while τ is the time step and $t_j=j\tau$ $(j=0, 1, 2, \ldots, N)$, where $L=Mh$ is a sufficiently large number. In this research, we replace the infinite boundary by $\left[-\frac{L}{2},\frac{L}{2}\right]$.

The nondimensional quantities are introduced for simplicity:

$$t\to\tau_1 t^*,\quad x\to\sqrt{D\tau_1}x^*,\quad \xi\to\tau_1\xi^*,\quad u\to\sqrt{\frac{D}{\tau_1}}u^*,\quad L=\sqrt{D\tau_1}L^*, \tag{8.164}$$

Eq. (8.161) and the corresponding initial and boundary conditions can be rewritten in the dimensionless forms (here we omitted the superscript * for simplicity):

$$\xi\frac{\partial^{2\alpha}P}{\partial t^{2\alpha}}+\frac{\partial^\alpha P}{\partial t^\alpha}+2u\xi\frac{\partial^{\alpha+1}P}{\partial x\partial t^\alpha}+u\frac{\partial P}{\partial x}+\left(u^2\xi-1\right)\frac{\partial^2 P}{\partial x^2}=0, \tag{8.165}$$

$$P(x,0)=\delta(x),\quad \left.\frac{\partial P(x,t)}{\partial t}\right|_{t=0}=0, \tag{8.166}$$

$$P(\pm L,t)=0. \tag{8.167}$$

Some useful definitions of the difference schemes are given as follows:

Definition 1. The first-order time and space derivatives can be approximated by the backward difference scheme, i.e.,

$$\frac{\partial P(x_i,t_j)}{\partial t}=\frac{P_i^j-P_i^{j-1}}{\tau}, \tag{8.168}$$

and

$$\frac{\partial P(x_i,t_j)}{\partial x}=\frac{P_i^j-P_{i-1}^j}{h}, \tag{8.169}$$

respectively.

Definition 2. The second-order space derivative can be approximated by the central difference scheme, i.e.,

$$\frac{\partial^2 P(x_i,t_j)}{\partial x^2}=\frac{P_{i+1}^j-2P_i^j+P_{i-1}^j}{h^2}, \tag{8.170}$$

Definition 3. The L1-approximation ($0<\alpha\le 1$) (Du et al., 2010) and L2-approximation ($1\le\beta\le 2$) (Huang and Liu, 2005; Sun and Wu, 2006) for the Caputo fractional derivative can be derived by:

$$\frac{\partial^\alpha P(x_i,t_j)}{\partial t^\alpha}=\frac{1}{\tau^\alpha\Gamma(2-\alpha)}\left(P_i^j+\sum_{k=1}^{j-1}(c_{j-k}-c_{j-k-1})P_i^k-c_{j-1}P_i^0\right), \tag{8.171}$$

and

$$\frac{\partial^\beta P(x_i,t_j)}{\partial t^\beta}=\frac{1}{\tau^\beta\Gamma(3-\beta)}\left[a_0\left(P_i^j-P_i^{j-1}\right)-\sum_{l=1}^{j-1}(a_{j-l-1}-a_{j-l})\left(P_i^l-P_i^{l-1}\right)\right], \tag{8.172}$$

respectively, where $c_0=1$, $c_k=(k+1)^{1-\alpha}-k^{1-\alpha}$ $(k>0)$, $a_0=1$, $a_l=(l+1)^{2-\beta}-l^{2-\beta}$ $(l>0)$.

For $\beta=2\alpha$, the definition of the L2-approximation can be changed as:

$$\frac{\partial^{2\alpha} P(x_i,t_j)}{\partial t^{2\alpha}}=\frac{1}{\tau^{2\alpha}\Gamma(3-2\alpha)}\left[a_0\left(P_i^j-P_i^{j-1}\right)-\sum_{l=1}^{j-1}(a_{j-l-1}-a_{j-l})\left(P_i^l-P_i^{l-1}\right)\right], \tag{8.173}$$

where $a_0=1$, $a_k=(k+1)^{2-2\alpha}-k^{2-2\alpha}$, $(k>0)$.

Through the difference schemes mentioned herein, for $0.5 \leq \alpha \leq 1$, the final discrete scheme of Eq. (8.165) can be expressed by:

$$\left(\frac{\xi u^2 - 1}{h^2}\right)P^j_{i+1} + \left[r_1 + r_2 + r_3 - 2\frac{\xi u^2 - 1}{h^2} + \frac{u}{h}\right]P^j_i + \left[\frac{\xi u^2 - 1}{h^2} - \frac{u}{h} - r_1\right]P^j_{i-1}$$
$$= r_1\left[\sum_{k=1}^{j-1}(c_{j-k-1} - c_{j-k})\left(P^k_i - P^k_{i-1}\right) + c_{j-1}\left(P^0_i - P^0_{i-1}\right)\right]$$
$$+ r_2\left[\sum_{k=1}^{j-1}(c_{j-k-1} - c_{j-k})P^k_i + c_{j-1}P^0_i\right]$$
$$+ r_3\left[P^{j-1}_i + \sum_{l=1}^{j-1}(a_{j-l-1} - a_{j-l})\left(P^l_i - P^{l-1}_i\right)\right], \tag{8.174}$$

where $r_1 = \frac{2u\xi}{\tau^\alpha\Gamma(2-\alpha)h}$, $r_2 = \frac{1}{\tau^\alpha\Gamma(2-\alpha)}$, $r_3 = \frac{\xi}{\tau^{2\alpha}\Gamma(3-2\alpha)}$.

By defining a new matrix G with the coefficients of the previous equation:

$$G_{il} = \begin{cases} \dfrac{\xi u^2 - 1}{h^2} & l = i+1, \\ r_1 + r_2 + r_3 - 2\dfrac{\xi u^2 - 1}{h^2} + \dfrac{u}{h} & l = i, \\ \dfrac{\xi u^2 - 1}{h^2} - \dfrac{u}{h} - r_1 & l = i-1, \end{cases}, \tag{8.175}$$

Eq. (8.174) can be simplified as:

$$GP^j_i = r_1\left[\sum_{k=1}^{j-1}(c_{j-k-1} - c_{j-k})\left(P^k_i - P^k_{i-1}\right) + c_{j-1}\left(P^0_i - P^0_{i-1}\right)\right]$$
$$+ r_2\left[\sum_{k=1}^{j-1}(c_{j-k-1} - c_{j-k})P^k_i + c_{j-1}P^0_i\right] \tag{8.176}$$
$$+ r_3\left[P^{j-1}_i + \sum_{l=1}^{j-1}(a_{j-l-1} - a_{j-l})\left(P^l_i - P^{l-1}_i\right)\right],$$

The discrete schemes of the initial and boundary conditions are given by:

$$p^0_i = \delta(x_i),\ \ p^0_i = p^1_i \ \text{ for } i = 0, 1, 2, \ldots, M, \tag{8.177}$$

and

$$p^j_0 = 0,\ \ p^j_M = 0, \quad \text{for} \quad j = 1, 2, \ldots, N, \tag{8.178}$$

respectively.

By defining $r_4 = \frac{u^2\xi - 1}{h^2}$, $r_5 = \frac{\xi}{\tau^{2\alpha}\Gamma(2-2\alpha)}$ $r_6 = \frac{1}{\tau^{\alpha}\Gamma(2-\alpha)}$, $r_7 = \frac{2u\xi}{h\tau^{\alpha}\Gamma(2-\alpha)}$, for $0 < \alpha \leq 0.5$, Eq. (8.165) can be simplified as:

$$\begin{aligned} r_4 P_{i+1}^j + \left[r_5 + r_6 + r_7 + \frac{u}{h} - 2r_4\right] P_i^j + \left[r_4 - r_7 - \frac{u}{h}\right] P_{i-1}^j \\ = -r_5\left[\sum_{k=1}^{j-1}(b_{j-k} - b_{j-k-1})P_i^k - b_{j-1}P_i^0\right] - r_6\left[\sum_{k=1}^{j-1}(c_{j-k} - c_{j-k-1})P_i^k - c_{j-1}P_i^0\right], \\ -r_7\left[\sum_{k=1}^{j-1}(c_{j-k} - c_{j-k-1})\left(P_i^k - P_{i-1}^k\right) - c_{j-1}\left(P_i^0 - P_{i-1}^0\right)\right], \end{aligned} \tag{8.179}$$

where $b_0 = 1$, $b_k = (k+1)^{1-2\alpha} - k^{1-2\alpha}$.

By defining a new matrix G^1 with the coefficients of the previous equation:

$$G_{il}^1 = \begin{cases} r_4 & l = i+1, \\ r_5 + r_6 + r_7 + \dfrac{u}{h} - 2r_4 & l = i, \\ r_4 - r_7 - \dfrac{u}{h} & l = i-1, \end{cases}, \tag{8.180}$$

Eq. (8.174) can be simplified as:

$$\begin{aligned} G^1 P_i^j = -r_5\left[\sum_{k=1}^{j-1}(b_{j-k} - b_{j-k-1})P_i^k - b_{j-1}P_i^0\right] - r_6\left[\sum_{k=1}^{j-1}(c_{j-k} - c_{j-k-1})P_i^k - c_{j-1}P_i^0\right] \\ -r_7\left[\sum_{k=1}^{j-1}(c_{j-k} - c_{j-k-1})\left(P_i^k - P_{i-1}^k\right) - c_{j-1}\left(P_i^0 - P_{i-1}^0\right)\right], \end{aligned} \tag{8.181}$$

The discrete schemes of the initial and boundary conditions are given by:

$$p_i^0 = \delta(x_i) \ \text{ for } \ i = 0, 1, 2, \ldots, M \tag{8.182}$$

and

$$p_0^j = 0, \ p_M^j = 0 \ \text{ for } \ j = 1, 2, \ldots, N, \tag{8.183}$$

respectively.

8.6.4 Comparison of Numerical and Analytical Solutions

By considering the parameter u as $u = 0$, Eq. (8.165) can be changed as the following form:

$$\xi \frac{\partial^{2\alpha} P}{\partial t^{2\alpha}} + \frac{\partial^\alpha P}{\partial t^\alpha} - \frac{\partial^2 P}{\partial x^2} = 0, \tag{8.184}$$

By using variable transformation, Laplace and Fourier transforms, and properties of H-function, the analytical solution of Eq. (8.184) for $t = 1$ and $\xi = 1$ was obtained by Qi and Jiang (2011):

$$P(x, 1) = \frac{1}{\mu |x| \sqrt{\pi}} \sum_{n=0}^{\infty} \frac{(-1)^n}{n!} \frac{1}{\xi^n} \left\{ \frac{1}{\xi} A + B \right\}, \tag{8.185}$$

where $A = H_{2,3}^{2,1}\left[\frac{\sqrt{\xi}|x|}{2} \middle| \begin{matrix} \left(1, \frac{1}{2}\right), (1 + n\alpha + \alpha, \alpha) \\ \left(\frac{1}{2}, \frac{1}{2}\right), \left(1 + n, \frac{1}{2}\right), \left(1, \frac{1}{2}\right) \end{matrix} \right]$ and

$$B = H_{2,3}^{2,1}\left[\frac{\sqrt{\xi}|x|}{2} \middle| \begin{matrix} \left(1, \frac{1}{2}\right), (1 + n\alpha, \alpha) \\ \left(\frac{1}{2}, \frac{1}{2}\right), \left(1 + n, \frac{1}{2}\right), \left(1, \frac{1}{2}\right) \end{matrix} \right].$$

Moreover, by considering the relaxation time parameter ξ as $\xi = 0$, Eq. (8.165) can be simplified as:

$$\frac{\partial^\alpha P}{\partial t^\alpha} + u \frac{\partial P}{\partial x} - \frac{\partial^2 P}{\partial x^2} = 0. \tag{8.186}$$

Using variable transformation, Mellin and Laplace transforms, and properties of H-function, Huang and Liu (2005) obtained the analytical solution of Eq. (8.186) for $t = 1$ and $\alpha = 0.5$, given by:

$$P(x, 1) = \frac{\exp\left(\frac{x}{2}\right)}{2\sqrt{\pi}} \int_{-\infty}^{\infty} e^{-\frac{1}{4}\sigma - x^2/(4\sigma)} H_{11}^{10}\left(\sigma \middle| \begin{matrix} \left(\frac{1}{4}, \frac{1}{2}\right) \\ \left(-\frac{1}{2}, 1\right) \end{matrix} \right) d\sigma \tag{8.187}$$

The curves of the comparison between the analytical solution and the numerical one of the previous two special cases are shown in Fig. 8.45. We can see that the curves of the analytical and numerical solutions are in very good agreement, which validates the correctness of the numerical one.

8.6.5 Results and Discussion

In this section, the nonlinear fractional differential Eq. (8.165) subject to the boundary and initial conditions is solved using the numerical discretization

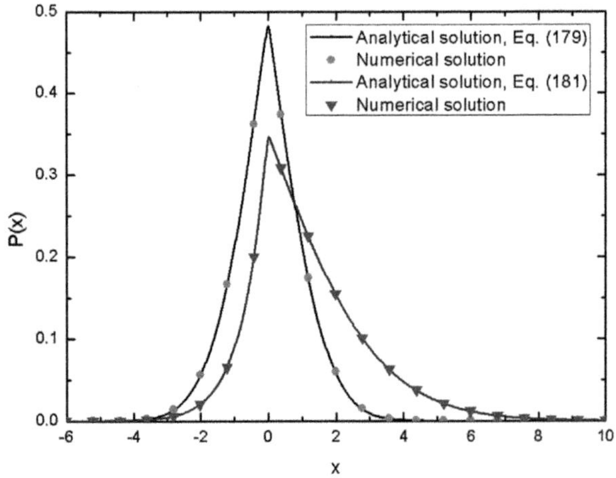

FIGURE 8.45 The comparison of the analytical and numerical solutions in the special case that $\alpha = 0.5$, $\xi = 1$, $u = 0$, and $t = 1$ for Eq. (8.179) and $\alpha = 0.5$, $\xi = 0$, $u = 1$, and $t = 1$ for Eq. (8.181).

method. The influences of different values of time, time fractional parameter, relaxation time, and convection velocity on the particles distribution are also discussed.

Figs. 8.45–8.48 show the particles distribution with different values of parameter t when $\xi = 1$ and $u = 0$ at the condition that the fractional parameters $\alpha = 0.5$, $\alpha = 0.7$, and $\alpha = 1$, respectively. For the time fractional parameter $\alpha = 0.5$, Eq. (8.165) corresponds a diffusion equation, which has a

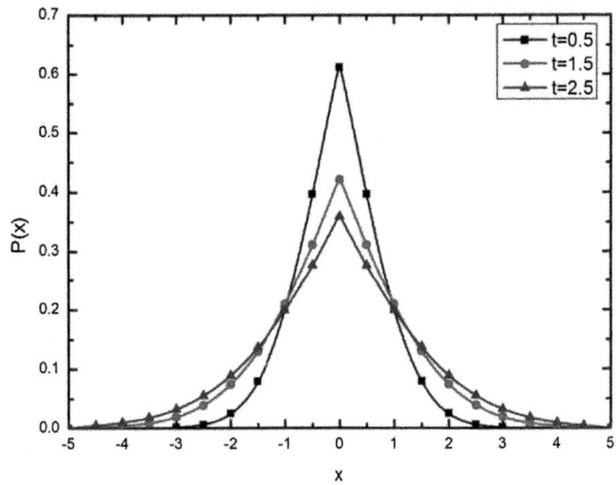

FIGURE 8.46 Distribution curves with different values of parameter t when $\alpha = 0.5$, $\xi = 1$, and $u = 0$.

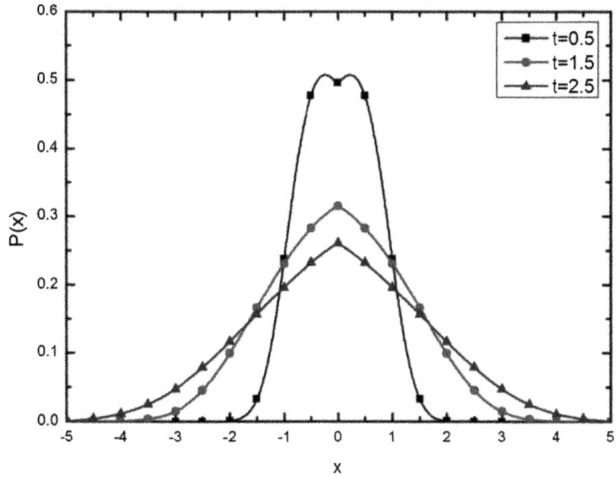

FIGURE 8.47 Distribution curves with different values of parameter t when $\alpha = 0.7$, $\xi = 1$, and $u = 0$.

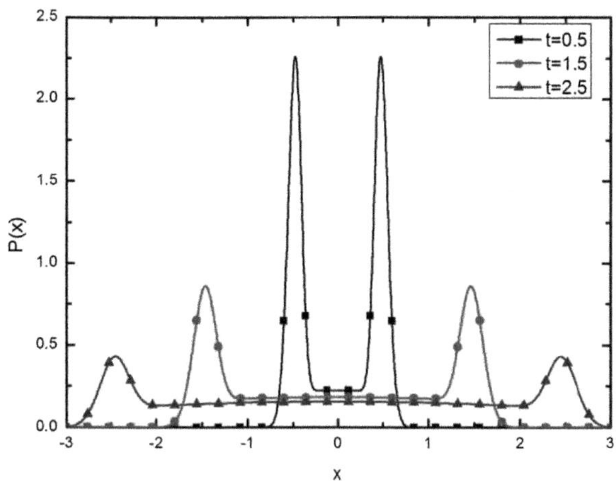

FIGURE 8.48 Distribution curves with different values of parameter t when $\alpha = 1.0$, $\xi = 1$, and $u = 0$.

parabolic character. Fig. 8.46 shows that the peaks of distribution curves become lower while both sides of distribution curves become higher with time increases, that is to say, the particles diffuse along both sides of x-axis gradually.

For the time fractional parameter $\alpha = 0.7$, the highest order of time derivative in differential Eq. (8.165) is 1.4, which coexists the characteristics of parabolic and hyperbolic. Fig. 8.47 shows that the distribution appears to be a hyperbolic curve at the parameter $t = 0.5$ and degenerates to the parabolic curves with time increase. We can conclude that the hyperbolic character weakens while the parabolic character enhances with the increase of time. For the time fractional parameter $\alpha = 1$, Eq. (8.165) corresponds a diffusion-wave equation. Fig. 8.48 shows that the particles propagate as wave form along the positive and negative directions of x-axis and the peaks of distribution curves become lower with time increase. Meanwhile, the value of the distribution between the two peaks is not zero and becomes smaller with time increases for the existence of diffusion.

Figs. 8.49 and 8.50 show the particles distribution with different values of fractional parameter α when $\xi = 1$ and $u = 0$ at the early time $t = 0.1$ and later time $t = 1$, respectively. The time fractional parameter refers to the memory of particles movement. It can be seen from the figures, with the increasing of the fractional parameter α, the peak of the particles distribution becomes higher at the early time $t = 0.1$ but this trend reverses at the later time $t = 1$ for the parabolic curves. The results indicate, for a larger fractional parameter, the diffusion velocity is slower at the early time but faster at the later time. In other words, the initial diffusion velocity is slower and actually the diffusion acceleration is larger. The results are important for addressing the electrotonic significance of decreasing spine densities (Liu et al., 2011; Henry and Langlands, 2008). For a larger fractional parameter, the distribution starts to appear with a hyperbolic character. With the increasing of the fractional parameter α, the peaks become higher for the hyperbolic curves at any time, namely the volatility of the curves becomes stronger.

Figs. 8.51–8.53 show the influence of different values of relaxation time ξ on particles distribution when $\alpha = 0.5$, $\alpha = 0.7$, and $\alpha = 1.0$, respectively. For the parameter $\alpha = 0.5$, the peak of the distribution curve becomes higher with

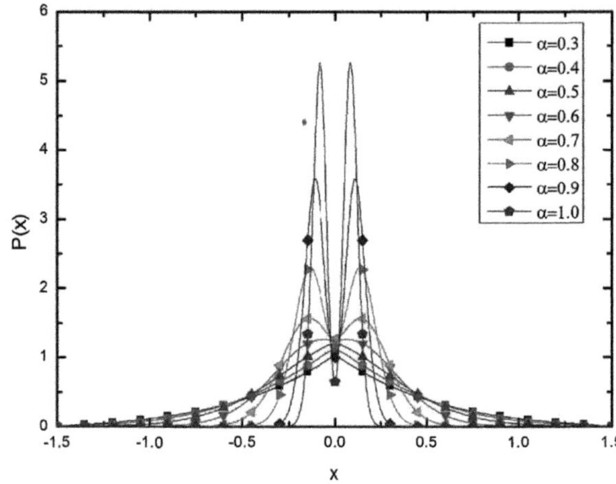

FIGURE 8.49 Distribution curves with different values of parameter α at the early time $t = 0.1$ when $\xi = 1$ and $u = 0$.

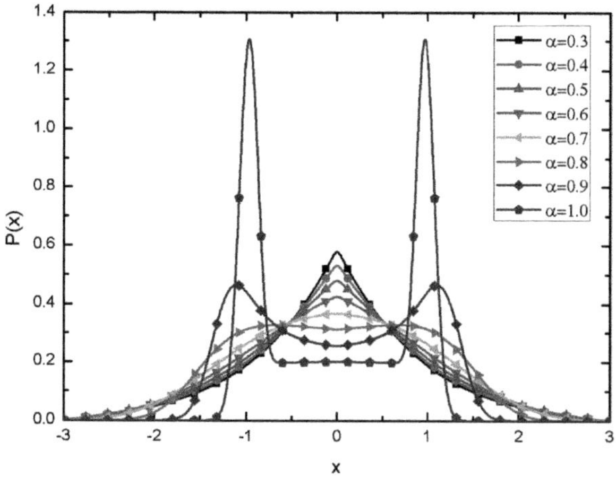

FIGURE 8.50 Distribution curves with different values of parameter α at the later time $t = 1$ when $\xi = 1$ and $u = 0$.

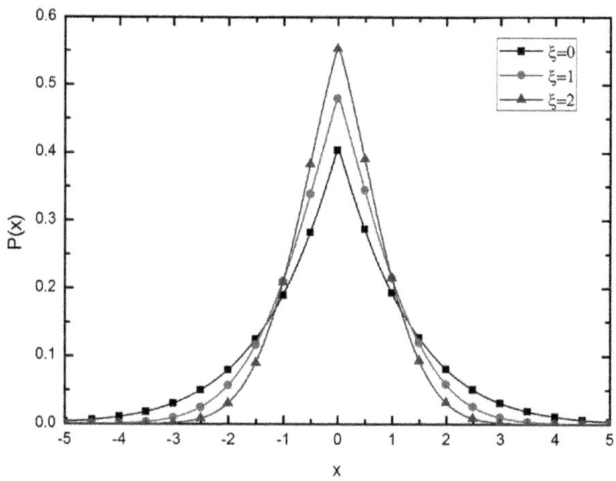

FIGURE 8.51 Distribution curves with different values of parameter ξ when $\alpha = 0.5$, $u = 0$, and $t = 1$.

the increasing of the parameter ξ. For the parameter $\alpha = 0.7$, the distribution curve appears hyperbolic character for a larger parameter ξ. For the parameter $\alpha = 1.0$, the peaks become higher and the distance between the two peaks becomes shorter with the increasing of parameter ξ. Comparing with the Figs. 8.46–8.48, we can conclude that the constant ξ has delay effect. The larger the coefficient ξ is, the stronger the delay effect will be, and vice versa.

The convection velocity has significant effects on particles distribution, which is shown in Figs. 8.54–8.57. Here we only discuss the convection velocity along the positive direction of x-axis. For the negative direction, its

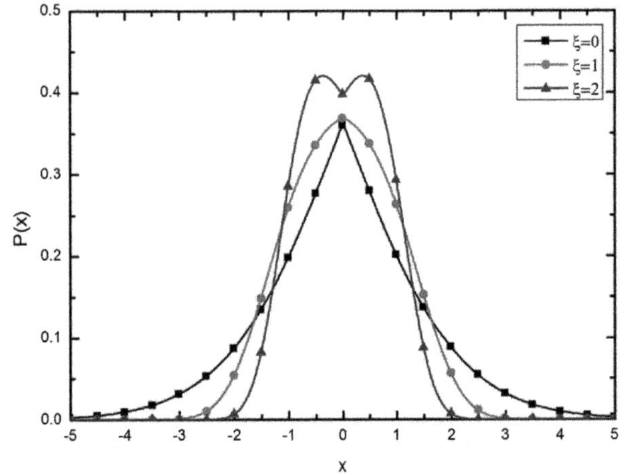

FIGURE 8.52 Distribution curves with different values of parameter ξ when $\alpha = 0.7$, $u = 0$, and $t = 1$.

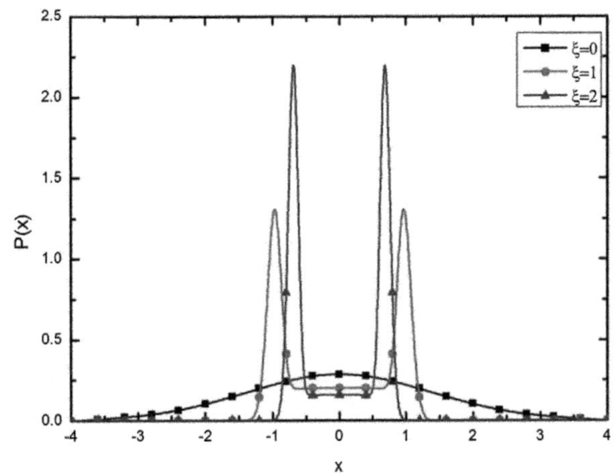

FIGURE 8.53 Distribution curves with different values of parameter ξ when $\alpha = 1.0$, $u = 0$, and $t = 1$.

influence on particles distribution can be obtained by similar analysis. It can be seen from the four figures that the convection velocity can influence the symmetry of the distribution curves.

For parameters $\xi = 0$ and $\alpha = 1.0$, Eq. (8.165) reduces to the convection-diffusion equation that is derived from Fick's law. Fig. 8.54 shows that the distribution curves move along the direction of convection velocity and the greater the velocity is, the father the peak will deviate from the initial position. Moreover, the peak becomes lower but not obvious for a greater velocity u. Fig. 8.55 shows the influences of different values of velocity u on the particles

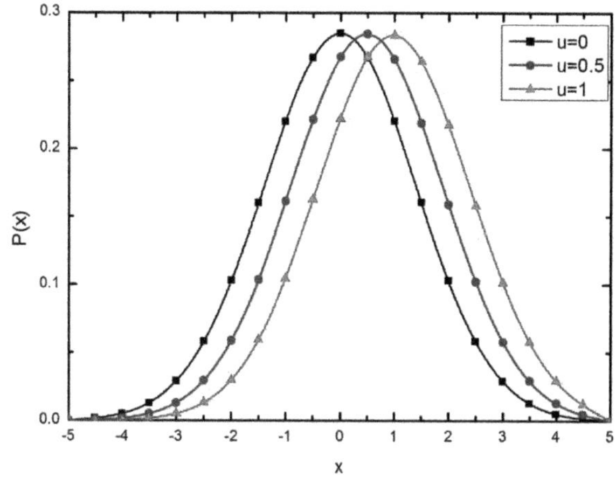

FIGURE 8.54 Distribution curves with different values of velocity u when $\alpha = 1.0$, $\xi = 0$, and $t = 1$.

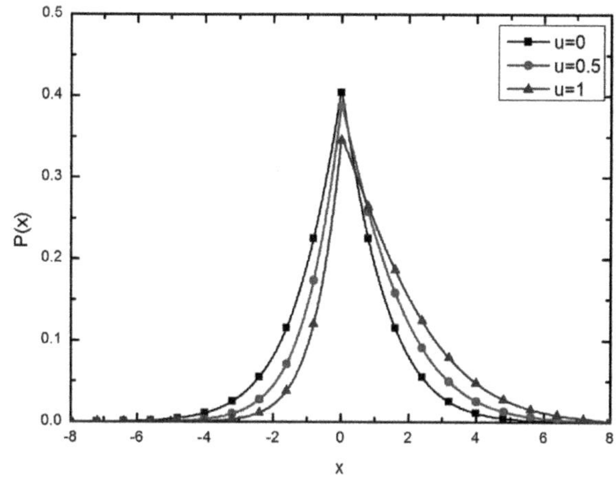

FIGURE 8.55 Distribution curves with different values of velocity u when $\alpha = 0.5$, $\xi = 0$, and $t = 1$.

distribution in the condition of the time fractional derivative of order 1/2, which corresponds to a parabolic equation. The greater the convection velocity is, the lower the peak of distribution curve will be. In addition, the particles diffuse faster along the positive direction but slower along the negative direction for a greater convection velocity. It is noteworthy that the peak does not move even when the convection velocity changes. Moreover, the distribution curves have sharp peaks because of the existence of time fractional operator.

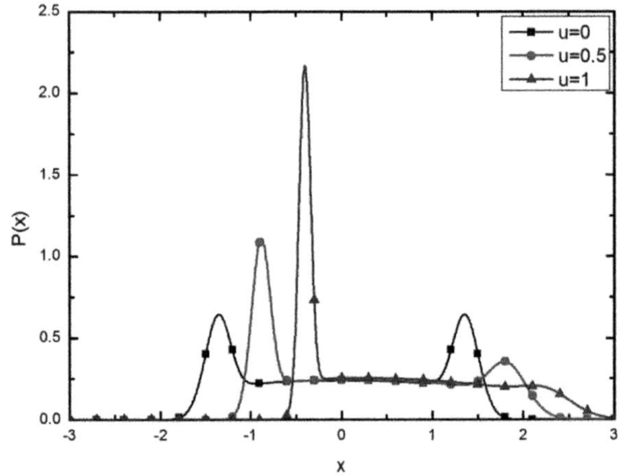

FIGURE 8.56 Distribution curves with different values of velocity u when $\alpha = 1.0$, $\xi = 0.5$, and $t = 1$.

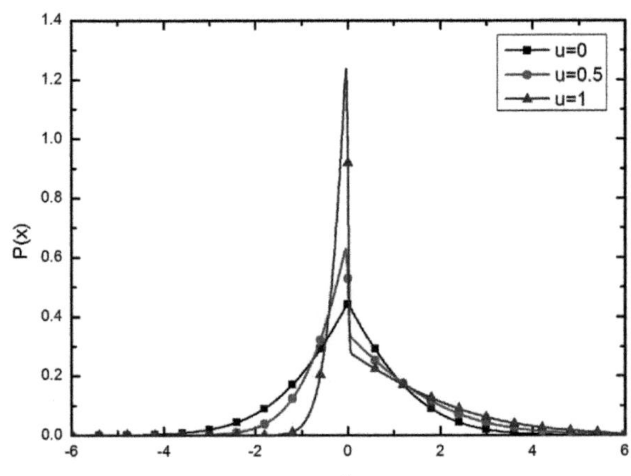

FIGURE 8.57 Distribution curves with different values of velocity u when $\alpha = 0.5$, $\xi = 0.5$, and $t = 1$.

The importance of time fractional operator is verified because of the largest difference at the initial position and time.

Figs. 8.56 and 8.57 show that the particles distribution presents as diffusion wave form and the governing equation has the coexisting characteristics of parabolic and hyperbolic due to the fact of the existence of relaxation parameter ξ and fractional parameter α. Fig. 8.56 shows the influence of convection velocity on particles distribution when $\alpha = 1.0$, $\xi = 0.5$, and $t = 1$. The larger the convection velocity is, the higher the left peak and the lower the right one will be. In other words, the curve peak becomes higher for the same directions of the convection velocity and wave propagation while lower for opposite directions. Moreover, the peak moves along the direction of convection velocity. Fig. 8.57 shows the particles distribution with different values of parameter u when $\alpha = 0.5$, $\xi = 0.5$, and $t = 1$. The cusps that are similar to the inflection points in Fig. 8.56 appear in the curves. They are the result of the existence of relaxation time ξ and fractional parameter α, which make the particles transport have the coexisting characters of parabolic and hyperbolic. For the existences of convection velocity, the peak becomes higher for a greater convection velocity. The right side of particles distribution appears in a general diffusion form and the particles diffuse faster for a greater convection velocity.

8.7 SUMMARY

We study some complex flow, heat, and mass transfer problems arising in non-Newtonian fluids. Four types of Fourier's heat conduction models are investigated for power law non-Newtonian fluids: Flow and heat transfer of power law non-Newtonian fluids over a rotating disk with modified Fourier's heat conduction law and fractional Maxwell fluid with modified Fourier's heat conduction law and modified Darcy's diffusion law; fractional Maxwell fluid unsteady natural convection boundary layer heat transfer over a vertical plate with modified Newtonian friction law and Fourier's heat conduction law; and the fractional convection diffusion in a comb-like structure with Cattaneo–Christov flux. The solutions are obtained numerically and the influences of pertinent parameters on the velocity and temperature fields are discussed in detail.

REFERENCES

Abel, M.S., Siddheshwar, P.G., Mahesha, N., 2009. Effects of thermal buoyancy and variable thermal conductivity on the MHD flow and heat transfer in a power-law fluid past a vertical stretching sheet in the presence of a non-uniform heat source. International Journal of Non-Linear Mechanics 44, 1–12.

Acrivos, A., Shah, M.J., Petersen, E.E., 1960. Momentum and heat transfer in laminar boundary-layer flows of non-Newtonian fluids past external surfaces. American Institute of Chemical Engineers Journal 6, 312–317.

Ames, W.F., 1977. Numerical Methods for Partial Differential Equations, second ed. Academic, New York.

Andersson, H.I., Korte, E., Meland, R., 2001. Flow of a power-law fluid over a rotating disk revisited. Fluid Dynamics Research vol. 28, 75–88.

Arkhincheev, V.E., 2007. Random walks on the comb model and its generalizations. Chaos 17, 043102.

Arkhincheev, V.E., 2010. Unified continuum description for sub-diffusion random walks on multi-dimensional comb model. Physica A 389, 1–6.

Arkhincheev, V.E., Kunnen, E., Baklanov, M.R., 2011. Active species in porous media: random walk and capture in traps. Microelectronic Engineering 88, 694–696.

Arkhincheev, V.E., Baskin, E.M., 1991. Anomalous diffusion and drift in a comb model of percolation clusters. Journal of Experimental and Theoretical Physics 73, 161–165.
Arunachalam, M., Rajappa, N.R., 1978. Thermal boundary layer in liquid metals with variable thermal conductivity. Applied Sciences Research 34, 179–187.
Atanackovic, T.M., Pilipovic, S., Zorica, D., 2007. A diffusion wave equation with two fractional derivatives of different order. Journal of Physics A: Mathematical and Theoretical 40, 5319–5333.
Attia, H.A., 2008. Rotating disk flow and heat transfer through a porous medium of a non-Newtonian fluid with suction and injection. Communications in Nonlinear Science and Numerical Simulation 13, 1571–1580.
Barletta, A., 1997. Fully developed laminar forced convection in circular ducts for power-law fluids with viscous dissipation. International Journal of Heat and Mass Transfer 40, 15–26.
Baskin, E., Iomin, A., 2004. Superdiffusion on a comb structure. Physical Review Letters 93, 120603.
Benton, E.R., 1966. On the flow due to a rotating disk. Journal of Fluid Mechanics 24, 781–800.
Carnahan, B., Luther, H.A., Wilkes, J.O., 1969. Applied Numerical Methods. John Wiley and Sons, New York.
Cattaneo, C., 1948. Sulla conduzione del calore, Some aspects in Diffusion Theory.
Chaim, T.C., 1998. Heat transfer in a fluid with variable thermal conductivity over a linearly stretching sheet. Acta Mechanica 129, 63–72.
Chen, C., Liu, F., Anh, V., Turner, I., 2010. Numerical schemes with high spatial accuracy for a variable-order anomalous subdiffusion equation. SIAM Journal on Scientific Computing 32, 1740–1760.
Cheng, C.Y., 2009. Nonsimilar boundary layer analysis of double-diffusive convection from a vertical truncated cone in a porous medium with variable viscosity. Applied Mathematics and Computation 212, 185–193.
Christov, C.I., 2009. On frame indifferent formulation of the Maxwell–Cattaneo model of finite-speed heat conduction. Mechanics Research Communications 36, 481–486.
Christov, C.I., 2007. On a higher-gradient generalization of Fourier's law of heat conduction. AIP Conference Proceedings 946, 11–22.
Ciarlet, P.G., 1978. The Finite Element Method for Elliptic Problems. North-Holland Publishing Company, Amsterdam.
Cochran, W.G., Goldstein, S., 1934. The flow due to a rotating disk. Mathematical Proceedings of the Cambridge Philosophical Society 30, 365–375.
Compte, A., Metzler, R., 1997. The generalized Cattaneo equation for the description of anomalous transport processes. Journal of Physics A: Mathematical and General 30, 7277–7289.
Crosan, T., Pop, I., 2001. Free convection over a vertical flat plate with a variable wall temperature and internal heat generation in a porous medium saturated with a non-Newtonian fluid. Technische Mechanik 4, 313–318.
Du, R., Cao, W.R., Sun, Z.Z., 2010. A compact difference scheme for the fractional diffusion-wave equation. Applied Mathematical Modelling 34, 2998–3007.
Ece, M.C., Buyuk, E., 2002. Similarity solutions for free convection to power-law fluids from a heated vertical plate. Applied Mathematics Letters 15, 1–5.
Ezzat, M.A., 2010. Thermoelectric MHD non-Newtonian fluid with fractional derivative heat transfer. Physica B 405, 4188–4194.
Fedotov, S., Iomin, A., 2007. Migration and proliferation dichotomy in tumor-cell invasion. Physical Review Letters 98, 118101.
Fetecau, C., Fetecau, C., Kamran, M., Vieru, D., 2009. Exact solution for the flow of a generalized Oldroyd-B fluid induced by a constantly accelerating plate between side walls perpendicular to the plate. Journal of Non-Newtonian Fluid Mechanics 156, 189–201.

Fletcher, C.A.J., 1988. Computationl Techniques for Fluid Dynamics 1. Spring-Verlag, New York.

Friedrich, C., 1991. Relaxation and retardation functions of the Maxwell model with fractional derivatives. Rheologica Acta 30, 151–158.

Friedrich, C., Braun, H., 1992. Generalised Cole-Cole behavior and its rheological relevance. Rheologica Acta 31, 309–322.

Ganesan, P., Palani, G., 2004. Finite difference analysis of unsteady natural convection MHD flow past an inclined plate with variable surface heat and mass flux. International Journal of Heat and Mass Transfer 47, 4449–4457.

Gómez, H., Colominas, I., Navarrina, F., Casteleiro, M., 2007. A finite element formulation for a convection–diffusion equation based on Cattaneo's law. Computer Methods in Applied Mechanics and Engineering 196, 1757–1766.

Gorla, R.S.R., Pop, I., Lee, J.K., 1992. Convective wall plume in power-law fluid: second-order correction for the adiabatic wall. Warme-und Stoffubertragung 27, 473–479.

Hady, F.M., 1995. Mixed convection boundary-layer flow of non-Newtonian fluids on a horizontal plate. Applied mathematics and Computation 68, 105–112.

Han, S.H., Zheng, L.C., Li, C.R., Zhang, X.X., 2014. Coupled flow and heat transfer in viscoelastic fluid with Cattaneo–Christov heat flux model. Applied Mathematics Letters 38, 87–93.

Henry, B.I., Langlands, T.A.M., 2008. Fractional cable models for spiny neuronal dendrites. Physical Review Letters 100, 128103.

Hossain, M.A., Munir, M.S., Pop, I., 2001. Natural convection with variable viscosity and thermal conductivity from a vertical wavy cone. International Journal of Thermal Science 40, 437–443.

Hossain, M.A., Munir, M.S., Rees, D.A.S., 2000a. Flow of viscous incompressible fluid with temperature dependent viscosity and thermal conductivity past a permeable wedge with uniform surface heat flux. International Journal of Thermal Science 39, 635–644.

Hossain, M.A., Munir, M.S., Takhar, H.S., 2000b. Natural convection flow of a viscous fluid about a truncated cone with temperature dependent viscosity and thermal conductivity. Acta Mechanica 140, 171–181.

Howell, T.G., Jeng, D.R., De Witt, K.J., 1997. Momentum and heat transfer on a continuous moving surface in power law fluid. International Journal of Heat and Mass Transfer 40, 1853–1861.

Huang, F., Liu, F., 2005. The time fractional diffusion equation and the advection-dispersion equation. The Anziam Journal 46, 317–330.

Iomin, A., 2005. Superdiffusion of cancer on a comb structure. Journal of Physics: Conference Series 7, 57–67.

Iomin, A., 2006. Toy model of fractional transport of cancer cells due to self-entrapping. Physical Review E 73, 061918.

Iomin, A., 2013. Fractional kinetics of glioma treatment by a radio-frequency electric field. The European Physical Special Topics 222, 1875–1884.

Iomin, A., Baskin, E., 2005. Negative superdiffusion due to inhomogeneous convection. Physical Review E 71, 061101.

Iomin, A., Méndez, V., 2013. Reaction-subdiffusion front propagation in a comblike model of spiny dendrites. Physical Review E 88, 012706.

Irvine, T.F.J., et al., 1988. Symposium on Fundamentals of Forced Convection Heat Transfer, vol. 101. ASME Publ. HTD, pp. 123–127.

Jamil, M., Zafar, A.A., Rauf, A., Khan, N.A., 2012. New exact analytical solutions for helical flows of second grade fluids. Communications in Nonlinear Science and Numerical Simulation 17, 141–153.

Jayanthi, S., Kumari, M., 2007. Effect of variable viscosity on non-Darcy free or mixed convection flow on a vertical surface in a non-Newtonian fluid saturated porous medium. Applied Mathematics and Computation 186, 1643–1659.

Kármán, T., 1921. Uber laminare und turbulente reibung. ZAMM-Zeitschrift fur Angewandte Mathematik und Mechanik 1, 233–252.

Kays, W.M., 1966. Convective Heat and Mass Transfer. McGraw-Hill, New York.

Khan, M., Ali, S.H., Qi, H.T., 2009. Exact solutions for some oscillating flows of a second grade fluid with a fractional derivative model. Mathematical and Computer Modelling 49, 1519–1530.

Khan, M., Arshad, M., Aujum, A., 2012. On exact solutions of Stokes second problem for MHD Oldroyd-B fluid. Nuclear Engineering and Design 243, 20–32.

Kim, S.C., 1995. Graetz problem solutions for a modified power law fluid over a wide range of shear rate. The Korean Journal of Rheology 7, 35–41.

Kumari, K., Pop, I., Takhar, H.S., 1997. Free-convection boundary-layer flow of a non-Newtonian fluid along a vertical wavy surface. International Journal of Heat and Fluid Flow 18, 625–631.

Langlands, T.A.M., Henry, B.I., 2005. The accuracy and stability of an implicit solution method for the fractional diffusion equation. Journal of Computational Physics 205, 719–736.

Lenzi, E.K., Silva, L.R., Tateishi, A.A., Lenzi, M.K., Ribeiro, H.V., 2013. Diffusive process on a backbone structure with drift terms. Physical Review E 87, 012121.

Li, B.T., Zheng, L.C., Zhang, X.X., 2010a. Numerical methods for solving energy equations of dilatant fluid flow. In: The Third International Conference on Computational Sciences and Optimization, Huangshan, Anhui, China, vol. 1, pp. 11–14.

Li, B.T., Zheng, L.C., Zhang, X.X., 2010b. Numerical investigation on heat transfer of power law fluids in a pipe with constant wall temperature. World Congress on Engineering 3, 1864–1866.

Li, B.T., Zheng, L.C., Zhang, X.X., 2010c. Unsteady forced convection heat transfer for power law fluids in a pipe. In: The International Heat Transfer Conference, Washington, DC, USA.

Li, C.R., Zheng, L.C., Zhang, Y., Ma, L.X., Zhang, X.X., 2012. Helical flows of a heated generalized Oldroyd-B fluid subject to a time-dependent shear stress in porous medium. Communications in Nonlinear Science and Numerical Simulations 17, 5026–5041.

Liu, F., Meerschaert, M.M., McGough, R., Zhuang, P., Liu, Q., 2013a. Numerical methods for solving the multi-term time fractional wave equations. Fractional Calculus & Applied Analysis 16, 9–25.

Liu, F., Turner, I., Anh, V., Yang, Q., Burrage, K., 2013b. A numerical method for the fractional Fitzhugh–Nagumo monodomain model. ANZIAM Journal 54, 608–629.

Liu, F., Yang, C., Burrage, K., 2009. Numerical method and analytical technique of the modified anomalous subdiffusion equation with a nonlinear source term. Journal of Computational and Applied Mathematics 231, 160–176.

Liu, F., Zhuang, P., Anh, V., Turner, I., Burrage, K., 2007. Stability and convergence of the difference methods for the space–time fractional advection–diffusion equation. Applied Mathematics and Computation 191, 12–20.

Liu, F., Zhuang, P., Liu, Q., 2015. Numerical Methods of Fractional Partial Differential Equations and Applications. Science Press, China.

Liu, F.W., Yang, Q.Q., Turner, I., 2011. Two new implicit numerical methods for the fractional cable equation. Journal of Computational and Nonlinear Dynamics 6, 1–7.

Liu, L., Zheng, L.C., Liu, F.W., Zhang, X.X., 2016. Anomalous convection diffusion and wave coupling transport of cells on comb frame with fractional Cattaneo-Christov flux. Communications in Nonlinear Science and Numerical Simulations 38, 45–58.

Lynch, V.E., Carreras, B.A., del-Castillo-Negrete, D., Ferreira-Mejias, K.M., Hicks, H.R., 2003. Numerical methods for the solution of partial differential equations of fractional order. Journal of Computational Physics 192, 406–421.

Makinde, O.D., Onyejekwe, O.O., 2011. A numerical study of MHD generalized Couette flow and heat transfer with variable viscosity and electrical conductivity. Journal of Magnetism and Magnetic Materials 323, 2757–2763.

Malekzadeh, P., Moghimi, M.A., Nickaeen, M., 2011. The radiation and variable viscosity eddects on electrically conducting fluid over a vertically moving plate subjuected to suction and heat flux. Energy Conversion and Management 52, 2040–2047.

Malik, M.Y., Hussain, A., Nadeem, S.B., 2013. Layer flow of an Eyring-Powell model fluid due to a stretching cylinder with variable viscosity. Scientia Iranica B 20, 313–321.

Méndeza, V., Iomin, A., 2013. Comb-like models for transport along spiny dendrites. Chaos, Solitons & Fractals 53, 46–51.

Milsaps, K., Polhausen, K., 1952. Heat transfer by laminar flow from a rotating plate. Journal of Aeronautical Sciences 19, 120–126.

Ming, C.Y., Zheng, L.C., Zhang, X.X., 2009. The flow and heat transfer of power-law fluid over a rotating disk. In: Proceedings of the 3rd International Conference on Mechanical Engineering and Mechanics, pp. 1242–1246.

Mitschka, P., 1964. Nicht-Newtonsche Flüssigkeiten II. Drehstromungen Ostwald-de Waelescher Nicht-Newtonscher Flüssigkeiten. Collection of Czechoslovak Chemical Communications 29, 2892–2905.

Mustafa, M., 2015. Cattaneo-Christov heat flux model for rotating flow and heat transfer of upperconvected Maxwell fluid. AIP Advances 5, 047109.

Nazar, M., Fetecau, C., Awan, A.U., 2010. A note on the unsteady flow of a generalized second-grade fluid through a circular cylinder subject to a time dependent shear stress. Nonlinear Analysis-Real World Applications 11, 2207–2214.

Osalusi, E., Side, J., Harris, R., Johnston, B., 2007. On the effectiveness of viscous dissipation and Joule heating on steady MHD flow and heat transfer of a Bingham fluid over a porous rotating disk in the presence of Hall and ion-slip currents. International Communications in Heat and Mass Transfer 34, 1030–1040.

Ostrach, S., 1953. An Analysis of Laminar Free Convection Flow and Heat Transfer about a Flat Plate Parallel to the Direction of the Generating Body Force. NACA Technical Reports Server, 1111.

Pal, D., Mandal, G., 2015. Hydromagnetic convective–radiative boundary layer flow of nanofluids induced by a non-linear vertical stretching/shrinking sheet with viscous–Ohmic dissipation. Powder Technology 279, 61–74.

Podlubny, I., 1999. Fractional Differential Equations. Academic Press, San Diego.

Pop, I., 1993. Boundary layer flow at a three-dimensional stagnation point in power-law non-Newtonian fluids. International Journal of Heat and Fluid Flow 14, 408–412.

Pop, I., Rashidi, M., Gorla, R.S.R., 1991. Mixed convection to power-law type non-Newtonian fluids from a vertical wall. Polymer-Plastics Technology and Engineering Journal 30, 47–66.

Qi, H.T., Guo, X.W., 2014. Transient fractional heat conduction with generalized Cattaneo model. International Journal of Heat and Mass Transfer 76, 535–539.

Qi, H.T., Jiang, X.Y., 2011. Solutions of the space-time fractional Cattaneo diffusion equation. Physica A 390, 1876–1883.

Rao, J.H., Jeng, D.R., De Witt, K.J., 1999. Momentum and heat transfer in a power-law fluid with arbitrary injection/suction at a moving wall. International Journal of Heat and Mass Transfer 42, 2837–2847.

Rashaida, A.A., 2005. Flow of a Non-Newtonian Bingham Plastic Fluid over a Rotating Disk (Ph.D. thesis). University of Saskatchewan.

Rogers, M.H., Lance, G.N., 1960. The rotationally symmetric flow of a viscous fluid in the presence of an infinite rotating disk. Journal of Fluid Mechanics 7, 617–631.

Rundora, L., Makinde, O.D., 2013. Effects of suction/injection on unsteady reactive variable viscosity non-Newtonian fluid flow in a channel filled with porous medium and convective boundary conditions. Journal of Petroleum Science and Engineering 108, 328–335.

Sahoo, B., 2009. Effects of partial slip, viscous dissipation and Joule heating on Von Kármán flow and heat transfer of an electrically conducting non-Newtonian fluid. Communications in Nonlinear Science and Numerical Simulation 14, 2982–2998.

Salem, A.M., 2007. Variable viscosity and thermal conductivity effects on MHD flow and heat transfer in viscoelastic fluid over a stretching sheet. Physics Letters A 369, 315–322.

Samarskii, A.A., 2001. The Theory of Difference Schemes. Marcel Dekker, New York.

Schowalter, W.R., 1960. The application of boundary-layer theory to power-law pseudoplastic fluids: similar solutions. American Institute of Chemical Engineers Journal 6, 24–28.

Seddeek, M.A., 2002. Effects of radiation and variable viscosity on a MHD free convection flow past a semi-infinite flat plate with an aligned magnetic field in the case of unsteady flow. International Journal of Heat and Mass Transfer 45, 931–935.

Smith, G.D., 1985. Numerical Solution of Partial Differential Equations: Finite Difference Methods. Clarendon Press, Oxford.

Sparrow, E.M., Gregg, G.L., 1959. Heat transfer from a rotating disk to fluid of any Prandtl number. Journal of Heat Transfer-Transactions of the ASME 81, 249–251.

Srinivasa, A.H., Eswara, A.T., 2013. Unsteady free convection flow and heat transfer from an isothermal truncated cone with variable viscosity. International Journal of Heat and Mass Transfer 57, 411–420.

Sun, Z.Z., Wu, X.N., 2006. A fully discrete difference scheme for a diffusion-wave system. Applied Numerical Mathematics 56, 193–209.

Tan, W.C., Masuoka, T., 2005a. Stokes' first problem for a second grade fluid in a porous half-space with heated boundary. International Journal of Non-Linear Mechanics 40, 515–522.

Tan, W.C., Masuoka, T., 2005b. Stokes' first problem for an Oldroyd-B fluid in a porous half-space. Physics of Fluids 17, 023101–023107.

Tan, W.C., Masuoka, T., 2007. Stability analysis of a Maxwell fluid in a porous medium heated from below. Physics Letters A 360, 454–460.

Vieru, D., Fetecau, C., Fetecau, C., 2008. Flow of a viscoelastic fluid with the fractional Maxwell model between two side walls perpendicular to a plate. Applied Mathematics and Computation 200, 459–464.

Wang, T.Y., 1995a. Mixed convection from a vertical plate to non-Newtonian fluids with uniform surface heat flux. International Communication in Heat and Mass Transfer 22, 369–380.

Wang, T.Y., 1995b. Mixed convection heat transfer from a vertical plate to non-Newtonian fluids. International Journal of Heat and Fluid Flow 16, 56–61.

Xu, H.Y., Qi, H.T., Jiang, X.Y., 2013. Fractional Cattaneo heat equation in a semi-infinite medium. Chinese Physics B 22, 014401.

Xue, C.F., Nie, J.X., 2009. Exact solutions of the Rayleigh-Stokes problem for a heated generalized second grade fluid in a porous half-space. Applied Mathematical Modelling 33, 524–531.

Zhao, J.H., Zheng, L.C., Zhang, X.X., Liu, F.W., 2016. Unsteady boundary layer natural convection heat transfer of fractional Maxwell viscoelastic fluid over a vertical plate. International Journal of Heat and Mass Transfer 97, 760–766.

Zandbergen, P.J., Dijkstra, D., 1987. Von Karman Swirling flows. Annual Review of Fluid Mechanics 19, 465–491.

Zheng, L.C., Liu, Y.Q., Zhang, X.X., 2011. Exact solutions for MHD flow of generalized Oldroyd-B fluid due to an infinite accelerating plate. Mathematical and Computer Modelling 54, 780–788.

Zheng, L.C., Liu, Y.Q., Zhang, X.X., 2012. Slip effects on MHD flow of a generalized Oldroyd-B fluid with fractional derivative. Nonlinear Analysis RWA 13, 513–523.

Zheng, L.C., Zhang, X.X., Lu, C.Q., 2006. Heat transfer of power law non-Newtonian. Chinese Physics Letters 23, 3301–3304.

Zheng, L.C., Zhang, X.X., Ma, L.X., 2008. Fully developed convective heat transfer for power law fluids in a circular tube. Chinese Physics Letters 25, 195–197.

Index

'*Note*: Page numbers followed by "f" indicate figures and "t" indicate tables.'

E

F

G

H

I

J

K

L

M

N

O

R

S

T

U

V

W

CPI Antony Rowe
Chippenham, UK
2017-09-12 17:33